# THE SEVEN MYSTERIES
# OF LIFE

# THE SEVEN
# MYSTERIES

*An Exploration in*
*Science & Philosophy*

HOUGHTON MIFFLIN COMPANY

# OF LIFE

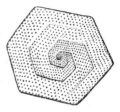

## GUY MURCHIE

*Illustrated by the Author*

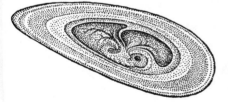

BOSTON

*Library of Congress Cataloging in Publication Data*

Murchie, Guy, date
The seven mysteries of life.

Includes index.
1. Life (Biology)    2. Life.    I. Title.
QH501.M87        574        78-2906
ISBN 0-395-26310-7

Printed in the United States of America

A  15  14  13  12  11  10  9  8  7

The seven small illustrations on the title page symbolize the Seven Mysteries of Life, clockwise, in order: *Abstraction, Interrelation, Omnipresence, Polarity, Transcendence, Germination* and *Divinity.*

Houghton Mifflin Company paperback 1981

# THE SEVEN MYSTERIES
## OF LIFE

# PRELUDE

Tᴴɪs is a book about life — all life in all worlds — and about life as the culminating celestial fact. I am writing it from the perspective of outer space as I did my last book, *Music of the Spheres*. That was about the material universe. and out of it the idea for this one sprouted as a sequel. But since the present subject perhaps over-ambitiously deals with things not only physical but mental and spiritual, it inevitably leads us, as you will see, deeper into philosophy than ever before. In fact the abstruse mysteries that soon cumulated and curdled in my notebook and which I eventually boiled down to seven in number became (to my wonder) the burden of the book.

Before introducing them, however, I will set the scene of observation so you may share in the discovery. Behold the body of Earth floating in living reality a thousand miles below us. See it, big as day, basking upon the black night of nothing, bedizened with stars — its mammose rotundity all fluted and furled in softest shades of bluish gray streaked with white.

Sometimes, looking at it out of the space station I imagine I'm in, I get the definite but indescribable feeling that this my maternal planet is somehow actually breathing — faintly sighing in her sleep — ever so slowly winking and wimpling in the benign light of the sun, while her musclelike clouds writhe in their own meteoric tempo as veritable tissues of a thing alive.

I was not brought up to think of the earth as a being. Such an impression has come to me only in this new outer perspective of space, which so provokes the thought that I cannot but wonder upon it. Could it? Just could the earth be alive?

Of course no one who lives down there would likely expect the terrestrial organism to be breathing with literal lungs. But, reflecting on it, lungs are far from essential to life. Insects and lesser creatures breathe easily without lungs. And obviously any being of planetary size, to live at all, must live in a very different way from the minuscule parasites that inhabit it, as do I in turn live so differently from my germs. Yet those germs — who are presumed never to have heard of me — remain as vital symbols and symptoms of my living. In fact they are enduring and tangible clues of my integral whole. And even more significant (as we will see in Chapter 14) is the growing evidence that Earth is a metabolizing superorganism who maintains her temperature, humidity and other characteristics within viable limits despite much greater changes in her celestial environment.

So I orbit here and dream about the rolling Earth, and wonder what music she is tuned to — what unseen ferment may already stir her geostrophic consciousness, what unimaginable tides of motivation may drive her evolution upon what yet unfathomed scales of time.

I am thinking of course of life in its broadest sense as embracing all kingdoms everywhere — and of mind and spirit as life's flowering in the large. More than presumably alive myself, however, I can neither escape nor deny an inevitable bias. For, although of course I must have my egocentric and geocentric prejudices, I yearn to savor the entire sweep of life (including me) from my new spatial vantage as best I may. I thirst to quaff it whole in full draught, undiluted, undoctored — withdrawing (in a less earthy metaphor) to cross-question it from without, from the rare detachment of my figurative watchtower among the stars. More specifically I seek to scrutinize the earth and her creatures down to their deepest marrow, to sense through their senses, read from their minds, pulse with their heartbeats — finally tune in, so far as I am allowed, to their profoundest spiritual potentialities.

So let us get on with the vital quest. I see unmistakable life down there, where the cloud systems flow and twist over the surface of Earth — where the fine threads called rivers braid themselves and glisten among the slower-pulsing mountain chains, their meandering loops now grinding gracefully downstream at so many miles a millennium, now overtaking each other until they short-circuit and fuse with a jerk, impatiently reacting and writhing, never letting their channels stay comfortable long enough to doze. Those slithering glaciers and squirming ocean currents are alive too, according to their respective natures — as are the fickle arms of forests advancing, withdrawing like dark flames, year by year — the sprouting lakes and islands and cultivated valleys, and now the cities almost exploding outward like frost stars on a windowpane.

Life indeed is splurging over Earth as never before, even bursting away from her altogether with rising frequency, in demonstration of her newly acquired self-consciousness. One cannot yet know how far away is Earth's zenith of life or evolution, or how long any optimum in bodily or spiritual development might endure. Yet against the tireless hustle of earthly rivers and the patient bustle of terrestrial hills, the spectacular manifestations of human life in our day have attained a startling acceleration and climax of development that is apparently natural and which surely must have a once-and-for-all uniqueness in the history of this and every other planet on which an equivalent phenomenon can occur. Our own generation in fact, by extraordinary coincidence, happens to come in the exact epoch of Earth's debut as a conscious planet — at the very moment in evolution when terrestrial beings have first begun to read the book of Earth, to measure themselves, to hang up their own "stars" and to guide (or misguide) their

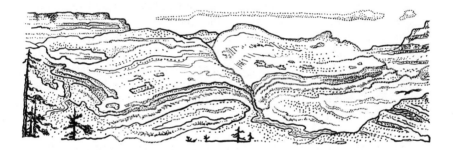

own evolution, advents symptomatic of an incipient blossoming of mind and spirit — a basic planetary germination that yet requires such a sophisticated explosion of pooled primary knowledge that the same sort of virginal springtide can hardly be expected ever to repeat itself on the same planet.

To help myself grasp the unique significance of such a planetary burgeoning, I like to refer occasionally to other ages and other worlds — to review our unfolding universe from less specialized perspectives — to remember, as did Lucretius two millenniums ago, in Mallock's charming paraphrase, that

*No single thing abides, but all things flow.*
*Fragment to fragment clings; the things thus grow*

3

*Until we know and name them. By degrees*
*They melt, and are no more the things we know.*

*Globed from the atoms, falling slow or swift*
*I see the suns, I see the systems lift*
*Their forms; and even the systems and their suns*
*Shall go back slowly to the eternal drift.*

*Thou too, O Earth — thine empires, lands and seas —*
*Least, with thy stars, of all the galaxies,*
*Globed from the drift like these, like these thou too*
*Shalt go. Thou art going, hour by hour, like these.*

*Nothing abides. Thy seas in delicate haze*
*Go off; those moonèd sands forsake their place;*
*And where they are shall other seas in turn*
*Mow with their scythes of whiteness other bays.*

It is probable, according to present knowledge, that most of the matter in our universe consists of very hot plasma, above a million degrees Fahrenheit, such as the simple, incandescent substances of far-flung suns and stars. But the even more scattered and rarer conglomerations of cooler solid stuffs, like the earth, may well have yet greater cosmic significance in that their ions have calmed down enough to offer hospitality to complete electron systems. For these have gradually permitted the diversification of the common planetary atoms, a few of which in turn have somehow managed to combine into such strange and potent complexities that they can recognize each other, compare impressions and, in some cases, even speculate on their own existence.

Thus, in the miracle that may spawn all miracles, I sense the surge of life upon a planet. But its entirety, of course, is very much too much for me to comprehend. Indeed, looking down with the perspicacity of my thousand miles of space, down upon the cloud-flecked paunch of a continent, even while remembering that I was recently confined to that thin veneer myself, it is a major effort to realize that that invisible biospheric surface is literally composed of millions of granulated films of life of innumerable but discrete varieties — not only creeping molds and soils and weeds and giant trees but buzzing flies and busy sparrows, darting fish and swarming germs, lumbering elephants and the finned vehicles of man amid invisible currents of pollen floating unnoticed up the moving halls of air, tuning in on each other magnetically, harmonically, corridor by corridor, stair on stair, scale for scale, note for note, *fa sol fa* — while mysteriously, ubiqui-

4

tously, the sun's rays penetrate the microcosmic keyholes of life to warm and activate the world — feeling, knowing, loving, creating . . . And all these teeming films of ferment and resonance somehow mingle and overlap and interact without cease, spatially, temporally, mentally, often fighting each other, ever competing for supremacy without losing their identities — indeed while continuously developing new species with new identities as evolution progresses.

This is incredible but literally true, and significance may be found in the fact that the films of life, like those of light and energy and matter, are not continuous but compartmented, with textures of interpenetrating units and groups of units: molecules, cells, organisms, plants, animals and their families, colonies and nations. The size of the units, however, has little relation to the thickness of the film networks, some of the thicker layers containing the smaller beings such as viruses and plankton, while various thinner ones may be made of large units like elephants or whales. Few of the films are obvious and many are virtually invisible, though of proven existence — like the established laceworks of rival songbird territories that cover almost every acre of the flowering land, a different cell texture for each bird species, the systems freely reaching and shifting seasonally through one another like untuned waves — amid other webworks of flitting insects as well as every kind of walking, crawling, slithering and burrowing creature there is. Not to forget the greatest numbers of all that swim through their equivalent precincts in the surrounding fresh and salty seas.

And if one would be thorough in considering life on Earth, one could not neglect the hierarchies of concentric systems of widely different size encased within each other yet so often dissonant in their conjunctions of scale and tempo. Like, for example, a soldier's stomach. While the workings of such a digestive organ go ahead under one motivation and plan, largely unknown to the soldier, other schemes are afoot within him on much smaller scales and with different motivations in the lives of his hormones, germs and viruses, even less known to him. And outside all these, the soldier's own conscious and subconscious motives may well be unwittingly at odds with the objectives of his army, which have not been divulged to him by his commanding officers, who, again, may be totally unaware of the true political or international pressures behind their campaigns.

If my explorations into such echelons of function and consciousness deepen my awe of the mystery of life, they may also make me aware of life's tenacity and pervasiveness. For these qualities are evident on every hand and it is obvious that, although no individual snail or　5 lily or sparrow can exist for more than a very few years, the genes within such creatures pass along their traits so steadfastly that snailery, lilyhood and sparrowness are always with us — immortal in their essence. It is through just such essences in fact that I am gaining the impression that life is probably as inevitable to the earth as earths are inevitable to the universe — the latter inevitability being a prime derivative of the terrestrial normality that is increasingly admitted by the astronomers.

What then is life's true nature? And is there any definable limit to life? Does it have measurable boundaries anywhere, or when? And how is life tied to consciousness? In simplest terms, what holds body to mind and to what degree may a mind take flight on its own? As for death: is it an end — or a phase of life? And of what is spirit made?

These are the kinds of basic questions I seek answers to, and try not to overlook as I gaze out at the warm maternal Earth and brood upon her uncountable creatures — creatures whose myriad forms and rhythms I see pouring out, not only spatially over the planetary surface and microscopically through every organism dwelling there, molecule by molecule, atom upon atom, but also coursing temporally through the whole length of evolution, mutation after mutation, generation on generation — in its entirety probably the most profoundly mysterious thing available to our consciousness.

To give you an inkling of where our search is going and what sort of mysteries it will turn up, let me say that almost immediately, as I wrote, I began to realize that there is something intangible behind the life in physical bodies — indeed behind all matter — and that this immateriality (energy, if you will) is revealed by the flow of time, which literally makes things into events. I find it convenient to classify all forms of this mysterious noumenon under the general heading of abstraction.

Next appeared the mystery of interrelatedness, which, geneticists tell us, is a measurable fact among all members of a species (including humanity in all its races) and, on deeper investigation, turns out to apply as well to whole kingdoms of creatures, not to mention the interrelations between kingdom and kingdom, or even between world and world, without end.

Third is the concept of the omnipresence of life, which denies that any impervious boundary has ever been found between any of the kingdoms or, for that matter, between life and nonlife, which leads inescapably to the conclusion that all rocks and seas and worlds, and consequently the entire universe, must in some sense be alive.

6 Fourth comes the polarity principle which recognizes the balance

and mutuality of the opposites we see everywhere: things like light and darkness, good and evil, male and female, predator and prey, matter and energy — all of which, by their contrast, give definition and meaning to life and make it work.

Fifth is something I call transcendence, which refers to the development of our perspectives on space and time as we grow older, as well as the progressive absorption of self into a wider awareness as one matures spiritually, all such factors ultimately revealing themselves to be, in effect, tools of learning in the inexorable drift from our present earthly finitude toward some sort of an Infinitude far beyond.

Sixth is the germination of worlds, a critical event that seems to happen once to every celestial organism and, after her billions of years of slow evolution, is occurring right now to Earth, as evidenced by many fundamental changes during what we call modern times — things that, as far as we know, never happened before and can never happen again on our planet.

And finally we come to the seventh and greatest mystery of all, the ultimate Mystery of divinity or whatever you choose to call the unknowable essence that leading thinkers have long believed somehow exists behind the creation and maintenance of all body, mind and spirit — not to mention behind every other known or unknown wonder of the universe.

# PART ONE

# BODY

# CHAPTER 1

# The Animal Kingdom

L ET US COME down to Earth now for a close look at life — a good, hard look at the nearest sample of life in the universe. I am thinking particularly of our cousins, the animals, it being only natural to begin with our closest kin. Besides, if animals aren't always the simplest creatures to understand, at least they are apt to be the most familiar.

Human knowledge of animals of course has a long history — as long as man's own — but it can reasonably be said to have started in a serious way only in the fourth century B.C. when Aristotle became the first scientific student of life on Earth as a whole. For it was he who earned an enduring reputation as the father of biology by doing more research in zoology than anyone else ever had, notably by cataloguing, describing and illustrating about five hundred kinds of animals. Naturally he had the help of assistants and students, including Alexander the Great, who used to send him specimens from his campaigns in

Asia and Egypt, but out of his assembled knowledge he managed to produce descriptions and explanations so accurate and modern in quality that they could not be improved on for twenty-two centuries.

"The catfish," he reported, "deposits its eggs in shallow water, generally close to roots or reeds. The eggs are sticky and adhere to the roots. The female catfish, having laid her eggs, goes away. The male stays on and guards them, keeping off all other little fishes that might steal the eggs or fry . . . In repelling the little fishes, he sometimes makes a rush in the water, emitting a muttering noise by rubbing his gills . . . He thus continues for forty or fifty days until the young are well grown . . ."

Although Aristotle's five hundred animals now seem pitifully few, the total number of varieties known did not appreciably increase for two thousand years, until Carolus Linnaeus of Uppsala, Sweden, in 1758 listed and classified 181 forms of reptiles and amphibians, 444 kinds of birds, 183 varieties of mammals and over 10,000 sorts of small animals and plants — giving each a generic and a specific name under his practical system of nomenclature that quickly became standard and has served biologists all over our planet ever since. As the modern method of classification of organisms into phylum, class, order, family, genus and species gradually developed in later years, the numbers of species identified and classified increased way past the million mark, so that now, for example, some 30,000 different species of protozoans are known, more than 10,000 species of sponges, nearly 9000 species of birds, on whom live (among the multitudes of other parasites) 26,000 species of feather-eating lice. Then at least 100,000 species of flies, a staggering 280,000 species of beetles and a meager 7000 species of mammals including man, who is merely one among some 250 in the order of primates, which is a small part of the mammalian class, which is but a fraction of the subphylum of vertebrates within the phylum *Chordata*, etc.

But there may still be more unknown than known species, particularly among the insects and microbes. Even sizable species are being discovered at a surprising rate: a new species of bird on an average of once a week, a new mammal nearly every fortnight, a new reptile or amphibian about twice yearly, a new insect about once an hour and every now and then a whole new phylum — the largest classification of animals — like that of a kind of worm with tentacles (*Pogonofora*) found on the ocean floor in 1964, then still another seaworm phylum (*Gnathostomulida*) in 1968. And if you think that anything as big as the legendary "abominable snowman" could not possibly exist yet remain unaccepted by science in this "advanced" age, remember that such hulking monsters as the square-lipped rhinoceros, largest of the rhino clan and second in size only to the elephant among land animals, was unknown until 1900, and also the huge brown bears of Kamchatka and Alaska (ten feet long and the biggest carnivores alive) unrecognized until two years earlier, while the mountain gorilla, largest of the apes, was only a myth before 1901, and the Komodo dragon, the tenfoot-long lizard known to kill and eat water buffaloes, was discovered by an airman downed in the East Indies in 1912. This list could be continued indefinitely through this century: the okapi, a kind of striped Congo "mule" added in 1900, the golden takin deer of China in 1911, the Yangtze dolphin in 1918, the giant panda first captured in 1936 . . . none of them small but all somehow overlooked and unrecorded until our own time.

To anyone with imagination, it should be clear by now that there could easily be, and almost surely must be, more such creatures still lurking shyly and unknown in the jungles of central Africa, the vast Amazon basin or perhaps Malaya. Why not? Or would they more likely be on some lush island like New Guinea where, as recently as June 1954, patrol planes turned up completely unknown stone age tribes estimated to number 100,000, some of whose cultures may be more primitive than those of Neanderthal man? There are innumerable halfverified legends around the still sleepy Earth, I hear — like that of "the spotted mountain lion" and the "Nandi bear" in Kenya, of the Loch Ness monster in Scotland, the "marsupial tiger" and the super otter called a "bunyip" of Australia, the "kongamato" or flying dragon of Zambia, some of which may prove to be real. As may also an occasional wary descendant of the presumably extinct giant swamp reptiles, or of the woolly mammoths in Siberia (reportedly seen alive last in 1918), the 15-foot sloth of Patagonia, the wingless 11-foot moa of New Zealand, the perhaps even larger Madagascan roc (of Sinbad the Sailor fame) whose 13½-inch egg has been found to hold 2½ gallons — all of which are intermittently rumored to be, and any of which just might turn out to be, still alive.

As we will see again and again in this book, just about everything imaginable has been, is being, or may yet be tried out somewhere, somehow, in this extraordinary world we live in. In evidence of which I testify that the recognized nonlegendary animal kingdom has long since adopted and perfected: movement without muscles, sight without eyes, hearing without ears, smelling without a nose, thirst quenching without drinking, eating without a mouth, digestion without a stomach or excretion, reproduction without sex, thinking without a brain and life without rest, sleep or death. And of course there are dozens of

13

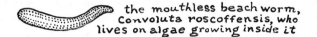

the mouthless beach worm,
Convoluta roscoffensis, who
lives on algae growing inside it

well-established senses besides the familiar five, not counting artificial ones (like the barometer), vicarious ones (like the watchdog), or most of the natural new ones that are continuously evolving.

And should you be one of those conventional persons who thinks of parasites as abnormal or perhaps goes so far as to drat the varmints, you may be surprised to discover, as I did, that parasites live both inside and outside most organisms in all the kingdoms, which makes parasitism thoroughly normal along with its derivatives: hyperparasitism (the phenomenon of parasites whose hosts are parasites in turn), hyperhyperparasitism, hyper$^3$-, hyper$^4$- . . . and on up to hyper$^n$-. And not only is the sun a parasite of the Milky Way as the earth is a parasite of the sun and terrestrial life parasitic on the earth, but you are a parasite of civilization (of which this book is a small token in evidence); and your dog, cat, canary, mice or others parasitic on you have each their own domestic fleas or lice or mites, etc., all of which (like you) must harbor itinerant bacteria which in turn carry viruses as passengers and, more than likely, stowaways still undiscovered . . . for it is not yet certain how many links such parasitic chains may have. Furthermore parasitism is part of the larger and even more complex subject of ecology or the interrelationship between all creatures and their total environments, which includes slavery, domestication, free sharing and a wide spectrum of bartering partnerships under the general heading of symbiosis. Parasitism may thus be a step toward brotherhood and even love, often becoming about as good in the long run for the host as for the parasite, both of whom tend to become dependent on each other under their unwritten but well-understood contract, as surely as you rely on your intestinal bacteria for digestion and vitamins or on your cat to suppress your mice.

In many such ways I see life deviously and imaginatively adapting itself to the world and to its own nature, shaping its form and behavior to the very Earth — to all the esoteric motions and melodies of the sphere. So it may be useful now to attend more closely the geography of animaldom — to observe more exactly how creatures evolve and function in relation to gravitation, to temperature, humidity, dynamic and chemical forces, light and the rhythms of days and seasons. Quite plainly, terrestrial life is part of the earth about as much as wheels and escapement are part of a clock. And, like a clock, the Earth turns, slowly but with order and cadence. Even more realistically, since

Einstein has demonstrated the nonfundamental and illusory nature of space and time, the physical essence of Earth life may be termed a spherical biofilm rotating in gravitational, electromagnetic and nuclear fields — a sort of gyrating bubble of evolving potency, a cosmic node of ferment.

While this bubble of life is materially concentrated in the fifteen-mile-thick earth skin of sea, soil and densest air, it is fueled and fed from energy in the planet's core and mantle and from the sun and stars. And it is bound together not only by gravitational, electromagnetic and nuclear cohesion but also by a kind of surface tension of interdependence, of chemical and psychological needs and intellectual accelerations interlaced with mystic, sometimes explosive, spiritual forces.

From a chemical standpoint alone, animal life without plants to eat would starve quicker but no more surely than most plants would die without animals. Even minerals would be drastically disturbed without both these higher kingdoms. It has been calculated that photosynthesis in present terrestrial vegetation would literally consume all the carbon dioxide out of the atmosphere within a year or two if it were not replenished by smoke from fires, engine fumes and the exhalations of animals and other consumers or decomposers.

## The Factors of Size and Gravitation

Although, at first thought, any correlation between the size of Earth and the size of her animal inhabitants is farfetched, a little reflection reveals a law of nature that says the creatures of Earth really must be approximately sized and shaped to the planet through the effects of gravity and chemistry, which vary as the creatures are large or small, fat or thin. It was Galileo who first articulated this law as the Principle of Similitude and it explains why no animal as little as a mouse need worry about getting hurt if it falls off a cliff or out of an airplane at any height since its body surface amounts to a parachute in relation to its meager weight, while an elephant is likely to be killed falling ten feet because his 10,000 times greater surface is made negligible by his 1 million times greater relative weight. This is because any two-dimensional surface increases as a square, while the corresponding three-dimensional volume or weight increases as a cube. And it explains why insects need no lungs and algae need no leaves, their simple spiracles and surfaces being ample enough to absorb all the oxygen or carbon dioxide their diminutive bodies need, while larger animals and

15

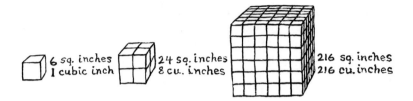

6 sq. inches
I cubic inch

24 sq. inches
8 cu. inches

216 sq. inches
216 cu. inches

plants would suffocate without their vast breathing areas in lungs and leaves. And, since digestion too is a surface function (of the surface of the intestines), it also explains why a ton of mice eats ten times as much as a ton of horses, and a ton of bacteria about ten times as much as a ton of mice.

Such inexorable disparities in size-surface ratios of course form a major geometric distinction between the microcosm and macrocosm, and their influence upon the forces and materials of Earth accounts for the size limits of its inhabitants, who could hardly walk on land if they were bigger than elephants or get enough to eat in the sea if much bulkier than whales or even stand still as trees if appreciably taller than a hundred meters. Furthermore, at the lower extremes, they could hardly be considered alive if their bodies contained less than a million atoms, this time not because of gravity but rather because of the uncertainties of random motion among individual atoms while their numbers remain too few to average each other into reliable order.

D'Arcy Thompson, who expressed similar concepts half a century ago in his classic book *On Growth and Form*, spoke of the form of any object as essentially a "diagram of forces" since "matter as such produces nothing, changes nothing, does nothing." Indeed he pointed out in some detail how physical forces create organic forms, from the hexagonal prisms of the honeycomb (which are merely cylinders under pressure) ⬡⬡⬡⬡⬡⬡ to the equiangular spirals of horns and shells, which permit drastic growth without a basic change of shape or even, in some cases, without so much as shifting their centers of gravity. In fact the very skeletons of creatures from microscopic radiolarians in the sea to the dolphin's skewed skull that may help him scull his way through the deep and to the cantilevered bones of beast and bird with their beautiful lines of tension and compression that are often mirror images of each other — all these are manifestations of stress upon growth by which nature creates form not so differently from the way the glassblower molds tubes and bulbs and vases by directing forces here and there in the space and time allowed.

The way in which earthly animals are molded by the earth in size or shape may be understood best of all perhaps by imagining what

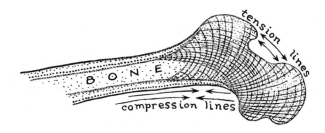

corresponding creatures would have to be like on other worlds. For instance, the gravity of a planet as big as Jupiter, with a force 2½ times stronger than Earth's, would require any Jovian donkeys to be as stout as terrestrial elephants in order to hold themselves up, chickens to have legs like ostriches, pigs either to crawl like alligators or swim in the way of porpoises (whose name, not inappropriately, derived from "pork fishes"). But on a moon with gravity much weaker than on Earth, an elephant might scamper about on spidery legs or a hippopotamus jump like a grasshopper and, if such a moon could have an atmosphere (as one of Saturn's is known to have), possibly beasts as big as wolves would have developed batlike wings on which they could flit with nary a qualm after their prey.

Could this be a being from the great ocean of Jupiter? Or the Saturnian Sea?

Or this a shrewd creature of Mars?

Actually this is a sea cucumber from the deep Pacific and this a tree-hopper from a forest, on Earth

Even on Earth as it is now, the ponderous elephant does not have any too easy a time of it, as is shown by the fact that of 352 known species of proboscideans (mastodons, mammoths, etc.) that in the last few million years have roamed the land, only two species survive: the African and Indian elephants. Significantly, their nearest, though remote, remaining cousins are the mermaidenly sea cow and the little woodchucklike hyrax — which suggests the capriciousness of natural evolution in its apparently heroic attempts to cope with a maturing planet.

The elephant's biggest problem of course is that he is so big. Far larger than any other animal that still walks the earth, he not only needs pillars for legs but literally an acre of lung surfaces to absorb oxygen, an eight-foot trunk to reach his food, a massive heart to circulate blood and hundreds of feet of guts and complex digestive organs to assimilate nourishment from the hundreds of pounds of foliage and grass he eats during his daily sixteen hours of browsing. These complications of course are not what makes the elephant so big. Rather is it his bigness that makes the elephant so complicated.

And among the more serious limitations of his size is his inability either to run, trot or jump, for experts claim that a trench seven feet wide is completely impassable to him, even if his stride covers 6½ feet. His top speed when enraged and charging has been clocked by a retreating jeep at just eighteen miles an hour.

Although the elephant's enormous size usually protects him from serious injuries by tiger, lion or buffalo and he has often been called "Lord of the Jungle," paradoxically some elephants have been panicked by Scotch terriers, ants and even by mosquitoes swarming upon their sensitive trunks. And they are notoriously susceptible to a great many afflictions, including nettle rash, seasickness and such human diseases as colds, pneumonia, mumps and diabetes. To cite perhaps their best-remembered case from history, of the fifty smallish "war elephants" that Hannibal drove from Cartagena through Spain, Gaul and over the Swiss Alps in 218 B.C., only eight lived to see Italy and fight the Romans. And of these but one (the only Indian elephant among them) survived the Apennines, then miraculously regained his health sufficiently to endure the long campaign of at least six great battles, amazingly returning after fifteen years in solitary triumph to the elephant training stables in Carthage. Since Caesar's time, however, the war elephant has been obsolete in Europe and Africa, his domesticated descendants (outside of Asia) relegated mainly to circuses and exhibitions.

Perhaps the elephant's destiny will lead him into a more aquatic life in future centuries for, as a vehicle of land transport, he has not found it possible to keep up with the truck or the water buffalo. Yet by erecting his trunk as a snorkel, he can march (not without some jitters) through rivers twelve feet deep, treading the bottom, and he has been known to swim continuously for six hours. The only serious setback suffered by an elephant swimming at sea to my knowledge was the case of a crazed one that thrashed about so much he attracted sharks, who chomped him into a skeleton in less than an hour. And there exists a convincing record of one Indian rogue who, seeming to adopt the slogan "Have trunk, will travel," one day left home and family to plunge purposefully into the sea, whereupon for twelve years he sashayed through a leisurely 200-mile island-hopping jaunt in the Bay of Bengal.

Certainly the buoyancy produced by immersion in water removes some of the disadvantages of excessive weight, as the dinosaurs knew and, more recently, the whales rediscovered when they evolved back to sea. The whales have adapted themselves to ocean life by stepping up their speed which, according to Froude's Law of the correspondence of speeds, tends to increase "in the ratio of the square root of the increasing length" — and by such changes as moving their nostrils upward to the top of their heads and disconnecting their lungs and windpipes from their mouths so they can swim underwater with jaws open to feed without fear of drowning. By such evolutionary advances bottlenosed whales have enabled themselves to submerge for two hours or to dive three quarters of a mile deep, as has been estimated from more than a dozen cases of whales getting their jaws tangled in cables. And the sea-lent speed has restored their power to jump so dramatically that more than once a blue whale weighing over a hundred tons has been seen to breach completely out of water. Yet when even a small whale comes close to shore he is in imminent danger of becoming cornered or stranded on a beach where, if the tide ebbs to leave him partly out of water, he is not only helpless but liable to suffocate when his lungs collapse under his no-longer-buoyant weight.

While the buoyancy of water is a blessing to whales, it can be a serious problem to some fish, for bodies suspended beneath its surface yet not resting on the bottom must cope with the inherent instability of their position. Most fish therefore are equipped with "swim bladders" containing fat or gas (mostly oxygen) which give them neutral buoyancy in the sea by making their average density equal to that of water, the bladders gradually absorbing or discharging matter as the fish goes deeper or shallower, thus enabling it to keep its balance in earthly gravitation. Indeed, if a fish wanders far from its normal depth, up or down, it has to face the very real risk of "falling" either upward or downward, the upward motion, if beyond control, eventually bursting its insides most unpleasantly out through its mouth. In this connection, did you know that the common cuttlebone (traditionally given pet birds to groom their beaks on) serves in life as the buoyancy tank of the cuttlefish, a squid cousin which by osmotic control can vary his cuttlebone density from .5 to .8, thus surfacing or submerging at will like a submarine. And the beautifully helical chambered nautilus uses gas in one of its storied and stately chambers to the same not-always-purely-poetic end.

But although the balloon principle can thus readily handle earthly    19 gravity in the sea, animals in aerial flight have not found buoyancy so easy to achieve. For some strange reason, no creature before man seems ever to have evolved anything approaching a bladder of hydrogen. And air, being about 800 times lighter than water, just will not support any very dense body unless it uses either some sort of a balloon or the active flying principle of deflecting the sky's substance to generate lift. Froude's Law, incidentally, covering all kinds of locomotion, explains why the ostrich does not fly, since, being about 25 times as tall and long as a sparrow, he would have to move $\sqrt{25}$ or 5 times as fast. That is at about the speed of a small airplane, which necessarily has many times an ostrich's "horsepower."

In addition to the upper size limit of earthly flight, moreover, there is a lower limit where tiny insects, too small to feel gravity, begin to be buffeted out of control by the so-called Brownian movement of random molecular motion that turns flying into swimming and ultimately, on the microscopic scale, swimming into digging. Competition with free molecules is not the only disadvantage in being small either, I notice, for the task of keeping warm and getting enough to eat becomes critical even sooner to the small. This is because heat is radiated only from a body's surface area, not from its total volume, and little creatures have relatively huge surfaces. So the food supply must be proportional to skin size, not weight — which explains why men may eat less than 2 percent of their weight daily on the average, while mice (despite the insulation of their fur coats) must tuck away a good 25 percent of theirs.

An extreme case in point is that of the tiniest and most numerous mammal of all, the shrew, who comes out at night and burrows largely unseen in every forest floor and through practically every garden in every continent but Australia and Antarctica, and is represented by more than 30 species in North America alone. Although some shrews are so tiny they weigh less than a dime and you seldom notice the traces of their diggings, their appetites are relatively enormous — inevitably so because such a minuscule creature, with so little mass per square inch of skin, metabolizes four times as fast as the smallest mouse (per gram of tissue) and therefore must eat up to three times his own weight in food every day, and not just vegetable matter (which

is good enough for a mouse) but also worms, grubs, insects, fish, frogs, and even mammal meat. The fact that he feels hungry enough to hunt almost every waking minute, a compulsion ultimately enforced by the lurking threat of starvation should he ever fast as long as three or four hours, gives him a ferocious disposition befitting the most terrible mammalian predator (gram for gram) on Earth. A water shrew has been known to kill a fish sixty times heavier than himself by biting out its eyes and brain, which is equivalent to a man killing an elephant barehanded. And all while holding his breath underwater which, with shrew metabolism demanding hundreds of breaths per minute, is impressive if it lasts five seconds. Imprisoned with another shrew and no other food within reach, a shrew has little compunction as to cannibalism either. In fact, when three of the beasties were left alone under a glass tumbler, in a typical example, two of them promptly killed and ate the third down to its last bone and hair, emitting a shrill batlike twitter the while. Then, a couple of hours later, the hungrier of the survivors suddenly attacked and polished off his remaining companion, whereupon he took time out to clean his whiskers, apparently feeling more than delighted with himself — for the moment, that is — after having so neatly converted two worthy colleagues into breakfast, lunch, some scattered droppings and a few unavoidably wasted calories of heat. The last act in this raw drama followed in about three more hours when the sole survivor's appetite had renewed itself to such a pitch that he finally seized the most accessible flesh still in sight, his own tail, and, working up from there, literally devoured himself to death — a dramatic demonstration that at least some creatures, driven to the extreme, actually would rather be eaten alive than starve.

It is reported that some shrews can dig themselves out of sight into firm soil in only one second. But a second, come to think of it, must be quite a spell for a shrew — perhaps equivalent to a minute for a whale — for there is evidence of a relativity in the consciousness of time in proportion to an animal's size. And the rapidity of life processes in such tiny beings includes not only metabolism but sense perception and the rates of everything from vibration and radiation to the conservation of angular momentum. If so, a second may be an even longer spell in the life of a hummingbird, since, of the more than four hundred known species (more than in any other family of bird), some weigh appreciably less than the smallest shrew and have the highest metabolism rate of any bird or mammal. According to the rule of thumb that a warm-blooded animal's metabolic pace must be in inverse ratio to its length, a hummingbird's metabolism is about a 21 dozen times faster than that of a pigeon or 100 times that of an elephant. Which may make you wonder why a hummingbird doesn't starve to death at night, since it cannot see to buzz among flowers for nectar in the dark or catch the minute insects it must have for protein — not to mention its need for rest.

The surprising answer is that the hummingbird's metabolism shifts into low gear at sundown, dropping to only one fifteenth the day rate. Thus, in effect, this tiny bird hibernates (or "noctivates") each night. It prepares for this by an hour of literally feverish feeding at top speed as the sun nears the horizon, its heart beating more than twelve times a second. Then, after roosting in the gathering gloaming and dropping daintily off to sleep, its temperature rapidly descends from perhaps 115°F. to nearly the temperature of the air, often 60 degrees cooler, its heart slows way down and it becomes torpid — so insensate it can be picked off its perch in your fingers like a ripe berry. In the high Andes, hummingbirds retreat into caves at night to avoid freezing in the chill mountain air, freezing obviously being a hazard to any such insect-sized creature. Its ease of flying, however, makes up for several of the wee hummer's other problems, for, most unpedestrian of birds, it has never been known to hop or walk a step, but flies on its shortest journeys, even down to an unpresumptuous jaunt of half an inch.

## LOCOMOTION

Efficiency in running and swimming as well as flying is influenced by much more than size or weight in relation to gravity. For although it is true that a large animal tends to move faster than a similar small one in proportion to the square roots of its linear dimensions (Froude's Law), there is no end to the ways in which the complex lever system of limbs and muscles can be put together. The badger, skunk and porcupine, for instance, not needing speed, have short legs suited to digging, scratching and climbing. But a hare or hound of equal weight has a much longer metatarsal joint or instep-lever, geared for fast running — and so does the horse, ostrich, kangaroo or other speedy animal, whether it specializes in sprinting like the cheetah (credited with 70 mph for a short distance) or loping like the camel (115 miles

in 12 hours). Keeping the heavier leg muscles close to the body (where they need not swing so far nor fast) plainly increases efficiency, as does the simplified and light hoof-shank system of the horse with its marvelously elastic ligaments that save energy by automatically re-straightening the fetlock joint and snapping the foot backward like a released spring every time it leaves the ground. Another speed factor

derives from the flexibility of the spine which not only greatly increases an animal's length of stride but can enable its foot actually to move in a smooth elliptical orbit in respect to the rest of its body — not so different from the cyclic path described by a planet or any point on the rim of a rolling wheel.

The wing motion of birds is somewhat circular too — mainly a propellerlike rotation which lifts as well as propels even during its upward stroke. Although only a minority of birds have succeeded in crossing the larger oceans, the way they achieved their flying efficiency is worth a minute's consideration. Through the evolution of wings, feathers, warm blood, powerful breast muscles, hollow bones and a remarkable respiratory and cooling system, they have worked out a difficult and delicate compromise between high power and low weight. Their excess baggage jettisoned by evolution has included such seeming necessities as teeth and heavy jaws, bone marrow, leg muscles, sweat glands, and much of their reproductive system, espe-cially during migratory seasons. Did you know that female birds make do with only one ovary (the left one) and that the sex organs of starlings, for instance, have been found to weigh 1500 times as much during the breeding season as in the rest of the year? That a flying pigeon uses one fourth of its air intake for breathing and three fourths for cooling? That bird air cooling systems include many air sacs and hollow bones (in some cases clear to the tips of wings and toes) through which their breath flows by efficient forced draft, the stale

upper wing bone of an eagle

exhalation going out passages that are largely separate from those entered by the fresh inhalation? That the bird is literally flying into breath so fast it can never run out of breath? This brand of super-charging of course is vital to the high-speed activity of flight, which also demands such "high octane" protein fuel as worms, insects, rodents, fish and seeds rather than the grass and leaves that suffice for cattle and elephants. And yet the skeleton of a frigate bird with a seven-foot wingspread has been weighed at only four ounces — actually less than the same bird's feathers!

Insects of course are Earth's primary flyers, since they were the first to take to the air, some 300 million years ago, probably in order to escape the "fish" whose fins were already turning into legs in their efforts to catch them on land. Birds could not have appeared for nearly another 100 million years, they having to await their lizard ancestors (of fish descent) whose front legs still had to evolve into wings as they leaped after the flying insects. Naturally it was relatively easy for the insects to fly, since the air's viscosity in relation to their negligible weight gave them a fair grip in it, a factor explaining how the smallest insects today can float on air with motionless wings, flapping only when they want to advance *through* it. Flying was so natural to them in fact that their wings evolved from their body wall, leaving all six legs intact to run with, at least four of which are normally on the ground at any instant. Twisting their wings like variable-pitch propellers as they beat them in figure-eight orbits at almost uniform speed (950 times a second in the case of the midge), they acquired an extraordinarily high muscular efficiency (calculated at 20 percent) which takes swarming locusts 300 miles nonstop at an average air speed of 8 mph, black flies almost as far, and enables even the gentle aphid to cross the North Sea. Most surprising of all is the monarch butterfly who, in stringent laboratory tests, has flown better than 6 mph for a continuous four days and four nights, a performance which, with the aid of a moderate tail wind, should easily get him across the Atlantic and no doubt is the explanation for his occasional appearance in England.

For sheer speed it is believed the big dragonflies and hawk moths are insect champions, who may cruise up to 24 mph, their wings beating alternately (front ones rising as the rear ones fall), reaching 36 mph with extreme exertion, while the giant paleozoic dragonfly of 300 million years ago with his 36-inch wing span probably could do 43 mph. Thus although insects have since been outflown by the birds, their still unchallenged position in the kingdom rests in no small degree on their ancient and honored ability without fuss or feathers just to

24

take off in any instant during 300 million years and buzz steadily ahead at 100 times their own length per second!

Speed underwater admittedly is neither as easy nor as important as speed in air, so one should not be too surprised that the fastest any fish can swim, even in a brief spurt, is only about ten times its length per second. Although a blue marlin has been credited with 50 mph at sea and a tuna with 47 mph, more reliable scientific measurements made in a special revolving laboratory tank give the fish speed record (by a 20-pound 4-foot barracuda) as but 27 miles an hour, while salmon, which proved unable to leap more than about 10 feet upward or 12 feet forward in the air, did so only after an extreme spurt that attained 18 mph in less than a tenth of a second.

It is now known that probably nearly all swimming creatures, and particularly sei and killer whales (who have been clocked at 35 and 34 mph respectively) and porpoises (up to 37 mph), achieve their speed through their remarkable capacity to move in water without creating turbulence. Their perfection of laminar flow is of course impossible to land animals or human swimmers, since it depends on extreme smoothness of skin (lubricated by special oils), as well as on a rare degree of legless, armless streamlining of form. And, interestingly, the surprising truth was revealed only when a thoughtful naval officer (who had been a biologist) one night observed a porpoise swimming almost invisibly through a sea in which boats and seals unavoidably stirred up fiery wakes of microorganic luminescence. Although squids use a kind of liquid jet propulsion and can zoom many feet into the air, they are not as fast as some other leapers, such as flying fish, while the beautiful silvery blue velellas, the iridescent Portuguese man-o-war and other jellyfish with sails, some of which can tack well into the wind, have too much drag (even with retracted tentacles) to attain even one mile an hour waterspeed.

Beneath solid earth, of course, quite different means of locomotion prevail. The skink literally swims in sand. But the common earthworm uses a kind of slow-motion ram-jet propulsion by eating in front and excreting behind as he goes, some of his species augmenting the alimentary flux by gripping the soil with retractable bristles which function in regressive waves exactly coordinated with alternate elongations and hunchings of his streamlined body. Then there is the burrowing snail's creepy advance by microscopic, rhythmic expansions and contractions of his single slimy foot, a technique that often enables him to overtake a worm and, if he is a worm-eater, lure it into his specialized worm-welcoming mouth, which turns out to be lined with one-way quills that permit no withdrawal. Still another kind of

25

motion is the flea's ability, with the aid of a boatlike bow, clean streamlining and six powerful legs, to "swim" up to 50 times his own length per second through a "sea" of animal fur, or to jump 120 times his own height at a peak acceleration of 140 Gs (30 times that of a moon rocket).

Below the world of worms and burrowing snails we may consider microscopic creatures like the ameba, who oozes along by pushing out lobes or ruffles of his single cell protoplasm and just flowing into them. His locomotion system (often called "ruffling") was only recently elucidated. It works in each lobe something like an advancing jet of jelly that spreads like the crown of a fountain at its forward end, turning outward in all directions, then back in the form of a surrounding cuff that condenses into a sleeve stiff enough to grip the ground until, turning once more in the rear, it liquefies and pours inward through the center. Biologists used to think that a musclelike squeezing of the sleeve's protoplasm at the hind end pushed the jelly forward like toothpaste out of its tube, but newer evidence strongly indicates that it is the continuous contraction of the jelly at the forward "fountain zone" (where it turns and stiffens) that literally pulls the central stream steadily ahead. So the consensus of opinion now is that the ameba advances somewhat as does an oriental king, by having his carpet swiftly unrolled before him as he walks, then as nimbly rerolled behind for further unrolling ahead. In fact he may travel thus half an inch an hour and, if he could keep going steadily in one direction, he might cover a foot a day or the length of a football field in a year.

Another method of microbe locomotion, up to several hundred times faster but still slow in human terms, is that of flagella or bacilli with tails. Some of them propel themselves by a recently discovered swivel motion, each flagellum lash rotating freely about its axis like the rigid propeller of a small airplane, generally pulling from the front end and changing course by reversing the direction of rotation. So far as I know, this is the only true wheel motion produced by nature before man invented the wheel about 3500 B.C. and it is powered by the equivalent of a reversible microscopic engine, something technological man has not learned how to make even today.

A different system again is that adopted by ciliates, common in mud and ponds, whose bodies are covered with thousands of rapidly waving hairs called cilia. Up to 20 times a second each cilium (one thousandth of an inch long) makes its stroke, much like a human swimmer's arm action, first reaching gently forward edgewise for minimum resistance, then sweeping rigidly backward broadside for maximum resistance, the beats coordinated in beautiful rhythmic waves of succession, like

pistons in an engine or stalks of wheat blowing in the wind. Even some visible animals use ciliated drive, notably two kinds of comb jellyfish the sizes of a gooseberry and a walnut, and often called respectively the "sea gooseberry" and "sea walnut," each of which has eight longitudinal belts of cilia (coordinated by a special balance organ) that steer and propel it like a spherical Caterpillar tractor.

Such techniques should not seem too exotic to us either for, little though we may be aware of it, our own bodies use similar methods, as, for example, our sperms that flagellate themselves forward with lashing tails, the cilia that line our windpipes and automatically convey unwanted fluids and dust particles out of our lungs, and not only real amebas in our watery fluids but white cells in our blood that move by the same ameboid motion.

This brings up a significant relativity in microbic activity since these tiny creatures, particularly ciliates, often work their wills not by going thither toward their objectives but rather by stirring up liquid currents that make their objectives come hither to them. Furthermore, they work their hundreds of limbs in a wholesale fashion, perhaps disbursing their energy on a random or probability basis. Certainly the use of extra limbs beyond an insect's six or a spider's eight contributes no more to speed than do extra wheels accelerate a train, and the centipede goes not a whit faster for all his 16 to 346 legs (depending on the species) nor does the millipede for his twice as many (which, in one Panamanian species, were actually counted as 784 legs) that carry him only half as fast. These multitudinous limbs do, however, give their proprietor a certain diffuse stability, or even what might be called a Lilliputian majesty.

The locomotion of snakes seems to be as mysterious as it is fascinating to most people but, I'd say, not totally unexpectable when you consider that a snake may have as many as 400 trunk vertebrae, each with its pair of ribs and at least a dozen bundles of muscles by means of which it may bend 25° to either side and 14° up or down. And, by a seeming miracle, young snakes never need slithering lessons, for they instinctively use a half dozen different techniques to slither their way ahead in grass, over pavement, up trees, through pipes, across sand dunes or, on the desert, even to hop along parched ground "too

27

hot to touch.'' The snake's common lateral undulation method works, of course, by pushing backward or partly sideways against grass stems, pebbles, etc., with continuously moving waves of his body. But such a technique doesn't work on a smooth surface like a tennis court, so a snake there usually reverts to one or a combination of other systems, like the concertina method (used by the inchworm) or the rectilinear method of advancing in a straight line (used by the earthworm). Although, inside a pipe, he can press outward to grip the walls and advance à la concertina, outside a pipe he must spiral, pressing inward and pushing against any irregularity. A tree is easier, usually permitting him to undulate among its branches. In soft sand, on the other hand, he may use the oblique loop-weave method of the sidewinder rattlesnake in which he weaves ahead with most of his body looping high off the shifty grains to reduce friction and heat.

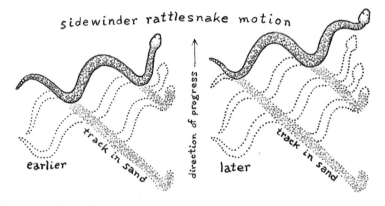

## ALTITUDE

The vertical limits of known animal life in the earth's biosphere are of interest here, as relating to air and water pressure, which derive from gravitation. Life thins down drastically by the 20,000-foot level above the sea, and mountaineers in the Himalayas report seeing only a few kinds of birds above that height — an occasional eagle or lammergeier (eaglelike vulture), perhaps a snow partridge, some crow-like choughs (ranging to 27,000 feet) or a string of bar-headed geese flying over Mount Everest on their annual migration to the lakes of Tibet. Although mammals lack the very efficient lungs of birds, not only yaks but sheep (including domestic ones) wander as high as edible plants will grow, and cushion plants have been collected at 20,130 feet. Mice and conies sometimes get about that high too, expectably pursued by weasels, foxes, wolves and snow leopards. Choughs by the way commonly alight on the backs of sheep and search their wool for insects, a service the sheep encourage by patiently holding still as long as the birds are near. And jumping spiders have been observed at 22,000 feet preying on flies and exploited by parasites. And aphids, moths, butterflies, beetles and centipedes about equally high, as well as semimicroscopic glacier fleas, mites and springtails, not to mention bacteria or smaller creatures. This level is known as the aeolian zone, for most of these tiny animals ride gracefully aloft on warm rising winds, along with pollen and dust, including spores and minute shreds of fungus or lichen on which they may graze as hungrily as birds and rodents feed on them in turn.

In light of the fact that the Air Force requires its flight crews to use extra oxygen above 10,000 feet, one might think that humans and other breathing mammals dwelling above that height could hardly live normal lives. Yet Tibet is largely 14,000 feet high and has many a bustling town above the common cloud levels. And I have seen reports of a permanent mining camp in Peru at 17,500 feet, where the air pressure is only half that at sea level, up from which men climb daily to work at 19,000 feet! There are also a few big cities with the pressure around 70 percent such as La Paz, capital of Bolivia, a metropolis of more than 375,000 people at 12,800 feet, the loftiest big city on Earth, with railroad stations, hospitals, department stores, a university and about everything else you would expect in a municipality the size of Miami. So life more than two miles up in the sky is sanctioned with a kind of normality — at least for an estimated ten million people here and there upon the scattered mountains of Earth.

Life in the abysmal depths of the oceans of course is much more difficult to observe than at comparable heights above sea level. But the greater volume of habitable, oxygen-rich room down there may make it all the more important, for the area of Earth covered by water 2½ miles deep is as vast as all the continents put together, and even the scattered abysses deeper than 4 miles total almost half the area of the United States.

It imposes more than a little strain on the imagination to visualize life in the sea-bottom world of utter and inky blackness, of passing mysterious shapes and unexpected encounters, of eternal cold (perpetually less than ten degrees above freezing) and of pressures thousands of times greater than in the air. But that there is life down there we know without doubt, for fine steel nets on cables several miles long have hauled its organisms up to the light of day and Jacques Piccard and Lieutenant Don Walsh who descended in their bathyscaph to the nadir nigritude of the Mariana trench in 1960, the deepest abyss known on Earth and more than seven miles straight down, saw flounderlike flat fish at the very bottom blithely swimming through water under a pressure of seven tons per square inch, their carefree relaxation certainly attributable to the pressure being equalized inside and outside the body and its cells.

Most abyssal animals, however, seem to be bug-sized burrowing forms such as bristle worms and brittle stars, mollusks, crustaceans and a few larger, crawling sea cucumbers. Fish with backbones remain very scarce below the top two miles because of the dearth of creatures they eat. And, even where they are plentiful about a mile down, their lives are made precarious by the fact that smaller edible creatures retreat into the darkness the instant they are sensed at a range of a few inches, while the fish themselves must flee in turn from bigger ones who want to eat them and, should they encounter still others of their own species but opposite sex, it becomes a rare opportunity for mating, with any effective choice between these alternatives requiring both discriminating decision and resolute action within a few seconds. And perhaps it was just such nerve-racking dilemmas that forced the evolution of the nightmarish denizens of the deep: the black swallowers and rattail grenadiers with needle-sharp fangs protruding through cheek holes even when their huge mouths are shut, and whose stretchable balloonlike bellies permit them to specialize in devouring unsuspecting fish twice their own size; the gulpers consisting of practically nothing but a swimming mouth trailed by a long lash ending in a red taillight that lures victims close enough

anglerfish full · gulper · lantern fish about to be eaten by an angler

so they can be lassoed by the lash and swished between the gaping jaws; and many other varieties of anglerfish with dorsal fishing rods from which hang lines with floats and luminous bait dangling in front of waiting teeth or, in a few species, displaying a tempting morsel even inside the already open mouth!

Below the two-mile depth there not only are very few luminescent fish or predators of any kind but virtually none that graze on fresh vegetation, which is absent because of remoteness from the sunshine and photosynthesis on which it depends. This may be why life is found to be slower in rough proportion to depth with the deepest fish breathing and metabolizing at only one third the rate of upper fish. In any case the only reliable food below two miles is the steady "snow-fall" of organic debris settling very slowly from above: crumbs of decaying seaweed, barnacle shells, shrimp skins, waterlogged drift-wood and half-dissolved excrement full of bacteria, taking years to get there. Seldom does anything like an edible carcass slip past the upper scavengers — only a rare and fast-sinking dead whale or maybe the forty-foot remains of a giant shark or humans entombed in a broken ship.

And not only are the creatures of the deepest chasms isolated by pressure and five miles or more of black water above them, but they are separated from all the other deeps of the oceans because of the vast subsea plains, plateaus and mountain barriers standing between these pockets with pressures far too low to permit them to pass over. Each abyss thus is like a solitary peak or ridge in reverse, a unique world that may have had no traffic with other such depressions for hundreds of millions of years. And the fact is shown also in the distinct high-pressure species dredged up from each trench, which in all cases have differed from trench to trench, each great abyss being the home of an entirely separate evolution, most of which are still so little known that it has not yet been quite ruled out that some one of them may have evolved a considerable degree of intelligence or con-ceivably even some sort of a sightless dexterity approaching that of man.

## TEMPERATURE

If pressure is a factor imposing limits on life's domains on Earth, temperature must be an even more critical factor. Did you know that a brook trout will die at room temperature or a frog when as much as one leg is placed in lukewarm water for a few hours, and polyps that build coral reefs may perish from temperatures only one or two degrees warmer than that of the ocean in which they have lived all their lives? Even we adaptable human beings will rarely survive a fever of 9°F. above normal, because the fluid cytoplasm in body cells coagulates in heat almost as does the white of an egg being cooked.

So animals have developed various ways of stabilizing their temperatures within tolerable limits. The one percent of them that are warm-blooded have the most elaborate physiological systems, which amount to remarkably efficient heat pumps and cooling radiators that use blood, lymph, sweat and air to transfer heat to and from the vital organs as needed — plus all sorts of supplementary devices, like the African elephant's wing-sized ears which, it is calculated, add an important 15 percent to his radiating surface when he holds them out from his head in hot weather. The much more numerous cold-blooded creatures such as fish, on the other hand, simply seek to stay in water that feels livably cool or warm enough, which can be a hazardous problem, as it is often difficult to know which way to swim up or down a gradient of only a tenth of a degree per mile.

Insects and land reptiles (not quite so "cold blooded") when exposed to the vagaries of the atmosphere, which is generally much less stable in temperature than water, must resort to many tricks of behavior to avoid roasting or freezing. Thus the greater earless lizard of the southwestern United States has been found, when active, to keep its temperature within 3.3°F. of its day mean of 101°F. 75 percent of the time. It does it by basking judiciously, by sensitively orienting its body at right angles to the sun's rays when it feels cold and therefore needs to absorb heat at the maximum rate. Later, as it warms up, it progressively turns more nearly parallel to the daylight until, attaining its normal temperature, it is directly facing the sun and exposing a minimal surface. Should its temperature rise uncomfortably higher, it will purposefully move into the shade of a rock or bush for a while to cool off, sometimes climbing into the branches to get away from the hot ground. And in the evening it is fond of burrowing into warm sand, tucking itself snugly under a granulated quilt for the night. In the morning, rather charily it will poke its head out into the sunlight, the while keeping the rest of its by-then-cooled body still under the covers, discreetly waiting until the sun has heated the blood coursing through a large sinus in its head enough to raise the temperature of its whole body. Only after it is thus prudently preheated all over does this wary reptile shake off its remaining bedding to venture forth primed for possible emergency speed and efficiency.

32     Snakes do better than lizards in some ways, their long flexible bodies

enabling them to expose more surface (per unit of mass) when they want to absorb or radiate their heat yet without forgoing their capacity to condense (by coiling) into a compact mass at noon or night or whenever they want to conserve either coolth or warmth. Another animal device for temperature control is having a skin that can change color or shade. By such means, many small reptiles and amphibians turn paler when hot, thereby reflecting more and absorbing less of the sun's rays.

Heat adaptation being vital to the diverse thousands of species of desert animals, it is not surprising that they have evolved a wide variety of drought strategies such as the spade-foot toad's way of sleeping through the nine driest months in a burrow sealed with his own jelly, the zebra's sniffing out and digging wells in dry stream beds, the tortoise's storing of a summer's supply of water in two sacs under his upper shell, the elf owl's quenching of thirst with wet spiders, the rabbit's sipping cactus water while radiating body heat from his giant ears, the deer's cooling himself with belches, desert birds nesting not according to sun declination but at the coming of rain, the bats that fly each spring to a cooler clime. And there are at least twenty species of desert fish swimming in the permanent oases, lungfish that sleep through long droughts caked in dried mud, eels that slither through wet grass at night from well to well, and dozens of kinds of snails and shrimps creeping and wiggling in brief puddles all over the desert after a rare downpour who, before drying up, lay eggs that can wait for decades, perhaps centuries, in the parched salty soil to hatch whenever the next rain comes.

But the animal most famous of all for its adaptation to drought is surely the Arabian camel, whose introduction as the ideal vehicle for desert transport, probably during the second millennium B.C., brought the incense land of Sheba into practical contact with Egypt and Babylon — a welcome and dramatic change after the age-old frustrations of trying to drive parched and emaciated donkeys from oasis to oasis, as did the Jews in their 40-year exodus. Camels have long been reputed to have a special water reservoir in their humps or, as some say, in their "fifth stomachs," but camel research in the Sahara in the 1950s by Knut Schmidt-Nielsen and others since has not revealed any such localized container. The fact that healthy camels can go without drinking for months, particularly in winter, and still refuse good water when offered it (as they sometimes do) does not prove they have been tapping a secret internal storage tank. For many animals from goats to desert rats can get along nicely without drinking, and even a human can abstain comfortably in cool weather if he eats plenty of juicy

33

vegetables and fruits and does not sweat. But in warm weather, particularly if he exercises, a man must lose water from his body so fast that he is likely to want to drink several times a day. This of course becomes inevitable in very hot weather if he is to keep his temperature within a couple of degrees of normal through the cooling effect of moisture evaporation, for a man suffering from heat cannot lose even 12 percent of his body weight in water and hope to survive. As he gets drier, his blood literally thickens and eventually becomes so sticky that his heart is overworked and his circulation too slow to conduct his internal heat to his skin, which throws him into a runaway fever that quickly accelerates into what is called "explosive heat death." If you "oiled" an engine with glue and neglected its cooling system, no doubt something similar would befall it.

The camel circumvents such a disaster, however, by keeping the loss and evaporation of his body water to a minimum. He does this through such devices as highly efficient kidneys that use much less water to eliminate the same amount of uric waste and a wonderfully specialized liver that sifts out a goodly portion of such wastes before they even reach the kidneys by shunting them back through the blood to the stomach for digestive reprocessing (along with incoming low-grade fodder) into new protein. He also largely seals off the watery plasma of his blood from the rest of his body water by means of albumin, so that he can be dehydrated to the extent of 25 percent of his normal weight without appreciably thickening his blood. And he saves a great deal of water by sweating much less than horses, cattle or men, which he manages by a delicate combination of heat insulation in his efficient camel's-hair coat and his hump that places reserve fat where the sun beats most fiercely, and less insulation where he has almost no fat on his flanks, belly, neck and limbs, from which in consequence any excess heat can radiate away most easily. Despite all this, however, the camel's temperature rises as much as 12°F. during a hot day. Yet the very fact that he can tolerate 105°F. internally without feeling feverish or losing strength saves him the evaporation of sweat or breath it would cost to keep him cooler. On the other hand, his ability to cool down to 93°F. on chilly nights — a kind of semihibernation — spares him both energy and the inevitable evaporation that would have been added if he had stayed warmer. This fluctuation may give you the impression that a camel's temperature is unregulated, but actually it is confined within strict limits, for his body will sweat hard to avoid rising above 105°F. and, if he has free access to water, he will seldom get warmer than 100°F., while 93°F. is his bottom limit, than which (until he is dying) he may cool no colder.

As for animals' ability to withstand heat without dryness, probably the palm should go to an odd fish called *Barbus thermalis* who lives in the warm springs of Ceylon at temperatures as high as 122°F. Or maybe to a pond snail who dwells in almost equally hot springs, despite the fact that some of them contain a lot of sulfur.

On the other hand there is a polar codfish named *Boreogadus* who swims about actively in saltwater at 29°F., several kinds of "ice fish" who do equally well in Antarctic seas (getting along virtually without red blood cells or hemoglobin) and penguins in Antarctica who routinely raise their families in air temperatures that sometimes drop below −100°F. And there is an insect known as *Grylloblatta* who looks like a cross between a cricket and a cockroach but dwells in the predominantly frozen soil of polar and subpolar mountains and is so far from being able to withstand normal insect temperatures that, if placed on a human palm, he will "burn" to death in a few minutes.

A crucial factor in an animal's adaptation to temperature is the fluidity of his fats and oily juices (as distinct from his watery juices) for it has been noted that cold-resisting species like the northern salmon have a more liquid type of fat than tropical fish — that is, a fat with a significantly lower freezing point that would qualify it as an antifreeze, four varieties of which have already been discovered in fish, all with a high content of the amino acid alanine. In addition there is the critical matter of enzymes which are very sensitive to temperature, slowing vital metabolism when too cool, enervating themselves by overactivity when too warm.

Easier to understand of course is simple insulation, a vital boon to large animals yet amazing in its variety. While insects are too small to carry enough insulation to effectively stabilize their temperatures, birds and mammals have evolved the enfolding of wonderful mantles of air in their feathers and fur. Eskimos do something similar in wearing their fur parkas extremely loose, so body-warmed air can flow about their limbs. Deer fur in winter not only holds air between the hairs but each hair itself is hollow and sealed at both ends, so a protective snow blanket may lie unmelted upon a sleeping doe's back for days if she remains still that long. The insulation value of fat and flesh also largely explains why Eskimos tend to be fatter than people of warmer latitudes, and why arctic animals such as walruses, polar bears and ptarmigan have stockier necks, legs and tails than most tropical ones. This effect indeed has recently been studied in laboratories where mice raised at 90°F. (close to their blood heat) grow up scrawny with fine long tails while others raised at 60°F. turn out fat with stubby tails. Then, some animals solve the temperature problem

35

by maintaining two internal temperatures: a tropical one for their main body mass and a semipolar one for such extremities as hoofs, tail, ears and nose, an expedient that not only reduces the heat loss of radiation from areas that do not need to be kept very warm but, for example, keeps arctic gulls' feet cold enough so they don't ever melt ice, which might risk their later becoming frozen in. It works through an extraordinary circulatory heat-exchange in which the warm arterial blood flowing toward an extremity is cooled by passing close to the returning cold venous blood, which, in the same exchange, is rewarmed preparatory to reentering the heated central parts of the body.

Beyond this multitemperature scheme there are the many ways and degrees of shifting the whole body into a lowered metabolism rate known as hibernation, which has been perfected by woodchucks, chipmunks and several other northern animals, and is known to take place to a slight degree even in the case of humans, such as Laplanders, Eskimos and Australian aborigines. It works because chilling is a general anesthetic that slows down all metabolic functions, reducing the body's need for food, water and oxygen to the point where it can sleep securely on its own fat until spring, breathing scarcely once a minute the while. Tests have shown too that a mammal's sight and hearing cease to function about when his temperature gets down to 94°F. or his pulse to 40 a minute. His breathing becomes undetectable around 80°F., his heart stops by 50°F., and oxygen is hardly needed in any form below 39°F.

## THE ELEMENTS OF LIFE

Now, having said something about the limitations imposed on animals by such factors as gravitation, altitude, temperature and humidity, we come to the question of chemistry — of the actual elements of Earth and the universe. Elements, after all, are what matters in matter, which is life's substance. And although we will later see strong evidence that material substance is a mere aspect of life, right now we may consider living bodies as material things.

About three fifths of all the hundred known elements of the universe have been found in the cells of living things. However, this does not mean that they are all "elements of life" or essential to life. The most significant and vital element in life on Earth may be said to be oxygen. Although oxygen atoms constitute only 1/18 of 1 percent of the atoms in the known universe, they include more than one out of every four atoms in the human body, one of three in the oceans and almost one of two (47 percent) in the solid crust of our planet with all its rocks, soil, forests, animals and cities. Hydrogen is closely associated with oxygen, especially in water ($H_2O$), and these two elements account for about eight ninths of the body, which may be two-thirds water. Carbon is likewise a prime element of life, composing about 10 percent of atoms in animals and man and a much higher percentage in vegetables, even though it is less than one fifth of 1 percent of Earth. After these in abundance for life come nitrogen 3%, calcium 2%, phosphorus 1%, sulfur .25%, potassium .2%, sodium .15%, chlorine .15%, magnesium .05% . . . But few of these elemental proportions of life correspond closely to those in the planet as a whole (even less so in the universe), for carbon is only the eleventh most abundant element on the earth's surface and even hydrogen forms less than a quarter of one percent of solid Earth, while silicon and aluminum (important bulk elements in rocks and soil) comprise 28% and 8% respectively of the terrestrial surface, yet are scarcely detectable at all in living tissues, though many creatures use silicon in feathers, shells, horns or in the skeletons of coral, sponges, radiolarians, etc. So it seems that life is quite selective as to the ingredients it chooses from those present on the planet. And it is in keeping with this principle that certain elements which life uses only in microscopic amounts are absolutely essential to it notwithstanding. For tests have conclusively proven that, unless animals and plants find and ingest these particular "trace elements," they must soon sicken and die. These vital traces, interestingly enough, include such metals as iron, copper, zinc, chromium and manganese, without at least a little of which neither animals nor plants can live — four atoms of iron being required, for instance, in every molecule of hemoglobin, which is the red in blood; copper and zinc in enzymes connected with breathing; etc. And they include cobalt, fluorine, selenium, molybdenum and iodine, all vital to animals.

A good example of a trace element that loudly signals its own absence is iodine, because a dearth of iodine in food and water brings out goiter in human beings. Only one atom of iodine in ten million in the bloodstream, however, is enough to prevent this swollen thyroid disorder by bolstering the metabolism-regulating hormone thyroxine with this vital element, which forms 65 percent of the thyroxine molecule. In evidence, just recall how easily the widespread adoption of iodized table salt has overcome the goiter problem in iodine-deficient areas such as certain inland parts of central Europe or mountains in the northwestern United States.

And the story is similar with iron, copper and other elements, which       37 are like microscopic keys linking the minerals of Earth to its creatures. A single ounce of cobalt, for example, which provides the central atom of each vitamin $B_{12}$ molecule, will sustain 800 sheep a whole year — but without it they quickly become anemic and start dying off. And sometimes trace elements have complex interrelationships, as exemplified by Australian sheep that get copper poisoning in pastures rich in copper but poor in molybdenum or inorganic sulfate — their deadly ailment being relieved only by feeding them a minimal trace of molybdenum together with sulfate, a combination that in some not-yet-clearly-understood way always manages to counteract the excess of copper.

Trace elements, like all other material substances and no doubt many immaterial ones, thus can become poisons when taken in large enough quantities. The purest water will kill you also — and not necessarily by drowning — if you take enough of it. And strong acids and alkalis of course have a violent corrosive effect, killing cells by the billions and often destroying a large animal by hemorrhage, shock or disintegration of some vital organ. But weaker poisons may be equally if more subtly lethal by merely nudging the sensitive acid-alkaline balance of the body beyond the tolerance limit of vital chemical reactions. Or, more subtly still, by barely interrupting some intricate chemical reaction upon which metabolism (therefore ultimately life) depends. Yet, despite the delicacy of organic chemical balance, a number of creatures have managed to evolve an amazing tolerance for particular poisons — among these the venomous animals and their parasites, or specialized ones like the brine fly who lives in the vats of saltworks, an oil-loving cousin who makes his home in petroleum pools, or the amazing drugstore beetle who can live for generations inside a bottle of belladonna, ergot, squill or any of a hundred other dangerous drugs. But of course even these animals must feel constrained and limited by their own exceptional, if not easily understood, adaptations.

## LIGHT

And if the chemistry of the earth, along with its gravity, temperature and humidity, can thus so severely discipline its life, one cannot afford to overlook the intangible but vital factor of light, the absence of which (as we have already noticed) makes it virtually impossible for any plants to grow in the black depths of the sea, and in somewhat different ways has a comparably profound effect in the black underground caverns of the land. In mentioning caverns, of course I am not thinking of the open areas near cave entrances inhabited by bears or mountain lions or men. Nor of the twilight zones farther in, where bats and owls roost and sometimes phoebes, jackdaws or hummingbirds, alongside of cave rats and all sorts of crawling things and a marginal vegetation of pale ferns, moss, mushrooms, lichen and algae. What I have in mind rather are the deep recesses of so-called absolute darkness, thousands of feet beyond the dimmest noticeable light. The only bird that penetrates this far is the fabulous *guácharo* or oilbird of Venezuela, which, in spite of its four-foot wingspread, has been known to congregate in raucous flocks exceeding 5000 in a single great cave, each bird navigating by its echo-location acoustic system similar to the sonar of bats (page 206), flying out into the open sky every evening at the gentle invitation of the gloaming, driven back to the cave each dawn by the gathering glare of day. Among those other creatures that do not emerge at all from the inner blackness even at night, there is little need for eyes and, as a consequence, any vision their ancestors bequeathed them seems to have devolved toward blindness. Indeed, as in the deep sea, the only call for sight in the innermost caves would seem to come from the occasional evolvement of luminescence, such as that of an exotic maggot of New Zealand which spins silken pendulums with beadlike globules of sticky mucus that glow blue-green in the darkness of certain grottoes and help lure and catch moths and other flitting insects that retain some vestige of vision. Among most of the mollusks, snails, slugs, crayfish, bugs and amphibians of such ink-black realms, palps are expectably much more developed than eyes, while hairs are used less for insulation than for feeling out one's prey or enemies. And there are innumerable special and looping food chains, such as those of insects that eat molds that grow on bat guano produced by bats that eat insects that eat molds . . .

While a few animals seem able thus to persist in pitch darkness for long periods, life in general on Earth both needs and appreciates the light. All the terrestrial food chains, for example, are firmly based on photosynthesis, and even the most addicted cave dwellers who "never" emerge must derive their ultimate nourishment from plants or seeds that originated in the light before they were washed or wafted or elsehow carried to the inner dark. And one should never forget that the influence of light is just as important at sea as upon land, since about three quarters of all the photosynthesis taking place on Earth occurs in the oceans. In fact, just as rabbits and deer graze on mountain meadows, in turn becoming meals for the prowling flesh eaters, so does all sea life pyramid upon the tiny one-celled algal "grasses,"

39

which are munched by plankton smaller than a pinhead, which are eaten by slightly larger fish, which later are devoured by still larger finned creatures, and so on up the scale of sizes in a seemingly ruthless but almost automatic winnowing process of eat and be eaten. And both these complicated food chains, on land and sea, are founded on the energy of sunshine that sparks the inexorable regimen of photosynthesis — something still quite mysterious yet so profound and important that you can be sure we will be looking at it with a much closer eye before we get through the next chapter.

Right now we only need to notice light's direct and extraordinary power to control almost every creature alive: through seasonal changes in the lengths of days, through rhythmic dawns and dusks, even through the reflected glow of the moon. How do birds know when to fly south in the fall? Or in what direction to head? Obviously whatever guides them can be no casual half-hearted influence when we consider the numerous authenticated reports of birds that have overcome their powerful parental instincts and actually abandoned their late unfledged broods to starvation rather than deny the irresistible call of the southland. And there is no doubt left that the shortening days and declining angle of sunlight are what mainly triggers this seasonal migration. Indeed a ponderable helping of recent research has established the reality of a surprisingly precise animal awareness of light, amounting to a celestial navigation sense, not only among bees and birds but extending to fish, reptiles, mollusks, worms and microorganisms. And this discipline of light, which includes starlight, moonlight, lightning, man-made illumination, animal luminescence and other forms, is already accepted as a key factor in the regulation of life in general — which naturally includes vegetable life — and most of the major rhythms of the body and its behavior are directly or indirectly geared to light. Not only are the sleeping and feeding habits of nearly all animals and man attuned to the 24-hour daylight-darkness cycle, but so are the entire life spans of some insects (mayflies for example), and innumerable lesser events from the diurnal color changes in fiddler crabs to the pulse rates of wombats and the metabolism of the twite and the smew . . .

I am told that snowshoe rabbits, for example, when blindfolded for an hour each autumn day, turn white earlier in the season than do unblindfolded rabbits, because visually shortened days act as potent "light signals" to forewarn their enzymes and prepare their fur for the coming of winter. Isn't it for the same basic reason that blind adolescent girls have their first menstruation later than girls who can see? And what about the crows who, after being illumined by artificial daylight before dawn on winter mornings, ripen their reproductive organs earlier and migrate sooner than their unenlightened brethren still roosting in the dark?

These are the merest hints at the influence of light upon life and its countless rhythms cadenced so inexorably to the motions of the celestial spheres. For, like gravity and temperature, light is as abstract as it is pervasive, and there is no known end to the constituents in its dominion. Nor are there definable limits to many of the other forces that shape life and its functions in the known worlds. So we must try to content ourselves as best we may with this sketchy overture to an immense subject, because, in a book like this that deals with all life, one must conserve and budget one's pages if one is to have room to investigate and write about life's numerous other and less-known aspects.

I therefore wrap up our space age bestiary of God's creatures without having yet discovered much more about the life of our sample planet than the scrimpy evidence already presented that animals are formed and limited in many unobvious ways by the Earth they are made of — an Earth indeed that seems to involve so much more than any man can hope to dream of that its total meaning is, for all I can tell, about as unfathomable as the Milky Way.

Yet, one wonders: is life also essentially good? Is planetary existence in any sense founded on wisdom, or love? Or are we rather drifting before the aimless winds of chance? If life turns out to be not wholly random, does that in itself imply a purpose to it, a design or possibly a spunk of something divine?

We have still to come to grips with these larger questions and, until we do, we cannot be fully aware of this node of flesh nor possibly realize what whole it is part of. Nor can we even learn whether a virus may potentially see the stars — nor in what billennium a coral polyp may at last discover the continent it created.

So let us listen to the quiet texture of the sphere. Let us sip our wine on the vine and whisper freshly with gentle grasses. Let us twine our toes into roots, surge with the sap and hearken to great trees. For theirs is a life that is older than animalness, deeper than thought, and that may well penetrate the primordial genesis of us all.

# CHAPTER 2

# Realm of the Vegetable

A S I LOOK DOWN almost casually upon the dark blue and cloud-flecked expanse of forest in northern South America, then upon the even darker jungles of central Africa, a curious nostalgia begins to pervade me. It derives from a sensation as strangely comforting and inexorable as the turning of the earth, which I can read like a clock from here, an instinctive realization that my flesh and bones are some-how parcel of that bulging mantle of vegetation that girdles the middle latitudes.

I wonder at the meaning of it — at the profundity of the relationship — as I recall, not without conscience, my most intimate arboreal affinities in two thirds of a century. During that time I have not only climbed many a tree but have consumed them in stoves and fireplaces in untold numbers. Though most often heedless, I've also been an admirer of trees and occasionally their midwife, nurse, godfather, pupil, trainer, experimenter, surgeon, conductor, executioner and undertaker. I wonder how much awareness in return any tree has had of being my staircase, watchtower, lumber, fuel, plaything, nursling, teacher, pupil, subject, patient, passenger, victim or corpse. However slight, I feel there has always been some degree of withy cognizance, which, in mathematical terms, must have had a value greater than zero. Or, if I may express it more humanly, is it not appropriate that I, a representative of mankind, should pause now and then in a forest to shake hands with a sociable bough? Or metaphorically to thrust my feet into roots until, wringing my empathy to its utmost, I imagine I can begin to sense just the tiniest inkling of the feeling of being a tree? Even guess that the tree may sense me?

I feel sure the venerable peasants of the Austrian Alps and other old-time woodsmen all over the world are not being unreasonably superstitious when they beg a tree's pardon before pruning it or chopping it down. Nor need any of us scoff at the beautiful symbolism, which is said to have originally inspired this tradition in the almost universal recognition of tree spirits, in concepts like the Persian Tree of Immortality, the bo tree under which Buddha received his enlightenment, the Hebrew Tree of Knowledge which opened the eyes of Adam and Eve, and in such powerful myths as the Scandinavian legend of the world tree of Yggdrasil, Odin's sacred ash of celestial dimensions, whose monstrous roots penetrated the subterranean kingdom of death, where the wily serpent Nidhögg gnawed at them ceaselessly and whose prodigious upper branches tossed among the stars of life (still commemorated in Christmas tree decorations) where four stags browsed daily upon their tender buds. The bards have it that, despite these dual consuming influences of death and life, the Yggdrasil tree's supernatural growth kept it constantly green, ever new, so that its heroic trunk bound Heaven to Earth and formed the imperishable structure of creation.

The appropriateness of such an archetype for our planet is striking, for vegetation in quite a real sense does bind the sky to the Earth. Not only is vegetation the land's chief organ for absorbing rain that falls from the clouds and (after using it briefly) evaporating the same rain from its leaves back into the sky to maintain the terrestrial circulation, but vegetation quickens and conserves the entire atmosphere by inhaling excess carbon dioxide and exhaling vital oxygen, the while holding up mountain ranges, stemming erosion with stems, rooting soil with roots, directing the courses of brooks and rivers and literally leashing stationary clouds to wooded slopes with invisible thongs of humid wind.

To understand the specific vegetable function, then, we need to    43 realize that trees are not just standing there twiddling their twigs. We need to look closer at what actually goes on inside a tree, at the workings and chemistry of plants, including the dynamic virility in the tree's driving, thirsty roots that spiral shrewdly downward with force enough (when needed) to split boulders weighing tons; in the breathing of its leaves that almost sniff their way upward in search of carbon and "glance" around for sunshine to cook their airy food; in the restless traffic of tenuous arteries that not only heft their daily tonnage of water to the sky but tote food and building materials from highest twig to deepest root hair, in subtle electrical discharges that transmit messages little less sophisticated than those sent through human nerves, and in the far-ranging migrations of spores that entrust a forest's future to the winds, rains, rivers and the wandering animals of the world.

One of the first things to be remembered about the tree is that, like all terrestrial life, it evolved from sea organisms. In fact it may be regarded as basically a seaweed that learned, after millions of years of desperate struggle, how to live on dry land. And, like the more mobile creatures that followed it, its body is still composed largely of water and depends utterly on water for life and nourishment. For these reasons the tree today has little choice but to make its home in soil, the most stable source of moisture on land. But it also needs the energy and building materials it can only get from sunlight and volatile air. Its architecture therefore runs to large surfaces both below and above ground, these systems perforce united by a trunk bulky enough to hold all the channels and pumps required for a continuous top-bottom interchange of materials, yet resilient enough to stand up to the fiercest winds.

What on Earth could fill such a bill better than a tree? And the beauty of its being develops apparently by random chance, while every root and branch makes its seemingly independent search for water and light, the whole tending to a highly efficient radial symmetry about the central bole, from which the foliage, according to actual mathematical analysis, literally disposes itself so the largest number of leaves receives the greatest amount of sunshine for the longest possible period. The fast upward flow of sap (more than 99 percent water) is through tubelike cells in the solid interior wood of the tree, while the much slower downward flow of syrupy gruel (only about 95 percent water) is through smaller cells in the outer layer, known as bast, just under the bark. The two systems normally move in opposite directions but are partially interconnected and, in the growing season, may go locally in the same direction for a while, particularly where new shoots are being supplied by both systems at once. It is helpful to remember that these two systems are separated only by the most vital part of the tree, the cambium layer (about an inch under the surface) where new "green" wood is continuously being born. Obviously most of this new wood adds a fresh ring each year on the outside of the layers of old rings that form the sapwood and heartwood, the main mass of the tree. A small portion of the growth, however, also produces a new yearly layer of bast inside the bark, the outside of which is inevitably aging, cracking, slowly splitting off and being discarded like animal scales, feathers or fingernails by the pressure within.

The reason I emphasize the cambium as the most alive part of the tree is that, although it is only one cell thick, it is the tree's only layer where cells thus divide and subdivide in birth, rapidly increasing their numbers, simultaneously augmenting their substance in two directions: at once building new rings outside the old wood and new layers inside the old bark all the way down to the deepest root hair and up to the highest fluttering leaf. And the fact that the level of the tree where the original seed sprouted is its oldest part naturally makes the trunk thickest there (just below ground) and causes it to taper gracefully and efficiently both upward and downward. The full three-dimensional form of each tree ring thus turns out to be a double cone, the main ring inside the tree trunk with its point upward, balanced by another inside the roots pointed downward, each complicated by branching and subbranching and created all over again each year as it is stacked outside last year's ring, like tomorrow's ice cream cones in a continuous nest — a symmetrical configuration that, engineers point out, enables the tree to reach twice the height of a cylinder of equal mass.

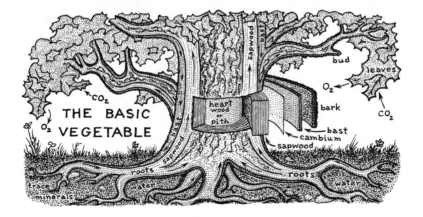

45

## Roots

If you are among those who think of roots as nothing but dull appendages sleeping peacefully in the stuffy dirt under a plant, you may be interested to know of their real adventures while aggressively hunting for water, air and mineral foods, which means fighting many a pitched battle against competing roots or animals, intermingled with making friendly, constructive deals with rocks, sociable molds, worms, insects and, more and more frequently, man. At the tip of each advancing thread of root is a root cap, a sort of pointed shoe or shovel made of tough, barklike, self-lubricating stuff that the root pushes ahead of it and replaces constantly by cambium cell division inside as the outside is worn away and turned into slippery jelly by passing stones, teeth, running water or other antagonists. But the tiny root cap is only the first of several specialized parts which, working together, enable the root to steer its zigzag or spiral course, skirting serious obstacles, compromising with offensive substances, judiciously groping for grips on the more congenial rocks, secreting powerful acids to dissolve the uncongenial ones, heading generally downward in search of moisture and minerals while ever careful not to run completely out of air.

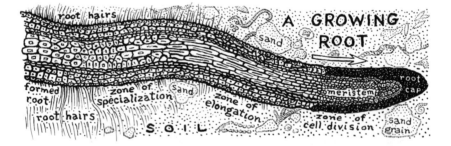

Close behind the root cap as it advances comes the meristem, the botanists' name for the concentrated terminus of the cambium layer (shown in the illustration) which literally never grows old. This is where the root cells divide at their fastest, and it is followed immediately by the zone of elongation where they gorge themselves with water and expand like hydraulic jacks, relentlessly forcing the root cap ahead of them. Then in turn comes the zone of root hairs, the unseen beard of microscopic filaments sprouting straight out of cellular protoplasm and providing the tremendous surface needed to absorb water out of dry soil fast enough to sustain not only root growth but the health and metabolism of the whole tree.

There is nothing fixed about these specialized root parts, mind you, except their relative order, for they are advancing continuously with the moving threads of root. And you may be sure the same cell that was born in the meristem cradle today will have grown enough to octuple in length by tomorrow, will sprout its first root hairs the next day, be covered with them the day after and probably lose them again before the end of the week as the advancing root leaves the cell farther and farther behind. The tree thus drinks in the manner of a very water-dependent beast, its total thirst actually far exceeding that of any animal on Earth. In fact, an ordinary cornstalk must drink and evaporate 50 gallons of water in order to reach maturity while a full ton of the same liquid asset is invested in the field for each loaf of bread.

Is not this the reason why many plants, like icebergs, remain about nine-tenths hidden beneath the surface where their roots distribute themselves in all directions as they systematically probe and search the soil, why a small alfalfa plant was found with its longest root groping thirstily and doggedly downward through dry sand for 31 feet, why mesquite on the desert is known to have reached a depth of more than 100 feet, why a random two-year-old tuft of grass carefully dug up by a scientist in 1937 had such a luxuriance of fine roots that their individual lengths, after being meticulously washed and separated out of 4 cubic meters of loam, added up to an amazing 315 miles? And more recently, and more astonishingly, why a botanist measured and calculated an innocent-looking little rye plant to have 14 billion root hairs with an end-to-end length of 6000 miles and growing at a cumulative rate of 100 miles a day while drinking from a total surface which, if it could have been spread out flat, would have exceeded the area of ten football fields! I know it is high-handed of anyone to rearrange roots into one long statistical strand, but doing it helps my wallowing mind grasp the incredible multiplicity with which even the gentlest plants steadily stitch themselves into the earth. Sometimes when I look at a blossoming shrub or tree, I imagine its myriad root caps needling their seemingly independent ways outward like a million tiny sewing machines, dutifully binding and weaving their central organism to the ground. And if one should project such a picture of a root system into accurate perspective with its upper body, it would be revealed as a sort of foreshortened mirror image of the visibly moving upper tree, the central taproot corresponding to the trunk though usually shorter, surrounded by great branchlike primaries often more

47

outspread than their counterparts above, along with lesser secondaries, slim tertiaries, twiggy "obliques" and finally the threadlike capillaries which hold the root hairs that correspond to the smallest veins in the upper leaves.

## LEAVES

Raising our attention by this comparison to the more familiar trunk and branches of the tree above ground, we find that things up here almost match those below also in the sequence of growth, the tough, slippery root cap being replaced by a protective bud or growing tip, followed in order by a meristem zone of cell division, a zone of elongation, then one of specialization, which, however, here needing to breathe rather than drink, sprouts leaves with nostrils instead of root hairs with throats. Certainly it is the leaf that is the most distinctive as well as the most vital part of most plants, which properly value it more than a root. For the leaf is to a vegetable what a lung is to a vertebrate and maybe more. If you stick a tiny piece of leaf from an African violet into moist soil, it will normally grow roots directly out of the leaf. And seaweed (in essence nothing but a leaf) can live and breathe indefinitely while drifting on the ocean's surface without any roots at all. By contrast a bare twig on a tree in winter can breathe only slightly through microscopic holes in its tender bark, doing so much less efficiently than an insect through its spiracles, for the twig without leaves is in effect asleep and, until it gets them, can hardly grow at all. Which may be why it is so eager — in some cases desperate — to unfurl its sprouts each spring. Failing in this indeed would be like a porpoise neglecting to surface for air, without which he must surely die.

The leaf thus turns out to be much more than a symmetrical flap of verdure. Indeed, surpassing the dreams of the greatest philosophers of old, it is revealed by the microscopes of modern science as not only a breathing organ but a bustling, automatic food factory full of tubes, retorts, chambers, valves and shutters operated by more timers, thermostats, hygrostats, feedback and catalytic controls (some of conflicting or unknown motivation) than anyone has yet been able to assess. Among its tubes are the upper ends of the same ones that raise sap through the wood all the way from the root hairs, as well as those that carry the gruel made by the leaf back down into the bast to nourish the whole tree. All such channels, by the way, not only convey vital liquids but also serve as structural bones in the leaf as in the tree. In a house it would be like having water pipes strong enough to double as joists and studs, to hold up the roof and bind everything together.

The retorts and chambers on the other hand are spaces between the microscopic cells and chlorophylled subcells, called chloroplasts, which are so arranged as to give the leaf access to the great quantities of light and air it needs. Most of a leaf's volume, you see, is normally air, which permeates the spongy green interior so completely that there is up to a hundred times more atmospheric exposure there among the chloroplasts than out upon the waxy, nonchlorophyll exterior skin. Botanists have calculated that there are about 600 square inches of surface inside a leaf for every cubic inch of its bulk and that a large elm tree has in all some 15 million leaves with an area, if spread out whole, of nearly 10 acres or, if unfolded into the sum total of air-breathing light-absorbing surfaces of all the internal chloroplasts, something like 25 square miles.

A tree, however, no more breathes at a constant rate than does an animal or a combustion engine, for it too has something equivalent to nostrils or intake manifolds that open or close as they want more or less air. These apertures may be described as innumerable invisible pores in the skin of the leaf, mostly on its underside, that look (through the microscope) rather like little vaginas and are controlled each by a pair of "guard cells" not unlike vulvae which automatically change shape under the stimulus of moisture, temperature and light. When the weather is very dry and hot these pores remain mere slits but begin to cleave apart as humidity makes the guard cells turgid, flaring wide open when it is cool and rainy and the chloroplasts have an excess of moisture to evaporate back into the sky, plus a related need for air and the nourishment they can derive by absorbing its carbon dioxide under the impact of sunlight. Pores in a leaf may number anywhere from a hundred on small arid plants to a hundred thousand on the large lush foliages growing in swamps or rain jungles. And the magnitude of their gas exchange is suggested by the calculation that

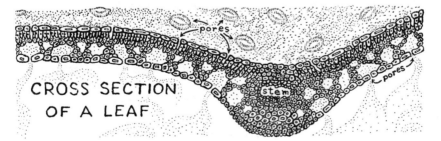

CROSS SECTION
OF A LEAF

pores

stem

pores

49

every day on the average a leaf requires as much carbon dioxide as would be contained in a column of air standing 150 feet high upon its surface, while in the same period the average tree in leaf evaporates more than a hundred gallons of water.

The pores of course are only one of many means by which the leaf adjusts to its changing environment. Another is the chloroplasts' ability to change their shapes and move around like monks within their cells, dispersing to intercept the maximum of light on cloudy days and lining up in each other's shadows to escape the sun when it has been bright too long. Sunlight penetrates right through leaves, you see, and is somehow sensed by the chloroplasts, evidently through submicroscopic lenses that amount to vegetable eyes — and the leaves' translucency or (conversely) absorptivity increases and decreases by as much as a third as the chloroplasts arrange and rearrange themselves. And often, it seems, leaves have to make decisions or compromises among conflicting motivations, such as choosing between closing their pores to conserve water (which slows their breathing) and opening their pores to snuff up carbon dioxide (which dehydrates them through evaporation). Usually the leaf settles such an issue by leaving its pores partly open like a man reconciling thirst and hunger by swigging drinks between mouthfuls of dinner. But experiments show that leaves on the whole are much more apprehensive of running out of water than out of air (actually just as vital), perhaps because, like many animals, they so often feel thirsty but almost never know suffocation.

## ENERGY

Having discussed how the vegetable functions from its roots to its leaves, it is time to ask such basic questions in botany as: What are plants made of? Do they breathe? What do they eat? Where do they really get their substance? And energy?

That the answers are far from obvious is shown by the long history of man's fumbling struggle to find them. Aristotle thought plants somehow ate their substance from the dirt, and it was not until two thousand years later that a Flemish alchemist-turned-chemist named Jan Baptista van Helmont tested the not-unreasonable assumption by accurately weighing the soil around a willow tree he kept growing in a pot for five years.

This classic first-ever scientific experiment in botany convinced van Helmont that plants do not consume earth, but it took many more tests before Joseph Priestley, an English amateur chemist, finally

proved that plants breathe as they grow. In 1772 he enclosed a growing plant in an airtight chamber and found that it suffocated and died
just as surely, if not as quickly, as an animal sealed in a similar
container. But much more surprising and wonderful was his famous
subsequent experiment of enclosing a plant and an animal together in
the same airtight chamber and discovering that both could live! — a
seeming miracle that was soon proved due to an exchange of chemically different kinds of breaths or gases shortly to be named oxygen
and carbon dioxide, the animal inhaling oxygen ($O_2$) and exhaling
carbon dioxide ($CO_2$) while the plant accommodatingly did just the
reverse.

This great discovery led to an unprecedented use of flowers in
sickrooms (in those days usually tightly shuttered against "harmful"
outside air), a custom that has continued ever since. But a Dutch
physician and chemist, one Jan Ingenhousz, was skeptical enough to
want to experiment further and by 1779 he had found that a plant's
exhaling of oxygen is done only by its green parts, particularly the
leaves, and only while they are illumined — never in the darkness of
night when the plant inhales more oxygen than it exhales. By 1796 he
had learned enough to postulate specifically that leaves by day decom-
pose the carbon dioxide they inhale from air, using the carbon (plus
some water) to build their bodies while they exhale the leftover oxygen. It was the first definite human awareness of the process we now
know as photosynthesis, which, because it involves some 300 billion
tons of sugar in vegetation each year, is today considered the most
massive chemical process on Earth!

To summarize the essential chemistry: when green plant tissue is
under strong light its water molecules ($H_2O$) are split into separate
hydrogen and oxygen atoms, of which the hydrogen (H) combines
chemically with part of the inhaled carbon dioxide ($CO_2$) to make
carbohydrate molecules ($C_6H_{12}O_6$), particularly sugars, starches and
cellulose to build the plant, while excess oxygen is exhaled.

That is a nutshell account of photosynthesis, this fundamental life
process that, when it was discovered, seemed so incredible and com- 51

plex that many researchers spoke of it as miraculous. And I doubt not they were right. In the first place, its prime source of energy is literally unearthly, being photons of light streaming at 186,282 miles a second out of the sun. Then, although light seems to have little effect on water *away* from plants, light striking water *inside* green plants receives a special tool for operating on it: the chlorophyll molecule. This curious, green, magnesium-centered, daisy-shaped molecule ($C_{55}H_{72}O_5N_4Mg$) with its long carbon "stem" and 12 "petals" surrounding the 12th element is the key to the whole operation. As it absorbs the light's energy, its hundreds of subatomic particles jump momentarily from their normal orbits in each atom to larger, more potent orbits. Although this leap of excitement only lasts about a ten billionth of a second, the jolt when they revert back to normal somehow (not yet clearly understood) splits hydrogen from oxygen in adjoining water molecules, whereupon the separated hydrogen is grabbed

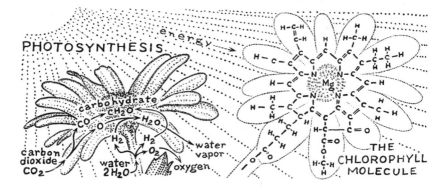

by big, spiral-shaped protein molecules called enzymes or catalysts, which seem specially prepared to handle it and promptly introduce it to freshly inhaled carbon dioxide (likewise in the hands of specialized enzymes) in such a way that the two substances immediately go into a kind of dance called a chemical cycle in which ingredients are intermittently added and extracted as they whirl and merge into new wood. To this I'll only add, at risk of whimsy, that the chlorophyll molecule's daisy shape is beautifully appropriate because the word "daisy" derives from "day's eye," meaning the sun, who not only energizes all vegetal growth (through chlorophyll) but is reflected in every sun-shaped daisy with the golden orb and white rays you see nodding in the meadow.

Of course photosynthesis is really very much more complicated than my brief portrayal suggests, because it also includes all sorts of less obvious molecules, such as carotenoids that help chlorophyll absorb energy of out-of-reach frequencies, other varieties of chlorophyll, various salts, minerals and trace elements from the soil, whole retinues of burly catalysts (such as ferredoxin, adenosine triphosphate, etc.), subcatalysts, acceptors, bouncers, cops, cooks, nurses, janitors and other chemical functionaries who keep things coming, going, stewing, growing — and literally bookfuls more . . .

When you think that all this is going on naturally and invisibly amid intricate valves and thermostats and regulators as if in a vast automatic factory inside every fluttering leaf, it is truly appalling — and you can hardly grasp the immensity by reminding yourself that the millions of invisible subcell chloroplasts in a leaf are huge football-shaped worlds in relation to the bustling industry within, each chloroplast packed with thousands of coin-shaped grana in neat piles, each granum in turn bristling with tens of thousands of bead-shaped quantasomes, each quantasome made up of about 200 daisy chlorophyll molecules, each of whose 137 atoms contains the mystery of creation!

## TRUNKS

Although we mentioned the vegetable's need for water in discussing roots and leaves, we must come back to water again now in order to describe how stems or trunks work. But this time the subject is not just finding water in the soil or later transpiring it into the air: it's the in-between problem of raising it as sap from many feet underground all the way up to the highest foliage — something that would seem a formidable feat in the case of a measured 368-foot giant redwood, which, counting its roots, could add up to a total height exceeding 400 feet. For, as water cannot be lifted more than 33 feet by earthly atmospheric pressure (which, believe it or not, is the force that "sucks" a drink up through a straw), something a lot stronger must be responsible. That something, according to a recent postulation, could be the pumping action of submicroscopic filaments that work like cilia. But the general consensus of botanists points first to the vertical pressure differential inside the tree, a force sometimes exceeding a quarter of a ton per square inch, though normally averaging closer to 350 pounds. In early spring it *pushes* the sap water up the tree, mainly with pressure generated from below by root osmosis (osmosis being the tendency for a permeable membrane to pass more liquid in one direction than the other), but as soon as the leaves sprout and begin chemically to consume as well as evaporate the upper    53 moisture, the negative pressure thus created in them *pulls* (by molecular tension) the tiny columns of sap all the way up from the roots and at speeds that sometimes exceed two thirds of an inch a second. This is possible because the chemical action (photosynthesis) in the leaf is immensely powerful and the hydrogen bonds in pure water ($H_2O$) give it great tensile strength (up to about 2¼ tons per square inch). Moreover it has been dramatically demonstrated through several kinds of tests on trees that in summer not only does the sap start rising every morning in the twigs earlier than in the bigger, lower branches, which in turn precede the main trunk and roots, but also tree trunks actually shrink to a measurable degree on hot days (due to the negative pressure within them) and, if punctured with a small hole, can be heard to hiss as they suck air inward.

If sap flow toward the leaves is hard to explain, the oozing of leaf gruel away from them is harder, roughly in proportion to its slowness, which takes minutes instead of seconds for each inch of advance. Some botanists say the gruel must be pushed by its production in the leaves because, unlike sap, it maintains a positive pressure in the growing season. But a few surmise there may be some sort of a physiological pumping action in the tiny bast tubes which, they suggest, may have already begun to evolve into organs that will ultimately be describable as rudimentary hearts.

## GROWTH

The first pioneer experimenter in botany to come to such a bold hypothesis, so far as I know, was Sir Jagadis Chandre Bose, founder and director of Bose Research Institute in Calcutta, who wrote about it in his *Plant Autographs and Their Revelations* in 1927, after inventing and developing the crescograph, an extremely delicate growth-recording instrument that revealed faint "peristaltic waves" in stems and tree trunks which increased whenever it rained and diminished during droughts. The telegraph plant (*Desmodium gyrans*), he particularly noted, has a kind of slow pulse and every three minutes its leaves nod slightly downward then gradually rise again. Whether such a rhythm, perceptible in some degree in many plants, may signal the beginning of evolution from the vegetable to the animal is a question only future research can decide.

Meantime we should not presume that plants, embedded as they normally are in all the nourishment they need, have neither motive nor capacity to move about. For move they certainly do, and some get

54

around actually faster than some animals, particularly such sedentary animals as barnacles, sponges and coral. A good example of a traveling vegetable is the slime mold that creeps and swarms over rotting stumps like an army of amebas. And there are the microscopic diatoms found swimming in all lakes and seas, who propel themselves with jets of protoplasm, some of which flow along slits in their sides, something like the paddles of sidewheel steamers. Also algae and ferns have spores or sperms that swim with whiptails, and a few species of these vegetable microbes attain a speed of more than half an inch a minute which, in proportion to their size, is impressive. If you compare a human sprinter running 7 times his height in a second with a jet airplane flying 50 times its length in the same period (at twice the speed of sound), the zoospore of a mere vegetable (aptly called Actinoplanes) has beat them both at an astonishing 100 times its own length in the same brief interval!

We will later consider some of a vegetable's other kinds of motion, such as opening its blossoms to the warmth of day, turning to face the sun, growing upright from a random seed, twining around a support, catching a fly, dodging a cow . . . but now let us behold in wonder the simplest motion of all: ordinary growth, which, in a plant, is required for just staying alive. According to Bose's high magnification crescograph, which automatically records a plant's growth on a graph on smoked glass, the average plant in India grows about one one-hundred-thousandth of an inch per second (almost an inch a day) with a slow pulsing motion, each surge upward followed by a slower recoil of about a quarter of the gain and the steps in the graph getting steeper and wavier at times of rapid sprouting, leveling out during drought and becoming completely horizontal at death.

Certain plants, however, "grow" much faster than the average, the fastest of all being probably bamboo. Indeed I saw a report in 1970 of a stem of *madake* bamboo that had grown a measured 47.6 inches in 24 hours. That is as fast as the minute hand of a watch and considerably faster than an ameba (a free animal) can move in any direction.

In Burma, during World War II, I recall that there were a number of occasions when soldiers by heroic efforts chopped a road through the jungle only to discover too late that they could not return over the same route a week after it was "finished" because of the tremendous volume and stubbornness of new shoots that had sprouted there as soon as the cutting stopped. I particularly remember hearing of an army truck, parked beside the road on a Tuesday morning during the monsoon, which would not budge the following Friday night because, as the sergeant driver explained, it had been "bamboozled" by the 3½ days' growth of bamboo that literally staked and wove its wheels and axles to the ground. If one could consider the kingdoms of nature to have been participating in the war in Burma, one would have to concede that it was the vegetables that generally emerged victorious over the humans and their mineral slaves.

Yet growth in such a plant as bamboo (really a genus of grass with exceptionally large elongation zones) is accomplished not so much by cell division as by mechanical swelling of cells. Indeed bamboo practically drinks its way upward, continuously extending its vertical pipelines by means of material they themselves have lifted. This lifting of course is accomplished through the pressure and friction of water in motion, for water is not just a drink, as any sailor can tell you, but at least as much a medium of transport — in vegetables even more than in animals, both kingdoms attesting thus to their evolutionary beginnings in the sea. In the cells of grasses and flowers, furthermore, water provides the vital pressure that gives them the stiffness to stand erect, without which they must immediately go limp or, in flower language, wilt. One can hardly be reminded too often indeed that water literally permeates all life on Earth, not only flowing daily through your body (of which it amounts to some 60 percent) but even more continuously through all vegetables (where it averages nearly 75 percent). And, besides water, there is the mysterious synthesis of vital substances from each of the three kingdoms of which vegetables are made — a synthesis that implements and coordinates everything from limb orientation to the processions of root hairs through Earth that resemble sheep grazing progressively farther and farther down a green valley, the complexity of their orderly interchange of materials being of the very essence of life.

This underground commerce furthermore accounts in large part for the basic order in any sort of vegetal growth. It hints at why baby plants first sprouting from seeds lose weight (like most babies at birth) for several days before they become adapted to their new freedom and, breaking the surface of the ground, gain the carbon dioxide and sunlight they need to photosynthesize their food and start increasing their total mass. It tells why something similar happens with transplanted saplings, which, like new kids in school, are awkward for a while as their roots and branches adjust to unfamiliar grounds and companions.

At first the sprouted seed, even while losing weight, multiplies the number of its cells at what a banker would call a compound interest rate with principal constantly increasing: the cells dividing and the new cell (though smaller) dividing again and again, virtually doubling their number at each generation. But as soon as the organism is mature enough to develop specialized zones in roots and shoots, its cell multiplication slows down toward a simple interest rate with fixed principal: new cells produced by the meristem zones elongating but (by then no longer meristem) not dividing again.

## FORMS

Each kind of plant of course has its own genetic character and attains its maximum growth rate according to its magnitude and capacity, the lupine at about ten days of age, the cornstalk in its sixth week, the beech tree after a quarter-century. And each one sprouts branches at angles and intervals of its own pattern, by which it may be recognized as a particular kind of tree or bush or flower from as far away as it can be seen. This basic order in branches is obscured, to be sure, by their more apparent randomness, yet the most careful measurements have confirmed that small nodes of meristem are methodically left behind the main meristem zones as they advance, and it is these nodes that later develop into knots and branches whose characteristic arrangement labels them as elm or oak, jasmine or jonquil in a subtle and beautiful harmonic periodicity. There is time as well as space in this vernal music too, for branches are born on different days as well as in different positions as surely as tree-ring calendars show both years and girths.

To be explicit, one of the simplest tree forms (found in maples, ashes, horse chestnuts and dogwoods) is the pairing of leaves, twigs or branches, two of which grow out from their mother stem exactly opposite each other, while the next pair (above or below) also grow opposite each other but at right angles to the first pair, giving maximum dispersion of foliage. A slightly more complicated form found in many more trees is the spiral. To see it clearly, tie a string to the base of a leaf, then extend the string along the twig and branch, looping it once around each leaf stem you come to, in as smooth a curve as possible. In the case of an elm or linden you will find the average leaf tends to be attached just halfway ($180°$) around its twig from the next leaf, so the string will spiral tightly at the rate of ½ turn per leaf. A beech tree, having leaves at only $120°$ intervals, yields a rate of ⅓ turn per leaf. An apple tree, oak or cypress with the common distribution of leaves at $144°$ averages ⅖ turn, a holly or spruce ⅜, a larch ⁵/₁₃, and so on. If you are mathematically inclined you may have suspected    57

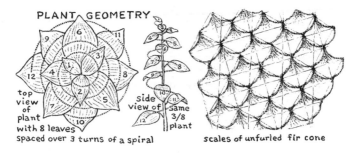

PLANT GEOMETRY

top view of plant with 8 leaves
Spaced over 3 turns of a spiral

side view of same 3/8 plant

scales of unfurled fir cone

by now that these fractions are not just random, for in fact each numerator and each denominator is the sum of the two immediately preceding it, both sequences of numbers forming the same simple and regular progression: 1,1,2,3,5,8,13,21,34,55,89,144,233,377 . . . This particular sequence was long ago named the Fibonacci series because a man by that name investigated it in Pisa in the thirteenth century. And it turns out to be the key to understanding how nature designs trees and is presumably a part of the same ubiquitous music of the spheres that builds harmony into atoms, molecules, crystals, shells, suns and galaxies and makes the universe sing.

Obviously in the vegetable kingdom it is not limited to the branching of trees. Its fractions beyond ⅜ have frequently been found in plants with compact seed or leaf systems, such as mosses, cabbages, or the florets in the center of petaled flowers with spirals going in both directions. Five thirteenths includes the white pine cone in which 5 turns of a spiral produce exactly 13 of its scales or seeds. Daisy centers use the three consecutive fractions 13/34, 21/55 and 34/89 and at least one sunflower head (22 inches in diameter) was recorded at Oxford University at an impressive 144/377 or 377 seeds to a spiral of precisely 144 turns!

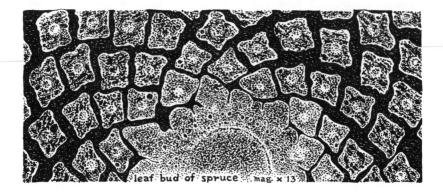

leaf bud of spruce  mag. x 13

If you suspect these are freak cases and that nature really has quite a carefree attitude toward mathematics, just try telling a pine tree to grow some way it doesn't want to. I mean, tie the tips of its branches so that the tall central one, the leader, is bent away from its accustomed verticality or the surrounding ones (in the case of a young white pine) from their stubborn 70° angles outward. No matter how you fix them, they will insist on growing back at their accustomed angles wherever and whenever they can reach, writhe or burst out of their bonds, for their geometric character is built into their very life. That is the abstraction that distinguishes one kind from another and it is geometric not only in the original Greek sense of basing its measurements on the earth (mostly through gravitation) but also on the earth's ancestors, the sun and to some extent the stars (through light and less obvious kinds of radiation), all of which give plants in general (as well as most animals) their bearings.

And the spirality of plants, when they have it, gives them the further identification of handedness, right or left, which may also (through molecules or inertia) be geared to the planet. At any rate the skewed tree or flower is most apt to turn clockwise going away from the observer (like a common corkscrew), which is called right-handed, to distinguish it from the less common left-handed or counterclockwise motion. This is particularly noticeable in helical climbing vines, some nine tenths of which, like seashells, are right-turning, as are the molecules of protein in all earthly flesh. Thus bindweed goes right with the majority, while honeysuckle twines left with the wayward few, a disparity that may have been at the bottom of Ben Jonson's observation in 1617 that "the blue bindweed doth itself enfold with honeysuckle" in a passionate embrace that to this day has never ceased to fascinate poets and geometricians.

If the turning of the earth were the direct or primary cause of the handedness of its vegetables, of course one would expect a variety that spirals clockwise north of the equator to turn counterclockwise when south of it, where the so-called Coriolis or torque effect is opposite. But so far there is sparse evidence that the equator is a definitive boundary between right and left vines, trees, shells or other creatures. Nevertheless, physical forces do definitely influence growth. Not only do mechanical tensions (presumably induced by hydraulic pressures) in their tissues keep grass and flower stems stiff enough to stand up in the wind (proven by the way stalks curl outward when cut and split in two) but, like animal muscles, plant fibers grow with exercise. D'Arcy Thompson reports an experiment in which several young sunflower shoots of a size that had broken in tests when loaded with 160 grams were left for two days bowed down by oppressive weights of 150 grams, after which ordeal they were individually strong enough to carry 250 grams. Then, in less than a week of further training, they developed so much stamina that each shoot could easily heft more than 400 grams!

Perhaps this is mostly a mechanical process of forcing fibers to line up parallel to the components of tension applied to them, as is known to happen with the chain molecules in molasses (a liquid vegetable) when it is boiled to a plastic solid and drawn into a rope of taffy, which can be continuously folded, twisted and pulled until the housewife finds it too tough to stretch any farther. If so, is form then created by the molecules that compose it? Or are the molecules rather produced by the form and function of the whole? Such philosophical questions come up again and again in the study of life and we will delve into them later on in this book. Here there is room only to make a preliminary inquiry into the cause and effect of growth which can but suggest a few macrocosmic factors that seem to influence what any particular plant is to be.

If you cut down a coniferous tree and examine its body in detail in an effort to find out what actually makes the lateral branches grow at such a different angle from the central stem, you will soon discover a reddish triangular segment of wood occupying the bottom quarter or more of each branch. Foresters call this "reaction wood" because it is wood that has had a reaction. Careful tests have shown that it is extremely sensitive to pressure. As soon as coniferous wood cells become noticeably crowded, as when a pine branch's weight compresses its lower parts, they start to redden and increase their rate of division. This reaction makes their side of a branch grow faster than the other side, thus bending the branch as a whole against gravity or whatever exerted the pressure. If a lateral branch is tied unnaturally high, reaction wood will form on its upper side and bend its free portion back down to its natural angle.

What enables the reaction wood to steer the branch so precisely? What keeps it from ever forgetting the exact characteristics of its particular kind of tree? These are not easy questions and they obviously involve the rather formidable subject of genetics, but last century Charles Darwin took a step toward the answer with his experiments in a plant's reactions when he found that a stimulus — say, a sunbeam — received in one part of the organism could elicit response in quite another part, indeed that the behavior of a young stem is directed by the tip growing at its top end in a way comparable to how a worm's whole body is controlled by the tiny brain in its front end.

Obviously plants do not have specialized nerve cells like those the animal kingdom evolved for high-speed transmission of messages from tongue to toe, yet a detectable "influence" somehow moves down a plant stem at a measurable rate. And, early in this century, that mysterious "influence" was proved capable of passing a cut in the stem if a film of gelatin was inserted to bridge the gap between the severed ends, suggesting that the "influence" was probably chemical in nature. But many more years of experimenting had to pass before Frits W. Went and others finally in 1928 isolated 40 milligrams of what turned out to be a potent growth hormone found in plants in several forms and which soon became well known as auxin. The potency of auxin moreover was so great that, as Went calculated to the amazement of scientists, if all the growth that theoretically could be produced by a single ounce of it were strung into one continuous line, it would encircle the earth.

Plants are now known to generate microscopic amounts of auxin somehow in their meristem tips at the ends of growing twigs, whereupon the substance flows slowly rootward through the bast channels along with the leaf gruel to stimulate growth in all the cells it reaches. Since most of it is concentrated in the meristem zones, that is where growth is fastest. Lesser amounts reach the elongation zones, and still less beyond them, but the gravitation of auxin to the lower sides of limbs and to other regions of relative pressure seems to be what boosts local growth there (exemplified by reaction wood), guiding and orienting the whole structure. And the fact that light repels auxin has the effect of concentrating the hormone in the shadowed lower sides of stems where its shade-biased growth steadily pushes the growing tips up from darkness toward light.

Other vegetable hormones were discovered soon after the auxins: the gibberellins, the cytokinins, abscisins, brassins and B vitamins such as thiamine and niacin, all either accelerating, decelerating or in other ways influencing the development of vegetation — and collectively providing more than a hint that plant growth must be a lot more complex than anyone had realized, being almost certainly controlled by an interplay of counteracting mechanisms not basically different from those involved in driving a car. At about the same time the direct effect of the duration of days and nights on blossoming was discovered, particularly the sensitivity of many plants to red light — a phenomenon that closely parallels the light sensitivity of animals.

And then a previously unnoticed intrarelation between different branches on the same tree began to attract the attention not only of botanists but even of a few philosophers. For it had been observed

61

that cutting off the topmost or leading stem of a pine tree induces the lateral branches just below it to bend upward in what seems to be an attempt to replace the lost leader. At least that is the ostensible purpose of the reaction, if one can believe there is any purpose in nature. The most remarkable part of the tree's response to the emergency, however, comes after the first scramble upward by the competing lateral branches, when the tree somehow singles out one (rarely two) from among these rising candidates and elevates it triumphantly in a few growing years to the vacant, vertical throne. The problem of succession is thus usually settled without a serious battle as the chosen prince of branches swings to the central supreme position with a seeming crown of authority while the other limbs drop placidly back to their accustomed places as if they recognized their new sovereign. How they recognize him, no one can say. It is known only that the cutting off of the top stem causes new reaction wood to appear at the base of the nearest lateral branches but how one of these is selected for honor and the others persuaded to forget it remains a mystery. The only thing reasonably sure is that some sort of a cooperative decision somehow gets to be agreed upon by the branches, if not by the whole organism. One might consider it the result of a kind of secret conference within the cells of the tree, perhaps involving a process analogous to voting in which proximity and responsiveness are at a premium and messages travel through the mysterious mediums of chemistry and electromagnetism.

The genetic memory behind such guidance in vegetable tissue, however, does not require a large piece to tell it how to grow. For any twig of certain kinds of tree, say a willow, can be cut off in spring and planted in moist soil and will sprout both roots and shoots. It is not only an organism: it is organized. It knows its top from its bottom. It feels a definite polarity and, even if you forget which end of a piece of willow was nearer the roots of its mother tree, the piece itself cannot possibly forget. If the wrong end is planted in the ground, the twig will in effect turn itself around by sprouting roots at the upper end which grow downward into the earth while shoots rise out of the soil from the bottom end to form a clump of willows, one of which a few decades later may become a beautiful big willow tree with no trace left of the misplaced twig.

Moreover polarity does not mean that any particular cell in the planted piece of willow is predestined to sprout a root instead of a shoot or vice versa, for the cell's action depends on its position in *relation* to the rest of the piece. If your cut puts the cell at the root end of the piece it will sprout a root. But if you cut it so the same cell

The mid-section of a willow twig will grow shoots or roots, depending on whether it is cut at A or B

comes at the shoot end it will grow a shoot. No matter how small the grains you cut willow into, they inexorably keep their root-shoot polarity, as surely as crumbs of magnetized iron keep their north-south polarity, and this will in willow asserts itself right down to hollow statolithic cells that contain dense, loose starch grains that keep rolling to the bottom of them in response to gravity, thus apprising the twig (which somehow "feels" them) as to which ways are up and down! Polarity is rather abstract, you see: a geometric directiveness, a built-in purpose like the homing instinct in birds. But we must defer further discussion of it until Chapter 18.

The only thing I still need to mention about plant forms is the extraordinary structure of cellulose, the basic stuff of the vegetable kingdom — generally called wood — which has long been (and will be yet a while) man's most versatile building material for anything from a bow to a barn. Most cells in a tree are tubelike and aligned parallel to the trunk and branches, yet frequently tending to curve as they grow, spiraling this way and that around the tapering, cone-shaped cambium sheath under the bast. Their walls, the toughest part of wood, are made of a fibrous matting in several layers, each one grained at a different angle (like the layers in plywood), the multitudinous, parallel, hairlike microfibrils that compose them spiraling now in a lazy many-looped helix, now in a steep corkscrew of few turns, winding around and around the tube cell, often mysteriously shifting course from clockwise to counterclockwise and back again as they weave themselves into their tight multi-mesh of incredible intricacy. And even the individual microfibrils have an internal structure no less amazing than the tapestry they are part of, each being a submicroscopic cable precisely woven of hundreds of long-chain molecules of cellulose, which stack together into continuous crystals — yes, crystals — of carbon, oxygen and hydrogen atoms of a texture and pagodalike lattice form that still far surpasses human comprehension.

## SIZES AND AGES

It is well known that some trees grow to giant sizes and almost unbelievable ages. In fact the tree is the largest, as well as the longest-living, kind of mortal organism on Earth. The tallest one ever measured, as I've mentioned, reached 368 feet which, added to the depth of its comparatively shallow root system, could well make its total height over 400 feet — and California timbermen swear they logged even taller redwoods in earlier days without bothering to measure them. Several such giants have been found to exceed 2000 years of age and their bulk can be judged by a contractor's estimate that any of them would provide "all the lumber needed for a couple of dozen five-room homes."

A very different sort is the Dragon Tree of the Canary Islands, genetically a member of the lily family, one giant specimen on Tenerife being reputedly around 4000 years old. And most extraordinary of all is the famous Montezuma cypress of Tule near Oaxaca, Mexico. It is only 110 feet high but an amazing 112 feet around the fluted trunk, which would make it perhaps the only large tree on Earth whose trunk circumference actually surpasses its height. Its age is adjudged to be over 5000 years, though this is hard to prove until the rings are revealed by coring or cutting.

Still another type is the banyan tree, a tropical species of strangler fig that continually drops roots from its branches, growing these into new trunks called pillars, until the whole looks like a dense grove. The world's most famous banyan, I think, is one in India said to have been described by Nearchus, the admiral of Alexander the Great, that today has a mother trunk 44 feet around and, at a recent count, 246 offspring trunks or pillar roots, some of them more than 10 feet in girth. They say it now spreads over more than an acre of ground and that a full brigade of 7000 soldiers can sit in its shade. If pillar roots were real trunks with sprouting branches there might be almost no end to the lateral growth of a single banyan tree, but somehow the pillar genes refuse to sprout, thereby putting a practical limit on a vegetable's span. Height, on the other hand, seems restricted only by the fact

THE
BANYAN
TREE

that vegetal weight, being three-dimensional, increases disproportion-ately to its 2-D supporting surfaces, as explained under Galileo's Prin-ciple of Similitude (page 15). In view of this you may wonder how, if land animals stop at elephants, a tree is enabled to reach a height twenty-five times greater. The answer appears to be that animal life, likewise 3-D and occupying perhaps 99 percent of its supporting bulk, cannot expand without soon outstripping its vital 2-D lung surfaces, while the life of a tree, essentially 2-D and residing mainly in the cambium layer (one cell thick), may occupy less than one percent of its body bulk, the rest being inert wood and bark that require neither air, water nor nourishment. This surface nature of living tissue in plants, moreover, not only gives a tree a tremendous reach, but the reach often spells the difference between life and death since the tree may not gain direct sunlight until it is hundreds of feet tall nor sufficient water until its roots have drilled scores of feet below ground.

When a plant does not have to hold itself upright against gravity, naturally it does not need to keep its girth in any particular proportion to its length, so thin vines that use trees for support can go to great lengths without overextending themselves — which explains how a jungle liana has been measured at an incredible 650 feet, and a single strand of sargasso seaweed, floating vertically in the buoyant Atlantic, at an unconfirmed, therefore even more incredible, 900 feet. But, curiously enough, many plants succeed at least as well by specializing in smallness, perhaps making up with their large numbers what they lack as individuals. The smallest any so-called normal plant (with stem, leaves, blossoms and roots) can be is about 1/5 inch tall, because it needs a minimum number of cells for this degree of organization, and cells, unlike computers, evidently cannot be further miniatur-ized. The tiniest plants of all (not counting viruses) are one-celled plankton, algae and bacteria, which move and feed so much like animals there is doubt in some cases as to which kingdom they belong.

## SOCIAL LIFE OF TREES

The animal-like capacity of banyans to strangle other trees here logically brings up the subject of plant aggression and the variety of ways vegetables barter, bargain or battle with animals, minerals and themselves. The most dramatic examples are to be found in the tropics, where there are perhaps twenty times as many species of plants (not to mention animals) as in temperate zones and where their vegetable parasites are not just gentle mosses, lichens or an occasional

65

fungus, as in Vermont or Germany, but much bigger and fiercer peppers, bromeliads with built-in water reservoirs, voracious orchids, armed cactuses, sword ferns and assaulting snakelike lianas.

In the rain forests there are more than thirty families of perching plants (called epiphytes) that begin·life innocently enough, sprouting out of a tuft of debris in some high tree crotch, silently dropping root threads till they penetrate the ground, slowly thickening as they drink until, years later, the dainty dangler may have become the vegetable equivalent of a giant python that stealthily coils itself tighter and tighter about its mother tree until it literally smothers and chokes it to death — then, after a few years of decay, stands smugly in its place until it too is eventually strangled into oblivion by younger, stronger rivals. Such is the slow, silent and terrible warfare of the jungle which, being observed so-to-say in slow motion, is much more apparent to the human eye than the swift, infrequent clashes between visible-sized animals, few of which take place in anything but utter darkness.

Poisons of course are used by many more plants than the familiar poison ivy, oak, sumac or toadstools (producing deadly muscarine), and there are so-called vicious trees, like the deadly upas of Java and the manchineel of tropical American shores with its luscious little "apples" that may kill anyone who eats them, and whose milky sap can blister flesh, paralyze muscles or blind the eyes of woodcutters. Less known are the plants that poison other plants, such as the aloof black walnut tree that stands by itself excreting a toxic chemical into the soil, the brittle bush of American deserts that sheds poisonous leaves, the jealous rubber-producing guayule of Mexico whose roots exude potent cinnamic acid to kill even most of its own seedlings, and not a few irascible grasses capable of retarding trees that threaten to drink away their water.

Vegetables can be benign as well as brutal, however, we must not forget, a good example being the Madre de Cacao tree (*Gliricidia sepium*) of tropical America, which, although its leaves, seeds and roots are apparently poisonous to rodents, so obviously helps neighboring cacao and other crops (perhaps partly through its nitrogenous root nodules) that it is widely known as the "cacao mama," "coffee mama," "clove mama," etc. Many northern trees likewise give shade and shelter to smaller neighbors. Indeed to such sun-shy ones as hemlock saplings in a forest or to wind-torn oaks on a stormy coast, this may make all the difference. Young white pines in an abandoned New England field naturally mother their rival black cherry seedlings (planted by birds perching in their branches), followed a decade or two later by maples and oaks that enjoy cherry tree shade in turn.

Many such sequences of tree species, each one shielding the next in a developing forest, are known to ecologists as important shelter chains in evolution. A somewhat more specific mothering relationship has evolved in certain northern forests where almost all spruce, hemlock and redwood seedlings get trampled or elbowed to death in the relentless bustle of ground plants, but where the rare lucky seed that lodges in a cranny of a rotting log several feet above the mob is thereby enabled to survive — perhaps literally on its grandfather's back. Thus fallen wood both enriches the soil and cradles infant trees whose roots eventually grope their way into the ground while their trunks slowly rise skyward in characteristic rows often remarked by woodsmen — rows that clearly commemorate the disintegrated ancient trunks that once lay down there to die and, dying, handed on the baton of life. Even logs floating in a lake or river sometimes sprout seedlings that, on being washed ashore, take firm root in solid ground or, as in a case I heard about, put out roots into the water which collect debris and gradually build up a floating island complete with soil, underbrush, earthworms and birds' nests!

## Special Environs

When plants learned to live on land, put down roots and breathe air, their main problem was how to obtain or conserve enough moisture to survive. Some found it easiest to remain partly underwater and so evolved air tubes down their stems, like the water violet, or gills like the pondweed, which can exist (if need be) totally submerged. And a few even evolved back to the briny deep, apparently without noticing that it's a lot brinier now than when life first evolved there. One of these is a real tree, the mangrove, which, like the seal and the walrus, is apt to be uncomfortable if it is not within easy reach of the ocean. Although the red mangrove cannot mature without taking root in earth or sea bottom shallow enough to let it reach the air and bear its fruit above the waves, it does not require either land or rain and has been known to exist unsheltered upon submerged shoals scores of miles from the nearest coast. It drops seedlings that normally get waterlogged at the root end and float vertically sometimes for thousands of miles on ocean currents as living driftwood, putting out secondary roots and budding leaves at sea, floating on and on until at last the roots touch bottom, take hold, and the plumule branches and bursts into flower at the top. Even if the floating seedling sinks before touching bottom, it has a chance to survive if it lodges in shallow    67 water, for it can be submerged a year and still shoot a long trunk to the surface, where it will joyfully catch its breath and bloom against the sky, its trunk eventually attaining a girth of nearly ten feet. These trees often build new islands too, for the advancing sand naturally piles up around their prop roots, making more room for seedlings, and the complex undergrowth harbors everything from driftwood, crabs and nesting cranes to oysters, which, at low tide, offer the makings of a stew ready for plucking right off the tree!

You might think hot springs an unlikely home for any vegetation, but green plants often grow in water as hot as 145°F. and pale ones without chlorophyll up to 162°F., while at least one hardy veteran, *Oscillatoria*, a blue-green alga perhaps little changed since the earth's early days, is reported to thrive in alkaline, silica-charged springs as hot as 194°F., and an extraordinarily tough breed of bacteria indigenous to deep oil-well brines sometimes lives even above the boiling point, 212°F.

Dry heat is a more normal problem, and plants survive in deserts mainly by either conserving their moisture or dying away to mere dry seeds that sprout only when it rains. Succulents store water in ingenious ways: the century plant in its leaves, the night-blooming cereus in underground bulbs, the cactus in its fat stem. Others, like the agave, hopefully leave their gutter-shaped leaves continuously open to catch the rare rain, the exact opposite of rainy-area plants like *Sarracenia minor* that wear perpetual hooded "umbrellas" to defend them from the too frequent downpours.

The giant saguaro of the American desert is perhaps the most dramatic example of a drought-resistant tree. Unlike the spindly mesquite, whose roots drill deep downward to sip continuously from whatever water may be under the desert, the portly saguaro puts its whole faith in a wide, shallow, lacy network of root that has been growing for perhaps 200 years but has no hope for a drink except when it rains. Since the saguaro is bulky, tight-skinned, sometimes more than 50 feet high and capable of holding 30 tons of liquid, it can wait comfortably through several years of drought if need be and, upon a sudden downpour, gulp up tons of rain in an hour, perhaps reaching saturation within half a day, which means being more than 90 percent water. American desert Indians not only found this bountiful plant a vital water resource and mashed it to extract the bitter liquid during extreme droughts, but they regularly ate its melonlike fruit, cooked it into a syrupy preserve, fermented the juice for "wine," pounded the seeds to "butter" and used the remaining long stems as lodge poles, the leftover scraps as firewood.

68     Quite insignificant by comparison appears the little patch of lichen

covering a weathered stone. It is scarcely a fifth of an inch thick and has no obvious rain reservoirs, yet it can survive drought for twenty times as long as the saguaro, stoically waiting in a dormant state — in one recorded case for 87 years! — until water instantly reawakens it to active growth. The only plants I've heard of that can get along completely without rain (or root water) are the American pygmy cedar (*Peucephyllum*) and the Saharan caper plant (*Capparis spinosa*), which keep green and healthy under extreme drought conditions just on the humidity they absorb from night air.

The way most plants cope with the onset of a cold winter is to retire into a state of hibernation. In doing so, trees that are not evergreen may almost completely dispense with chlorophyll, photosynthesis and evaporation by shedding their leaves, breathing only feebly through bark and buds, and hardly metabolizing or growing at all. Although severe cold spells may have frozen their sap solid and its consequent expansion combined with wintry winds may have cracked their brittle wood to fill the sap channels with air bubbles, happily the coming of spring has a way of correcting the damage by creating new sap tubes and priming them with fresh bubble-free sap or, in the case of certain trees and vines (birch, maple, grape . . . ), vernal root pressure has been known to force new sap into some of the old air-filled tubes as well.

Many smaller plants on the other hand, evidently finding hibernation too difficult or wasteful as a winter strategy, take the more drastic step, in a greater or lesser degree, of dying. Thus they save the energy it would require to keep going all through the cold months and, if they are annuals, their seeds can still sprout and carry on in spring or, if perennials, their roots expectably "come alive" with the season. This way of wintering, in case you like the philosopher's perspective, is a demonstration of one of the advantages of death, a subject we will explore in some depth in Chapter 20.

## STORAGE

Do plants store up any reserves of nourishment to tide them over a drought? Do they have anything to compare with an animal's stomach or stomachs or surplus layers of fat? Yes, plants often stash their food or fuel in special reservoirs in their leaves, bulbs and stems. An onion, for example, is really a head of leaves that adapted itself to hoarding plant food. A potato is a stem that evolved a belly for starch. The acrocomia is a tall palm that has done the same with a bulging paunch up to 17 feet in girth. The kapok can be a portly tree

20 feet high under ideal cultivation, although it grows skinny and 100 feet high when it must compete for sunlight in a rain forest. The baobab is the fattest of all trees, often exceeding 100 feet in circumference though only 40 feet high and, being soft and pulpy inside, is a favorite food of African elephants who, if they get the chance, will knock it down and consume it completely! The creeping cashew of the Amazon valley, on the other hand, looks like a shrub, but actually has a huge secret trunk that it conceals underground as it circumvents its rivals by living like an iceberg: eight ninths out of sight.

Many flowers, even dainty ones, have also made their place in the world by evolving into giants, increasing both their appetites and their storage capacities over millions of years until they have turned into hungry monsters, like a rain forest "violet" that now sometimes hulks as big as a plum tree, a "milkwort" that twines upward 100 feet and more, a rose from which arose hawthorn and apple trees, a pea that evolved giant locust trees, lilies that turned into the Joshua trees of California, and the aforementioned Dragon Tree of Tenerife. Then there are the 120-foot bamboos of southeast Asia, which are really overgrown members of the ancient family of grasses, a sunflower 30 feet high in the Galápagos, a bignonia that became the catalpa tree, a giant "verbena" that resembles a horse chestnut tree, a species of 60-foot tree ferns of Tahiti, tree lobelias, tree heathers, tree cabbages, tree geraniums and finally the gargantuan groundsels of Mount Kilimanjaro that are in fact 20-foot cousins of daisies supported on stems 4 feet thick. Although all trees must ultimately have evolved from something much smaller, it is more than possible that these big brothers of our little friends may today only be in passing phases of gigantism that will eventually evolve back toward temperance and lowliness, just as the heirs of club moss trees that towered 90 feet high and 5 feet thick in the coal ages now stand only a few inches tall.

OVERGROWN TREES AND FLOWERS

acrocomia  baobab  (a lily) Joshua tree  (grass) bamboo  tree fern  lobelia  tree heather  (a daisy) groundsel

I suppose the simplest and smallest of vegetables (not counting viruses) are bacteria. But they, not being given to paunches, have solved their storage problem differently by evolving tastes for foods that would likely be plentiful even in a drought or a famine. Thus some kinds of bacteria have acquired an appetite for hydrogen sulfide (rotten-egg gas) and now use it instead of water in photosynthesis, which makes them exhale sulfur instead of oxygen. Certain other bacteria meantime have learned to live on ammonia found in decaying tissues, on hydrogen, methane or even on iron, oxidizing each of these inorganic foods to build themselves carbohydrates not with light energy, as in photosynthesis, but with chemical energy in the alternate process known as chemosynthesis.

## FUNGI

No doubt the most successful near relatives of bacteria are molds and yeasts, which are classes within the subphylum called fungi. But as fungi have no chlorophyll of their own, they rely (as we do) on the photosynthesis of others, and their tastes have proliferated accordingly. So, starting with leavened bread and fermented wine, they can live on almost any vegetables (including fungi), meat and fish (dead or alive), cheeses, jams, sauces, etc., then paper, ink, books, cloth, leather, bone, dung, soil, wax, wood (including living trees), wool, upholstery, girdles, glue, paint, creosote, rubber, asphalt, plastics — and a few even on things like copper salt and glass — almost anything but metal.

In other words, fungi are about as unfussy as eaters can get and therefore are among the best adjusted organisms on Earth. It is true that humans are inclined to think of them as coming from the other side of the biological tracks but, before we drop them from the register, let's remember that we too depend utterly on the chlorophyll of other creatures, we too ferment the food we eat and we too stop at almost nothing in our search for diverse edibles from spaghetti, corn flakes, seaweed, opium, coffee, vodka, bird's-nest soup and blazing plum pudding to quinine, vitamin pills, canned rattlesnake and chocolate-covered ants. While man is a new and single, experimental species of unproven stability who cannot refrain from flirting with massive self-destruction, the fungi have been solidly established on Earth several hundred times as long and are now sensibly diversified into nearly 100,000 species that live in the ocean, throughout the soil, on all lands (including glaciers) and whose invisible spores float everywhere in the air and to some degree through space.

Such success of course is partly due to simplicity and to modesty of aim.  Fungi avoid headaches, stomachaches, toothaches and heartaches by just not having any heads, stomachs, teeth or hearts.  Instead of private or internal digestion, for instance, they use a public, external system and rather than eat first and digest afterward, as we do, they digest first before even deciding whether the meal is worth eating — the "proof," so to say, coming *before* the "pudding"!  This is possible because the body of a fungus is primarily a branching meshwork of microscopic filaments called mycelia, whose long cells grow at their tips so fast you can actually see them snaking across the field of a microscope, and they are continuously exuding enzymes that disintegrate or digest everything edible (often 100 percent of solid matter) within reach, later consuming what they can of it at leisure.  The filaments may be as fine as one one-hundred-thousandth of an inch in diameter, but they put out side branches about every half hour, and thus a single spore multiplying into cells can add up to a total length of hundreds of miles in two days' growth.

Although most fungi prefer a moist climate of around 80°F. and get sluggish when it is dry or cold, you can't kill them by freezing, because they readily hibernate and wait confidently for warmer times.  A few species, on the other hand, like it hot and grow so lustily in summer that they generate heat above 130°F. in grain elevators, sometimes stimulating heat-loving strains of bacteria to multiply so fast the temperature rises to the bacteria's limit of about 170°F., whereupon chemical oxidation, if conditions are right for it, will cook the fusty, fuming ferment to the smoldering point at which it may "spontaneously" burst into flames.

## LICHEN

Where fungi have made their most important mark, I would think, is in their constructive associations with other organisms.  Lichen, for example, is a very ancient and proven partnership between fungi and algae.  Evidently it began casually some 350 million years ago when algae were trying to broaden their beachhead on dry land and fungi were casting about for handy green vegetables to eat, or it may have happened earlier when both were still completely at sea.  Either way, the tough active fungi seem to have taken over the gentle algae cells like farmers who herd cows, gradually discovering how to weave tight corrals of mycelium around them while "milking" them of their sweet, photosynthetic juices, at the same time rewarding them with protection from sun, weather and enemies while munificently bestowing the obvious benefits of travel without removing the basic comforts of home. Although some 15,000 "species" of lichen are already recognized, each is more specifically classified as an established combination of a species of fungus and its own completely domesticated species of alga. Experiments have shown that it is possible for the fungus and alga to live apart in a fashion when forcibly separated, but the operation is about as ticklish as depriving a farmer of his favorite cows. The fungus is apt to go hungry and the alga risks being dried up or washed away with the changing moods of the weather while both partners obviously "yearn" to get together again. The fungus generally assumes the role of "husband" in this remarkable symbiotic "marriage," but there are also a good many cases (as among humans) where "she" manages to dominate "him" both in size and strength. Whoever wears the pants, however, it is virtually always a satisfactory union even to the point of having offspring, who are begotten not sexually but by simple division as small crumbs of lichen (alga cells enclosed in mycelium) break loose from the parent body, often to be wafted aloft on the wind and to float great distances in a more or less dormant state before alighting on some bleak crag or arctic scree, where a little moisture may induce them to shoot out new filaments to secure a grip before they are blown away again to a fate unknown.

The only other symbiotic partnership of fungi and green plants I can think of that may be more important than lichen is the curious union between fungi and the roots of evergreen trees. It's known as mycorrhiza, and occurs when the roots go exploring, almost calling, through the ground for vital minerals, an act that provokes the ground in return to arise and answer the roots in the form of the invisible threads of fungus containing those very minerals — all because the fungus too is hungry, in this case hungry for the energy obtainable only from the roots of those same sunlight-absorbing evergreen plants above them. Which is why nurseries sell pines and spruces with burlapped balls of earth around their roots: so they will be assured the fungi partners they need to survive.

THE MYCORRHIZAL RELATION

## Seeds

We cannot end a chapter about the vegetable kingdom without saying something about seeds. Because although the kingdom got along without them until about 350 million years ago, when evolution devised its first one, the seed is now well established as nature's best all-around invention for starting vegetation in new places. It also possesses a world-changing power that is truly mystic in the sense that it can potentially reproduce not only the complete tree or plant it came from, but all the trees that can descend from that one, indeed (if conditions are favorable) whole forests of them, including those that will eventually mutate into new species and evolve into higher forms of life, ultimately (in billions of years) permeating all the kingdoms without any known limit.

In other ways too, the seed is not the simple thing it seems. Nor is it the beginning of a new plant, which normally is conceived when an ovule is fertilized by a grain of pollen. For the fertilized ovule has to grow for many days to become a seed. A seed thus is a sort of vegetable egg, a partly developed organism in a protective shell with all the food it expectably needs until it can forage for itself in the outside world.

On the average seeds are the size of a grain of wheat or rice, but a few of them go to extremes and become hulking 80-pound coconuts on the Seychelle Islands or invisible specks of orchid "dust," of which it takes 8 million to weigh an ounce. They may be of almost any color, and of a great variety of shapes, from the sphere of the black nightshade, the ovoid of the bean and the lens of the lentil to the disk of tulip, the star of puncture vine, the javelin of oat, the torpedo of mangrove, the tusks of devil's-claw, the propeller blade of maple, the glider of zanonia, the feather of clematis and the parachute of dandelion.

It is understandable that much of a plant's energy goes into seed production, for upon the seed depends the whole future of its kind and, to a large degree, of all the animals that feed on (and therefore distribute) that seed. The flower that preceded it obviously evolved to make sure the ovule got fertilized to start growing the seed. The fruit that followed evidently appeared as a home for the seed and often also a vehicle to carry it to its destination. Inside the seed meanwhile a tiny, new, unseen plant takes shape, forming a stem, a root and the beginning of leaves, all neatly stowed with nourishing starch in the armored and sealed container. The twin halves of a peanut are a familiar example of the two first leaves of an unborn plant. When the seed has attained the fullness of seedhood, however, it stops developing and goes to sleep to await "G-day," its germination date, its mysterious time of appointment with the outside world. In the vegetable kingdom, you see, the gestation period for seeds is not regular and they rarely hatch open on a predictable schedule but instead must await a number of conditions, presumably to allow time for dissemination.

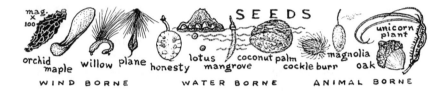

WIND BORNE          WATER BORNE          ANIMAL BORNE

Dissemination means literally "outward sowing," which well expresses seeds' evident eagerness to travel. I am not thinking just of the modest ones like violet seeds, whose pod bursts like a tiny popgun, shooting them a few feet to possible fresh ground, but also of the tufted seeds of milkweed, thistledown or willow silk that ride the breezes, tumbleweed included, and of those who prefer to hook themselves a ride on a passerby, like the grapple plant that catches hold of moving paws and hoofs, like the burdock that clings to any fabric with its now-patented (Velcro) brand of fastener action, and the ones that simply stick with one kind or another of gummy glue to the fur of passing animals or the feet of birds. In all there are at least 5000 different sorts of migrating birds and bats, by the way, who quickly spread seeds from country to country, often by eating them in fruit, then voiding them next day hundreds of miles away, and later perhaps dying in time to let them germinate in rotting flesh.

Many other seeds have taken to using man to disseminate them, mostly without his permission, some going in for chancy modern tricks like getting picked up in tire treads of cars or airplanes, some stowing away as "weeds" among more popular or ticketed travelers, some just riding unbeknownst in old-fashioned ships, trains, wheelbarrows or shoes. Most remarkable of all from an evolutionary viewpoint may be the increasing numbers who gain willing human cooperation, even to being presented with gay raiment like seed packages that serve as blossoms to catch the eyes of those who may plant them in exotic gardens of unwonted, if not unwarranted, congeniality. A few have even surmounted some sort of pinnacle of spiritual aspiration by be-

75

coming the choice of such heroes as young John Chapman of Massachusetts, who in 1806 took two canoeloads of apple seeds down the Ohio River for his now legendary planting tour of the West as "Johnny Appleseed," or merchant seaman Aloysius Mozier who in the 1950s personally handed out more than a million seed packages to needy folk in Asiatic ports.

If dissemination distributes seeds in space, however, it is germination that distributes them in time, a dimension no less critical to a seed seeking a season suited to its survival. Germination naturally requires moisture, free oxygen and moderate warmth, but it would be much too simple, in fact suicidal, if all seeds could germinate at the first warm sprinkle of rain, for desert plants would often sprout into a deadly drought only momentarily relieved by those few drops, or an apple shoot would respond to a December thaw barely in time to be cut dead by a January freeze. So a great many seeds are provided with special insulated or time-locked coats to inhibit germination until the optimum moment. The waterproof "varnish" on a cherry seed remains undissolvable until ground by a gizzard and corroded by digestive acids and bacteria, ensuring that if the seed attains optimum sowing (by a bird) it will not only be alive but cocked for triggering by the next rain. Many such seeds are adapted to the digestions of particular animals, such as quandong seeds in Australia, which, after the prunelike fruit is eaten by an emu, germinate in his droppings; and the seedpod of the camel thorn in Africa, favored by the elephant, from whose dung the seeds are commandeered by dung beetles to be buried in seed beds specialized for rapid sprouting.

Apple seeds stubbornly resist germination until several months of cold have convinced them that the succeeding warmth must really be spring. Certain orchid seeds also hold off until a friendly "infection" of mold has eaten away their raincoats, while a few larger seeds, curiously, have evolved what amounts to a yearning for burning, notably jack pine, whose scorched cones opportunistically snap open during a forest fire to reseed the ground when it is bare of competition.

A great many seeds, it has recently been discovered, are extraordinarily sensitive to light (particularly red light), some (like mistletoe) germinating only by day and others (such as onion) only in the dark, though always other factors complicate the outcome. Some seeds also contain remarkably accurate natural "clocks" as proved by their inherited 24-hour rhythms and their surprising sensitivity to precise lengths of days and nights that not only trigger germination but the opening of buds and blossoms later on. Although long periods of dormancy tend to make a seed *less* viable, they also often make it *more* germinable, a distinctly different thing. The record for unfrozen dormancy is now held by two ancient lotus seeds found in a peat bog in Japan in 1951. Affirmed by carbon-14 dating to be 2000 years old, the seeds were cut open, put in tepid water and sprouted four days later, soon blossoming into beautiful pink flowers, just as they would have if they had been allowed to germinate before the birth of Christ.

If they are frozen, however, seeds will retain their viability almost indefinitely, like the arctic tundra lupine seeds found in 1954 after 10,000 years of lying deep in frozen silt in the upper Yukon valley and which, when planted on wet filter paper in 1967, germinated within 48 hours to produce normal healthy flowers. Certain kinds of seeds, on the other hand, will not germinate before a set date no matter what you do to hurry them. Even boiling them for hours does not disturb their slumber, for evidently their "clocks" have not struck the appointed hour. It seems that a sort of digestion process goes on in acorns and other seeds, slowly converting fats into carbohydrates, which may be what works their "clocks."

When "G-day" finally arrives and germination is triggered, the first noticeable change in the seed is that its shell has become permeable and it drinks water, its crystals unlock and it swells up somewhat, as an animal will stretch itself after a long sleep. Thus wheat and corn kernels absorb about half their weight in water and the garden pea drinks its full weight. Next the increased pressure of the swollen seed bursts the shell open, perhaps at prepared seams (as in the walnut) or "doors" (as in the coconut), while the released embryo, inhaling oxygen like an inward yawn, starts dividing its cells in rapid growth, usually sprouting first a root downward, then a shoot upward, like a baby hunting for a nipple (corresponding to earth) before he opens his eyes (corresponding to leaves).

The buds that form on a tree every winter are worth mentioning here too because, like seeds, they are a strikingly successful means of carrying on life from one season to the next. As soon as dead leaves drop off in autumn, their stems parting at a "tear off" line plainly perforated for the purpose, new buds start forming at that line in preparation for next year's greenth. And, like a seed, each bud is an embryo organism, perhaps an unborn limb, with its own tiny stem, leaves or petals and a stock of sugary food neatly folded in the little waterproof shell that protects it, not so much from the cold (which keeps it asleep) as from drought at a season when the tree's water system may be too icebound to replenish any evaporated liquid.

Every tree and plant, it is found, has its own special budding inducements: a warm day for the pea, a warm night for the tomato, a 77 particular day-night ratio for the cocklebur, a holding thaw after pro-longed cold for the lily, a combination of warmth and long daylight for the strawberry. And most of them are ineffably complex and mysterious.

Also, like seeds, the leaves when they appear develop their own distinctive shapes and "purposes," in peas some becoming tendrils for climbing, in the barberry some turning into thorns for defense, others producing hard-to-fathom aerodynamic effects. And of course whole trees take on their characteristic shapes too: the vaselike, wide-crowned American elm, the oblate spreading oak, the tapering fir, the graceful tresses of willow, the tall quill of poplar, the candle flame of cypress, the feather duster of palm . . .

One of the collective functions of plants that impresses me strongly is the rarely mentioned but rhythmic, almost peristaltic, swallowing action of their seasonal changes: the sprouting upward of needlelike shoots in spring, the reaching over (sideways) of foliage in summer, the valvelike drop of dead leaves in autumn covering everything on the ground, the burial and decay in winter while awaiting the next upward penetration of spring that will actually pierce and pass the old growth, sewing it securely into the earth. Every year this sequence of stitching and pumping is repeated relentlessly, inexorably, up-over-down-under, up-over-down-under, like a global sewing machine, 1-2-3-4, 1-2-3-4, slibaroom, slibaroom, slibaroom — a heart throbbing in the breast of nature herself — devouring the waste of the world, binding up the shroud of death, digesting the past, winding the clock of life.

It reminds one of man's increasing debt to the older kingdoms and emphasizes the eternal dependence of animals upon vegetables. It raises questions of evolutionary purpose and ultimate planetary poten-tialities. Does a tree in any sense have a self or a consciousness distinct from the cells inside it, from the forest around it or the earth under it? If men may be looked upon as trees walking, may not trees in turn be animals standing? So far as I can see, we have no reason to suppose consciousness requires nerves or, if it does, that trees are utterly nerveless. And while their feelings may be exemplified by their silent tuning in on radiations from Earth, moon, sun and stars, by their apparently leisurely adaptation to the forest, still they are due some share of our courtesy and consideration. Should they not, like the animals, have their basic rights in a human court of law?

Long ago Walt Whitman asked, "Why are there trees I never walk under but large and melodious thoughts descend upon me?" Somehow the trees' melodies seemed to come to him in sylvan syllables, leading him to hearken to a live oak as a voice singing in the wilderness, perhaps even as a whispered word of God, as when he wrote: "It grew there uttering joyous leaves of dark green."

Descending thence with Walt's melodious thoughts down the weathered trunk to the gentle grass at its feet, let us ponder anew the little things, the earth itself, the stirrings inside all creatures. For what does life finally spring from? Which of us has seen the world of his forefathers of a million years gone by? Who will transpose the conversings of blood with mud, the dainty dialogues in dew? And who can say, when one cherishes a rose, how little or much the rose may cherish in return?

CHAPTER 3

# The World of Little

THE TOWN OF DELFT, five miles southeast of that bustling new port called The Hague in Holland, was prosperous but rather quaint in the seventeenth century. Among its thousands of inhabitants there was nothing about the stocky, bright-eyed little draper and alderman named Antony van Leeuwenhoek with his magnifying glass for examining fabrics and his off-hour hobby of grinding lenses to indicate that he was on the verge of a world-changing discovery. Even now it is easy to forget, as we look down on the struggling planet from our perspective of space and time, that Holland was then probably the most progressive country on Earth, an imperial republic leading the planet in naval power, commerce, science, art and letters. Yet among his creative fellow townsmen like Jan Vermeer the painter and Regnier de Graaf the physician, Leeuwenhoek at forty struck no one as much of a pioneer in any branch of knowledge. Not only was he unconnected with Delft's university, he had not even finished elementary school, spoke an ungrammatical Dutch and his only assets were dexterous fingers, excellent eyesight and an extraordinary curiosity abetted by a streak of the proverbial Dutch stubbornness. He had also picked up the rudiments of mathematics, mostly from doing occasional surveying jobs and keeping the books in his drapery shop.

Lens-making in those days was one of the newer do-it-yourself fads introduced by the Renaissance, perhaps comparable to building a homemade telescope today. Although its forgotten history went back at least to the ground and polished magnifying crystals of ancient Nineveh and had entered the realm of commerce upon the invention of eyeglasses in Florence in 1286, its biggest triumph had come only three hundred years later, shortly after the appearance of the telescope (probably in Naples in the 1580s), when Galileo made himself a series of double-lens tubes in 1609 that enabled him to discover the moons of Jupiter, the rings of Saturn and to resolve the stars of the Milky Way. Leeuwenhoek probably did not know about Galileo's subsequent construction of a compound microscope (5 feet long) through which flies looked "like lambs" and walked "on glass while hanging feet upwards."

Leeuwenhoek was content enough just with grinding out hundreds of tiny, single, convex lenses, usually around one eighth of an inch in diameter, which he would shape as fat as beads (for high magnification) and polish with great skill and mount between two plates of brass, copper, silver or, in a few cases, gold (smelted in his own forge) with a little peephole like the aperture of a stopped-down camera. And he usually arranged an adjustable pin on the back side, so that any small object stuck on its point would be exactly in focus. He would thus hold biological specimens like the eye of a gnat, a fly's brain or a spider's spinneret up for inspection against the open sky or, illumined by candles at night, against a dark background for clarity. He did not notice anything very startling at first, but when he magnified droplets of stagnant pond water in the summer of 1674 with lenses that enlarged more than 100 diameters, he was amazed and thrilled to see "very many little animals," as he put it in a letter to the new Royal Society in London, some round, some oval, some with "paws" and tails, darting every which way through this liquid that the world until then had presumed to be just an inert mineral. Later, carefully measuring these microbes in relation to a cheese mite (barely visible to the naked eye), he wrote: "as the size of a small animalcule in the water is to

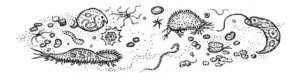

that of a mite, so is a honeybee to a horse, for the distance around one of these little animals is less than the thickness of a mite's hair.''

Soon Leeuwenhoek got a phial of seawater and discovered the teeming microscopic life we now call plankton: first, ''a little animal that was blackish, having a shape as if 'twere made of two globules'' and jerking about ''after the manner of a very little flea . . . displaced, at every jump, within the compass of a coarse sand grain.'' Then some ''animalcules which were clear'' with ''oval figures and snake-wise motion,'' a third kind, ''mouse-colored . . . very slow'' with ''stingers'' at both ends, and a very tiny variety (presumably proto-zoans) ''that, whenever they lay . . . out of the water, would burst and flow . . . into three or four very small globules . . .''

In rainwater he discovered a kind that seemed to be composed of from five to eight clear globules loosely clinging together with a long tail ending in another globule. This animal (now identified as a vor-ticella) fascinated Leeuwenhoek by sticking out ''two little horns'' (actually eye stalks) which it continuously swiveled around ''like a horse's ears,'' but it also aroused his pity as ''the most wretched creature I've ever seen for when, with its tail, it touched any particle it stuck entangled in it, then pulled itself into an oval and did struggle by strongly stretching itself to free its tail, whereupon the whole body snapped together again leaving the tail coiled up serpentwise. . .'' Leeuwenhoek was appalled to behold ''hundreds of such animalcules caught fast by one another . . . within the span of one coarse grain of sand.''

He almost certainly saw thread bacteria as ''most exceeding thin little tubes'' to the number of ''tens of thousands in a single drop of water,'' swimming slowly ''like eels'' but ''as well backward as for-ward.'' And another faster kind that ''would oft-times shoot so swiftly forward . . . for half a hair's breadth . . . that you might think you saw a pike darting through the water . . .'' The smallest lively objects he saw, so minute he could not ''assign any figure to 'em,'' were ''a thousand times less than the eye of a full-grown louse,'' but just may have been inert particles activated by the random, thermal bombard-ment of molecules that was to be described in 1827 by the English botanist Robert Brown (who also assumed the particles alive) and today is commonly called Brownian movement.

Between looking at such oddities as an embryo oyster, a whale's eye, an elephant's tooth, a pig's tongue, a rye germ and a ram sperm, Leeuwenhoek improved the magnifying power of his lenses up to 275 diameters, by which time, one might say, he really had got his teeth into his hobby. I mean by that that this was when he made one of his biggest discoveries by examining a rotten molar from his own mouth, which, he was staggered to realize, was literally crawling. "I can't forbear to tell you, most noble Sirs," he wrote the Royal Society, that "I dug some stuff out of the roots of one of my teeth . . . and in it I found an unbelievably great company of living animalcules, amoving more nimbly than any I had seen up to now. The biggest sort bent their body into curves in going forwards . . . I must confess that the whole stuff seemed to me to be alive, with the number of animalcules so extraordinarily great that 'twould take a thousand million of some of 'em to make up the bulk of a coarse sand grain . . . Indeed all the people living in our United Netherlands are not as many as the living animals I carry in my own mouth this very day."

Thus did our little draper give the world its first solid evidence of the teeming bacterial life within us, which had been only intuitively suspected by such earlier philosophers as Marcus Terentius Varro of Rome, who in the first century B.C. wrote (perhaps after reading Lucretius on the nature of epidemics and pestiferous atoms) that "near marshy places . . . live certain minute creatures which cannot be seen by the eyes but out of the air enter the body through the mouth and nose and there cause serious diseases." Leeuwenhoek, moreover, actually examined the air with his lenses, closely scrutinizing innumerable "earthy motes" that he caught floating "commonly in the air" and which must have been "given off by dustsome things. Indeed," he explained, "you can't so much as rub your hands together when they are dry nor stroke your face, without thereby imparting a multitude of tiny scaled-off particles to the air; and 'tis even so with wood, earth, smoke . . . Furthermore," he opined, "I'll not deny that there can be in the air living creatures so small as to escape our sight . . . Indeed likely they would be begotten in the clouds where, in the continual dampness, they could remain alive and so be conveyed still living to us in mist and rain. I fancy I have seen something of the sort in the early summer of this year [1676] on two occasions when there was a heavy mist here . . . and I can well understand that in all falling rain, carried from gutters into water-butts, animalcules are to be found . . . for, along with floating dust, these creatures can be carried in the wind."

It is not only the scope of this unpretentious man's revelations that is remarkable. So is the degree to which his accuracy has been confirmed in succeeding centuries, for the findings of modern meteorology have established that every cubic centimeter of air within some ten miles of the earth has at least several dozen invisible particles of dust floating in it and, if the air happens to be over a smoggy city, the

83

number can reach into the millions. Many of these tiny nuclei are dry salt granules tossed into the winds as fine ocean spray whose liquid evaporates quickly enough to release a microscopic residue of solid crystal usually encrusted with a little living tissue. Other suspended dust particles are bits of earth raised by friction from such things as shoes, hoofs, wheels, brooms and wind. Some are invisible flakes or spicules from smoke or exhaust fumes. Some are ashes left by vaporized meteorites, billions of which daily stab into the top of the atmosphere from space. And a goodly percentage also are aerial plankton in the form of wafted spores of algae, fungi, bacteria or other plants, grains of pollen, even a few actual seeds and viable fragments of lichen.

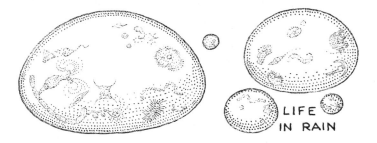

LIFE IN RAIN

Almost any of these motes of dust, particularly the ones that originated in soil or water, may include animal microbes that are dormant while dry or cold but awaken and start swimming when a cloud droplet happens to condense around them, perhaps eventually conveying them to Earth inside a raindrop. Indeed it is the clouds that are the most fertile parts of sky and every puff of cumulus or nimbus is unquestionably alive with tiny animals and plants that live there for generations (a microbe's generation often lasting less than an hour), eating, breathing, excreting, floating, swimming, competing, reproducing . . . Most of them are less than one five-hundredth of an inch long and shaped in a hundred different ways, as the illustration (somewhat out of proportion) suggests, the larger ones ranging from algae, seeds and pollen to springtails and mites, often picked up by a cloud that slides up the windward slope of a mountain. The tinier ones, like bacteria and diatoms, may be lifted directly out of the sea by rolling fog, even bearing their own stores of phosphate, methane, carbon and sulfur dioxides and other nourishing compounds.

The fact that moisture in the air condenses only around solid nuclei of course makes dust a vital factor on Earth, without which there would be neither clouds nor rain nor fresh water nor life much beyond the sea. And the presence of many of Leeuwenhoek's creatures actually inhabiting the wind and rain reveals that life permeates the sky about as extensively if not as densely as it permeates all bodies of water and virtually all land.

## MICROBE CLASSIFICATIONS

Something not generally realized about the microcosm even today is that it not only bustles faster than our visible macrocosm, but also, as more and more is revealed about it, it seems steadily to deepen in profundity until no distinction between animals and vegetables is left, for the good reason that the microcosm preceded both kingdoms in evolution and apparently remains Earth's prime proving ground for the basic inventions of life: growth, metabolism, sense, response, locomotion, reproduction, sex, death . . .

The main divisions of the microbe world, in order of diminishing size, are first, the creatures of borderline visibility (mostly animals of the largest one-celled dimensions such as amebas, slipper-shaped ciliated paramecia, bird-beaked cladocerans, anteater-nosed thacheliuses, goose-bodied lionotuses, heart-shaped ostracods, wormlike nematodes, ova and the larger plankton of sea and sky, averaging about $1/100$ of an inch in diameter), second, the one-celled algae, fungi such as yeasts, small plankton, sperms and blood cells (ranging to about $1/5000$ of an inch), third, such tiny plant life as bacteria, rickettsiae and primitive blue-green algae (averaging $1/25,000$ of an inch) and finally viruses (around one $1/1,000,000$ of an inch).

Such a vast complexity of animals, vegetables, animal-vegetables and possibly (in the case of viruses) animal-vegetable-minerals, does not make up a very neat hierarchy. Indeed no two biology books I have seen have classified microbes exactly alike — and how could they with nature herself apparently undecided as to how to rate her proliferation of primordial but constantly evolving miniature models? Yet, if for no other reason than that the total mass of microbes on Earth has recently been estimated to be a good 20 times greater than the mass of all earthly animals (including all animal microbes), we certainly cannot afford to neglect the microcosm.

Have you ever seen a paramecium? Although a single-celled protozoan discovered by old Leeuwenhoek in pond water, it is not what you would call really primitive like the simple and almost formless ameba for, in the course of its evolution (which has taken as many millenniums as our own), it has happened to develop an extraordinary complexity. Neither is it the biggest of protozoans, one unruly variety of which (called *Chaos chaos*) attains to a length of one fifth of an inch, but it is just visible and swims commonly in wet earth, often inside the stalks of green grass, and in dry weather it may sleep encysted in hay or dust almost anywhere. It is shaped like a slipper, but glides "heel" forward except when maneuvering around an obstacle, propelled by its thousands of cilia that row in eight or ten flashing waves of continuous coordinated stroking down its body (page 27), some of them at the same time whisking food into the funnel-shaped mouth, the while gyrating in a corkscrew course due to the oblique angles of the cilia and perhaps the skew of the oral groove that leads to the mouth. When it wants to back up, the creature instantly "shifts gear" and all its cilia paddle in reverse, and it can turn in any direction.

Like all living cells the paramecium has a nucleus to direct its growth, and the bubblelike vacuoles that enclose its digesting food can be seen under the microscope slowly circling around this floating center in the transparent body, the undigested residue eventually arriving at a fixed anal pore to be evacuated. The animal breathes oxygen by diffusion through its surface and, after using it for the combustion of food as do all animals, it exhales or excretes leftover carbon dioxide and nitrogenous wastes along with excess water, which served mainly as a vehicle for the food. It may be worth mentioning too that the paramecium uses two special blossom-shaped organs for getting rid of these fluids (the "petals" of which are canals converging on the central bladder), and it is believed to be a world's champion in elimination, having been known to void its entire body volume of liquid in half an hour: 10 times faster than an ameba and 1000 times faster than a man. Other features are its "defense" arsenal of "poisonous harpoons," which it can shower against an attacker, and the extraordinary fact that some of its species have been observed to diversify into eight sexes, a subject we may look forward to delving into in Chapter 5.

Another common ciliate is called didinium and is a kind of shrew of the watery microcosm, specializing in preying on paramecia. It looks like an oil can decorated with two belts of flickering cilia and a snout on top, with which it pokes about at high speed. When it meets a paramecium it jabs into the soft body and easily sucks it into itself. Even though the paramecium is usually bigger than the didinium, it doesn't faze this pirate, whose snout gapes open into a voracious mouth while his sides bulge to accommodate the meal, soon turning him into a pulsing ball several times his original size. One might suppose an entire paramecium at one sitting would satisfy the didinium for at least a week, but an hour to such a microscopic creature must be equivalent to 100 times as long to a man, for the average didinium has been observed to waste away toward starvation if he eats less than half a dozen paramecia a day!

An animal that feeds on a similar principle but in a more vegetal manner is the suctorian, who usually attaches himself to something solid, then extends his dozens of tentacles straight out from his rounded center, like an octopus with open arms. As soon as any soft-bodied victim blunders into a tentacle, even if it is 10 times bigger than he, it is grabbed in several places and sucked "dry" in about 15 minutes, while the suctorian balloons beyond recognition.

But to get down toward the essence of life we had better look closer at that simplest of all creatures that are indubitably animals: the ameba. This variable one-celled blob of protoplasm lives almost anywhere life exists, from the bottom of the ocean to the topmost crests of sandstorms swirling over the Sahara, including the inside of your body, and is sometimes visible to the naked eye as a white speck one fiftieth of an inch across. The most essential of its parts is the nucleus that drifts about inside the surrounding jelly (made of protein, sugar, starch, fat, etc.) called the cytoplasm. The latter is enclosed in an almost imperceptible membrane, without which the creature would quickly lose itself by diffusing outward. Yet the membrane lets water and gases flow through it easily, and, if the ameba is cut in two, both halves will immediately seal their unprotected surface with new membrane.

Being without a fixed form, the ameba takes naturally to "walking" on temporary "legs" and eating with a momentary "mouth," these appendages often serving both purposes at once. We have already described its strange oozing locomotion (page 26), which entails using nearly every iota of its cytoplasmic jelly as a leg at some stage of its gait. And, curiously enough, it eats with only a slight variation of the same oozing motion, in effect swallowing its prey through a kind of outflanking maneuver of its leg-lips, like an army surrounding an unwary enemy. If this suggests that an ameba habitually puts its foot in its mouth, at least it is deft enough to create a throat with the same encircling gesture, and as the throat carries the prey inward it becomes in turn a "food vacuole" or wandering stomach of water that was swallowed with the meal and in which the struggling prey will soon be digested — that is, if it doesn't first break loose again, a common dénouement among amebas, who can never be sure any mouthful is really "down" until it has been drugged into submission by digestive juices. Superfluous liquid of course is voided (as with the paramecium)     87 from a bladder vacuole that periodically piddles it through the outer membrane.

Another branch of the microworld I should mention is a completely different phylum of animals, called rotifers because they seem to have rotating heads. Though smaller than paramecia and most other protozoans, they are vastly more complex for they are metazoans (multicellular), which means that their bodies contain not just one but hundreds of cells wonderfully specialized and combined into such organs as eyes, brain and developed muscle and nervous systems. And, like most other microbes, rotifers come in many varieties, including some that float on ponds as transparent bubbles, some that look like worms and some that live in graceful "ivory towers" built of their own molded fecal "cobbles." But the most familiar sort has the amusing form of very proper old ladies standing wrapped in furs and tight skirts with elaborate ciliated coiffures and, as Leeuwenhoek put it, "all aquivering and ashivering." Some of them look ridicu-

lously like dowagers standing gossiping outside the opera house in a snowstorm while awaiting a taxi, balancing on their little feet, intermittently doubling up in fits of mirth or leaning toward each other all atwit over a juicy anecdote, and you can see in the microscope the orbits of the swirling "snowflakes" stirred by their gyrating heads. Reinforcing the illusion is our certain knowledge that these fluttery figures are actually females, for rotifer males are few and ironically much smaller than their "wives." Yet either sex can telescope by pulling feet and heads into their barrel-shaped mid-regions preparatory to encysting themselves against drought or danger, in which state they look like tired poodles beside their uncysted dowager companions. Their ability to seal themselves into their "barrels" while awaiting better times is obviously important in their evolutionary success, for they are extraordinarily cosmopolitan and travel constantly upon the feet of birds, on other animals and vehicles, including microscopic dust carried by the wind, which results in their being found on every continent, in every desert and ocean, even frolicking about in hot springs and sleeping upon glaciers in Antarctica. When not stowing away they can also travel either by walking like inchworms, using both ends for feet, by swimming individually with their cilia crowns serving as propellers, or swimming collectively, a dozen or more of them joined by their feet and radiating outward to form a revolving sphere. When feeding, however, they keep their regular feet firmly planted by gripping any solid object with glue supplied by special glands, while, at their head ends, everything they suck in and swallow passes through the "jaws" of a chewing mill with muscle-operated teeth that, in some species, resemble a row of bird beaks that can reach out to peck reluctant crumbs.

The roster of odd beings in the little world below our normal vision thus seems as endless as it is fantastic, and I imagine Leeuwenhoek could not have been more fascinated if he were discovering the fauna of Mars.

## PLANKTON

Where the microcosm is most dramatic in scale, if not significance, is in the ocean, where plankton live everywhere in numbers far exceeding the stars of the Milky Way or even of all the billions of Milky Ways now known to exist beyond our own. The word "plankton" appropriately derives from the Greek for "wandering" (just as the related word "planet" came from "wandering" as applied to the "stars") and is now used to designate all the minute swimming and drifting life of the fluid seas. A good way to introduce oneself to the incredible quantity of planktonic beings might be first to try to visualize the 100-odd scattered specks of shrimplike, rounded or wormy animals in a cubic foot of average seawater as they swim about browsing on some 10,000 invisible crystalline plants among an estimated 100 million much smaller bacteria, algae and protozoan creatures, many of them inside the bodies of their larger companions. Although the most fertile plankton regions are a lot more populous, this probably is pretty accurate for a random bucketful of ocean, 147 billions of which are actually present in each one of the 340 million cubic miles of the earthly seas!

It is hard to decide what is the commonest kind of plankton animal because every phylum in the kingdom has at least some members living in the sea and very likely all of them are to be found somewhere among plankton, one certainly being the phylum of *Chordata*, which includes fish and man. Crustaceans, however, seem to be the most frequently mentioned class of plankton animal. I mean particularly the ubiquitous shrimplike copepods that range from the size of a rice grain down into invisibility and, under a magnifying glass, look iridescent, now blue, now green, now yellow, as they turn their humped little bodies, the while twiddling their legs or rowing jerkily along with feathery oars. Sometimes they and their companions make one think of buffalo grazing restlessly on an ancient Wyoming prairie in spring — wee pirouetting bison amid a smattering of odd-shaped, prancing horses, elk, sheep, giraffes and kangaroos. The analogy is apt because they really are grazing, though on diatoms instead of daisies amid green algae instead of green grass. But copepods could hardly be more exotic, trailing graceful scarfs in pink, blue and purple, as they jostle among new-hatched finny fry with staring silvery eyes, here dodging a ghostly globe of fish eggs, there a slithering, bristle-nosed arrowworm or a pteropod "sea butterfly" that corkscrews upward, looking like a cross between a spaceship and a bumblebee.

Some plankton sprawl with multigesticulating tentacles like tiny octopuses, some throb as breathing lampshades with tasseled fringes or glow at night while flashing red, white and blue lights around the rims of their umbrellas. One kind, the polychaete worm, is as transparent as glass and swims with forward undulations while paddling his 30-odd double flippers, but, when frightened, rolls into a tight ball to plunge elusively down to a "safer" level of society. Now and again along will come a bustling baby barnacle, an infant mackerel or perhaps a two-inch shrimplike krill, favorite fare of whales and seemingly a whale himself beside the microscopic copepods and diatoms. Up, down and on every side one beholds a hundred animal fantasies through the lens in as many seconds of watching.

There is seemingly no end to the unfolding details revealed by increasing magnification. Even vegetable plankton such as diatoms, initially too small to see, appear in the magnifying glass as beautiful skeletons shaped like spheres or triangles or diamonds or boats, with hundreds of "holes" in them arranged in curious symmetric patterns and serving as geometric sieves that strain the teeming sea. But in a powerful optical microscope the "holes" themselves turn out to be hexagonal sieves full of much smaller "holes," which, through an electron microscope, are revealed as still smaller sieves made of even smaller "holes," the descending series continuing down to molecules in the form of crystal lattices composed of atoms that are made up of elementary particles that act and react as abstract waves of energy . . .

Such microvegetables dwell in the upper levels of the sea where sunlight penetrates and "cooks" their meals by photosynthesis, just as in grass and trees. Plankton animals, on the other hand, do not have to stay so high, for, although they climb to graze in the lush upper pastures (mostly at night), they can also browse below upon the spicy, if less green, bits of dead and decaying plankton that perpetually settle through the black depths, like a slow snowstorm laced with vitamins and spiked with phosphates, nitrates and iron compounds that are continually being washed off the land by millions of little streams before being transported down hundreds of big rivers to add their share to the oceans' salts. Some of the larger plankton animals can swim upward as fast as 500 feet per hour, and even the microscopic ones will climb perhaps 25 feet in the same time for their nightly greenery. Some are thought to sleep during the winter in the cooler waters, a kind of microcosmic hibernation. A few manage to migrate vast distances in the bilge water of ships, thus significantly propagating their kind throughout the seven seas.

Some are barrel-shaped and eat, breathe and swim all with one continuous pulsating motion, sucking in water at the open head end, straining it, then expelling it out the open tail end on the ramjet principle. Some live in snail shells buoyed on bubbles. Some are bell-shaped medusae with pendulant elephantine trunks that ingest tiny animals picked off their tentacles, the tentacles having first captured them with stinging cells called nematocysts. The nematocyst, by the way, is an extraordinary but widely used automatic weapon (found in various forms on the tentacles of all fresh- and saltwater polyps and jellyfish) that explodes when any animal touches its hair trigger, first scissoring a tiny gash in the skin of the victim, then instantly shooting a long microscopic syringe through the breach with such high hydraulic pressure that the jerk it undergoes at full extension bursts its tip, flooding the deep wound with a paralyzing poison!

Perhaps most amazing of all plankton animals, however, is a microscopic flagellate called *Oikopleura*, some species of which range down to the size and shape of a human sperm, who builds a "fish trap" (for catching bacteria-sized food) so elaborate and efficient that nothing in man's vast array of fishing gear can approach it. This contraption first appears as a kind of second skin secreted out of his own body pores, a clinging envelope around him that is both transparent and elastic. Then, having neatly pared himself from it with his long tail, he undulates the same tail to pump water into it, rapidly inflating it about him like a balloon, the while spinning thread inside it and weaving a set of webs beyond the capacity of any spider: two coarse grids across the double front doors to bar intruders above edible size, and two cone-shaped fine-mesh nets within to funnel food from the entering streams directly into his open mouth. The whole "machine" is thus a sophisticated, continuous and almost automatic trap, streamlined to facilitate its jet propulsion through the sea, as its pilot-owner with waving tail keeps its internal currents flowing and presumably a careful eye on its five hatches, one of which is an escape port under the bowsprit, just in case —

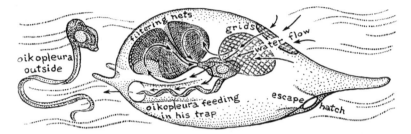

Plankton living near the bottom who haven't learned how to make houses or traps by exuding them whole from their bodies often do about as well by secreting a sticky mortar with which they pick up sand grains, sponge spicules, mica flakes or other available hard materials, gradually compounding them into shells around themselves. The many spiny or shelled plankton have indeed thus evolved hundreds of slightly different techniques for construction of the thousand varieties of their skeletons, shells and shelters, most of them automatic methods, such as the absorption of mineral elements into their bodies along with food to be later excreted ready-mixed for unconscious building. And a typically ingenious method of preying is that of *Foraminifera* or "sea spiders," so called because they sprout hollow bristles fine and numerous enough to crisscross into webs which enmesh microscopic prey that the bristles digest on the spot by exuding gastric juices, then sucking the nutrients into their central calcareous shells that are often attached to seaweed. The while myriad crumbs and skeletons of dying older generations of sea spiders hail slowly but inexorably downward to the ocean bottom to cumulate in 100 million years such massive deposits of chalk as the white cliffs of Dover. The procedure is different of course with coral polyps, but the excreted cumulation just as sure and the building material much the same.

Virtually all planktonic sea creatures range thus through the earth's watery expanses and in various mediums react upon each other in ways that demonstrate them to be essentially parts of one organism. Indeed as the atmosphere circulates water, dust, spores, pollen, seeds, lichen, microbes, insects, spiders, birds, bats, man and other odd beings and baggage, acting somewhat as the planet's breath, which is occasionally infectious as well as refreshing, so is the sea a kind of blood to the Earth's living body that continuously conveys not only all the obvious creatures but, along with the already mentioned invisible ones, immeasurable quantities of germ cells, eggs, hormones, enzymes, vitamins, chemicals and other mysterious, subtle secretions that influence and coordinate the whole in space and time.

## DIRT

Most concentrated of all the earth's mediums of life, though less vast and volatile than air or ocean, is the solid land — particularly the soil, which normally contains more living protoplasm and is of a higher level of organization than all that is in the roots embedded in it. I am speaking of the microcosmic earth we are made of and which obversely, to some degree, is made of us.

As you probably suspected, it is a dirty story — in the most literal sense. Of course you know I'm not talking about filthy dirt but rather the earthy kind. Even filthy dirt, however, is primarily a mental abstraction, since a bread crumb becomes "dirt" only when it falls away from its loaf, say, onto an otherwise clean floor. "Dirt," in other words, is something that is out of place. But the only dirt in the fundamental sense is soil or earth and the microscopic airborne and seaborne dusts that are largely composed of it. Ultimately, of course, all these kinds of dirt derive from space dust, which is thought to have aggregated the whole earth over billions of years and whose most significant ingredient turns out to be graphite, stuff sticky enough to collect such wandering interstellar atoms as hydrogen and oxygen and combine them into all the familiar chemical compounds of our world.

One of dirt's most vital characteristics, moreover, is that it is highly absorptive, for nothing is too dirty to be absorbed by dirt. Indeed its metabolism is quite indispensable in pollution control, while it supports not only the visible, macrocosmic plants like trees and flowers but also the microcosmic ones like mold and bacteria that rot and disintegrate them, making it the natural and ultimate refuge of refuse, of dead bodies, sewage and waste. If dirt, in this sense, is an agent of death, it is even more (and paradoxically) the stuff of life. For God    93 was not speaking figuratively when He told Adam, "Dust thou art, and unto dust shalt thou return." In fact dirt is literally what you and I are made of, what we sprang from, what we walk on, what our food was raised in, what we will soon be buried in and must ultimately diffuse into. And we are the heirs of uncountable generations that passed the same stuff down from one to the next: "earth to earth, ashes to ashes, dust to dust . . ."

The human "clay" that forms our bones and tissues is thus actually molded into man (as we will see in Chapter 6) by a temporary (genetic) rearrangement of the same elements to be found in any common clay or soil: ordinary atoms of carbon, oxygen, hydrogen, nitrogen, silicon, sodium, chlorine, phosphorus, sulfur, aluminum, manganese, calcium, potassium, iron, copper, magnesium, zinc, cobalt, etc., which are arranged in both organic and inorganic molecules in the living body and, to a surprising degree, similarly in the living ground. This means that dirt is something basically congenial to us — which is to say: friendly stuff. In fact, next to flesh, dirt turns out to be about the friendliest stuff imaginable. And as a vital ingredient of the sky and the nuclei of rain, its airborne form (dust) is part of our very breath and blood. It is likewise a prime factor in the ocean that evolved our ancestors and, according to modern genetics, has so many ancestors in common with ours that it is more than our kith, being to a measurable degree our actual kin.

But what exactly are dust and soil? Do they in any sense have a definable structure? One might not think so to hear farmers bemoaning the fact that no two fields are the same or even recognizably similar, yet soil scientists agree that the most varied earths have a lot in common. For all soil is more or less porous, being composed about half of air, which may be largely replaced with water after a rain, and half of rock particles that serve as a bony skeleton to which clings organic matter teeming with microbes and almost as active chemically as living flesh. The rock portion is primarily made of tiny grains of quartz, the hardest part of most rock, which survives the slow erosion of its softer components of feldspar, mica, etc., under the impact of sunlight, wind, rain, ice, flowing streams, chemical digestion, groping lichen, moss, the roots of plants and the hoofs of animals.

During the first few million years of this grinding and wearing process the skeletal grains are sand. At least geologists define them as sand down to a diameter of about one thousandth of an inch, then as silt from one thousandth to one ten-thousandth of an inch and, below that, clay. While not usually considered alive, they effectively serve as the bones of dirt and, in their way, are about as complex and active as the bones of an animal. This is particularly evident in the molecular

structure of clay, which comprises at least 30 elements whose crystals tend to align themselves in alternating sheets of silica and hydrated alumina like club sandwiches, the layers of which can be only one thirty-five-millionth of an inch apart, and electrically charged so that rather intense chemical interactions take place between them. In one typical form of mica clay the alumina sheets have hexagonal holes that line up opposite similar holes in the silica, and the sheets get riveted together with potassium ions that exactly fit the holes. But the greatest potential in clay probably results from the very smallness of its particles, on the geometric principle that cutting any solid into pieces increases its total surface area proportionately to the smallness of the pieces (page 15), the effect becoming dramatic when these get down to the caliber of fine powder. Thus the finest clays have surfaces as vast as ten acres per ounce, which makes them strongly colloidal, notoriously sticky when wet and chemically potent enough to be molded and baked into bricks, tiles and pottery.

Another vital part of soil is called humus, which comes from rotting vegetable and animal matter and is the mucky protein that helps hold the skeletal grains of quartz together, along with many other compounds of carbon, oxygen, nitrogen, phosphorus, etc., that add up to the basic living substance of Earth. And all these parts of soil, both organic and inorganic, are mixed together none too evenly while, except in sandy places, they tend to form crumbs up to about one eighth of an inch in diameter, which are each a tiny sample of the local earth. These crumbs are familiar to anyone who gardens or handles dirt, and seem to be tranquil little clods of inert, mellow tilth. But their apparent quiescence is almost completely illusory, for they are not only teeming with individual vegetable and animal life but are in a real sense alive themselves. They actually inhale oxygen and exhale carbon dioxide, and tests show that normally the air in soil down to about 5 inches deep is completely renewed every hour. And for many feet below that the soil breathes, though progressively more slowly as the moisture and carbon dioxide content of the air increase with depth. If it seems incredible that hard clay could be breathing, just remember that crevices only one thousandth of an inch wide, much too small to see without a microscope, are as much bigger than an oxygen molecule as a valley 120 miles wide is bigger than a man.

95

## Soil Creatures

Of course the breathing of soil is to a ponderable extent the collective respiration of the myriad individual lives within it, which are rather staggering in their interrelated profusion and, as we shall see, may quite logically be considered cells so inherent in their corporate organism that, without them, it becomes not only infertile but hardly deserving of the name of soil. In order to find out just what or who populates the soil, various teams of scientists have marked out random sections of wild forest land in New York State, grassy meadows in England, farms in Sweden, etc., and sliced off the top layers of earth for detailed examination. Not counting the larger animals like foxes, rabbits, mice, moles, shrews, snakes, turtles, toads, ground birds, that here and there had burrowed into it, the scientific census of all visible smaller inhabitants washed and sieved out of the dirt showed an average of about 40,000 living visible animals per top cubic foot, including some 25,000 barely visible mites and 10,000 tiny springtails, with the remaining 5000 divided between about 100 kinds of bugs, beetles, bristletails, lice, spiders, ants, wasps, worms, millipedes, centipedes, crickets, flies, moths, slugs and other forms. And when it came to microscopic life, the numbers were literally billions of times greater, since a single ounce of average fertile soil has been found to contain not only well over a mile of fungus mycelium and comparable lengths of other roots, but about 20 times as many bacteria as all the men, women and children on Earth, with progressively greater and lesser numbers of viruses and other microbes in roughly inverse proportion to their sizes.

If you visualize this immense population in three dimensions, densest at the surface just under the soil's roof of grasses or leafy litter, its continuous bustle increasing as the temperature rises in spring, decreasing as it falls in autumn, you will begin to grasp the real nature of earth — its vast hordes of eight-legged mites crawling in the dark between the black crumbs of loam each of which is itself a seething ferment of microbes around a charged crystal, often meeting nervous little springtails, slithering nematode worms, tardigrades, odd lice, baby spiders and occasionally a huge milling millipede or a giant earthworm hunching its way through the tangle of thirsty root hairs and creeping fingers of fungus. It is a world full of ambush, stealth, surprise, instant reaction and fierce fighting for survival, yet there is harmony and contentment too, even love of a sort, where fungus and probing root embrace each other and mating worms intertwine.

But the predominant interrelation seems to be predatory, which, as elsewhere, assumes the form of chains: with burrowing owls hunting shrews, shrews digging worms, worms eating fungi and amebas, both of whom eat worms in turn. The worms eaten in turn, moreover, are not earthworms but tiny threadworms or nematodes, most species of whom are thousands of times smaller than earthworms and glide snake-like between the microscopic crumbs of soil, now sipping juice from root hairs with their dainty stylets, now swallowing bacteria or worms smaller than themselves. A census of the soil around grass roots has counted nematodes in numbers exceeding 10,000 per cubic inch, and they are well known to form the principal diet of certain ingeniously predacious fungi which may constitute more than half the total weight of living tissue in the same soil; when these fungi "smell" nematodes nearby, they are known to grow sticky knob "worm traps" upon their mycelium threads as well as loop snares, through some of which, sooner or later, worms will try to pass. One of the simpler of these amazingly varied types of snares contains about three sensitively triggered cells arranged in a single loop which, within one tenth of a second of when they feel pressure inside the circle, swell enough to close it or constrict anything inside it in a resilient but relentless stranglehold, which almost always subdues a caught nematode, despite its frantic writhings for freedom. Thereupon the snare starts sprouting shoots of mycelium into the victim which in a few hours grow the full length of his juicy body and rapidly digest it from within. Even if the lashing of a snared worm breaks the noose from its parent mycelium and he crawls away with a fungus collar, shoots of mycelium will still grow into him from the collar and he will usually be dead in an hour. The sticky knob traps work differently, being evidently as attractively flavored as are lollipops to a child, but when a nematode puts his mouth around one he discovers too late that it not only sticks like glue, anchoring him to the spot, but soon shoots mycelium down his throat to digest him from inside, just like the loop snares.

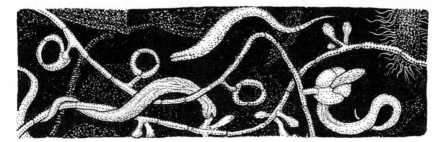

It seems fantastic to realize that such desperate warfare is being fought perpetually between armies of animals and vegetables inside almost every flower pot or swatch of mud that sticks to your shoe or shovel. But more than murder is going on all through the dirt, and the friendly bargaining and bartering there may be even more prevalent than the killing. At least a great many mycorrhizal threads of fungus (as we mentioned in Chapter 2) are now known to grow toward neighboring tree roots and when the fungus meets the root hairs it seems to be welcomed into the tree's body where it has been actually seen (under the microscope) to invade the root, cell by cell, until parts of it are completely consumed or digested. At the same time some of the bast gruel from the outer tree is drawn down into the fungus that enfolds the root hairs where its sun-derived energy is obviously welcomed in return. But all the underground dealings and conversations between plants are not expressed through such close embraces, because some roots are known to exude hormones capable of diffusing outward for several feet to where they find and activate the spores of other plants, causing them to release spermlike flagellates with whiptails that swim confidently straight back through the soil to the plant roots like homing pigeons. It is thus that the vegetable kingdom nourishes and fertilizes its members. And thus, among other ways, do vital traces of minerals get around: magnesium atoms slipping stealthily into chlorophyll, iron into blood and wandering atoms of copper, zinc, cobalt, molybdenum, iodine and cadmium into almost any organisms that will have them — these and all such exchanges exemplifying the great principle of universal interrelatedness that we will be coming back to in Chapter 13.

## GERMS

Meantime, to return now briefly to Antony van Leeuwenhoek, I must point out that he was so busy and excited for almost two decades in discovering and reporting the endless wonders of his little animals who "hopped like magpies" and "swam with their hair" that he could hardly find a moment to reflect on their significance for humanity and the world. Indeed it was only in 1692 that he got around to attributing his good health to his drinking every morning the new Ethiopian beverage called coffee so steaming hot that it "scalded the animalcules" in his mouth and, he was confident, finished off at least most of those on his front teeth.

This is about as close as he got to proposing a germ theory of disease, for he was probably too much of a realist to want to add to the numerous vague and ill-founded speculations about invisible "fever carriers." These had been recorded through history, certainly as far back as Aristotle, who theorized on the subject in the fourth century B.C. In most of the ancient world, moreover, despite Lucretius and Varro (page 83), illness and affliction had generally been considered divine punishment for sin and their treatment was primarily religious, a frank appeasement of angry gods. Only very occasionally did more advanced theories arise, like the Hindu concept of the "five winds" that activate the body or the Greeks' Hippocratic thesis that life is a precarious balance of the "four humors," of white phlegm, red blood, and black bile and yellow bile.

So far as anyone now knows, ancient diseases were generally similar to present ones, and I have found evidence of at least forty familiar ailments that were rife in ancient Egypt and almost as many in each of Babylon, India, China, Greece, Mexico and Peru. These include scores of afflictions like ulcers (treated with mercury in ancient China), mumps (salved with music in Greece), the common cold (soothed with sage in Egypt) and constipation (helped by sarsaparilla in Aztec Mexico, by corn silk in Inca Peru, rhubarb in China, oxgall enemas funneled through a bull's horn in Egypt or through a medicinal flute in Babylon).

The really serious diseases, however, those that obviously affected mankind the most, were the plaguey sort such as smallpox, which may have originated in China — where it was described in 1700 B.C. — and spread westward to India, the Near East and Egypt, where King Ramses V is known to have suffered from it half a millennium later. The great plague of Athens described by Thucydides in 430 B.C. is now believed to have been smallpox, the virus of which may well have decided the outcome of the Peloponnesian War, as the germ of a similar epidemic, reported by Diodorus Siculus to have ravaged the Carthaginian army besieging Syracuse in 396 B.C., set the Mediterranean stage in favor of Rome before the Punic Wars.

Bubonic plague is another such scourge of mankind whose origins are dim in ancient history, but it may have been crucial in stopping the Persians after Thermopylae in 479 B.C. and is thought to be the organism behind the so-called great plague of Justinian that hit Egypt in A.D. 540, spread swiftly to Palestine and shattered Byzantium to the tune of 10,000 deaths a day, undeniably delivering the coup de grâce to Roman civilization. Bubonic plague evidently also erupted in the hot summer of 1204 in Asia Minor to such lethal effect that the Fourth Crusade was stopped in its tracks and forced to abandon all hope of reaching Jerusalem. But it was in the fourteenth century that this frightful disease, then called "the Black Death," attained its zenith, appearing suddenly out of the east in 1348 to wipe out more than half of Europe's population of over 100 million. It struck again in 1361, not quite as deadly but killing nearly half of the survivors and their children; again it showed up in 1371, this time afflicting only about a tenth of the remaining people, many of whom recovered; and afterward yet a dozen times, gradually diminishing in force until 1820, when it seems finally to have faded out as a serious danger to the Western World.

Comparable careers of other deadly contagions include typhus, meningitis, anthrax, scarlet fever, yellow fever and, in their various ways, malaria, filariasis, cholera, tuberculosis, typhoid, influenza, diphtheria, measles, leprosy, gonorrhea, syphilis, rabies and many more — all being the collective names for colonial organisms recently discovered to have flourished at one time or another inside individual humans, but also recorded throughout history to have been imperialistic super-organisms exploring and competing irregularly about the surface of the planet, at once invisible and as big as the world. Like other organisms, they are kept by evolutionary forces in a perpetual state of change, now looming out of "nowhere," now vanishing like smoke, now aiding or hindering a fellow disease for no known reason, ever probing, adapting, competing, evolving. There is evidence, for instance, that leprosy was all but extinguished by the medieval waves of bubonic plague that swept through the huge leper colonies of Europe, a case of one disease eliminating another.

Man, however, may now be influencing diseases more than they influence each other, not only by conquering or controlling many such killers as bubonic plague, typhus, tuberculosis and polio, plus totally extinguishing smallpox (as he seems to have done), but roughly seventy-five percent of present human ills, especially in the wealthier countries, are attributable to man himself. Called the iatrogenic diseases, many of these have arisen directly out of man's adoption of artificial drugs (like embryonic deformity after a pregnant woman has taken thalidomide), some sprang up indirectly (like a new strain of bacteria that feeds on penicillin) and some (such as high blood pressure, heart disease, cancer and asthma) can be largely attributed to the adverse effects of new human habits and occupations.

Returning now to Leeuwenhoek, we can see that it was his epic discovery of the reality of microscopic hordes of living organisms that made possible the increasingly plausible germ hypotheses of the eighteenth century, outstanding among them the closely reasoned treatise on germ theory published in 1762 by Dr. Anton von Plenciz of Vienna. And then came the pragmatic discovery in 1796 by Edward Jenner, a country doctor of Gloucestershire, who, after hearing that milkmaids who'd had cowpox were immune to smallpox, made a few experiments and then publicly demonstrated that nearly anyone could be immunized against smallpox by rubbing matter from a cowpox pustule into an open scratch on his skin. Of course no one could explain why it usually worked and there were vociferous doubters and a good many cases of persons thus "vaccinated" (from *vacca*, cow) coming down with infections (not necessarily a pox) as a result of some unwitting contamination.

Half a century later the causal relation between a microbe (fungus) and a vegetable disease (potato blight) became pretty well established during the terrible Irish famine of the 1840s, but it remained for a young French chemistry professor named Louis Pasteur, who had done research in the physics of crystals, to prove the connection between microbes and human or animal disease. Studying the fermentation of beet juice leavened with yeast for an alcohol maker in Lille in 1857, he noticed a certain optical property in the fermented crystals that resembled what he remembered having seen in biologically grown crystals and this made him suspect yeast of being a live organism rather than just a chemical, as was the common assumption of the day. Testing the idea, he soon realized that all of the various kinds of fermentation in milk, wine, bread, cheese, etc., might be caused by living organisms, perhaps different ones in each case, and that some of the invisible living creatures Leeuwenhoek had found floating around on dust could well cause diseases too. It was relatively easy to prove that a sterile liquid like freshly boiled bouillon would not ferment if kept free of dust-laden air but it took Pasteur (with the help of others) many years to work out the organic chemical processes by which bacteria convert milk sugar into the lactic acid of souring milk, by which yeast changes grape sugar into the alcohol of wine, by which certain bacteria turn wine into the acetic acid of vinegar, and so forth. Yet this he eventually did, mostly during the 1860s, and by his great work the germ theory — that invisible creatures are capable of conquering and killing men or even elephants while swimming through their blood — preposterous on the face of it, became generally accepted as fact. And the fantastic concept was further established by such perspicacious pioneers as the English surgeon Joseph Lister, who quickly and successfully applied it in his operating procedure by sterilizing open incisions with a spray of carbolic acid; a decade later by the young German medical researcher Robert Koch, who, having learned to photograph bacteria, identified the bacilli that cause anthrax, tuberculosis, cholera and others; followed by increasing numbers of other biologists with their growing lists of positively identified disease-carrying germs.

It was not until 1887, however, that anyone actually saw one of the ultramicroscopic microbes we now call viruses (which had been suspected by Pasteur and others). The Scottish surgeon Dr. John Buist, who beheld the barely visible reddish specks of cowpox virus (about 1/100,000 inch long) in his microscope, supposed them to be "spores" of bacteria or fungi. But a decade later a gruff Dutch botany professor named Martinus Willem Beijerinck discovered that the juice of a diseased tobacco leaf, after being passed through a porcelain filter to eliminate all bacteria, would still infect healthy tobacco leaves and the mosaiclike infection could spread indefinitely. He could not see any microbes in it with his microscope, so he concluded that the agent of the disease must be a "filterable virus" too small to see and somehow living without any solid structure in the liquid. While his "liquid life" concept never was taken seriously by other biologists, a third of a century passed before an English medical researcher, Dr. Willis J. Elford, in 1931 obtained conclusive evidence (with ultrafilters which stopped viruses while letting liquids through) that viruses must be solid particles.

And then in 1935 a young American chemist, Wendell M. Stanley, succeeded in purifying a solution of tobacco mosaic virus (TMV to virologists) into the strange form of a white, sugary, crystalline powder that seemed to be nothing but inert mineral yet was very potent in infecting tobacco. It was incredible to biologists that any purely mineral solid, chemically cooked and rigorously reduced for more than two years until obviously "dead," could still carry and spread a disease, but the dry crystals turned out to be about 95 percent protein and 5 percent nucleic acid and perfectly capable, in the presence of tobacco leaf cells, of reproducing themselves scores of times in an hour.

Before we get ahead of our story, however, I must point out that the resolution of this major mystery in what seemed to be the ultimate microbe will have to await Chapter 6.

## CELLS

Here is where we need to take a look at what is coming to be recognized as the basic unit of life, the cell. Cells have not been easy to investigate, being generally invisible and therefore quite unsuspected by man until A.D. 1665 when Robert Hooke happened to notice their compartmented structure while examining pieces of cork under his low-powered pre-Leeuwenhoek "microscope." The segments presumably reminded him so much of monks' cells in a monastery that he naturally named them cells also. But although he drew detailed pictures of them, he did not realize they were present in all plants and animals. In fact it was not until 1839 that the botanist Matthias Jakob Schleiden and the zoologist Theodor Schwann propounded the startling theory that the cell is the "vessel of living matter," a new idea for which Rudolf Virchow was to win wide acceptance in 1859 by demonstrating that all cells, in both vegetables and animals, originate from the divisions of earlier cells. In other words, the cell at last was revealed as the natural unit of life, not only capable of independence as a complete one-celled animal like an ameba or a paramecium, but surprisingly independent even when part of a large multicelled organism, feeding, excreting, reproducing and in many cases moving about and making responsive decisions. In a nourishing liquid (known as a "tissue culture") almost any body cell can now be kept alive *outside* the body, like the free creature many of its ancestors must have been millions of years ago, and this discovery in the early 1930s convinced biologists that cells must be much more than the simple blobs of jelly they appeared.

If you have never seen a body cell floating by itself in a tissue culture, I might say that at first the ovoid speck usually seems inert and helpless, drifting idly off after separation from that greater organism, its multicelled body society. But as soon as the lonely cell touches a solid object it responds. Perceptibly it bulges toward it. Then a protoplasmic finger forms on the cell, pointing and reaching out in the direction it wants to go. If the solid object is the inside surface of a test tube, the fingertip usually flattens against it, gluing itself to the glass. Then the finger contracts, pulling the rest of the cell toward that glued spot. The cell then normally takes a second step as another finger reaches forward, and thus it creeps with apparent purpose on its way.

Startled to find such unexpected deliberation in human flesh, biologists have been pursuing the cell with more and more powerful microscopes ever since, eagerly exploring its mysterious spots and shadows to learn how this curious unit organism functions. In the 1940s they discovered a faint but persistent stringiness in the protoplasm near the nucleus that eventually proved to be a complicated and beautiful network of tubes and necklaces, through which the teeming cell populace of porters and messengers could be detected flowing like street traffic in all directions, delivering supplies and orders to every part of the cell and of course to a great extent beyond it among adjoining cells and the outside world. The beads on the necklaces (some of them loose beads) in turn were found to be individual chemical generators, soon named mitochondria, which are endlessly turning out the dynamic fuel adenosine triphosphate, now commonly abbreviated ATP, a very ancient kind of bio-explosive that powers all of life's material activity from growth to muscle contraction and is, as you may remember, a vital step in the chemistry of photosynthesis.

Between and all around the tubes, necklaces and other special parts is the real interior space of the cell, a sort of storage place that serves also as a lobby and reading room, where fat vacuoles loll about full of

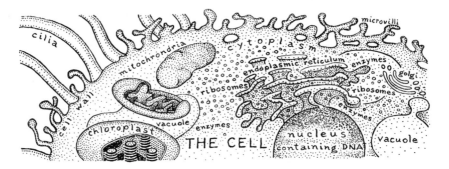

water, oil or gruel and curious enzymes are stacked like books and periodicals on library shelves. The most useful "volumes" are continually being passed and circulated about the cell corridors while an important percentage are confidently dispatched abroad like overseas mail in the form of explicit messages, in effect written, coded and posted by one cell to a fellow cell it has never seen but somehow manages to correspond with, even while they are worlds apart.

The key to this enzyme library is located, as long suspected, in the cell's nucleus in the form of a master code, a code literally made of genes and guarded like a state secret. The nucleus that encloses it turns out to be a sort of cell manager's office built like a vault, out of which painstakingly accurate copies of the cell's construction plans are rolling off the duplicating machines at speed for immediate distribution to ensure that every part of the cell knows what is expected of it and that not only all parts of the cell but all cells of a whole multicelled organism conform to the same standard. The very difficult job of cracking the cell's genetic code was only undertaken in the 1950s after scientists practically blasted their way into its nucleus and extracted the key substance of deoxyribonucleic acid which, by popular demand, was quickly abbreviated to DNA. The heart of the nucleus of every cell, they found, is essentially a tapelike coil of DNA that carries the code of life in it as surely as a magnetic tape carries a speech or a tune. DNA however is not only more indelible than magnetic tape but there is something unquestionably mystic about it. In fact, while ruling the cell by regulating its chemicals and dispatching its enzymes, DNA subtly conceals the ultimate origin of its power. In a way DNA acts as the cell's god, a designation appropriately spelled out in the Latin word *deo*, which forms the first three letters of deoxyribonucleic acid. And, godlike, it broadcasts its omnific decrees at electronic speed through a technique so intricate and awesome we must defer looking deeply into it until Chapter 6.

I can say here, however, that DNA's decrees are translated into action with an elegance and perfection evolved over billions of years, a micro-majesty perhaps best exemplified by the cell's oft-repeated but deliberate division into two daughter cells. This kind of dividing is called mitosis and is a normal act of growth that is happening in millions of places in your body at every moment. If you've ever wondered why big animals, say whales, don't multiply this way, splitting down the middle to produce two new half-sized whales, thereby circumventing the problems of babyhood, it's because the whales would have to be genetically the same. They would inevitably be identical twins and identical not only with each other but also with their parent. Which would be disastrous from the standpoint of evolution because evolution needs continuous variation in order to be flexible enough to adapt to changing conditions.

Mitosis, on the other hand, is just right for cells. Being in the microcosm and therefore very numerous, cells don't need to be flexible individually but only in statistical masses. Besides, the heirlooms of cells, unlike macrocosmic davenports and teapots that would be damaged by being cut in two, can safely be split into identical halves right down to the chromosome and gene levels. And that actually is how their bequests are bequeathed. It is an orderly, even stately, performance taking about an hour in each case, displaying a discipline, beauty and wisdom hard to imagine as springing solely from that microscopic cell. First there comes a sort of secret decision that whispers forth a quiet mobilization order that rapidly grips the whole cell in a state of pseudo crystallization. This begins just outside the nucleus at a spot called the centriole, from which tiny lines are seen to shoot out like frost feathers across a windowpane, lines (probably representing microscopic fibers) that in a few minutes have organized the whole cell into two halves around two poles that appear as mirror images of each other. This polarity of course includes the nucleus, which, following the rest of the cell, progressively straightens and regiments its coils of

DNA into duplicate forms, each paired member attached to an opposite centriole by threads that converge on it as if wound on spindles in a textile mill. When the whole cell is thus completely regimented into the mirrored double-crystal form, an equator of cleavage appears halfway between the poles, a bulbous waistline that begins to pinch inward as the upper and lower hemispheres pull apart, eventually thinning to a tight waspish shape before they finally separate into two new round cells (illustrated on page 164).

You might think these twin baby cells would rest for a day or two after their birth, but no, their nature is to move and, as long as they are free, they keep moving at an average speed about equal to the hour hand of a small watch. Furthermore each one is under some mysterious compulsion to make a crucial decision within its first two or three hours: it must either begin to specialize, after which it may never divide again, or start to divide and subdivide further, in which case each time it cleaves into children it will also "conceive" grandchildren in the children's centrioles. By such rules do cells in general — vegetable as well as animal ones and including the 50 trillion cells in your body — live and renew themselves, diversifying the while into flesh, leaf, bone, wood, blood, muscle, nerves, skin, bark, hair, seed, eyes and all the other substances needed. And despite, or because of, their restless activity, each one somehow always seems to be in an appropriate place at every moment of its life, a life whose duration depends largely on the role it has chosen to play, the role in turn normally conforming to whatever specialized cells it has been associated with. Thus unspecialized vertebrate cell tissue, if exposed to a piece of spinal cord, will nearly always start producing cartilage tissue, a first step toward becoming a spine. Yet the same unspecialized tissue, embedded instead in muscle, will in more than 99 cases out of 100 rapidly specialize into muscle cells.

As to what actually makes a cell begin to specialize, very little is known, but biologists have noticed that body cells on the loose commonly sprout sensitive whiskers they call microspikes or pili, which whisk about and feel what is near them and, if it is congenial, penetrate and cling to it. Presumably it is these tentacles (more than any other factor) that enable cells to spot and scrape acquaintance with their neighbors, whereupon, assuming the neighbors "taste" good, they flock together with them into the dense masses familiar to us as flesh and bone, at the same time exchanging genetic material and perhaps exuding short-range chemical bonds for greater cohesion, like a kind of living mortar.

There is no reason to suppose the individual cell has any real choice as to whether it will specialize and settle down as part of a gut or a leg or an eye in this social development, which may resemble a well-rehearsed army mobilization, but, as I said, it obviously makes a lot of difference in the expectable life spans of its descendants. For an epithelial cell in the gut lining lives only a day or two before it dies and is sloughed off, but white blood cells can last two weeks and red ones four months. Nerve cells, however, which cannot normally be replaced at all, may survive more than a hundred years — a decidedly opportune span in the case of centenarians, who depend, like the rest of us, on irreplaceable nerves.

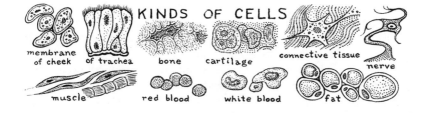

KINDS OF CELLS

membrane of cheek | of trachea | bone | cartilage | connective tissue | nerve | muscle | red blood | white blood | fat

Although specialization commits cells to particular functions and life expectancies, it does not necessarily limit them to fixed locations, as can easily be seen in the travels of blood cells. Moreover, when flesh is bruised or cut, it is now known that millions of deeper-lying body cells (other than blood cells) apparently eagerly migrate upward to replace those lost in the injury, and skin cells detach themselves from all sides to spread individually over the wound's surface, floating upon the fluid film that automatically wets it while strewing it with scarcely perceptible fibrous "guidelines," along which the free cells have been seen to move and coagulate. In a way such semi-imaginary guidelines are like faint cattle trails on a Texas range, over which the cells amble forward about as obediently as steers steered by a proficient herdsman and, like them, almost surely playing a part in some hard-to-fathom overall plan. One can even imagine them enjoying their share in this mystic migration of mercy that is the healing of a wound.

Although muscle cells (like many other specialized ones) cannot replace themselves by dividing, new muscle cells are constantly being assembled from microscopic old fragments of muscle floating about in cell fluid. It is a process curiously like crystallization, in which the flowing bits manage to combine into filaments, which somehow bundle together into fibrils, which by the hundreds align to form fibers of striated muscle which, although more dynamic than any vegetable cells, turn out to look (under the microscope) remarkably like a piece of wood. This crystalline structure evidently is vital to the sophisti- 107 cated coordination that enables many millions of muscle cells simultaneously to contract as a single muscle, gossamer layers of filaments sliding past each other inside the tiny fibrils, in some cases (as in midge flight) going through the cycle of contraction and relaxation up to 950 times per second, each muscle exactly balanced by its opposing countermuscle for perfect control. Muscles in a real sense are the body's engines, with their filaments pumping back and forth like pistons in the fibril cylinders, fueled internally with ATP generated by the hundreds of mitochondria in every cell. Mitochondria are remarkably like the chloroplasts in leaves (page 49), especially in their alternating layers of protein and fat that synthesize ATP, but of course are powered not directly from sunlight (as are chloroplasts) but indirectly from food (originally vegetable carbohydrates) created through the sunlight. And they resemble chloroplasts also in their ability to move about fairly freely within their cells and to reproduce themselves, behavior suggesting to some biologists that they may have once been independent organisms, which somehow got domesticated into their present specialized symbiotic role.

Nerve cells are also intimately involved in muscle cells, which they control more precisely than your car's distributor controls its engine, and they regulate many other types of cells while bestowing on an organism a continuous awareness through its senses. Fact is: it takes more nerve activity than you probably realize just to maintain human consciousness, for researchers have discovered that the average of billions of nerve cells sends reports to the brain at a rate that can never drop as low as one report per second per cell while the brain's owner is alive and well.

And bone cells, blood cells, germ cells, vegetable cells, etc., all have their own special natures. Skin cells begin to die almost as soon as they are born by stiffening or "crystallizing" into a callus, which may eventually become as hard as wood, bark or horn. Heart cells, even when separated from each other, are still hearty enough to continue faintly beating as individuals, apparently never quite forgetting their chosen purpose. And, if given half a chance, they will catch and cling to their fellows again, incorporating gradually into a hollow ball while synchronizing their rhythms, from there on redeveloping as best they can their full single throb and function as one vital organ in an organism they cannot help (at however primitive a level of instinctual being) feeling part of as long as they live.

Thus does the cell exhibit its mysterious propensity for transcendence from the microcosm into the macrocosm. Even as does life itself transcend (as our Fifth Mystery will explain) in a still more universal sense.

# CHAPTER 4

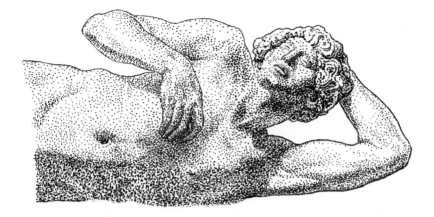

# The Body

HERE LET US TURN the eye of perspective from life's invisible elementary units, the cells, to what cells can assemble and become: a visible body. Which is to say that, having just examined the functioning of these constituent parts, we now want nothing so much as a good look at our corporeal whole, to take its measure and see how it works.

## PHYSICAL LIMITS

I was surprised to discover that, among adult humans, some weigh 200 times more than others and that variations in the human species cover as wide a scope as in the dog species, which includes both Chihuahuas and Saint Bernards. In evidence, the generally accepted heaviest human who ever lived was Robert Earl Hughes of Monticello, Illinois, who died in 1958 at the age of thirty-two a few months after being carefully weighed before witnesses at 1069 pounds, which is considered about normal for a six-year-old elephant. While the lightest human on record seems to have been Zuchia Zarate, an emaciated nineteenth-century Mexican midget 26½ inches tall, who weighed 4.7

109

pounds at the age of seventeen yet managed to survive until she was twenty-six. Literally, she could have stepped out of a balloon a mile in the sky without much risk of injury if she had had an ordinary umbrella in her hand and knew how to open it up as a parachute. The tallest human ever reliably measured was Robert Pershing Wadlow, credited with 8 feet 11 $^1/_{10}$ inches, who died in 1940 when he was twenty-two, and the shortest, a Dutch midget known as Princess Pauline, who was only 23.2 inches tall at the age of nineteen. As for the human capacity for drastic physical change, the record probably goes to an American circus "fat lady" known as Dolly Dimple, who at the age of fifty-eight weighed an almost lethal 555 pounds but, to foil the undertaker, quit her job and in 14 months under close medical care slimmed down to 120 pounds, the while reducing her "vital statistics" from a formidable 84"-84"-79" to a fetching 34"-28"-36".

If extremely large or small people do not seem entirely normal to you, I must point out that total humanity includes a complete spectrum of monsters that are definitely even less normal, though at least forty kinds of them occur often enough to have standardized names. Indeed I hardly think it would be possible to imagine any sort of creature, even one (or is it more?) with multiple or fused heads, extra limbs, tails, horns or other appendages, that is not on record somewhere sometime as having actually been born to some real animal or human parents. And that verifiable fact raises the interesting abstract question of whether any meaningful boundary can ever be drawn between one individual and two or, for that matter, between 0 and 1 or 1 and 1½ or any other numbers.

It is in the supposedly empty gap between a single whole human and two whole humans, I notice, that most monsters find their logical designation, which may turn out to be anything from a single inoffensive individual with a slight suggestion of an extra person about him (a sixth finger, say, or some small appendage) to two individuals who are barely merged by a narrow peduncle of flesh (the bond connecting Siamese twins), which hardly needs make them unattractive. But midway between these borderline cases there recurrently appear all sorts of genetic nightmares that represent more than one person without ever becoming entirely two — obvious mutations or mistakes in development that express themselves in scrambled multiple features, sometimes repeating heads or eyes sufficient for several people while omitting legs or other parts entirely.

In duration of living also, human beings have a remarkable record, even without swallowing the claims of Methuselahs or accepting centenarians who can't quite find their birth certificates. I mention this hazard, because there has probably been more self-deluded bragging and deliberate fibbing in the allegations of longevity (sometimes for patriotic or political reasons) than in any comparable branch of statistics. In fact scientific researchers, I am told, have discovered a suggestively stable ratio between the numbers of illiterates and of claimed centenarians in the major countries of Earth, which seems to be due to the tendency of very old people, particularly ones who keep no written records, to pad their ages at a rate that fairly consistently averages 17 years per decade. Notwithstanding all such foibles, humans very likely live longer than any other mammals (including whales, one of whom with identifying markings was sighted in an Australian bay over a period of 90 years), but they certainly have serious warm-blooded rivals among vultures and possibly condors and certain eagles, the only doubt stemming from the fact that no one is known to have kept systematic birth records of animals long enough yet to prove precisely how long the longest-living ones can live. The oldest irrefutably documented age attained by a human being appears to be 113 years 124 days by an eighteenth- and nineteenth-century Quebec bootmaker named Pierre Joubert; and the rarity of verified centenarians in general is suggested by the failure of the most reliably pedigreed large group of people anywhere, the British peerage, to produce a single 100-year-old peer in ten centuries — until Lord Penrhyn "proved the rule" by dying in Towcester, England, on February 4, 1967, at the age of 101.

Animal longevity seems to vary roughly in direct proportion to body size and in inverse proportion to reproductive capacity, with cold-blooded creatures usually outliving warm-blooded ones, but there are so many deviations from these rules that zoologists have hardly begun to explain the phenomenon. Cats live longer than dogs, for example (known limits being 35 for cats to 27 for dogs), while zebras outlive lions in zoos (38 to 35), but presumably not in the wild state where they serve as lion food. In accord with the fact that a 580-pound giant clam has been estimated (from growth layers) to have lived about 100 years, you might think animals with the fast metabolism of birds would burn themselves out and die young, but something about the flying life must give them durability, for birds live much longer than most other animals their size, including cold-blooded ones, with the exception of a few insects (queen ants have lived 19 years, cicadas regularly 17), many small reptiles (30 years or more) and toads (up to 54 years). Even bats live three or four times as long as mice their size. But it is the large tortoises that evidently hold the record for age among animals, the oldest documented patriarch of all being Tu'imalila, the    111 tortoise reported to have been given to the Queen of Tonga by Captain James Cook in 1772, which died in Tonga in 1966, probably after living more than 200 years.

Trees, on the other hand, usually outlive animals, and some of them last more than 20 times as long, evidenced by the ring count of a recently cut bristlecone pine in the White Mountains of California that proved it had lived over 4600 years, having started from a seed just before the pyramids of Egypt were begun. Of course if one regards the resproutings from old stumps or roots as a continuation of the life of the same organism, vegetables can be considered practically immortal. And this is even more true of most animal and vegetable cells when you count the twin offspring from a divided cell as continuations of the old cell that formed them. By the same reasoning, a strain of bacteria recently revived from a Permian limestone formation in Germany is probably 250 million years old, while similar pre-Cambrian cells, if they prove viable, could have lived two billion or more years, roughly the age of such a star cluster as the Pleiades, and evidently only possible because (on Earth at least) there seems to be no long-range aging process in the successive generations of cells.

If bodies are far exceeded in longevity by the cells that compose them, the bodies can take heart in the fact that they are now gaining relatively. The life expectancy of prehistoric man is estimated to have been only 18 years, but under the civilizing influence of ancient Rome it rose to 22 years, in England in the Middle Ages to 33, in America in 1900 to 47. While India and other poor countries are still floundering under a life expectancy in the 30s, western Europe, Japan, Australia and the United States by now have raised theirs to 70 years, mostly by improving the art and science of resisting disease. Even dogs in America have doubled their expectancy (from 7 years in 1930 to about 14 now) at a rate faster than man's, theirs having been comparatively much lower when man finally acquired an easy means of helping them.

## Bones and Muscles

So far medicine has discovered little about how to control the seemingly inexorable process of aging and biologists do not even agree on its definition. But just as metal is said to "fatigue" while its molecules gradually collect in knots, so does a similar coagulation transform bone, muscle and flesh as the years go by. Body cells tend to lump together, while dying off faster than they can be replaced. And, after the age of about thirty, water and reserves noticeably decrease, bodily efficiency declines roughly one percent a year, tissue wastes away, enzymes disappear, mutations damage genes and organs wear out, one after another. Typically, in the body of an eighty-year-old man 50 million of whose cells are dying off each second, while perhaps only 30 million new ones replace them, muscle has lost 30 percent of its former weight, the brain has shriveled 10 percent, nerve trunks have shed 25 percent of their fibers, each breath uses 50 percent less air, each heartbeat pumps 35 percent less blood, the blood absorbs oxygen 60 percent more slowly, and the kidneys (curiously like loyal members of a team) have sacrificed their efficiency by half to help other organs worse off than themselves.

Yet the fact that a body will continue functioning through all such adjustment speaks for the wonder of this corporation of 100 organs, 200 bones, 600 muscles, trillions of cells and octillions of atoms that are physically you, a human cosmopolis, a mysterious going concern that is so adaptable that a contortionist recently squeezed himself into a 19″ × 19″ × 19″ box, and an emergency specialist on the New York Police Force commented that "the average human body can be compressed into a space 4 to 6 inches wide without serious harm." And I'd say the roots of this resilience permeate the body's microcosmic structure with such elegant perfection that no one could critically examine it without a respect amounting to awe.

Consider collagen, the little-known main ingredient of animal protein. It is the tough, fibrous connective tissue of body parts as diverse as skin, ligament, tendon, gristle and bone, and it accounts for 40 percent of all human protein, including the webby material between muscle fibers and around cells in many organs. It has the texture of leather and the guts of glue. In a tendon it has the tensile strength of a light steel cable. In the eye's cornea it is stacked like plywood and appears as transparent as glass. In bone its fibrils are arranged like bridge girders for maximum resistance to expectable stress. Its basic molecule is a left-handed polypeptide helix and three such helices are twisted around each other to form a right-handed superhelix, whose three parts are evidently held together by hydrogen bonds while their wholes lock into various larger crystalline patterns according to their function. Among other qualities, collagen can adapt to changing requirements while being dissolved and recrystallized, a capacity to transform that makes it particularly effective in healing wounds. However it increases its cross-links with age, presumably through normal molecular movement, and the result· is obvious in the stiffening of joints, the hardening of arteries and the cracking of leathery skin with advancing years.

The only thing that may hold the body together more than collagen is the skeleton with its muscular covering. The skeleton of course is a living engineering structure: the pelvis an arch into which the lower spine wedges like a keystone, while the vertebral column stands above it, supported by muscles as a ship's mast is secured by shrouds, the foot a small cantilever bridge, the knee a pulley wheel on a crane that swivels in the ball-and-socket joint of the hip, the elbow and knuckles simple hinges . . .

At first thought bones seem not really alive, but this illusion is quickly dispelled by a microscopic study of their growth, which turns out to be a highly organized activity comparable to the building of a subway in a bustling city without disturbing it. All the time a young bone is lengthening, it is sealed at its ends by disks of collagen-reinforced cartilage that correspond to the shields behind which men drill a tunnel. The disks grow steadily ahead at their front surfaces, while cells bearing such minerals as calcium, phosphate and carbonate arrive in a continuous stream behind them to convert their rear parts into bone, almost as if men were coming up with concrete to replace the dirt excavated by the tunnelers. The construction is in no way crude, but follows microscopic specifications in a plan of exquisite design, for not only is the cartilage composed of parallel columns of cells in a kind of granulated honeycomb, but the much harder bone turns out to be a hydroxyapatite crystal of still finer grain whose mosaic of tiles is laid in a carbonate and citrate mortar with collagen for a binder. And while the bone grows in length it also grows in girth, accomplishing this through the combination of bone erosion (by wrecker cells inside the marrow cavity) and bone building (by mason cells on the outside), the latter accretion serving also as a kind of mineral bank in which reserve calcium, phosphorus, trace metals, etc., are conveniently stored for the body's future need.

Muscle seems so different from bone that it is hard to believe at first that it too is basically crystalline. Yet recent work at the molecular level has established the fact beyond question. The common muscles that men and animals use to move their limbs and bodies at will are called striated, because under magnification they show cross stripes at right angles to the fibers that run the length of them, like the grain in wood, also crystalline in form. Indeed these cross stripes exist because the fibers are not only perfectly parallel to each other but all their parts are exactly matched crosswise, giving the whole a carpetlike warp-and-woof texture that is vital to its function. If you doubt this, you should see a muscle fiber under a series of microscopes. First, in low magnification, it looks like one of hundreds of silken violin strings that are bundled together while stretching from the tendon at one end of the muscle to the tendon at the other several inches away, and it is usually less than one thousandth of an inch in diameter. Then, under higher magnification, you see that the fiber is made up of a thousand or so smaller and parallel fibrils, generally aligned into striped ribbons. Finally, under still higher magnification, each fibril turns out to have a grain of hundreds of even tinier parallel filaments of two distinct kinds, one of them twice as thick as the other, arranged in a beautiful hexagonal pattern reminiscent of cartilage and other honeycomb forms. But the special thing about the filaments in muscle fibrils is that the thick ones (made of long, golf-club-shaped myosin molecules) are chemically very different from the thin ones (made of chains of round, golf-ball-like actin molecules) and the two kinds, lying alternately side by side, can be made to react powerfully upon each other in such a way that they are forced to slide past each other.

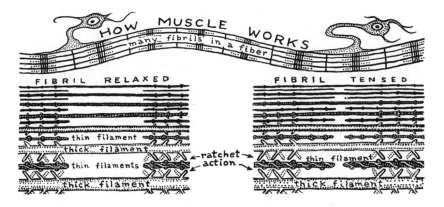

This, in fact, when done by millions of filaments in unison, *is* the contraction (or relaxation) of a muscle. And although every aspect of how the filaments exert force upon each other is not yet fully understood, the evidence shows it to be a kind of ratchet action played by chemical cross-bridges between these two kinds of protein. Contraction is triggered by a nerve impulse that hits the fiber and depolarizes the membrane enclosing it, discharging the fiber's normal negative electric charge of a tenth of a volt to release calcium ions throughout the muscle, shortening it at the rate of a tenth of its length per hundredth of a second under a tension of 40 pounds per square inch of cross section. Moreover, the ions are dispatched and routed to the filaments by means of a marvelous network of microscopic tubes and sacs that simultaneously handles both telegraphy and plumbing, controlling everything from the swimming muscles of a whale that pitches five times a minute to the flight muscles of a midge that beats its wings a thousand times a second!

## NERVES

What I mean by telegraphy in a muscle is of course the wonderful communication system of its nerves. But nerves are also in all the body's organs, in the skin, the spinal column and in unbelievable masses in the brain. In fact through the human body as a whole there stretch something approaching a hundred billion neurons or nerve cells, most of which look under the microscope rather like octopuses, with bulbous central bodies from which extend long tentacles. And when you get to the ends of these axons, as the arms are called, they characteristically turn vegetable in form by spreading out into uncountable vinelike branches that would make a botanist think of the glory-lily with its tuber and far-probing tendrils.

Individually such neurons are, in a sense, the threads of our intelligence, impalpable threads that collectively weave themselves into the tapestry of thought. At the least, they have a nonmaterial aspect as waves of abstract logic and visualization, whose ever-shifting relationships shape our consciousness. At the most, they possess unimaginably weird and esoteric properties. When appropriately triggered, they fire invisible barrages of electrochemical pulses whose lightning-swift patterns encode our every motive and whim. More astonishing, each of them is a sophisticated living computer capable of evaluating not only thousands of competing signals per second but, in the same interval, making decisions in response to them all.

The technique of neuron response and function, it turns out, depends primarily on the junction points or synapses between these cells, which are where their branches intermingle and effectively touch each other in thousands of places per cell, where each cell can send and receive messages from hundreds of other cells at once. These quadrillion points of contact then are like the points in spark plugs, if you can imagine a living engine with a quadrillion spark plugs, each of which is flashing from an irregular but endless fusillade of electrical pulses.

Sometimes the synapses close and let their pulses pass through to the next neuron, an action that involves the flow of chemical transmitters that fit like keys into locks on both sides of the gap, and sometimes the synapses remain open and stop the nerve impulses. It all depends on the message. Inhibitory synapses in fact are about as common and vital as excitory ones, serving to define the delivered signals by a kind of electrochemical pruning, as well as to protect the brain from being swamped with trivialities that, if they were not rigorously screened, could turn a normally stimulating act like opening one's eyes into a deadly convulsion.

The physiology of the nerve impulse itself has always been very hard for researching scientists to comprehend, being at least as mysterious and complex as lightning, which interconnects clouds and Earth, and, if anything, even more abstract. Like lightning, the impulse in effect generates itself, forming a transient wave of electrical excitation that advances the length of a nerve fiber, which is comparable to an insulated wire yet contains a low-resistance core of potassium ions enclosed in a high-resistance, porous membrane surrounded by sodium ions. The insulating membrane, I must explain, does not block the flow of electricity completely but rather serves as a variable filter, a sophisticated and sensitive switch that, in a precise pattern, allows current to leak between the negatively charged potassium ions within and the positively charged sodium ions without. In fact it is the advancing front of this current (exchanging potassium ions for sodium ones) that constitutes the moving wave of the nerve impulse, the abstract something that conveys its message at an average potential of .13 volts and a speed that has been clocked all the way from 2 mph in the case of a one-twenty-five-thousandth-inch-thin, eighty-year-old, visceral nerve to 285 mph in a one-hundredth-inch-thick, young, spinal cord nerve cell. And there is even evidence that a comparable impulse can travel in vegetable tissue under certain conditions, and possibly in minerals — these being just two more examples of the universal capacity of matter to respond to the world, to react, relate and be alive.


## SKIN AND BREATHING

One does not usually think of skin, which one sloughs off relatively easily, as a vital organ. But skin, in fact, is the largest as well as the    117 most versatile organ of the body. It holds our liquid flesh (two-thirds water) in one piece and shields it against crippling blows, microbial and chemical invasions, heat, cold, searing ultraviolet sun rays, and at the same time regulates blood flow, excretes waste, houses the sense of touch and is one of the most important means of sex attraction. Further, the same tissue that becomes skin differentiates in other places and in other animals into hair, nails, scales, claws, hoofs, horns, quills, beak and feathers, all using the collagen molecules in the required crystalline patterns, the softer varieties regenerating damaged parts so quickly and automatically that we take the act for granted as healing and usually forget that without such self-repair we might have great difficulty in surviving even a year. The full subtlety of skin, moreover, cannot possibly be appreciated without studying it microscopically, when every part of the body is seen to be covered by skin of a different specialized texture, every individual possessing his unique fingerprints and pore patterns, his two million hair and sweat glands, his surface nerves, blood capillaries or, in small lungless creatures like the earthworm, a sievelike velum that "breathes" oxygen directly to the blood.

Oxygen being not only vital but by far the most prevalent constituent element in the animals of Earth, as it is in the surface of the planet itself, a terrestrial body must have ample access to the oxygen it requires. In microscopic creatures like protozoans, therefore, oxygen diffuses readily through the delicate cellular integument, roundworms (as I've said) breathe through their whole skin, insects have networks of tiny windpipes called spiracles that convey oxygen to all their tissues (none more than half an inch from the surface) and larger animals, if living in water, have evolved gills to filter oxygen from that liquid or, if on land, lungs to extract oxygen from the atmosphere by a rather incredible air-conditioning system analogous to an inverse tree of vapor whose branches spread inward within the body from a trunk connected to the sky outside. The "leaves" of this lung tree are called alveoli, microscopic air sacs that sprout from "twigs" or bronchioles that stem from bronchi branching off the trunklike windpipe that carries air through the nose and throat. There are some 300 million alveoli in the normal human lung, which tremendous number is needed, as are leaves on the tree, to provide enough surface to enable the proprietary organism to breathe at a livable rate. If spread out flat, the total area of a man's alveoli would add up to 40 times that of his skin: some 750 square feet, or all the floor space of an average house. Unlike birds, mammals breathe both ways (in and out) through their tubes and windpipes and they have a remarkable filtering system
to clean the incoming air. Nose hairs and bone convolutions entrap

almost all dust particles bigger than one twenty-five-hundredth of an inch, while smaller ones down to one ten-thousandth of an inch usually settle on the walls of bronchi or bronchioles, where they sink into the mucus lining that moves continually upward, propelled by microscopic cilia whose hairs lash back and forth 12 times a second, yet enough faster in one direction than the other to keep the mucus escalating. Still smaller dust particles reach the alveoli in large numbers, where they are likely to be engulfed by scavenger cells or carried off by flowing lymph, although some percentage remains to age the lung by causing the growth of fibrous tissue.

The purpose of breathing of course is to bring oxygen into contact with blood, which eagerly absorbs it from air in exchange for carbon dioxide and other wastes and circulates this vital element throughout the body. The exchange takes place almost entirely in the alveoli, which are made of cobwebby nettings of blood capillaries so fine that the red blood cells literally must slither through them in single file, yet normally a red cell lingers only three fourths of a second in a capillary or, during hard exercise, one third of a second. Moreover the gas swapping is enabled to go on continuously despite the intermittency of breathing, because most breaths transport only a pint of air while the lungs hold some one-and-two-thirds gallons, leaving them with plenty of reserve even at the end of a deep sigh. In consequence professional divers have learned after long training how to hold their breath for minutes even while doing strenuous work deep underwater, as was demonstrated by a Greek sponge diver named Stotti Georghios who, in 1913, without any equipment except a weight for descent and a rope for ascent, went down 200 feet to put a line on the lost anchor of an Italian battleship. Techniques of inhaling oxygen beforehand were later developed as means of greatly prolonging such submersion, the record for which, last I heard, was held by a student at Wesleyan who, a few years ago, voluntarily held his head underwater for an astounding 21 minutes, time enough for a cross-country runner to run more than 4 miles. Further explanation of how this is possible came when experiments with aquatic mammals as well as humans revealed that the body's way of adapting itself to oxygen deprivation is immediately to reduce its metabolism, as in hibernation, a reaction known in the case of seals to slow their heartbeats to one tenth the normal rate, with corresponding retardation of blood flow, particularly in the extremities. Breathing water without gills is also possible, even to humans, provided the water is enriched with the right amounts of oxygen and salt, and recent experiments offer hope that this discovery will eventually make deep-sea diving a lot simpler and safer.

## BLOOD

If the body system that draws oxygen to the blood is remarkable, the companion system that pumps blood to the oxygen is astounding. In fact the capillaries that expose the blood to the air for oxygen in the lungs and the even more numerous capillaries that distribute the blood's oxygen to all the body's tissues add up to a total length of about 60,000 miles in the average man. This can be so because capillaries are like garden hoses beside the river of an artery, each one less than one thousandth of an inch in diameter, with more than 3 million of them threading through each square inch of the cross section of a muscle, each capillary carrying its swift but invisible trickle of blood of a volume that individually amounts to scarcely two drops an hour, yet in the whole body totals a good 100 gallons in the same period, as the heart patiently pumps the body's 6 quarts of blood again and again through all its tissues at an average pace of one round trip a minute, though in strenuous exercise this may be accelerated to a circuit every 15 seconds. Blood circulation is continuous, you see, but not at all constant in an active body, for there are local muscular, electrical and chemical controls that perpetually regulate and change the pattern of flow, like farmers opening and closing ditch gates in an irrigated valley, flooding first one area then another, bypassing obstacles too big or heavy to be flushed away, even reversing the current in certain channels. Operation of the system as a whole has been found to be supervised by special centers in the brain, which receive continuous information from sensory monitoring devices located at strategic points and which send back orders to the heart and to thousands of arterial, venal and capillary control stations.

The heart, relatively speaking, is a new invention in the life of Earth and not basically essential. The most primitive creatures have no hearts, although, as we observed in the case of the tree, they seem to be hinting that they are about to start evolving something of the sort to aid their circulation. Microscopic and larger animals up to the complexity of flatworms hardly even have circulation, preferring to leave it to their bodily motion to stir enough diffused oxygen into their juices. But the more complex roundworms have tubes to channel their "blood" through part of its course. The first "hearts" seem to have been nothing but faint waves of peristaltic motion (like the waves that nudge food through intestines), which gradually became localized and developed into swellings with a pulse. As circulation was mostly open and unconfined by blood vessels (as it still is in clams, shrimps, insects, etc.), heart action then was more comparable to gently stirring soup with a spoon than to anything that could be called pumping — which may explain why the squid needs three hearts, the grasshopper six and the earthworm ten. And even when the heart evolved its valves with completely channeled blood flow, it still awaited a future history extending from the single-loop circulation of fish to the loop with a side (lung) branch of amphibians and finally to the now well-perfected double-loop circulation of mammals, which uses a two-chambered heart to pump blood first to the lungs to absorb oxygen, then to the whole body to distribute it.

The amount of blood the heart can pump through a modern man may be judged by the size of his trunk artery, the aorta, which has been known to reach 1½ inches in diameter where it leaves the heart, actually exceeding the girth of the water pipes supplying an average house. To keep such a channel busy, the normal heart does daily work equivalent to lifting a ton from the ground up to the top of a five-story building and in a lifetime may beat four billion times, or once for every man, woman and child on Earth, each time squeezing, twisting and literally wringing the blood out of itself with four complex sets of spiral muscles triggered by a kind of electrical timer that is directly connected with the brain, while its rubbery one-way valves automatically flap, click and rest with their familiar song of lubb-dup, lubb-dup, lubb-dup just beneath the front ribs.

Blood itself is an extraordinary and complex protein fluid one cannot live without, yet it can be swapped back and forth between any of countless persons who have blood of the same type, because it has no nuclei limiting it to one person. Its liquid part, called plasma, is much like the primordial body juice of amebas and other small creatures. It is the prime ingredient of life's colloidal soup, the stuff of sweat and tears, salty and transparent, which evolved directly from ancient seawater. Its function is mainly to transport food, oxygen, vitamins and minerals to cells in exchange for waste matter. Blood's solid part consists almost entirely of coin-shaped red cells that are remarkably elastic, so flexible they can elongate and fold up and sneak through a capillary of barely half their own diameter. They are manufactured automatically in the marrow inside bones, and their principal component is hemoglobin, a protein molecule practically identical with chlorophyll, except that chlorophyll's central atom of magnesium has here been replaced by four atoms of iron, giving hemoglobin a curious chemical yearning to combine with oxygen. In fact it is the same yearning that oxidizes iron into rust. This trait of red blood cells of course is vital, being what makes blood absorb oxygen in the lungs and transport it all over the body, to where the relatively low pressure in the capillaries unloads it again, replacing it with carbon dioxide and other unwanted matter to be evaporated as exhaled breath.

Other ingredients in blood include some thin plate-shaped cells that are essential to clotting, which they accomplish wherever there is a blood leak by releasing a chemical that crystallizes one of the plasma's proteins, fibrinogen, into a kind of network enmeshing the red cells together into a solid plug; and there are a few white cells that act like cops among the red ones, being big and bossy but usually numbering only 1 to 750 of the reds, although, when there is an invasion of bacteria or any other kind of insurrection threatening the established order, their numbers rapidly increase as they engulf and destroy the enemy. There are also various hormones, enzymes, odd molecules and metallic atoms drifting about on their bloody errands. And I should mention the blood's companion system of lymph (watery stuff largely composed of plasma) that circulates more or less separately, rather heartlessly, but with valves and filtering nodes that serve to isolate and eliminate dangerous infections.

## DIGESTION AND ELIMINATION

Another part of the body, with the biggest elimination function of all, is the digestive tract, which no animal can do without in some form. It starts with the familiar mouth with its prodding tongue, its jets of saliva, its teeth for cutting and grinding. Then comes, in humans, the delicate but resilient throat that swallows thousands of times a day despite being subjected to temperatures from ice cream at 10° to soup at 170°F., that talks tens of thousands of daily words, interrupted by an occasional 200-mph tornado of a sneeze or a cough, the while patiently transporting 3000 gallons of daily breath, laden with a billion dust and smoke particles, not to mention countless microbes and viruses, a small but unknown number of which can be deadly dangerous.

The stomach, next in sequence, is probably the most misunderstood of organs and one of the hardest to describe, because it is constantly writhing, kneading, throbbing and changing its form, particularly when its owner is under emotional stress. Empty, the stomach hangs from the bottom of the gullet like a deflated balloon, some 16 inches from top to bottom, but when it gets a message that food is coming, it begins to squirm in anticipation. Full, it tightens into a plump kidney-bean shape perhaps 10 inches high, yet undulating with waves of contraction. Its juice (mainly hydrochloric acid) is so corrosive it can dissolve zinc and would blister your palm in an instant, but the mucus lining is so impervious that virtually no food, drink or gastric juice can reach the stomach walls. Between the muscular churning of the organ as a whole and the sting of its juice (felt in heartburn when it rises into the throat), food is effectively softened up for the intestines, into which it is doled in teaspoon-size doses by the judicious opening and closing of its pylorus or back door.

Once in the small intestine, longest section of the coiled 30-foot intestinal tract, the acid-laced food is neutralized by alkaline digestive fluids from pancreas and liver, then bombarded with powerful enzymes that disintegrate the proteins into amino acids, the carbohydrates into sugary glucose, the fats into fatty acids and glycerol. This is manipulated in detail by millions of hairlike projections called villi, lining the intestinal walls like nap on a towel, which dominate the chemical and frictional work of milling, absorbing and sorting the useful nutrients so they can be quickly fed into the bloodstream, at the same time letting the waste solids drift on into the large intestine and out the rectum. This is how it happens with most omnivorous and carnivorous eaters like man, but the process is considerably modified in herbivorous animals, like the cow with her four stomachs, in the sea cucumber who cures his bellyache by literally discarding his belly before swimming away to grow another one, in the spider or the starfish who digest their food outside their bodies.

Elimination, by the way, does not absolutely require an intestine or anus, for some creatures don't live long enough to have an elimination problem (death or pupation doing the eliminating) while others merely slough off waste from their outer surfaces like bark from a tree. The anus too is apt to have a different and less unsavory connotation to animals than to people. When the sea cucumber (see illustration, page 17), who had been evacuating his indigestibles out of his mouth for a hundred million years, finally evolved a separate anus for the purpose, thereby gaining a choice as to which opening to breathe through, he chose the anus! And of course this choice may have been influenced by the parasitic little pearlfish who traditionally inhabits his rectum, using his anus for a door, and who undoubtedly feels, like so many of us, that "there's no place like home."

Once we accept the rectum as a haven, it naturally follows that we accept its produce as something less than obnoxious and, in the case of primitive animals, it can turn out to be good to eat, handy to build a house with, even sculpturally delightful. Indeed where larger creatures merely get rid of their ill-smelling dung, small ones often package 123 or mold it into something useful, sometimes into dwellings of extraordinary beauty, like the tiny rotifer's beaded tower, into dramatic landscapes, like the glistening white guano isles of Peru or the limpid South Sea atolls of exquisitely excreted coral. All shells in fact are either secreted or excreted, the distinction not always clear, while beach wastes in general offer perhaps the most appealing of all forms of ordure: the architectural fecal pellets of mollusks that turn out such housetop shapes as tiles, double gutters and even a little oriental temple roof with its bottom edges turned upward toward the sky. But it is on the ocean floor, ultimate bowel of the deep, that excrement is most pervasive and, surprisingly, ofttimes graceful. More than once its convolutions have made me think of ancient oriental script, as if the worms who ejected them were trying to convey some sort of weird, benthic intelligence.

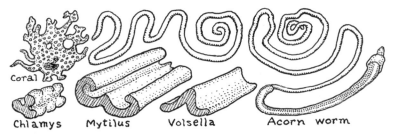

Coral  Chlamys  Mytilus  Volsella  Acorn worm

## KIDNEYS AND SECRETIONS

The liquid side of digestion and elimination inevitably centers around the kidneys, those twin organs that are indispensable to life and have to do almost entirely with the watery two thirds of the body. Kidneys are organs that evolved when fish moved into the fresh water of rivers, where they began to need something that would pump out the excess of the strange unsalted fluid they found themselves increasingly absorbing by osmosis. By the time real freshwater fish appeared, the kidney had developed into a sensitive filtering system that not only forced the unwanted water out, but in the process absorbed and salvaged most of whatever salts and valuable minerals would otherwise have departed with it. The kidney's key unit, the nephron, in modern form is smaller than a pinhead and, under a microscope, looks like a worm with a round skull — a skull that turns out to be full of blood capillaries continuously exuding watery plasma, but, at the same time, absorbing from it vital amino acids, glucose, minerals, proteins and (as an evolutionary afterthought on moving ashore) retaining more than 98 percent of the fresh water. The small residue of course is urine, and it dribbles out of tiny, wormy tributaries which, joining together from some million nephrons in each kidney, ultimately drain into a bladder to be excreted at opportune moments.

But the main function of the kidneys is maintenance of an exact proportion of water in the blood (which they clean and reclean to a total of 40 gallons a day), plus a strict mineral and chemical balance in the salty, colloidal sea that still surges inside us, involving a close coordination with the lungs, heart, liver, pancreas, etc., not to mention the sugar concentration, the acid-alkali balance and the temperature, all of which are intimately monitored by the nerves under the central supervision of the brain.

Glands, although minor organs, are nonetheless essential parts of the body. Consider the female breast with its system of some eighteen little rivers of milk, each with thousands of tributaries, combining into a common delta at the nipple. It is something like an extension of the lung, only its millions of alveoli exchange not carbon dioxide for oxygen but rather blood for milk, assembling (with the aid of hormones) long casein molecules out of short amino acid ones. Without the breast few babies (outside modern civilization) could live. And it is made possible by a single microscopic droplet of the hormone estrogen secreted at puberty by the ovaries, which transforms a skinny girl into a shapely woman ready for pregnancy. The adrenal, thyroid, pituitary, gastric, salivary, sweat, tear and numerous other glands also exude comparably potent hormones that carry their specific messages and cast their mysterious influences over every part of every creature endowed with them.

Hormones, created as needed, fortunately seldom seem to accumulate in bothersome amounts, and oxygen, though vital, is something bodies on Earth do not normally need to store in quantity since they can simply inhale it as they need it from water or air. But almost all other important body substances, including trace metals, being not breathable, must be tucked away somewhere for ready availability. Of these the most storable of all is fat which, being 88 percent carbon and hydrogen, is chemically close to gasoline and other hydrocarbon fuels, its favorable energy-per-pound ratio enabling birds and insects to fly and other animals to go long distances in search of food. As it is usually stored just under the skin, it also serves as insulation against heat and cold, not to mention acting as a protective cushion and support for delicate structures, such as joints, kidneys, eyes and sex organs.

But the most important thing we haven't yet gotten to in discussing the body is its system of replacing itself in evolution through reproduction — something that unavoidably (but not unwillingly) leads us into the ever-fascinating subject of sex.

125

# CHAPTER 5

# The Complement Called Sex

As I gaze pensively out of space at the pale, plump Earth, Venus seems so close behind me that I could almost be looking out of her eyes. That would be appropriate too, because she is a symbol for fertility, which is what this chapter is about. Indeed the continuum of life on any planet depends on fertility, which of course means continuous renewal, usually termed reproduction. And the most obvious mechanism of reproduction is founded on the complement called sex.

Exploring this subject in evolution, one of the first things we discover is that sex evolved very early on Earth, in fact long before death, which, as we will see in Chapter 20, is mainly associated with multicelled organisms. This means that sex goes back to such simple life forms as amebas, bacteria and viruses, among whom everything seems to have been tried from "birth" by sprouting buds to "sex" without gender and "death" by fissiparous propagation. Does that leave "love" an art? Or is it just something one falls into?

Starfish, you know, get offspring by occasionally releasing an arm.

And, in most very primitive creatures, sex is not absolute but relative, the differentiation into maleness and femaleness appearing in a wide graduation of degrees. It is enough to make one wonder what sex, in its essence, really is. And whether sex plays half so important a part in reproduction as most humans are brought up to think.

## RELATIVITY OF SEX

The general notion in the eastern Mediterranean region before the emergence of the Hebrews or the Greeks was that sexual attraction is due to the two sexes having originally been one — and this was reiterated from Genesis, wherein woman was created from part of man, "therefore shall a man . . . cleave unto his wife, and they shall be one flesh," to Plato's *Symposium*, in which Aristophanes observes that man's joining woman reunites "our original nature, making one of two." Wasn't this after all just one application of the long-accepted concept of gravity as caused by the tendency of like to seek like, of the yearning of free stones to hug the stony ground, of smoke to rise up and embrace the cloudy sky?

Dr. Alfred Kinsey threw some scientific light on this theory in the 1940s, when he discovered that, in humans, there is actually less difference between the sexes than among individuals of the same sex. And yet, even before birth, curious disparities appear. Girls on the average develop faster in the womb than boys and the female heart beats faster. After birth, a girl seems more sensitive, reacts more strongly to removal of a blanket or to a touch. Also she learns to talk earlier and better and is more interested in people. A boy, on the other hand, is more independent and self-reliant, more interested in inanimate objects. Confined alone in a strange pen, the girl is apt to cry, while the boy tries to find his way out.

Good evidence for the closeness of the sexes can be found in the successful treatment of several cases in this century in which a baby boy's penis was accidentally cut off — after which, with the help of reconstructive surgery and hormones, he was "reassigned" as a girl and brought up to live a happy and "normal" life. As long as such reassignment takes place before the age of about four years, it seems to work well, as "she" doesn't remember being "he" and the psychological conditioning almost takes care of itself. And should the reassignment be made while the embryo is still sexually undifferentiated (during the first 16 weeks of pregnancy), it can be done with hormones alone.

127

In the case of animals, especially cold-blooded ones, changing sex is relatively easy even when the young one is half grown, because such measures as increasing feeding may turn a male into a female. So will a change in temperature. Or, in the case of a warm-blooded creature, it can be done by extracting ovaries to switch, say, a hen into a cock. As a matter of fact, sex reversals happen naturally in many species. Quahogs, I am told, are born and grow up male, but later half of them turn female, undergoing a sort of sexual demi-death, perhaps when the activity of youth shifts toward the passivity of age. Slipper shells do this too, and cup and saucer shells, commencing every season as boys, but almost all of them saucily switching later through a phase of ambisexuality into adult females — leaving only a few stragglers to linger on as solitary male bachelors.

Sex among these lowly folk seems to depend a great deal on food, since the best-fed individuals turn female the earliest, while the poor scrawny ones get left behind as males (although the opposite happens in the case of oysters). In some species, such as the marine worm *Ophryotrocha*, if the portly young females are later underfed they revert back into males again. Indeed among most primitive creatures of the sea and practically all insects it is a general rule that the smaller individuals are males and the bigger, fatter ones females, the basic reason being that the essential female function is to produce and feed young, while the only important thing expected of a primitive male is to dart blithely about fertilizing every egg in reach with no ensuing responsibility.

It may surprise you to know that fish have evidently evolved the easiest and quickest sex-reversing capacity of any animal, some species not only changing from male to female as they grow but a few, like groupers and guppies, developing the ability to switch sexually back and forth within seconds, almost as readily as you shift gears in your car. If two girl guppies meet while feeling amorous, an ichthyologist told me, one is likely to start turning into a boy so he can mate the other. But occasionally it happens that both shift at the same moment and find themselves still stymied as two boys ten seconds later. And this sort of flipping to and fro can theoretically continue several times, as with people frustrated in trying to get by each other on the sidewalk. In practice, however, the experts say, such an impasse with guppies usually results in a furious fight, with the winner, oddly enough, emerging as a female who somehow forces the other to stay male.

This brings up the interesting point that the male by no means always dominates the female in nature, doing so mainly where intromissive copulation gives him the unilateral potentiality of rape. But, as that male advantage is lacking among most birds and simpler animals, mating is often initiated by a bigger, stronger female. Furthermore, many primitive creatures (including plants) are hermaphrodites who possess both male and female organs, so mating may go both ways at once. In fact the guppy's flexibility comes from having both testicles and ovaries with some sort of valve that switches the flow from milt to roe. Yet true hermaphroditism (offering simultaneous sperm and egg flow) is entirely normal among most plants and many animals, from snails, who make love with their feet, to earthworms, who commonly emerge from the soil to court one another at night on lawns, spending happy hours adjusting and aligning themselves head to tail and tail to head, with the aid of a sticky mucus they exude, so that the sperm pores on the fifteenth segment of each worm exactly coincide with the egg pores on the tenth segment of the other worm.

Most such hermaphrodites thus have little difficulty begetting enough offspring to keep their population up, but, in times of famine or stress, when they don't meet each other so often for cross-fertilization, each one still has the possibility of fertilizing itself by uniting its own sperm and ova. Indeed a dynasty of laboratory snails has actually been kept going on self-fertilization for 90 consecutive generations (during 20 years) without noticeable loss of vitality. And some kinds of arrowworms have even been found to *prefer* self-fertilization and apparently use it exclusively in their natural environment in the deep sea.

Hermaphroditism and sex reversal are not the same thing as homosexuality, as is evidenced by their comparative rarity among the more highly evolved creatures all the way to man, who, nevertheless, is occasionally known to shift gender and has actually (in very rare cases) managed to function as both a male and a female at the same time. The most striking example of a well-adjusted human hermaphrodite I've heard of is the individual reported in Lisbon in 1807 as having normal sperm-producing testicles and "some beard on the chin" but otherwise the appearance of a slim-hipped, attractive girl with graceful figure, adequate bosom and charmingly feminine voice, not to overlook the usual female organs which functioned well enough that she-he had already been pregnant twice between presumed affairs in which he-she may in turn have impregnated others!

Easier to understand, I think, is the mental, emotional and spiritual side of the shift of gender — something beautifully explained by the English writer James Morris ♂ who, despite his warm relationship with a loving wife and four children, recently became Jan Morris ♀     129 after ten years of taking female hormones and undergoing surgery.

"From the age of three I always knew my gender was feminine," he allowed. "Not my sex but my gender. Gender is more important than sex, and more subtle. It's not a matter of body or organs. Or even chemistry. It's mystical . . ."

After the operation "she" first felt astonishment, then joy. "She" literally sang. On reflection "she" described it as a "rebirth." "I had a loss but also a discovery and a gain." Asked how "she" now felt about "her" former wife, "she" said, "It made no essential difference. We will always love each other no matter what. If, instead of a woman, I had turned into a horse, I'd love her just as much. And how interesting it would be!"

## Organs

The organs and techniques of "love," like other aspects of life, are surprisingly various, and new ways are apparently being tried out in an endless sequence as evolution unfolds. Both the kangaroo and the opossum, for example, have forked phalluses that more than conveniently fit into their mate's forked vagina, while snakes and lizards have completely double ones (called hemipenes) that erect "inside out" like fingers of a rubber glove and work separately, each in its own time sliding into the female's single anal-vaginal passage (called a cloaca). The gray squirrel, on the other hand, has a hooked penis, the mole shrew an S-shaped one with flanges for taming his own little shrew, and a number of animals including the spider monkey, wild cat, lynx and puma all have barbs on theirs, indeed horny ones that point backward and latch upon complementary "projections" in the vagina that did not evolve there likely by pure coincidence. In each of these cases the locking mechanism tends to hold the mated pair together, much as dogs are sometimes locked by the knot midway on his organ, and, despite all their caterwauling, perhaps this is nature's way of accomplishing for cats some of what the Dyak people have sought for humans by embedding silver knobs (called ampallang) in the penis to increase friction, or the Celebesian custom of encircling it with goats' eyelashes for enhancement of feminine delight. The only animal I can think of who has gone further is the Abyssinian bat whose member is densely coated with bristles. On which I've seen a meticulous zoological report saying that it works like a bottle brush.

A number of mammals have stiffeners inside their male organs, ranging from the mink's cartilage to actual bone in the wolf and fox (as Aristotle first recorded), and which have served humanity as natural tools, from the walrus's formidable rod, traditionally used to club seals, to the raccoon's dainty prong with a slight hook at the end that an eighteenth-century dandy would carry on his watch chain for a toothpick mounted in gold, and that backwoods tailors still claim is just the thing for yanking out basting threads. Although aging humans may occasionally have fancied they themselves could do with something of the sort, it has proved anything but a boon to the few abnormal men actually born with a bone in the penis. Nor have any of the very rare semi-monsters with double penises or double vaginas ever managed such an extravaganza as the theoretically possible double-barreled marriage (opossum style) or derived any other noticeable benefit from their superfluities, which evidently can no more impart double ecstasy than can a millipede outrun a centipede.

Maybe if you are male with an imagination sufficiently lubricious you would prefer to descend into the microcosm and be a roving tom rotifer, one of those curious little pond beings (page 88) who are smaller and rarer than their bustling females and appear to have quite a merry sex life, the chief feature of which is that their relatively enormous phalluses can be poked into any part of any female rotifer, which should make rape (if there were such in the microcosm) as easy as tag. This sort of "hypodermic insemination" is also used by leeches and some arthropods, but often requires a certain amount of hustling about. So if you like to loll perhaps you'd rather be a *Schistosoma haematobium*. That's a kind of microscopic fluke who lives in human

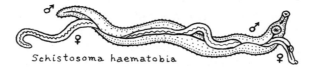

*Schistosoma haematobia*

blood and gives people "snail fever" in the wet tropics. The threadlike female grows up already encased within the burlier male. And in certain other microbe species their sizes are reversed; she is bigger and, in effect, has him securely zipped inside her nightie. Such affairs illustrate what is known as "perpetual copulation," requiring no mate hunting nor even maneuvering into position, and I guess the ultimate example is a kind of double hermaphroditic animal appropriately called *Diplozoon paradoxum*, who lives in the gills of carp and matures in pairs that soon grow together into Siamese twins permanently riveted in two places: where the phallus of each penetrates the vagina of the other in an un-untiable double splice.

As for the mating problems of larger animals, some seem quite baffling — though obviously animals who haven't become extinct must have solved them somehow. How can a porcupine, let's say, cope with the 25,000 quills that defend his bride, each with its dozens of barbs and their tendency to expand when moistened by flesh, which would be deadly against his exposed belly and genitals? The fact is: he succeeds by playing it cool from the start, cautiously grunting and sniffing her over, then, fully aroused, rearing up on his hind legs and tail to waddle forward expectantly. If she also rears up, he will likely begin spraying her with urine, which seems to be accepted by the porcupine maiden as a kind of personal bouquet, and, if it keeps up, soon gives her all the thrills of a bridal shower with the ultimate disarming effect of inducing her skin muscles to pull her quills down flat like grounded spears so he can safely mount her from the rear.

One might suppose there must be a section under Galileo's Principle of Similitude or Froude's Law of the correspondence of speeds expressing the relation between the tempo of love and body bulk, the larger beasts presumably taking more time at it than the smaller, but the pace of intercourse turns out to be geared more closely to complex factors of mechanical and chemical physiology than to overall size. Among birds, the swiftest of known animals, the ostrich and the duck are relatively slow in coition because they have intromissive organs, while most other birds can mate in a flick by merely touching their cloacas together, frequently (as you've probably noticed) in flight. Elephants may seem clumsy, but they not only are highly sophisticated in courtship, with large repertoires of provocative gestures, proddings, nippings and subtle erotic teasings with the trunk, but sometimes they complete actual copulation within 18 seconds. Whales 20 times bigger than elephants also have elaborate courtships featuring wild leapings and very tender full-length fondlings with flipper and fluke, and have been known to finish their coition in less than five seconds, despite the prodigiosity of a phallus up to ten feet long and a foot thick. Cattle too are quick on the draw, bulls sometimes winding up the job in two seconds, while an Indian antelope (the nilgai) appears champion of mammals, finishing in "a fraction of a second" (according to Heini Hediger of the Basel zoo). Yet stallions with their longer organs take minutes, lions and tigers usually a quarter hour, pigs and bears most of an hour and minks, martens and sables commonly over an hour. Slower still (without considering the "perpetual copulators") are many small creatures like snails or earthworms who often take several hours and aquatic ones from diving beetles to sea turtles who have been known to cling together for days.

Some of these may be slowed down by the cold like the remarkable Alaskan stone fly, who has been seen to mate rather stiffly yet successfully in a temperature of 32°F., with its body hardly a degree warmer. Some are deterred by the heat like elk, who often have to ascend to windy heights between times to cool their testicles back to potency. Some are delayed by inexperience — the chimpanzee, for instance — because, unlike mice and simpler creatures who make love by instinct, he must learn how. Some are almost indifferent to sex like the male gorilla, who waits until a female makes a floozyish pass at him before giving her so much as a glance. Some suffer from roughness like the rhinoceros, who can't seem to get a real bang out of his inamorata until she literally bashes and gores him bloody and staggering. And female rodents being courted are often vicious to the point of being murderous. In fact wood rats get killed about as often by shrewish females as by rival males, usually during the long sparring period of preliminary coition that is required to make her ovulate.

Which brings up the interesting general rule that, in species where the female is eager or at least cooperative, the penis is rather short, but in species where she is shy, reluctant or violent, it is proportionately much longer, with of course a correspondingly deep vagina in the female — measures that appear to be nature's quaintly persuasive way of keeping her from changing her mind once she has succumbed. Other male tools for restraining females range all the way from the stag beetle's huge mandibles and the frog's love thumbs to the duckbill platypus's spur that not only firmly grips the female but, some think, injects a sedative to soothe her. And the aforementioned male diving beetles have curious suction disks on their forelegs for getting their long-lasting vacuum hold on the slippery bride, while a few kinds of flies (*Dolichopodidae*) have cuplike blinders that neatly fit over their loved one's large and rather starry eyes during mating which may well help to keep her attention on what she is doing.

The position of mating makes a difference too of course and varies widely, most mammals using the familiar rear approach of dogs and cats, a few (ranging from millipedes to beavers to whales) mating belly to belly (often vertically in water) and surprisingly some (like the rabbit) back to back because the male organ aims rearward. This rear-to-rear position, however, is commonest among insects, notably bed-bugs and flies (including butterflies and mosquitoes), where in some cases it is offset by the male's slender abdomen making a U-turn, except that certain unlucky males (notably the horsefly) while mating get flipped over backward by the female's buzzing wings, then dragged ignominiously and upside down through the air behind her. With most birds and insects where the sexes are of nearly equal size, it matters little whether the male mounts the female or the other way around because sex organs of the cloaca type need only be touched together to pass the sperm. Indeed the aggressive females of some species, such as mole crickets and saber grasshoppers, have made themselves quite a reputation for mounting and apparently dominating the male.

A few animals have even evolved a definite handedness in their mating. I mean the male specializes in either right- or left-handed sex organs that require the female to have a complementary handedness for the meeting to be successful. The most striking example is a family of small fish living in tropical American waters called the *Poeciliidae* among whom guppies, mosquito fish and the four-eyed anableps are best known. Instead of laying eggs like other fish, these give birth to swimming minnows, but their most unusual feature is the male sex organ, which evidently evolved from a ventral fin and can be half as long as its owner. In erection it enlarges and swings forward until, in some species, its tip is almost even with the fish's nose yet pointing perhaps 30° to the right or left. In several species this fishy phallus has fingerlike appendages that one can imagine must be delightfully handy for feeling its way into the female and it is sometimes also abetted by two sets of comblike retrorse spines (apparently evolved from side fins) for clasping her the while. But she definitely must have her orifice on the correct side, right or left, to receive the male, else the whole match is off.

## PRODIGALITY

The oyster, when it comes to mating, seems to have a knottier problem than the anableps, because he not only is handed, being permanently attached to a solid surface to the "left" (deep) half of his shell while the "right" (flat) half is a movable lid, but he has no legs, flippers, fins or tail and therefore no chance to go courting or embrace a mate. So what does he do but compensate with prodigality. He alternately pours forth sperms and (shifting sex) eggs in immense numbers to the open sea. Indeed if the estimate of one malacologist (that an oyster can lay a septillion eggs a year) is anywhere near correct and all the issue survived, in less than five generations our world quite literally would be his oyster.

Another sort of exuberance amounting to a molluscan orgy is practiced by the oyster's near cousin, the sluglike sea hare, a kind of shellless snail sometimes two feet long that grazes on seaweed and is well named both because of its long earlike tentacles and its promiscuity, which is completely hermaphroditic. Curiously the sea hare's phallus is on the right side of its head and, when playing the male, it puts its head between the finlike fans of a companion playing the female, gradually oozing its member into the genital opening. In spite of the sluggishness of its movements — which we ought to forgive since it is, after all, a kind of slug — it doesn't seem to miss any tricks, for at the same time it is a male to one sea hare it will be female to another on the other side, and as many as fifteen of them have been seen linked thus in a chain. In a few cases such a chain's ends (understandably lonely) had somehow worked their way around to meet and join, forming a continuous loop, with each animal playing a double role in a sort of subsea version of what the French call *la ronde* — which makes me wonder how any sea hare, once plugged into such a sexual merry-go-round, ever gets out of it!

The fish in general seem sexually more carefree than any other classes of large animals, not only roaming the ocean with few territorial restraints but (with some exceptions) doing their fertilizing externally and wholesale. The great fish schools are therefore mostly co-ed and the individual boy and girl fish do not bother looking for partners, nor do they date or mate, because both sexes are so densely mingled that many more than enough germ cells are cast together upon the waters to fertilize and hatch all the progeny that can live. In fact one might say the whole school is "going steady" with itself all the time.

Naturally such prodigality requires an enormous output of sperm and eggs, almost all of which will inevitably get eaten or lost even when they have the luck to meet each other and become fertilized. Yet nature seems to delight in just such production splurges, even in the case of internal fertilization, where it seems hardly necessary, for it is estimated that the human male ejaculates over a hundred million sperms in an average orgasm, the stallion 50 times as much and the boar an unsurpassed 85 billion just for one shot at a jackpot of baby pigs. And if you think one shot is a fair day's output for even a tame farm animal, you should hear the plaint of a sheep farmer who kept a big flock of ewes in a fold by themselves so they couldn't be bred out of season. Somehow one night his lone and lonely ram rammed his way through the bars and made the most of it with the flock. The farmer never could find out how many ewes the rambunctious ram mounted during his maximum of eight hours inside the fold (some ewes no doubt more than once), but five months later, as a result of that one night, 114 of the ewes gave birth to lambs!

The astronomical multiplicity of sperms does not mean that these cells are individually unreliable or necessarily ephemeral, for it takes only one — virtually any one — to fertilize an ovum, and it may have great powers of endurance before it gets its chance. Bats, for instance, who mate each autumn in midair (quite a trick with their intromissive organs), leave the viable semen stored in the female all winter to fertilize her ova shortly after she wakes from hibernation in spring. And turtle brides, who are adept at the waiting game, have been known to store sperms for more than four years. A lot more diversiform than ova, these male cells come in a wide variety of shapes, as the illustration shows, propelling themselves with everything from lashing tails to the spore ooze of worms.

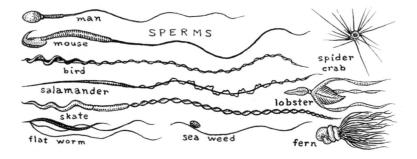

## VEGETAL SEX

I don't think there is much need to separate the animal and vegetable kingdoms in their sexual traits, for there is no incontrovertible line between them. Neither do vegetables such as blue-green algae, who are thought to have invented sex long before animals existed on Earth, owe anything to the animals in resourcefulness of lovemaking. Even though plants are less completely divided into males and females, they are just about as varied in shape, in their hermaphroditic and conjugal devices, and I suspect that, at some level of awareness, they may really enjoy their gentle version of conjugation, which of course they have evolved in many cases by ingeniously exploiting (sometimes actually enslaving) the animals.

As a matter of fact, divining the subtleties of vegetable reproduction has been no easier than divining those of animal reproduction, and even the great Aristotle argued *a priori* that plants could have no sex. It was the Babylonians who first noticed that date palms were of two kinds, only one of which bore fruit and then only when at least one of the other kind was also present in the area. Later some imaginative Babylonian surmised that the bearing tree must be female, the other male, whereupon progressive date farmers began to help the two get together. It took many centuries for this dating idea to be generally accepted.

In our own day it has been proven that the wind can pollinate date palms when trees of the two sexes are more than 50 miles apart and it is certain that thousands of different flowers and blossoms, male and female, are specifically shaped and adapted to fertilization by butterflies, moths, bees, beetles, birds, bats, snails and other creatures. One bee or butterfly in fact can attend about 20 flowers a minute, or 20,000 of them in a long summer day, so if there are any bees or butterflies active in a garden, however few, there is little chance of any suitable flower not being fertilized. Meanwhile let me say that, like mollusks and a goodly number of other animals, the many hermaphroditic plants usually separate their sexes by time if not by space in order to avoid fertilizing themselves. Thus while the primrose has a male flower on one stalk and a female on another at the same time, the mallow and sage start off male but at a later time turn female, and the *Aristolochia* makes the opposite time shift from female to male. Others, such as the buttercup, merely have a chemical barrier against self-fertilization, and others again, like the dandelion, have rejected sex altogether and simply dispense their familiar tufted seeds that require no fertilization. Should the appalling prospect of the abolition of sex strike you as having merit, however, may I point out that, successful though the common dandelion is in our day, it has for some reason sacrificed its future for its present, by denying itself the variability it once had through constantly recurring new combinations of male and female cells (the evolutionary "purpose" of sex) and therefore it has become quite inflexible. Some botanists go so far as to say it is *de*volving instead of *e*volving since, unlike sexier flowers, it is likely to get wiped out by a future environmental change it cannot adapt to.

If the wind in the willows is a poetic phrase to you, it is assuredly even more so to the willows and many other trees, to most grasses, mushrooms and a good tenth of all flowering plants. A breeze is in truth a link in life itself for, like the exuberant oyster, these graceful creatures have long found it vital to broadcast their spores wholesale to the open ocean of air around them, entrusting their descendants to the invisible rivers of wind which waft them as surely as any watery currents below — and certainly more swiftly. Under the microscope pollen grains appear amazingly varied, often tiny spheres with studs or warts like those of campion, or bristles in the case of mallow, either of which kind is nicely designed to grip the hairs of the messenger

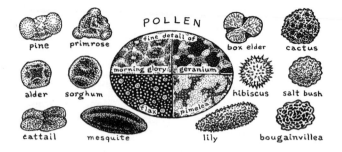

POLLEN

pine — primrose — fine detail of — box elder — cactus

morning glory — geranium

alder — sorghum — hibiscus — salt bush

flax — pimelea

cattail — mesquite — lily — bougainvillea

bees that carry it. Or it may be still tinier and lighter, like the ovoid pollen of the chestnut tree or the bubble-bearing pollen of pine — just right for floating on air. Wind-borne pollen is understandably smaller than animal-borne pollen because it not only has to be supportable by airy molecules but must be produced by the billions to compensate for the enormous expectable wastage in being strewn at random through the sky. A single ragweed plant has been measured to generate 1600 million pollen grains per hour, which, seen over a summer landscape with other such pollen in vast clouds, will produce a faint blue haze of astounding fecundity.

From the viewpoint of each microscopic grain of pollen of course, it makes an extremely hazardous journey, as scarcely one in a million can hope to alight precisely upon a waiting stigma (female organ) of exactly the right kind of plant to spunk it into a seed. Yet while the vast hordes are drying up, drowning or merely withering in frustration, the lucky few will each touch a bit of stigma, usually sticky and sweet with just the chemical complement needed to germinate it.

Germination is expressed by the pollen's sprouting a kind of phallic root that in a few hours elongates to worm its way down the surface of a papilla (hair) of the stigma, which is significantly coated with a potent chemical aphrodisiac. Once it penetrates the papilla's base, it draws further strength from the very nutritious tissue now around it,

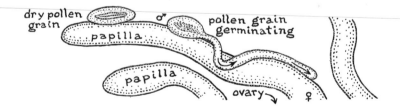

dry pollen grain — ♂ — pollen grain germinating — papilla — papilla — ovary — ♀

which also enfolds it snugly like a vagina, seeming to seduce it deeper and deeper until at last it bursts triumphantly into the cavity of the ovary and seizes one of the ovules, pouring its contents into her until their two substances fuse into one and she is fertilized. This is the key genetic act in vegetable reproduction, the union of male and female germ cells that conceives a new plant.

## BIRTH

The mammalian equivalent of seed germination or egg hatching we know as birth. I will not dwell long on it, for it is comparatively easily understood as a critical step in reproduction. But I must mention that it varies about as much as any other function in life, the ease of delivery among mammals alone ranging from that of the opossum baby, who weighs one twenty-fifth of an ounce or one ten-thousandth as much as his little mother and may be born only eight days after intercourse but already able to climb hand over hand up into her pouch, to that of the newborn blue whale who may tip the beam at fifteen tons or over one tenth the weight and two fifths the length of his huge mother. Although a fetus is a relatively easy burden when the uterus around it is supported in the sea, the very real risks in whale birth are shown by the many known cases of mothers who did not survive it. The uterus nevertheless has proven a remarkably adaptable organ on land as well as sea, having been measured in some pregnant mammals to stretch to 400 times its original size, a flexibility obviously important in bringing forth a youngster big enough to get along independently his first day out as is true of the agouti, a rodent born with a hairy coat and open eyes.

You might think the porcupine mother would be in for trouble in giving birth to a prickly baby, especially in cases of breech presentation, but actually it is easy, for the infant (as with so many animals) comes wrapped in a smooth membranous sac. On the other hand one would not expect mice, who beget as many as seventeen litters a year, to have any of the whale's trouble delivering them, but, surprisingly, at least one European species of mouse called *Acomys caharinus* with relatively large offspring has been observed to need "midwives" and uses them in two thirds of all births, the experienced female mice helping with difficult confinements, biting off the umbilical cord and cleaning the little ones as they appear. Some degree of midwifery is practiced, incidentally, by a good many other animals, from male frogs to female elephants.

A more unusual kind of birth assistance was evolved over 200 million years ago among the first land mammals to return to the sea. I am thinking of Stenopterygius, a kind of pre-porpoise of the Triassic age whose young always emerged very slowly tail first, taking weeks to be born, during which time the baby-to-be learned to swim at first just with its tail, then gradually worked its body and flippers into the act until, when at last its head came free, it was well able to rise to the surface for air. Modern porpoises, using a newer, presumably safer, system, are born 100 times faster, but afterward usually find themselves "tied" to their mother's "apron strings" for several days as the tough umbilical cord remains intact while they are led around getting used to breathing, swimming and nursing. More primitive animals such as fish, some of whom produce live young, can afford more carefree birth, as witness the great manta ray, who is born in midair while his mother (sometimes 20 feet in wingspread) leaps clear of the waves. A few kinds of fish in a reverse amphibian tactic crawl out of the water to give birth on dry land. And there are curiously handed "births" like those of the starfish and sea urchin, who develop in the left-hand side of their larval "mothers." However, since they almost totally consume her body before emerging, this sort of debut might more properly be considered a metamorphosis.

It is in the human sphere that some of the strangest births have been recorded, revealing the extremes of which nature is capable. The size of surviving human babies ranges all the way from a ten-ounce girl born unattended in England in 1938 to a 24-pounder in southern Turkey in 1961, and a few unconfirmed reports of ones up to 33 pounds. The numbers of babies that can share a birth have an interesting frequency schedule, human twins occurring statistically about every 86th confinement (when no ovulation-stimulating hormone, such as gonadotropin, has been used), triplets roughly once in 10,000, quadruplets once in a million, while about every year one mother somewhere on Earth produces quintuplets, about every four years sextuplets, more than once every decade septuplets, and approximately every three decades octuplets, the two octuplet cases thus far this century having been reported in Tampico, Mexico, in 1921 and in Kwoom Yam Sha, China, in 1934.

As significant as multiple births in a single delivery, I should think, are multiples of deliveries and the total numbers of offspring per mother, so it is interesting to read of the unconfirmed report of a Brazilian woman who begat 44 children in her life, all single births, to achieve which she could hardly have missed any chances in averaging a baby every eleven months from puberty to menopause. Then there is the authenticated case of Madame Feodor Vassiliev of Russia, who bore 69 children, with sixteen pairs of twins, seven sets of triplets and four sets of quadruplets. Her peasant husband deserves mention too, I'd say, for, after her well-earned demise, his second wife bore him eighteen more children in six pairs of twins and two of triplets, making him the father of 87 children, from only two wives and with no single birth in the lot!

## EGGS AND INCUBATION

The ovulation and gestation rates of terrestrial animals are quite unmistakably influenced by the earth and her motion in relation to the sun or, in the case of tidal creatures like oysters, by the moon as well. Indeed large wild mammals generally bring forth their young only in spring, thereby alloting them three quarters of a year in which to get ready for their first winter. Even eggs are seasonally regimented so the young have the maximum chance to survive. Starting as a germ in every female before she is born, the egg remains a single cell for a long time, but, on being fertilized in the mating season, it suddenly starts growing in distinct layers. A typical bird's egg yolk (food for the embryo) has twelve layers created in six days, as in the first chapter of Genesis, strictly at the command of the sun: a yellow layer each day and evening followed by a thin white layer each night. Next the egg's albumen appears, its white layers accompanied by the watery fluid that supports and protects the central yolk. Then come two tough membranes plus four final layers of chalky shell that harden in less than a day. The outer one of these is patterned and colored in the exact design evolved by that particular species of bird.

Eggs vary almost as much as seeds, their shapes ranging from the nearly round ones of fish and turtles to the long pointed ones of the cliff-dwelling murre which, if disturbed, have the life-saving tendency to roll in tight circles. Some are laid like beads on a string or stuck together in jelly, in pods, biscuits, cups, spiral coils or vermicelli threads. Some flying fish and also the ocean's only insect (the water-striding *Halobates*) anchor their eggs to drifting feathers shed by sea birds. Certain moth eggs are cuboid and laid in courses like bricks. Shark eggs have horns. Wasp eggs may be injected into living animals, who will serve as food after they hatch. Others won't hatch until they have been chewed, swallowed and partly digested. And some ant parasites lay polyembryonic eggs which, like MRV missiles, separate into many. The eggs of insects are probably the most diverse of all,

as the illustration suggests. But even birds' eggs range in bulk all the way from the pill of the hummingbird to the bigger-than-football ovoid of the extinct "roc" of Madagascar, and in color from robin's-egg blue and egret green to the emu's black and the blood red of a newly laid ptarmigan's egg.

Some birds with few enemies, such as the albatross or the petrel, lay only one egg a year and, even if it is lost, will refuse to lay another. Some, known as "determinate layers," in effect count and remember their eggs and stop laying at four or some other definite number whether or not any eggs have been removed. Others, called "indeterminate layers," will keep on if someone takes their eggs, a case in point being the flicker whose newest egg was experimentally removed daily until she had laid 71 of them in 73 days. Champion in this category, however, must be the domestic duck who laid 363 eggs in 365 days! And perhaps equally impressive is the performance of a wild ruddy duck weighing one pound who in about two weeks laid a single clutch of fifteen eggs weighing over three pounds or three times her own weight.

Most birds incubate their eggs by sitting on them, but a few have devised rather intriguing alternative means. I'm thinking of the turkeylike megapode, who lays hers on a huge pile of fermenting, steaming compost, then covers them with just enough sand to keep them at 92°F., or the maleo of Celebes, who specializes in hot springs and fumaroles, sometimes literally incubating her eggs by volcano! Meanwhile inside the shell during this critical period (which, depending on the species, can last anywhere from 11 to 90 days) the homogeneous yolk with its microscopic ovum miraculously turns into a hatched bird. In the first two or three days the ovum duplicates and reduplicates itself into a visible embryo with rootlike blood vessels reaching thirstily outward into the nourishing yolk. After two fifths of the incubation time has passed, the embryo has separated itself from the shrinking yolk except for a kind of umbilical blood stalk that feeds it. Halfway, most of its organs are visible, particularly the dark, shining eyes. At the three-quarter mark, the chick is fully formed, yet

142

has still to absorb the final third of the yolk via its cord. In the last day or two before hatching, it starts to breathe from the expanding air chamber at the large end of the egg which is replenished through the shell's pores, the shell steadily thinning as its lime flows into the growing bones. And associated with the breathing it makes faint peeping sounds, by means of which, scientists have discovered, adjacent eggs can talk to each other and their mother and coordinate their hatching, ensuring that all the chicks of a clutch start pecking their shells to get out at the same time. This is important for giving them all an equal chance to survive, especially in the case of such helpless and naked hatchlings as baby songbirds who seem geared to grow at a furious rate. In some cases such youngsters have been known to multiply their weight by 50 in three weeks from eating mostly proteins conscientiously brought to them every minute or two all day long. A pair of great tit parents in England indeed were recorded in one test as making more than 900 feeding trips to their brood in a single day.

Mammal incubation, on the other hand, while it occurs mostly in the womb instead of an eggshell, works in the same basic way except that the mother's body directly supplies the young mammal with food for a much longer period, including serving it milk after birth. Evidently it was for this important purpose that the great mammalian invention of the nursing breast was evolved with its own wide range of models. Did you know that the familiar pair- or multiple-pair-breasted physiognomy of man, dogs, cats, mice, etc., is only one of many designs and that the number of breasts or nipples can vary widely even in the same species. I guess the official maximum is the 22 teats in two rows on the tenrec of Madagascar, who has litters up to 36, but there are all sorts of odd cases, like the opossum with 13 teats (sometimes more) close together inside her pouch where the tiny young clamp on for more than five weeks, with quick starvation the hard lot of the late-born, who commonly find all stations already occupied. Baby bats have a better chance and perhaps a merrier time hanging to their mother's teats while she zigzags intricately through the night sky, grazing on insects every couple of seconds. The platypus mother, believe it or not, serves her milk from no teats at all, and the young, after hatching from eggs, must lick it off her fur. The whale mother also eschews teats in favor of slits, which are more streamlined, and has them way back on her tail so she can lift them clear of the water while her air-breathing calf is getting his nourishment pumped into him with special milk muscles. The dolphin's mammary glands are often colored like targets, so the bumbling baby, even while tied to her "apron strings," can't miss them. And targetlike also, come to think of it, is many a human breast with its dark red disk signaling the way to the nipple.

And this brings to mind the further fact that the variability of breasts in number and location applies to humans almost as much as to animals, for medical anomalies include practically every abnormality one can think of from amazia (complete absence of breasts) and cases of only one breast to all sorts of supernumerary ones in odd places from under the arm to the back, even the thigh, sometimes turning a figure into a panorama. Handily enough, most of these function too. Anne Boleyn is reputed to have had three breasts and there is a famous painting in the Louvre by Rubens showing a woman with four. That this anomaly is more prevalent than most people realize is suggested by one researcher's report of finding 60 instances out of 3956 persons examined, which is 1.56 percent — the men, surprisingly, sprouting extra nipples about twice as often as the women.

Of course nursing is only one side of mothering, which is quite general on Earth, even among such animals as turtles and alligators — and by no means excludes the vegetable kingdom (page 66). Centipedes and scorpions are particularly maternal toward both their eggs and young, and the Surinam toad mother carries her eggs on her back until they hatch as fully formed toadlets. I could go on, but there seems no point in further belaboring this large and well-accepted subject.

## ORIGINS OF SEX

We have by now looked at many aspects of sex from its abstract relativity to its concrete workability. Yet I cannot conclude without saying something about sex's origins in life and evolution, its reason for being.

So let us take a glance backward into the older, simpler, littler, lower world, where procreation is more and more a matter of chemistry and more orderly — where, beneath the large-scale, emotional drama of love, there unfolds the basic need that makes it happen. We find ourselves then back in the habitat of cells and molecules, the microcosm where reproduction and sex began. Obviously we must sift and sort some elementary facts. What in essence is sex? And is it the same thing as reproduction? No. These phenomena are not only *not* the same (witness the dandelion's prolific reproduction without sex) but, in a basic way, they are opposites. I mean, if the main requirement of reproduction is to multiply the numbers of individuals, turning the singular into the plural, then sexual activity or the uniting of two or more individuals is contrary to it in reducing multiflesh into one flesh or, to use biological terms, conjugating the separate sperm and ovum into a single embryo.

To clarify our thinking on this not-so-familiar aspect of the subject, let's first consider reproduction, which of course is more fundamental than sex. Reproduction is basically a kind of growth — and it is a growth that almost always leads to separation into new units, as when a bit of the parent body splits off and grows into a new parent body, which in time splits again. All life somehow must reproduce itself (whether or not via sex) if it is to continue beyond its individual organisms, which obviously cannot go on forever just as they are. And the cell, generally regarded as the unit of life, has evolved three principal ways of reproducing, only the newest and most complicated of which makes use of sex.

The first and simplest method of cell reproduction, already described in some detail (page 105), is division of the cell into halves or, in a few cases, smaller fractions. Second is the budding or sprouting upon the cell of a small, specialized sporelike appendage, which soon breaks loose and grows into a new complete cell. Yeast propagates itself this way. Third is the specialization of the cell into two or more genetic strains, which, when they attain enough differentiation, become unable to split and reproduce themselves until they can get together and exchange genetic material. This most ancient form of copulation is the exception, not the rule, in single cells, which generally reproduce by simple division (one into two), but it nevertheless seems to have become important in the long run for maintaining genetic vigor, and is known to be practiced at rare intervals by both bacteria (vegetables) and protozoans (animals), usually starting when they are under stress, say from drought, famine or temperature change. It also occurs among viruses, which are hardly more than minerals and where the strains (genders) are labeled only plus and minus or (at the molecular level) right and left. It may even, as the illustration suggests, have begun between identical crystals that are indisputably minerals!

EVOLUTION OF SEX

| MINERALS conjugation of equal crystals | VEGETABLES mating of equal cells | mating of unequal cells | ANIMALS full sex differentiation |

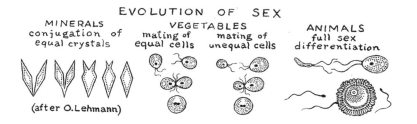

(after O. Lehmann)

145

The interesting thing about this primordial version of sex is that it has little to do with reproduction and not only was imposed upon it but, in fact, actively opposes it. Its function seems rather to be to reinvigorate the species as with a shot in the arm after a long, hard siege of redundant stagnation — and, in the case of paramecia, this genetic swap definitely slows down reproduction by consuming over an hour's time, enough for three generations of ordinary division, which division obviously must be postponed the while. Thus 2 protozoans making "love" of a morning (sexually so to speak) will end their session as 2 protozoa instead of 16, which otherwise would have been the number of their great-grandchildren procreated (by simple division) in the same period. So don't assume sex necessarily promotes a family.

Yet sex has its usefulness, for it increases variableness and adaptability by its exchange of genes (significant details of which we will take up in the next chapter). In essence sexual conjugation is a randomizing process, a shuffling of the genetic "cards" that, in the hundreds of millions of years it has been going on, has been largely responsible for the extraordinary variableness and proliferation of nature. It is also a liberalizing process as compared with the conservatism of plain reproduction. It keeps trying out new combinations of genes, partly by chance meetings (as between boy and girl), partly by selection (as when a girl with many suitors decides whom to say "yes" to), the results, if any, appearing in subsequent offspring.

In the very beginning of course sex was without gender. There was literally no such a thing as male or female anywhere on Earth in that pre-Cambrian time about a billion years ago. No creature had sex organs, and the organisms that mated were more similar than any two brothers or two sisters you ever met, with the possible exception of identical twins. Under the biologist's microscope one can see the same thing still going on among many single-celled plants and animals, where it is an enduring trend of elementary life. As these cells divide and multiply and divide again, or sometimes sprout buds that cast off on their own, once in a long while, particularly under stress conditions, a new and intrusive urge seems to pervade them. Some flagellate, say, instead of splitting in two as its ancestors did for 100 generations, will hesitate and probe about. It may be getting a giddy, reckless feeling of sociability it never knew before. It will show interest in some other flagellate and, after only the briefest of introductions and perhaps a moment to screw up its "courage," the two will touch and melt and merge completely into one, as if they were two minds with a single thought.

If this happened every time two cells met, of course their number would steadily diminish — but it does not; it can be deadly dangerous to both cells (which they seem to sense), and most of them go on dividing and disregarding their neighbors, only resorting to this risky population-reducing merger at rare intervals. Although the two merging cells are clearly almost identical when they first try this "exciting," liquid hug, evolution gradually brings on greater complexity, a proto-hermaphroditism followed by increasing polarity and disparity between them, permitting them eventually to specialize, which is the real beginning of sex. Since some traveling must be done to get them together, and some food must be provided to keep them nourished afterward, their specializing takes the logical direction of dividing these duties, one kind of cell going in for mobility by growing a tail, the other kind concentrating on food storage by growing fat. The first kind, we soon realize, is evolving into a sperm, the second into an ovum. The first therefore emerges in time as the male, whose essential function is to be active, to hunt and travel. The second becomes the female, whose essence is passivity, who stays home and nurtures the young.

Of course many variations of the basic idea of sexual polarity as demonstrated in merging or swapping of genes are tried out. This is nature's wont and, while some free cells become simple sperms and ova that merge into one cell before they start their reproductive multiplication, others just touch or conjugate long enough to exchange material, then separate and one or both may reproduce later, depending on their degree of maleness, femaleness, hermaphroditism or other sexuality. The paramecium (page 86) is a one-celled animal that briefly conjugates to swap genes, using this primitive protosexual technique to ease the stress of life, on the principle that "two do better than one." The mating pair express themselves in what might be described as the oldest known language of love, a sort of chemical conversation in which each partner puts forth a substance that influences the other, both of them seeming to feel better and more relaxed for this discourse evolving into intercourse. It may not be as limited a dialogue as you would think either, for there is no fundamental reason why there can't be more than two sexes and biologists already have discovered eight sexes just among paramecia. On hearing this I once asked an authority on sex at the Harvard Biological Laboratories (whose name not inappropriately turned out to be Dr. Raper) whether these eight sexes could have any sort of mechanical manifestation because, in my innocence, I had imagined lovemaking among octosexed paramecia as analogous to a kind of baseball game in which each player had to touch all four bases twice in order to make a home run. But he said, "No, the sexual differences in parameciums are only chemical."

Yet the complications of multisexed life remain both fascinating and baffling. Just think of the dilemma of a hostess faced with the etiquette problem of seating guests of eight different sexes harmoniously and congenially at the banquet table. Then consider some of the higher varieties of fungi which, according to Raper, have as many as 24,000 sexes — only, as he points out, they aren't really sexes as the word is commonly understood but rather mating types, properly called syngams, each of which is intrasterile within itself and intersterile with certain other syngams, while interfertile with the great majority of its own species. Incidentally this is a more liberal arrangement than it may seem — for, I was given to understand, if one were a fungus and should happen to meet another fungus of one's own species, say in some shady nook, one would have about a 98 percent chance of being able to mate with it successfully. Which is a much higher batting average than anybody, including Jacques Casanova, ever attained as a human.

Some of the multiplicity of sexes may perhaps be explainable through the relativity of sex, an aspect one doesn't ordinarily become aware of much, outside homosexual circles. In saying relativity, I mean that sexual characteristics, notably in primitive creatures, seem to depend on how much of this or that chemical the organism contains, with the implication that there are many degrees or even gradients of sexuality. If true it would explain why a microbe is so often observed to be male in relation to one companion at the same time it is female in relation to another. For all I know, there could be individual organisms with a whole arsenal, repertoire or spectrum of sexes, each sex chemically (if not mechanically) specific to one of dozens of potential mates it might someday feel a yen to get intimate with. Almost anything appears possible. Bacteria are now known to swap genetic material at random or by accident, as when two or more of them are mangled by an intruding object, by the action of viruses adept at lugging genes around or by "deliberate" conjugation. Cutting is one of the most familiar means of reproduction among worms, almost any species of which, no matter how you slice it, seems able to grow its parts into whole new worms. And since multicelled organisms, unlike single-celled ones, do not need their whole bodies for reproduction, most of them soon evolve the more efficient system of keeping just a relatively few special cells for the purpose: like sperm and ova. Since sperms need liquid for swimming, however, when life spread from the sea to the land (where the more variable environment particularly demanded the adaptability of sex), it had to bring a little sea along with it or find a substitute in order to get to the ova and reproduce. That is why animals still either lay their eggs in water or retain wet internal ducts for the sperm.

Primitive plants naturally evolved something similar, most of them living in wet places or, if crowded away, perforce waiting for rain, as do ferns and mosses, whose sperms still swim, but the majority of their descendants eventually found this dependence such a fatal drawback that they "taught" their sperms to fly or hitch rides, mostly by progressively miniaturizing them into pollen. And not a few species (of animals as well as vegetables) even compromised between sex and nonsex by alternating generations between sperms and spores.

It would be only natural for humans to think of nonsexual reproduction or virgin birth as limited to the most primordial or simple creatures, but actually it is found here and there through practically all levels of life, including our own. Sometimes it is a phenomenon of only half the year, as among aphids, water fleas, etc., where males are unknown in spring and summer but normally show up every autumn. But there are at least two species of lizards and four of fish that have no males ever, their embryos developing unfertilized by what is known as gynogenesis. Frogs' eggs, moreover, can be "fertilized" without any sperm at any season by pricking them with a needle, or sea urchin eggs by adding sugar, salt or other substances to the water, and they grow up into real frogs and urchins.

In such cases, as you would suspect, it's the mother who is the virgin. Yet if a sperm is "fertilized" by being allowed to enter a fragment of egg that has no nucleus, the young that develop from out of it have no genetic mother — so in this instance, oddly enough, it's the father who is the virgin! Animals as big and advanced as turkeys have recently been raised by the U.S. Department of Agriculture without any fathers at all, their ova evidently produced by mitosis instead of meiosis (page 164), and have later begotten their own offspring. So the principal biological evidence that the Virgin Mary had to have experienced a miracle — if not an immaculate misconception — to have given birth to Jesus is his maleness, and a human race of Amazons is theoretically possible. Besides, has it not long been known to science that there exists such a thing as human sexless reproduction: notably in the case of the embryo that, without aid from the opposite sex, somehow divides into two and becomes identical twins?

Winding up this chapter on sex, we might revert for perspective to
149

the question of the relative importance of space and time in all of life's reproductive methods, for there is obvious significance in the observation that encapsulated spores (which are tough and have no sex or fertilization problems) can go almost anywhere in space but are not very adaptable over a period of time, while sperms (which can survive only where they find an ovum) are very limited in space yet can adapt to almost any environment if only they (and their descendants) have time enough.

Such an issue is admittedly too unfamiliar to us humans to be easy to grasp but, I can imagine, it might seem a lot nearer fetched to the mind (if any) of a sperm or an ovum, or any other kind of sex cell. Sex cells, you know, are practically immortal — potentially so anyway — and for that reason, if for no other, it would seem worth the effort to try to visualize their extraordinary outlook. Generation upon generation they live on and on without dying, being passed "down" from parent to child, repeatedly conceiving and sprouting out a body in which to shelter, feed and renew themselves, indeed diverging again and again into many bodies at once, tantalizing them with wild lusts so they will be sure to free them before they have lost the power to find their way and sprout the next body — and so on indefinitely.

Thus, in a sense, you and I are nothing but a sperm's (or ovum's) way of duplicating itself and staying alive. And what is a chicken but an egg's way of producing another egg? Come to think of it, what really evolves in evolution may not be so much the obvious vegetables and animals we see around us as their hidden cells of sex — not the mortal individuals so much as their immortal lifelines of genetic continuity. For evolution is essentially a kind of perpetual living flux, a turbulent continuum of interrelatedness that as a whole must be much greater than any sums of its parts.

We will be returning to the subject of interrelations in Chapter 13, but, before that, we need to delve deeper than sex and its cells. I mean we must dig into the very core of whatever it is in these germs of generation that enables them not only to preserve everything our ancestors evolved during billions of years but whatever, in each reproductive cycle, still makes it possible to move each of their descendants one more step forward into the unknown future. And this, should our endeavor meet with success, means simply that we shall have come that much closer to the ancient and elusive elixir of life.

# CHAPTER 6

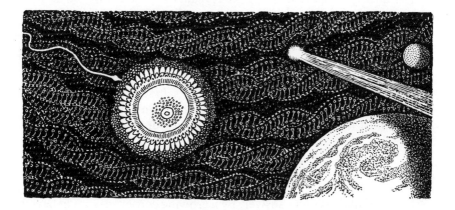

# Secret Language of the Gene

W HEN I LOOK upon the earth and moon and other worlds so patiently going their rounds, each sticking to its own well-defined beat like so many territorial animals, I can't help but wonder at the order and discipline of it all. If these rotund organisms are alive, as I am more and more convinced they are (at least in a cosmic sense), there must be some factor regulating their growth and behavior, just as is true of raindrops, bubbles, apples, seeds, eggs and other globoid forms that come and go — something that directs the motion not only of whole entities but even of the atoms, molecules and other substances that enter and leave them in the endless exchange of life.

So I ask myself: if life is a flow, isn't it because only the fluid state permits matter to maintain enough molecular motion to give it the chemical complexity that spells life? Of course I know the solid state of bones includes metabolic drift, but far surpassing this movement are the torrents of blood, lymph, water and air that keep the body alive by supplying it with vital secretions plucked out of the passing world. The processes involved are naturally abstract in essence, like an ocean wave that continuously advances through vast populations of molecules, shaping them as it goes — or like a candle flame fed by liquid wax and air so that its fluctuating form persists as a pattern in the invisible currents of oxygen, hydrogen, carbon, nitrogen . . .

The abstract aspect of life thus revealed is as inescapable as it is elusive, since it leaves no room to doubt that life is essentially only a pattern in a metabolic current of elements and not really dependent on any particular atom or molecule or cell. What, then, should we think of as the physical basis of life? And from what, if anything, is a body ultimately derived?

It is an ancient enigma. Like Chopin's first prelude (sometimes called "The Question") it does not expect an immediate answer. Rather must it glean satisfaction from being truly heard.

## HEREDITY'S UNIT

The first inklings of an answer, by a growing consensus, are believed to lurk somewhere in the nucleus of the living cell — particularly in the genes that direct the cell's growth and behavior. But what is a gene? Perhaps the surest explanation one can give is that it is a letter or a word in nature's message to a seed or an egg telling it how to be a tree, a bird, a turtle, a snake, a platypus . . . A gene is one step in the secret recipe for growing up, for living. It is a wave of the unseen wand that turns a tadpole into a frog, a caterpillar into a butterfly. It is a basic unit of heredity. It is what controls the mysterious automation that heals a wound, regenerates a lizard's tail or guides a migrating bird with the help of the stars.

Such automation is called genetic because it is presumably provided by genes in each cell, but just how they provide it is still in process of being unfolded. Last-century biologists, who knew little of the microcosm, theorized that whatever hereditary controls are in a fertilized egg inevitably must be parceled out as the egg divides and grows so that each new cell can be directed as to what it is to do. One of the two cells created in the egg's first division, they reasoned, contained directives for all the structures on the right side of the body, while the other held the plans for those on the left. The second round of division (at right angles to the first) in turn separated the instructions for the upper half of the body from those for the lower half, while every subsequent division in like manner reduced the number of determinants per cell until finally, at the birth of the animal, each of its cells carried only about one unit (gene) of hereditary information. In other words each fingertip contained only genes with fingertip information, each eye contained only eye information, and so on.

Though such a hypothesis had the appeal of logic and simplicity, it was proved untrue in a classic experiment early this century by a German zoologist named Hans Spemann, who operated on an embryo salamander, using semimicroscopic glass threads and tubes of his own devising. Most of the cells of the unborn creature (still attached to its mother) apparently had not yet committed themselves to specialization, for the tiny animal looked as blank as a baked bean, with barely perceptible nubs where the head and limbs had begun to form. Yet Spemann knew from experience exactly where an eye would appear in a few more days' development, and carefully cut a square piece of prospective-eye flesh out of the emerging head and exchanged it for a piece of the same size and shape dissected from the emerging tail, immediately transplanting both into their respective new positions to see how they would grow. Would the eye appear in the tail, as it should if each cell had been given instructions only for itself? Within a week it was clear that the answer was no — for the eye showed up in its regular place in the head, notwithstanding the fact that it fashioned itself out of cells that would have become a tail had they not been moved. Obviously, then, all body cells must have genes fully qualified to make any body part required of them — like men building a house, each of whom has been given blueprints for the whole job, so that, no matter what section he finds himself in, he can build correctly in relation to the whole. When a body composes itself, therefore, it is immaterial what flesh is used to form its features. Any of its flesh is good enough to make an eye, even as God the Potter may mold any clay to His creative wish.

This revelation of a complete genetic blueprint in every cell, extravagant and wonderful though it seemed, inevitably left a lot to be explained as to how life grows. Students of embryology had long been aware that when a cell begins to specialize it generally commits itself to a particular task and ones with such different functions as eyes and tails had never been known to tolerate an exchange after they were established. However, Spemann continued his imaginative experiments, and it was not long before he realized and proved that development is controlled not just by genes but to a great extent by a continuous "interplay" between them and the rest of the cell. He did not fail to take into account also the ceaseless influence on a cell's growth of all its surrounding cells and every manner of far-ranging hormones, chemicals, viruses and other intrusions. The process in which a fertilized ovum thus divides in order to multiply itself into 2, then 4, 8, 16 . . . cells came to be recognized as more and more subtle and complex — an unfathomably sophisticated natural increment of cells, in which each organism diverges into its own pattern, the cells       153 by the time their number approaches 50 forming one of various models of a hollow sphere known as the blastula. This round and growing speck sooner or later, for some little understood reason, finds itself compelled to draw its single-layered surface inward at a certain point like a rubber ball sucking a dent into itself, usually thus becoming a cup-shaped two-layered body, the gastrula, which will later shape the dent into the beginnings of a gut. And, along with the gut, the outer gastrula will fold inward to make a groove that, if the embryo is to become a chordate, will steadily work itself into a spinal column, while other foldings, unfoldings and sproutings of increasing intricacy form the head, limbs, muscles, organs and detailed features of the being-to-be.

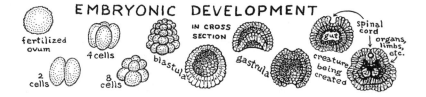

EMBRYONIC DEVELOPMENT

By such a process all animals literally organize their organs and themselves. So do human embryos, dramatically increasing their weight ten-thousand-fold in the first month. And I note with interest that the body's original single opening, the one that formed when the blastula dimpled inward to become a gastrula, first served as a mouth that was increasingly expected to double as an anus until, this crudity becoming intolerable, a new mouth broke through at the other (now forward) end, at last permitting sensible throughway digestion. Later feelers, nose and ears appeared (also forward), followed by the eyes. At the same time undoubtedly appendages of locomotion were developing (aft) through relentless trial and selection — tail, fins, flippers, legs, toes, wings — to produce the million-odd different creatures we know today. One can readily recognize the main phases in human embryos, whose heads bulge like bladders in their early liquid period, only gradually solidifying, and who, like others, have gills and flippers before they develop lungs and hands. By the time he is halfway to birth, obstetricians point out, the human fetus has learned to move and turn and kick, to drink his surrounding fluid and, in some cases, to suck his thumb, practicing for the outside world. When he is hurt he will cry, strengthening his lungs with liquid breathing. At eight months he can hiccup, sneeze and begin working his kidneys and bowels. Although he sleeps most of the time in the womb, he intermittently wakes and hears continuous throbbings, rumblings, occasional voices, scary bangs and soothing music from another world. There is even evidence that he may see changes of reddish light and darkness through briefly opened eyes. And a doctor in Toronto, watching unborn twins through a fluoroscope, saw them actually fighting.

# THE ORGANIZER

The mechanisms hidden within such a miraculous luxuriation of flesh and function were almost completely unreachable until Spemann, still experimenting, made his greatest discovery in locating specific growth spots he named "organizers," which at every developmental stage proved to be the actual inducers of the next step. The salamander eye, for instance, begins as a tiny swelling upon its brain sprouting a stalk with a cup at the end that soon reaches nearly out of its skin. The stalk does not have to go farther because the cup is a potent organizer that promptly induces the nearest skin cells to dip down into it and form an eyeball: transparent lens, cornea and all. Spemann even demonstrated the cup's power by embedding one from an embryo salamander under the skin of its belly, with the result that the following week he had a baby salamander with a belly eye!

The same kind of induction, as he called it, is known to multiply a fertilized egg's cells through the blastula stage, in which the "dorsal lip" of its indentation becomes the "primary organizer" of the future body by defining its spinal axis. If any very young embryo tissue is grafted upon such an organizer, Spemann found, it will be organized into a new organizer, which in turn can induce new specialized growth, including the sprouting of an entirely new embryo. In certain situations even organizer tissue that has been established as "dead" by all the standard tests will still induce some such development in a graft. In others a normally living organizer may kill its brother cells by the thousands in order to shape a living body — producing, for example, a frog by degenerating a tadpole's tail, a development sometimes called ungrowth. A striking example of ungrowth is *Pseudis paradoxa*, a species of frog in Trinidad who ungrows to such lengths that when he is an adult frog (measuring two inches) he can be literally less than a quarter as long as he was just after he lost his tail when a juvenile frog (almost ten inches).

All kinds of cases could be cited of the ways of organized development, for there are whole hierarchies and successions of organizers that step by step turn the gastrula's outer layer of cells into skin and nerves, its inner layer into lungs and digestive tract, its ensuing inter-

155

mediary layer into bone, muscle, blood. The organizer seems to give orders that a basic structure be made, trustingly leaving it to local influences to produce the details, as was pointedly demonstrated when Spemann grafted a piece of prospective brain from a frog embryo into the prospective mouth region of a newt, which then obligingly grew a usable mouth in the right place. Yet, surprisingly, it turned out to be not the toothed mouth of a newt, but unmistakably the gaping mouth of a frog. All this suggests that the genes, whatever they are, act something like tones in a tune. Certainly they have a time axis of some sort, each following and fulfilling the gene before it as if they can somehow sense that they are unit parts of a sequential whole, footsteps in a procession, notes in the melody of life.

His experiments understandably gave Spemann's laboratory something of the reputation of a "monster factory," but they were only the most prominent among several similar early research efforts in the field of growth which continued to make significant discoveries, such as that if you cut off a shrimp's protruding eye, leaving the nerve ganglion in its stalk, a new eye will grow out of it — but if you amputate the ganglion along with the eye, an antenna will sprout in place of the eye. Or if a mantis's antenna is cut off, two sections from the head, a new antenna will regenerate, while, if only one section of stump is left, a leg will grow instead. Findings of this kind reinforced the conclusion that body cells have full repertoires of potentialities within their genes and that they depend on local organizers only to decide which potentiality to materialize. This points to an interesting genetic variant of the ancient Mosaic Law of an eye for an eye and a tooth for a tooth, which may have to yield place among some creatures to a wider range of possibilities, perhaps even more fantastic than this automatic substitution of an antenna for an eye or a leg for an antenna.

## Genetic Puzzles

It may be worthwhile to take a quick look at regeneration in general, which seems to be a fundamental organizing property of living matter. Why can a salamander regrow an amputated limb when a mouse cannot? The significant difference between these animals must be in the degree of their evolvement or specialization. A primitive creature such as an ameba is, so to speak, a blob of unspecialized goo, more than 95 percent of which, as we have seen, can become a foot, mouth, nostril, stomach or anus as occasion demands. Its protein and other protoplasm seem as flexible and uncommitted as a batch of fresh-mixed cement. But a slightly higher form of life — say, a hydra — is specialized at least to the degree of maintaining a tentacled form. Although here again almost any sizable piece of flesh cut from the hydra will regenerate into a whole new hydra, the tentacles turn out to be an exception and will not grow into hydras because they have become too specialized, having in some genetic way committed themselves beyond the capacity for that much adaptability. The age of the individual organism of course is a major factor, for its early growth swiftly repeats its ancestors' evolutionary history, with the primitive, simple, embryonic stages retaining much more regenerative potential than will later be possible to the mature, specialized, complex animal. In consequence any creature, including a human, may regenerate any part if called to do so early enough; identical human twins, triplets, etc., are nothing less than a single embryo that broke into two or more pieces, each of which then regenerated into a complete human being. In the case of the famous Dionne quintuplets, the original female embryo is known to have split into five parts in 1933, every one of them developing into a consummate woman by 1950. That is, five full skeletons, five human brains, five pairs of seeing eyes and hearing ears and the complete bodies and minds that go with them — all springing from one disintegrated embryo under what was left of its own guidance.

The older or more developed an embryo is, of course, the less it can regenerate any lost part. Yet it has been found that in some mysterious way the inhibiting effect of age can be largely offset by artificially increasing the number of nerve fibers at the site of required regrowth, which seems to indicate that the nerve impulse conveys not only information but a stimulus actually vital to growing. Electricity is an important factor here too, working specifically through the natural flow of electrons in skin over a stump. Even artificial electric current applied from outside is known to have accelerated the healing and regenerative process in many cases. There is a limit to it though for, beyond a certain stage of development, the only hope of recouping a loss has been found to be by grafting — and grafting in turn has its own narrow limits attributable to the long-known reluctance of any flesh to accept any other kind but its genetic equal, which has traditionally meant flesh from either the same body or that of its identical twin.

The most obvious reason for flesh's inhospitality to strangers is the need for a defense against abnormal growths, such as tumors or invasions from outside, particularly the subtle intrusion of germs and viruses. But immunity is not all that simple. In fact the new science

157

of immunology has sprung up mainly to deal with its spreading complexities that involve the body's production of defensive protein molecules called antibodies. For, although the antibodies usually do a good job of disarming dangerous invaders, they cannot unaided discriminate between good invaders and bad, and naturally tend to oppose helpful ones such as a transplanted kidney even when that kidney represents a last hope for life. That is why modern researchers have made a major effort to learn how to control the antibodies by typing, studying and comparing them. From their viewpoint the immune system is one of the most important organs in the body — and it weighs two pounds.

Another and even more fundamental approach to this basically genetic problem has resulted in the experimental procreation of a black-and-white-striped mouse who genetically had two mothers and two fathers . . . It was done in 1964 by mixing and transplanting genes from the fertilized ova of two newly pregnant mice, one black, the other white, to assemble the first four-parent creature in history. Later other experimenters went a step farther to produce a series of previously "impossible" hybrid creatures by using viruses to integrate the genes of species as classically disparate as mice and men and, very recently, of a chicken and a tobacco plant to create the first artificial cross between a plant and an animal, a so-called plantimal that lived for five hours! The viruses that made this possible were of infectious strains that, for some reason, are wont to dissolve cell membranes wherever two cells touch, at the same time fusing what remains of the membranes into a single composite membrane around what has thus become one compound cell with two nuclei. Since there are no antibodies or other mechanisms for recognizing incompatibility within a single vertebrate cell, such corralled nuclei do not reject each other, and even larger hybrid cells with three entirely different nuclei have been created, which produced daughter cells that thrived and continued duplicating themselves.

## THE VIRUS

Part of the explanation for these phenomena is in the nature of a virus (page 102), now known to be the smallest biological structure possessing all the information needed for its own reproduction, and whose discovery finally filled the perplexing gap between molecules and organisms. In a sense a virus is a gene with a coat on, out wandering about the world, for its core is made of either deoxyribonucleic acid (DNA) or its subsidiary ribonucleic acid (RNA), carrying hereditary information and wrapped in protein. Yet it is far from complete unto itself — in fact it is a very special but fundamental kind of parasite that cannot reproduce itself without entering a cell and using the cell's own genetic and reproductive machinery. For this reason its protein coat is often equipped with chemicals enabling it to dissolve cell membranes (as mentioned above) presumably to assure it of admittance to a host when it finds one.

Most of the time a virus just lies around as an inert crystal, "lifeless as a rock" and perhaps staying that way for centuries or millenniums. Yet, unlike a rock, it may "wake up" at any moment. All it needs is the warmth and moisture of some vulnerable cell that it can swiftly enter and infect, in the same motion reproducing itself hundreds of times within the hour — in some cases (apparently by chance) breeding a new type of offspring that may spread fast enough in a year to kill 20 million people, something a new, deadly influenza virus actually did as recently as 1918. The submicroscopic size of most viruses of course is the most obvious factor in their ability to slip inside cells. For notwithstanding the fact that they are gigantic compared to the simple molecules of water, air, etc., their atoms numbering in the millions, they are still so unbelievably tiny that one quintillion of them could fit inside a Ping-Pong ball. And if this spatial comparison makes little impact on you, you can make it a temporal one by realizing that, if the viruses had commenced pouring into a Ping-Pong ball at the supposed beginning of the universe (18 billion years ago) at a steady rate of one virus a second, the ball would now be only half full.

Of course not all viruses are the size of the polio virus used in the above calculation. In fact some are a thousand times bigger, yet would still need 30 million years to fill the Ping-Pong ball. And others are eighty times smaller and couldn't fill the ball in two trillion years.

Significantly it took only thirty years for science to figure out the structure of the newly discovered viruses through electron microscopy, x-ray diffraction analysis and other sophisticated methods. It is now known that these infectious particles are assemblies of identical protein subunits stacked symmetrically around cores of nucleic acid into the shapes of gemlike polyhedrons, including some of Plato's five famous regular solids, plus prisms, spirals and a few more complex forms. The smallest virus yet discovered is made of nothing but a core of nucleic acid (RNA), while another, a little larger and known for its beauty, is the polio virus, a dodecahedron of 12 perfect pentagonal facets, probably made up of 30 spherical subunits. Another is the polyoma virus of rodent cancer, an icosahedron of 20 equilateral triangles composed of 42 hexagonal prisms. The tobacco mosaic virus is like a long tubelike pine cone with 2130 protein "seeds" in a continuous spiral, the mumps virus a spheroid full of endless helical RNA "spaghetti," the influenza virus perhaps something of a bristling "sea spider" (page 92) around a coiled core, the T4 virus a kind of hexagonal mosquito with 6 leglike triggers and a beak for injecting bacteria.

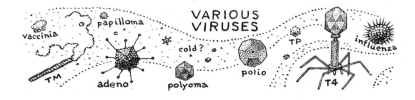

One is apt to think of viruses as swimmers like amebas, propelling themselves where they wish. But actually they are generally as inert as grains of sand and get to their destinations only by being swept along willy-nilly in their surrounding tide of liquid, occasionally following magnetic gradients but continually buffeted by the agitations of molecules collectively known as Brownian movement, in trees wafted by sap, in animals by blood and lymph, in the sky by wind and rain . . . Nevertheless, through patient willfulness or blind chance, they get wherever they are going and a percentage of them manages to attack their prey like a pack of dogs around a bear, as in the case of the T type (also called the bacteriophage or "phage" for short) that punctures a bacterium with its syringelike beak to pour in its DNA. Such an attack, interestingly enough, is as much rape and suicide as it is murder, for it injects genetic material that merges with the bacterium's own genes, while neither the virus nor his victim survives the act. Once the long thread of viral DNA has been shot into the bacterium, you see, the virus's residue of protein is but an empty husk, and the injected DNA disintegrates in the very act of insinuating itself into the bacterial genes, somehow deceiving them into accepting it piecemeal as parts of themselves. Although there is no remaining sign of the virus in the bacterium, since it has fused with the nucleus already there, about 20 minutes later new viruses begin to form by the hundreds and within 30 minutes the bacterium pops open like a burst balloon, liberating a horde of viral offspring identical with their virus "father," who only half an hour ago raped their "foster mother," whose torn corpse now is all that remains of "her."

Although such a virus attack on bacteria is very common, it is not always so one-sided. Indeed there is at least one species of bacterium, called *Flavobacterium virurumpens*, that lives in mud and can kill viruses, specifically the tobacco mosaic virus, which it attacks with enzymes at certain points in the spiral protein coat, literally dissecting it into irregular clusters of one or more of its 2130 "seeds." Not many kinds of viruses are yet considered exactly "good" by man, but one kind "infects" tulips with beautifully colored patterns that increase their value, and another stopped the midcentury plague of rabbits in Australia with frightening efficiency.

Undoubtedly man's greatest benefit from viruses, however, is what they teach him about genetics, because viruses are the simplest of all reproducing creatures. In fact geneticists already have such a detailed concept of viral genes that one of them states that the difference between a mild virus and a killer could be the replacement of only 3 of 5,250,000 atoms in a particular species. All sorts of viral cross-breedings can be made experimentally by letting different varieties of viruses simultaneously infect the same cell, the offspring often inheriting traits from more than one "parent" through this most primitive level of "sexual" intercourse. Also, since it has been proven now that viruses can cause and cure certain kinds of cancer, evidently by changing the genes of cells (a phenomenon called mutation) and, since science is learning to influence mutations (by radiation, etc.) as well as to manipulate viruses, it looks as though viruses may soon be made to order by man, possibly including some that will destroy cancer cells but leave normal ones alone.

## DISCOVERY OF THE GENE

To round out the genetic story, we need now to look back to the mid-nineteenth century when the first great discoveries about genes were made. It was in Brünn in Moravia (now Brno in southern Czechoslovakia) that a young Augustinian monk named Gregor Mendel, who had studied mathematics at the University of Vienna, experimented methodically with peas and other plants, trying to find out the laws of inheritance. The accepted belief of the day was that animal and human heredity was directly transmitted through blood, while vegetable heredity presumably worked through some element in plant juices, the progeny in either case receiving a mishmash of all their ancestors' characteristics. On the rare occasions when black-haired parents unexpectedly had a baby with red hair it was supposed to be because    161 the baby somehow miraculously happened to derive a few drops of relatively unmixed blood from a red-haired ancestor further back. If a red flower that had been crossbred with a white flower produced pink progeny it was because any red liquid mixed with a white liquid must yield some blend of pink. Charles Darwin, aware of the naïveté of these notions, which, if true, would have progressively eliminated the natural variations in organisms on which (he had discovered) evolution depends, had been experimenting quietly with plants for years in a serious effort to learn precisely how colors and other traits passed from generation to generation, but his results were inconsistent and confusing, eventually leading him to abandon the project in favor of more promising lines of endeavor.

Then in 1858 Mendel started his classic experiments with hundreds of plants in a section of his monastery garden, carefully labeling each with fertilization data as he played the part of a bee or butterfly in artificially breeding it to others by transferring pollen on a brush from male anthers to female stigmas in a planned and systematic pattern. With single-minded patience in minding his peas and cues and a combination of intuition and luck in his choices of plants and traits, he soon found extraordinary evidence that traits do not mix like stirred paint but rather like shuffled playing cards, always retaining discrete "factors" hidden somewhere inside every organism, including the organisms of animals, such as a few light and dark mice that he experimentally bred in addition to his plants.

In the case of the flower *Mirabilis jalapa*, commonly known as four-o'clock, to take a simple example, two of its many varieties come out in all-red and all-white petals. When pollen from either variety is used to fertilize the other, the resulting seeds grow into pink flowers with nothing visibly red or white about them, yet when these pink second-generation flowers are fertilized in turn, either with each other's pollen or their own, the third generation produces not only more pink flowers but also red and white varieties, just like their grandparents, in the proportion of 25 percent red, 50 percent pink and 25 percent white. It thus became clear that whatever factors initially made the flowers red and white were not all used up by being mixed into pink, for the pink flowers always kept their power to revert back to red and white in another generation. And not only that but, as the mathematically minded Mendel could readily see, they did so according to orderly rules, most obviously the well-known mathematical law of probability under which, if you flip two coins, each red on one face and white on the other, they will fall so as to show only their red faces 25 percent of the time, both a red and a white face 50 percent of the time and

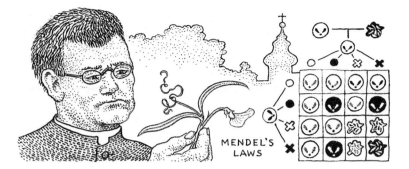

MENDEL'S
LAWS

only their white faces the other 25 percent of the time.

This simple case involving red and white "factors" (which we now call "genes") was a lucky discovery, giving Mendel his main clue. But inheritance usually is more complicated than this and far less obvious because genes seldom are so evenly balanced and so clearly definable. In fact many genes go in pairs (each member of which is known as an allele) like the genes for roughness or smoothness in a guinea pig's fur, one allele or the other being apparent but not both, and one of them often being dominant over the other, so that when both are inherited only the dominant allele shows, like the black eyes of children begotten in the marriage of a blue-eyed Swede to a black-eyed Turk. When Mendel had hypothesized dominant and recessive alleles he was elated to discover he held the key to prediction of the way many traits will be inherited, including the occasional procreation of a blue-eyed child by black-eyed parents when both these parents have inherited hidden recessive blue-eyed alleles.

Genetics is a science that, ever since Mendel struck the light, has been rapidly developing into what is probably the most complex if not the most baffling of biology's many branches. Certainly it is beset with the mathematics of hard-to-visualize, submicroscopic factors and their seemingly infinite interrelationships. I cannot avoid mentioning a few main points, however, such as that Mendel, in working out his laws of inheritance, realized that every sexual coming together of two germ cells (pollen and ovule, sperm and ovum . . .) must double (in the new combined cell) the number of genes that were in each old separate cell, and that this doubled number inevitably has to be halved again somewhere — presumably wherever germ cells are created by dividing the body's ordinary cells — in order to prevent each generation from inheriting twice as many genes as their parents did, which would make the gene number increase more than a thousandfold every ten generations, a manifest absurdity. Mendel did not work with a 163

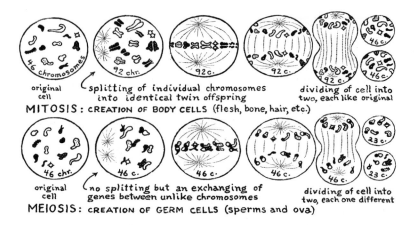

original cell — splitting of individual chromosomes into identical twin offspring — dividing of cell into two, each like original

MITOSIS: CREATION OF BODY CELLS (flesh, bone, hair, etc.)

original cell — no splitting but an exchanging of genes between unlike chromosomes — dividing of cell into two, each one different

MEIOSIS: CREATION OF GERM CELLS (sperms and ova)

microscope and therefore could not see or count the chromosomes (as strings of genes are called) in cell nuclei, but he made a reasonable assumption of what happens both in mitosis, the normal reproduction of body cells by division into two complete new cells (page 106), and in meiosis, the more special production of germ cells from body cells, in which, as I have just suggested, the division separates the original chromosomes with their genes into two groups (not exactly alike but normally paired into alleles), each containing only half as many genes as the body cell they came from. While Mendel knew no way to check whether this genetic halving really happened, his logical surmise turned out to be remarkably accurate and eventually made him the acknowledged father of genetics — even though none of the few German-speaking scientists who glanced through his revolutionary article published in 1866 then had the independence of mind or imagination to realize its importance.

Furthermore, the two kinds of male germ cells (sperms) made by dividing their parental cell's chromosomes into the two groups (one containing so-called X and Y chromosomes, the other double X chromosomes) proved to be capable of fathering respectively only male or only female progeny, showing why the birth rates of boys and girls are so evenly balanced, each embryo's gender being determined entirely by which kind of these equally numerous sperms (XY = ♂, XX = ♀) succeeds in penetrating the ovum first. From a mathematical or probability viewpoint meiosis thus shows itself comparable to halving a pack of cards (genes) at random, both half packs then forming germ cells (sperms if inside a male body and of two kinds; ova if inside a female, but of only one kind) — while sexual fertilization is like bringing two such complementary half packs (necessarily from differ-

164

ent original packs) back together more or less by chance to form a new whole pack (an embryo) whose cards (genes) thus originated half from a random half of its father's cards and half from a random half of its mother's cards (genes). This constantly repeated shuffle indeed is nature's main way of varying the offspring of every generation of life to ensure that all their potential traits will ultimately be given their chance for a trial in the evolving world.

If Darwin could have seen Mendel's meticulous account of his genetic discoveries when it appeared in the *Proceedings of the Natural History Society of Brünn*, both men would surely have been stimulated by the contact, and the science of genetics would have hatched a third of a century earlier. As fate decreed it, however, Darwin's puzzlement about the mechanism of inheritance was matched by Mendel's frustration that other scientists thought he was only an eccentric bookkeeper indulging his whims with a new garden hobby — so Mendel's work slept unrecognized until 1900, when, in one of those mystic historical coincidences that certainly spring from more than chance, three men (De Vries in Holland, Correns in Germany and Tschermak-Seysenegg in Austria) simultaneously but independently rediscovered Mendel's reports, repeated his experiments and confirmed his conclusions. By then Mendel had been dead for sixteen years.

With the Mendelian laws of inheritance now suddenly established, genetics blossomed spontaneously and attracted increasing numbers of imaginative pioneers, including Spemann. One of the earliest was Thomas Hunt Morgan, who discovered in experiments with fruit flies that genes can cross from one chromosome to another and that this appropriately happens when the chromosomes go into a kind of love dance just before meiosis, pairing and entwining each other in a kind of fond farewell as they separate to begin their new life as sperms or ova. Another pioneer, Archibald Garrod, correctly conceived that genes operate through enzymes, an idea almost as far ahead of its time in 1908 as Mendel's "factors" had been in 1866. And by 1913 accelerating research indicated that the chromosomes, familiar under the best optical microscopes as wormlike specks in cell nuclei, almost surely store their genes in one-dimensional order. No one yet had the faintest idea what genes looked like or how they worked, but a zoological count of chromosomes had been started which eventually revealed a curious hierarchy: the cells of a round worm, for instance, having 4 chromosomes, of a fruit fly 8, corn 20, a mouse 40, man 46, certain butterflies 62 and, curiously, the one-celled paramecium about 200.

But research has been accelerating ever since then, and by now not only have functioning genes been synthesized in the laboratory but literally hundreds of different kinds have been identified by their exact function and labeled with a number. A single gene theoretically can control the creation and assembly of a 60-subunit icosahedron which, perhaps for that reason, is the most usual shape of small viruses, whose subunits seem as uniform as building tiles and so tenoned and mortised that they can only fit stably into one simple symmetry. The only one-gene "organism" so far discovered in nature seems to be a potato virus. Larger structures naturally need more genes, usually including a few for "nuts" and "bolts," or "buttons" and "zippers," not to mention those of temporary function like "hammers," "wrenches," "funnels" and "scaffolding." More than 100 genes have been mapped for the complex T4 virus, for example, of which gene no. 23 makes the major subunit or building "brick," gene no. 20 closes off the icosahedral shape, gene no. 27 is a key to the tail, no. 18 assembles the tail's sheath, no. 15 "sews" a kind of "button" to the core, no. 31 adds solubility, no. 66 is an elongation factor, etc. Although a few rare genes have been found that will function effectively at any stage in the development of an organism, like salt in cooking a stew, genes generally work only sequentially and must apply themselves in an exact order if they are to succeed. Thus they create protoplasm not in the simple way knitting needles knit socks (by starting at one end and adding similar stitches until they reach the other) but rather through an assembly-line process of assigning jobs to specialists, most of whom build essential parts for others to put together into complete, working wholes.

## The Secret: DNA

After World War II, when the electron microscope suddenly made it possible to magnify a pinhead to the size of the Pentagon building and new x-ray equipment started "seeing" chromosomes as fibrous, twisted cords, the gene molecule at last came into focus, resolving itself into a beautiful and intricate spiral, now deservedly famous, if not exactly familiar, as DNA (page 105). In fact it proved (on close analysis) to be a double helix ladder, something like a spiral staircase with curving but parallel sides cross-linked every millimicron of the way by steplike rungs embedded at both ends. And the detailed order of these chemical rungs comprises the information actually dispensed by the cell's nuclear management as new cells are conceived and fabricated into all the specialized body materials of bone, blood, gut, nerve, hair, skin and sinew.

It is not easy to determine exactly where an individual gene begins and ends in the spiral stair — even the definition of a gene is in dispute — but it is generally understood that a gene is a continuous and complete sequence of DNA that, when active, progressively replicates either more DNA or a slightly different molecule called RNA (ribonucleic acid) out its side chains, as you can see in the illustration. DNA can contain anything from a few rungs to many thousand of them, depending on the complexity of the function involved. And a very simple chromosome, say in a bacterium, is a stair made up of several small genes strung in line like beads on a necklace. But a larger chromosome in a more complex creature like a human being would likely consist of several such strings wound around each other something in the way a large steel cable is twisted from smaller cables, and the larger and more complicated the organism, the larger the number of strands of DNA that are wound into its chromosomes.

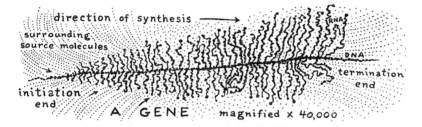

The universality of this extraordinary material is quite beyond comprehension, but a slight idea of its scope may be gleaned if you can realize that on Earth alone DNA is widely presumed to have existed since "life" began some three billion years ago, and it now directs in detail all cellular growth in every kind of organism, from viruses to mold to houseflies to trees to whales. In your own body, for example, DNA is distributed into your fifty-trillion-odd cells and, as suggested earlier, it remains uniquely your own and identical in atomic arrangement from head to toe, whether in bone, nerve, muscle or other tissue. The only cells that don't have it are red blood cells which, without nuclei, are impersonally exchangeable between all members of your blood group. If all the tightly wound DNA in even a single cell nucleus of your body were uncoiled and the pieces laid end to end, it has been estimated the invisible genetic thread would extend five feet, which would make the DNA in all your 50 trillion cells stretch out 50 billion miles or enough to reach to the moon and back 100,000 times.

In Chapter 3 I likened DNA to magnetic tape of a mystic origin, which seems an apt analogy for this extraordinary "one-dimensional"  167 molecule that holds all the essential data of a lifetime. Yet, amazingly, DNA is made of nothing but common elements put together as twin backbones of repeating groups of sugar and phosphate with cross-rungs of nitrogen compounds whose numerous variations spell out the ineffable code of lif . You might think that the four standard chemical rungs of adenine, thymine, cytosine and guanine (usually abbreviated A, T, C and G) would be much too simple to convey all the sophisticated information and intricate instructions needed to assemble an entire organism, but a leading geneticist has calculated that if we were to translate the coded messages of a single human cell into English, they would fill a thousand-volume library. So the four rungs somehow seem to manage to express themselves. And increasingly overwhelming evidence shows that they do it in stages, even as do the simpler dots and dashes of Morse code compose letters that make words that write books that fill libraries. It is significant, moreover, that all phonetic alphabets contain about twenty letters, which, not by chance, is the number of different amino acids that surge through protoplasm and its genes. And the very A, T, C and G rungs in those genes, operating like linotype machines, are what pick up the amino molecules, one by one, to mold from each the equivalent of letter type, whose sequences spell out words in the 20-letter amino alphabet. Indeed this is the source of the long polypeptide chains that stream from DNA like ticker tape, the hundreds of amino words in a typical one composing a complete genetic paragraph that exactly delineates a particular protein molecule so it can be assembled and sent forth to live a few weeks or months as part of an arm, a liver or an eye.

When the genes are asleep or dormant (which they are perhaps 90 percent of the time) it is noted that the chromosomes embedding them in their cables of DNA remain slim and inconspicuous, but when they are awake and "turned on" — that is, actively replicating themselves into more protoplasm in the cell's protein factories (called ribosomes) — the chromosomes puff out in noticeable swellings at all their busy points. For decades geneticists could not figure out what was going on inside these chromosome puffs but, now that DNA's genetic code has been cracked, they realize that the DNA chains in these spots (either genes or small groups of genes) somehow have "got the word" that they are to produce and they respond by suddenly stiffening and straightening out their coils, literally untwisting themselves so the twin strands of the DNA double helix can come apart. This stretching and undoing inevitably takes extra room and is what creates the puff (as the illustration shows), the full beauty of which we have scarcely begun to appreciate.

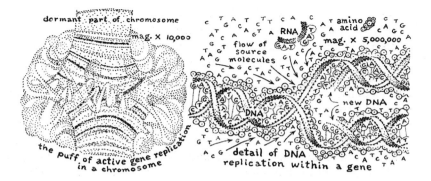

dormant part of chromosome

mag. X 10,000

the puff of active gene replication in a chromosome

flow of source molecules

RNA

amino acid

mag. X 5,000,000

new DNA

DNA

detail of DNA replication within a gene

What we have learned, however, is that this is the first stage of cell growth, of reproduction of a new cell that is usually a replica of the old one, but also often modified, more specialized or, in the particular cases of meiosis and mutation, fundamentally different. What happens is that, as the two strands of the DNA separate, each is bereft of part of itself, leaving it feeling incomplete, like a bride-to-be at the church door. But it has been unzipped down the middle of each rung, so to speak, precisely between the two letters that form the rung (these being always either AT or CG, or their reverse orders of TA or GC, a total of four alternatives) so that each A is now yearning for another T and each C for another G and vice versa. Moreover, since the medium surrounding the DNA in the cell's nucleus is full of millions of identical nitrogen compounds (A, T, C and G) and many other chemical fragments, all milling freely about in the unbelievably rapid hurly-burly of molecular turbulence, each lovelorn "letter" almost instantly finds a new partner that fits or suits it just like the old one. It is probably an almost automatic process of high-speed high-pressure trial and error, but it results in each strand of the separated DNA replacing its missing twin out of their common surroundings, including not only the amputated rungs but the whole absent strand, so that the DNA chain has been completely duplicated and each half of the next cell division can be assured of having exactly the same DNA as its fellow in the new generation.

Alternatively when DNA is not just reproducing but is serving as chief administrator for its cell, directing internal growth and other cellular activities, it creates (instead of more DNA) the aforementioned different kind of molecule called RNA (ribonucleic acid) that serves as a sort of lieutenant to which it delegates specific responsibilities. RNA is thus a special breed of genetic catalyst or enzyme (page 104), and it assumes various forms depending on the task at hand, mostly 169 doing errands outside the DNA boss's office, presumably so that the vital DNA itself need never risk venturing abroad.

RNA's main job is manufacturing protein, because proteins are the common molecules of life, providing not only the substances of blood, muscle, bone, nerve, etc., but also acting as the mobile chemical intermediaries or enzymes so indispensable to practically all life's physical processes. Every cell indeed, just to stay alive, has to be continuously and simultaneously grinding out hundreds of different models of protein. And a bacterium, which will sometimes reproduce itself in 20 minutes, may fabricate and distribute any of 1000 different kinds of protein in that short time, most of them enzymes geared to turning out membranes, ribosomes, mitochondria and hundreds of cell details, each in its proper sequence relative to the others. Protein in all its variations is the same miraculous molecule we have been describing as a long complex chain made up of twenty kinds of amino acids, each protein being a particular combination of them.

When DNA makes RNA it evidently unwinds its double strands and invites in complementary nitrogen compounds, etc., from the surrounding medium, as if replicating itself, but it does it somewhat differently (accepting more oxygen atoms for one thing) and so many kinds of RNA stream out of DNA — some conveying messages, some acting as brokers — that I couldn't possibly describe or explain them, even if I knew how.

The place where all this happens and where proteins are manufactured is inside the ribosomes, those thousands of globules floating like submicroscopic cherries about the cytoplasm just outside the cell's nucleus (page 104). It is not yet known exactly how each ribosome receives its RNA messengers when they arrive from the nucleus, but we can deduce that it does so somewhat as a player piano receives a roll to play a tune with. At least there is no doubt that the ribosome "reads" the RNA tape and "learns" the message, which it then expresses in a sequential production analogous to playing music. But naturally this abstract melody of life is not literally heard. Perhaps it is physically more analogous to the linotype already mentioned, since it produces the long polypeptide chain molecules we know as protein. And it is known that the ribosome scans the messenger-RNA with the close assistance of so-called transfer-RNA in precise sections: three chemical rungs at a time.

Such a section of code is called a codon and it corresponds to a three-letter "word" written in the four-letter RNA alphabet: A, U, C, G. The codon may be AUU, for instance, which specifies the amino acid tyrosine, or UUU which means another, called phenylalanine.

170

Sixty-four different three-letter combinations or "words" are possible in this alphabet, far more than ample for designating all the 20 amino acids. The codon is inspected somehow by the little transfer-RNA brokers, who continuously bustle about the ribosomes searching for the combinations that represent their clients, like schoolboys looking for their family names on a bulletin board. Each broker holds the complementary half rungs of three letters that stand for one particular amino acid, probably reading the successive codons by the Braille method, literally feeling the raised "letters," testing each codon until (by trial and error) he finds the one in twenty that fits — and so gives life its form and substance.

But this genetic code or DNA-RNA language in which specifications for all living systems are written with only four letters and a maximum of 64 three-letter words is neither as simple nor as foolproof as it appears. This is shown partly by its mutations or "typographical errors" during transmission, which, while at first they seem to be accidents, from the broader viewpoint of evolution and the long-term need for new species to cope with changing environs, are eventually revealed as vital factors in the way of life. Like other "accidents" mutations are individually much more likely to be destructive than constructive, for the same reason that hitting a balky watch with a hammer at random hurts it thousands of times more often than it helps it. Yet out of many thousands of mutations, inevitably a rare one now and then will bring a beneficial attribute: perhaps an unprecedented capacity to tolerate a drastic drop in temperature that, at the onset of the next ice age, will enable the species that received it to survive. And it is just such rare helpful mutations that lend life its vital flexibility.

Another remarkable thing about mutations is that they gain their effect on evolution in a subtle, cumulative way that until recently escaped notice. Mutagenic changes, as I have said, are much like misprints in the typed page of a manuscript, misprints that might be made by a typist who averages one error every time she types a page of 2000 letters. Such a rate may be considered reasonably low or normal. Yet if the typist starts with a perfect copy and keeps copying it perfunctorily without thinking of its sense or correcting mistakes, each time copying her latest copy, including its latest misprint, by the time she has finished the thousandth copy it will contain something like 800 errors (allowing for errors in copying errors which, in rare instances, actually correct them) and the page's meaning will have become unrecognizable. That is about how DNA copies itself, automatically and blindly, cell division by cell division, mutation upon mutation, generation after generation, and how even the lowest mutation rate will in time transform and evolve life.

Although the genetic code has been deduced mainly from studies of the colon bacillus or bacterium living in our intestines, it appears to be a universal code, at least as far as the earth is concerned, applying equally well to every organism from tobacco plants to human beings. Cracking the code may turn out to have been the most important scientific accomplishment of the twentieth century, surpassing the dissection of the atom and its nucleus or the dramatic conquest of space, for its potential consequences in eugenics and evolution are incalculable. It was done by the coordinated efforts of thousands of researchers in dozens of nations, progressively fitting together the pieces of evidence, mapping genes and mutations, rendering amino acids radioactive one after another to trace their separate movements and deduce their individual parts in genetic function.

## THE ABSTRACT ASPECT

The abstract nature of life is made clearer as we study the gene, for it becomes obvious that the gene is essentially a catalyst, not a creator, materially speaking. Indeed when parents create a child, they do not create substance out of nothing but use the material of their environment, only giving it a genetic key or direction for growth, a pattern for synthesis, an abstraction. The atoms that form the child are any old atoms of the earth that just happen to be passing through the mother's body as the child needs and absorbs them — and the essence of the child is in its genes, the mysterious blueprints of growth and development, of physical, mental and spiritual unfoldment. In fact there isn't a very fundamental difference between mixing *inorganic* cement in a cement mixer and mixing *organic* cement in one's digestive system. Either way, one takes available stuff from one's environment and, through a mixing or stirring process, builds one's abode with it, shaping the masonry into walls or bone or hive or shell or flesh as occasion requires.

In some sense, I have little doubt, genes know what they are doing, for they are memory incarnate, letters of living purpose, the script of life in a material universe. Stretching my imagination a little, I can think of them as grains of mind or even psychic feathers or scales, as veritable units of thought "made flesh," as St. John put it, to "dwell among us" or, if you like, as the flowing texture of mortality metabolizing its body in space and time. And a gene is not necessarily bounded, even abstractly, for it amounts to a composed pattern of simpler units, each of which subtly interacts with its sister genes on either side, so that not only may a given gene have several effects but an effect may come from several genes. Furthermore there are genes that control other genes and a few that trigger disintegration and death in ostensibly temporary opposition to growth and life, thus showing death to be, biologically speaking, a detail of life.

Interestingly also, genes have been found to have the material molecular structure of aperiodic crystals and it is thought that they very probably evolved in the same basic way as other crystals, which are well known to be the structural basis of rock, as well as of wood, bones and flesh. And this may explain not only why radiolarians in the sea seem to have about as many thousands of species as there are possible geometric structures to hold them together, but also why there are "genes" of a sort in snow crystals, their rarely seen nuclei being the microscopic dust particles or germs that the ice forms around and whose countless variations in shape logically account for the illimitable "species" of snowflakes, as suggested by D'Arcy Thompson in *On Growth and Form* in 1917.

These concepts of course more or less transcend the material aspect of life and the body, which have been the subject of Part One. Indeed they introduce an abstract quality into this book that naturally leads us to the intangibilities of Part Two. And that is where we may hope to discover that life's range of senses is wider than man has been aware of until now — also where human vision may be permitted a peek into an invisible mirror, deep within which, if we are fortunate, we may hope to see — what else? — the mind.

# PART TWO

# MIND

# CHAPTER 7

# Eleven Senses of Radiation
## and Feeling

I T IS SURPRISING that we so rarely feel lonely out here in space. In my case, it may be only because I am more than normally aware of being a family member of the universe. In others, it often seems something else, perhaps in a way like being on the stage. If that idea is puzzling, may I remind you that we here have completely relinquished the sheltering rondure of the globe, each point on which is hidden from every other point except its immediate neighbors. And our resulting loss of privacy makes this station a kind of celestial goldfish bowl, indeed a habitation that is in the direct line of sight of more and more people the higher we go. Furthermore, the station's resemblance to a stage is striking, as our eyes are dazzled by the unfiltered spotlight of the sun, the lesser glare of the moon and the footlights of planets and stars, while everywhere before us, below and above, yawns the black void of an unseen audience whose true gamut of feelings we can hardly begin to surmise.

If the challenge of our view from space is basically a challenge of mind, this first page of Part Two is the place to mention it. Here we leave body to start our investigation of mind, which, even more than body, is a criterion of life. Besides, as mind's interaction with body must come primarily through the senses, these next two chapters on sense will serve, I hope, as a fair introduction to mind.

In any case, while now and again some susceptible astronaut may experience something akin to stage fright on his debut here, they all get more or less used to it and most of them seem to enjoy the stimulus of the unearthly sensation. I call it a sensation rather than a sense but, come to think of it, I wouldn't be surprised if it eventually evolves into a new sense. Certainly it adds something to the remarkable spectrum of senses already known on Earth.

A lot of people seem to think there can be none but the five traditional senses of sight, hearing, smell, taste and touch. In a way they are right, I suppose, if you assume that only the ones most obvious to humans are to be included. But surely there are more senses in Heaven and Earth than you or I have dreamed of. And I have increasingly had the feeling that the time has come when someone should pioneer into the subject as a whole with a fresh, untrammeled outlook. So, out of more than idle curiosity, I've jotted down a list of all I could think of and it came to 48, not even counting the stage-in-space "sense" previously described. Then, by combining the most closely related ones, I trimmed the number to 32. Of course a lot depends on how one defines a sense, and on arbitrary choices, like whether you decide to lump the sense of warmth and coolness or the sense of dryness and dampness in with the sense of feeling, and whether you want to include the senses (or are they instincts?) that animals, plants and (conceivably) rocks have but most humans evidently don't.

Here is my list of the principal senses of all creatures:

## The Radiation Senses

1. Sight, which, I should think, would include seeing polarized light and seeing without eyes, such as the heliotropism or sun sense of plants.

2. The sense of awareness of one's own visibility or invisibility and the consequent competence to advertise or to camouflage via pigmentation control, luminescence, transparency, screening, behavior, etc.

3. Sensitivity to radiation other than visible light, including radio waves, x-rays, gamma rays, etc., but omitting most of the temperature and electromagnetic senses.

4. Temperature sense, including ability to insulate, hibernate, estivate, etc. This sense is known to have its own separate nerve networks.

5. Electromagnetic sense, which includes the ability to generate current (as in the electric eel), awareness of magnetic polarity (possessed by many insects) and a general sensitivity to electromagnetic fields.

## The Feeling Senses

6. Hearing, including sonar and the detection of infra- and ultrasonic frequencies beyond ears.

7. Awareness of pressure, particularly underground and underwater, as through the lateral line organ of fish, the earth tremor sense of burrowers, the barometric sense, etc.

8. Feel, particularly touch on the skin and the proprioceptive awareness of intra- and intermuscular motion, tickling, vibration sense (such as the spider feels), cognition of heartbeat, blood circulation, breathing, etc.

9. The sense of weight and balance.

10. Space or proximity sense.

11. Coriolis sense, or awareness of effects of the rotation of the earth.

## The Chemical Senses

12. Smell, with and beyond the nose.

13. Taste, with and beyond the tongue or mouth.

14. Appetite, hunger and the urge to hunt, kill or otherwise obtain food.

15. Humidity sense, including thirst, evaporation control and. the acumen to find water or evade a flood.

## The Mental Senses

16. Pain: external, internal, mental or spiritual distress, or any combination of these, including the impulse and capacity to weep.

17. The sense of fear, the dread of injury or death, of attack by vicious enemies, of suffocation, falling, bleeding, disease and other dangers.

18. The procreative urge, which includes sex awareness, courting (perhaps involving love), mating, nesting, brooding, parturition, maternity, paternity and raising the young.

19. The sense of play, sport, humor, pleasure and laughter.

20. Time sense and, most specifically, the so-called biological clock.

21. Navigation sense, including the detailed awareness of land- and       179

seascapes, of the positions of sun, moon and stars, of time, of electromagnetic fields, proximity to objects, probably Coriolis and other sensitivities still undefined.

22. Domineering and territorial sense, including the capacity to repel, intimidate or exploit other creatures by fighting, predation, parasitism, domestication or slavery.

23. Colonizing sense, including the receptive awareness of one's fellow creatures, of parasites, slaves, hosts, symbionts and congregating with them, sometimes to the degree of being absorbed into a superorganism.

24. Horticultural sense and the ability to cultivate crops, as is done by ants who grow fungus, or by fungus that farms algae (page 72).

25. Language and articulation sense, used to express feelings and convey information in every medium from the bees' dance to human literature.

26. Reasoning, including memory and the capacity for logic and science.

27. Intuition or subconscious deduction.

28. Esthetic sense, including creativity and appreciation of music, literature, drama, of graphic and other arts.

29. Psychic capacity, such as foreknowledge, clairvoyance, clairaudience, psychokinesis, astral projection and possibly certain animal instincts and plant sensitivities (page 308).

30. Hypnotic power: the capacity to hypnotize other creatures.

31. Relaxation and sleep, including dreaming, meditation, brainwave awareness and other less-than-conscious states of mind like pupation, which involves cocoon building, metamorphoses and, from some viewpoints, dying.

## The Spiritual Sense

32. Spiritual sense, including conscience, capacity for sublime love, ecstasy, a sense of sin, profound sorrow, sacrifice and, in rare cases, cosmic consciousness (page 591).

To sum up, I've grouped these senses into five categories. First, the radiation senses (1–5), which include not only vision and electromagnetism but temperature awareness because that comes from infrared radiation. Second, the feeling senses (6–11), including not only touch and pressure perception but hearing because that means "feeling" the pressure of successive sound waves vibrating in the ear. Third, the
chemical senses (12–15), which encompass not only smell and taste

but hunger and thirst because these are largely controlled by the chemistry of the body. Fourth, the mental senses (16–31), which appear to develop mostly in the mind, among which are included a few like the urge for procreation and the navigation instinct which generally involve several other senses, not to mention having variations roughly in proportion to the number of species using them. And finally the spiritual sense (32), which is presumably newly evolved and appreciably developed only in man.

It is perhaps more than we should have done even to try to enumerate these thirty-two senses — for obviously they overlap a good deal, are often controversial, arbitrary, ambiguous and vague, if not ill defined. And many of them could, with fair justification, be termed instincts or capacities rather than senses. But it seems to me useful to list them, if only to demonstrate the complex nature of the broad subject of sense and whatever other channels there may be by which the mind relates to the body.

Since all these senses evidently arose during the past few billion years of earthly evolution, it seems expectable that more will evolve in the earth's next few billion years — and indeed that some must be in process of evolving now, presumably the kind often referred to as occult or "sixth" senses, because they are mysterious, immeasurable or found only in a few rare or divinely endowed individuals. All the senses, I would think, have passive and active aspects. I mention the passive first because that is usually the better known. It is the reception by eyes, ears, noses, etc., of radiation, of vibrations, of particles sent forth from elsewhere. But the sources of these waves, emanations or messengers and the source viewpoints are vital factors to the senses too of course, and they are the active end of the sense exchange, which often amounts to a throw-and-catch feedback.

The chameleon's eye, for instance, is not only actively bombarded by photons of light out of the sun and by the glances of other eyes upon it from outside, but it possesses two kinds of passive vision through one of which it sees the world around it (via those same photons) and, through the other, its own body, in effect visualizing what is seen by those same outside eyes that look upon it, then influencing its body to shift the color pigment of its skin accordingly. This is proved by the fact that, if you blindfold a chameleon, he can no longer camouflage himself by matching a changing background. The songbird likewise has a sense of singing as well as of listening, the two being intimately attuned. And the skunk possesses a perceptive sense of broadcasting scent as well as of smelling it, even though these involve opposite ends of his body. Sonar and radar are only two 181 out of many sense systems that must send before they can receive.

Next come the vicarious senses, the percepts borrowed or handed on from others, such as a watchdog whose nose sniffs an approaching stranger so his master's ears will hear the warning bark and his master's eyes look out for whoever is coming. And these include the tempo of the cricket's chirp, the droop angle of the rhododendron's leaves and the space-density of animals in a flock that tell you the temperature like a living thermometer.

Then there are the artificial senses, starting with the stick, the shovel, and calendar, the clock, the weathervane, the abacus, the lightning rod, the telescope and the smoke signal, and reaching to the camera, the speedometer, the telephone, the taped color TV program, the pacemaker, bionics and the computer. We don't count these vicarious and artificial systems in our main roster of senses, however, because they are uncountably numerous, particularly the artificial ones, evolving and multiplying constantly.

## SIGHT

How far science has taken us can be suggested by the surmise of Empedokles in the fifth century B.C. that "perception is chiefly in the blood, especially near the heart, our organ of consciousness. For we think mainly with our blood in which, of all parts of the body, the elements are most completely mingled." Other ancient philosophers wondered whether a bell might be consumed by frequent ringing or musk all used up on a journey because of perfuming a hundred miles of countryside. And sight, according to a theory still prevalent in the seventeenth century, was produced by light inside the eye flashing forth to illumine whatever was seen, like the glance cast by a man at an attractive woman, which seems to involve something moving actively to her, something positive she can feel. But the man's active glance is now understood to be positive mainly in a mental or spiritual sense, while the most significant material involved (a beam of photons) originates in the sun or a lamp and is reflected primarily from the woman into the man's eye rather than the other way. Indeed, as some people who put on dark glasses to aid their vision seem not yet to have learned, the eye is a two-way organ, a "window of the soul" that actually may serve its owner better by being looked *into* than *out of*.

Yes, the eye is the prime achievement of sensory evolution on Earth, an ellipsoid of optical revolution, whose chief source of energy is 93 million miles away in the sun! There remains a good deal of mystery about how it evolved, but it has become probably the most important and widely used of all sense organs, its rudimentary forms being found among the "lenses" of leaves that are vital to photosynthesis and even among translucent and transparent minerals, where its significance is still but dimly understood.

The vision of simple forms of marine animals like the one-celled flagellate *Euglena* is similar to that of plants, consisting primarily of light-sensitive eyespots that are microscopic and may have evolved directly from chloroplasts, the organs of photosynthesis that aim the leaf toward the sun (which gives it energy too) or, should the glare get too hot, turn it toward the shade. Light sensitivity, however, may be a general attribute (or at least a potentiality) of all cells, for experiments show that a flash of light elicits a measurable response in the brain cells of insects, in the heart muscles of certain snails and in numerous other muscle cells, even some in mammalian skin.

When the ocular lens finally evolved, using various conglomerations of translucent cells, nature experimented rather lavishly, sometimes sprouting the lens on the body's outer surface (as in the scorpion), sometimes at the ends of stringy arms (as in the stylophthalmus),

STYLOPHTHALMUS

sometimes sinking it in a pit (as in the limpet), sometimes enveloping it completely in protoplasm (as in the snail), at the same time trying out all sorts of eyes of simple, ingenious, weird, radical and multiple function. Scallops, for example, developed as many as 200 eyespots (each with its lens and protoretina), which respond to darkness and are vital to these mollusks as they swim about by jet propulsion, an activity in which they can easily be caught and eaten should they fail to notice the occasional shadow made by an approaching predator. But the first camera-type eye to evolve seems to have been that of the pinhead-sized *Copilia*, probably the smallest animal capable of seeing actual images of things around it. It has a transparent body with two sets of double, microscopic lenses geared to L-shaped "retinas" that sweep to and fro like scanning radar dishes, flashing slow signals through optic nerves to a microscopic brain found able to reconstruct a complete image after each scan, which takes at least a fifth of a second and sometimes ten times that.

The next kind of eye to evolve on Earth seems to have been the cluster of eyespots that became the compound eye, common to a few marine animals and many insects. The supreme example is the compound eye of the male horsefly which arrayed about 7000 lenses in crystalline rows like a microscopic honeycomb. Unlike human eye lenses, these are rigidly fixed in focus and therefore rarely form sharp images or distinguish a bee from a pebble, but they register the movement of any visible object passing from lens to lens with such efficiency that a fly may accurately judge the speed of anything from the minute hand on a watch to a swooping bird or a lashing tail, a perception which often enables it to escape in time. This also explains why honeybees are particularly attracted to flowers swaying across their line of sight.

In the seemingly endless course of earthly evolution, eyes have produced bewilderingly varied answers to the visual needs of their owners, among which a number of general solutions have emerged. A swift animal moving about a large or complex environ, for instance, is apt to have big eyes (relative to body size) for wide, sweeping vision: the squirrel, the dragonfly, the eagle. A nocturnal creature has even bigger light-receptive ones, almost absurdly conspicuous, as in the owl and the tarsier. But an underground animal, an internal parasite or a cave dweller, has small eyes or none at all: the mole, the worm. And various water animals have periscopic ones: the hippo, the frog, the fiddler crab or the stylophthalmus larva who waves his about like arms.

As to specialized optical systems, predatory animals who pursue and catch elusive prey of course require keen forward vision, with at least two eyes coordinating stereoscopically, while the quarry, habitually in danger from unexpected directions, naturally have bulging eyes on both sides of their heads to ensure 360° vision. Which is why the prowling cat and the owl look straight ahead and the timid rabbit and the deer see sideways and all around, even backward. The owl in fact has eyes fixed in their sockets like headlights, so he has to turn his whole head to shift his gaze, the doing of which for millions of years has evolved a neck so flexible it can swivel in a tenth of a second more than a full circle: approximately 400°. And most game animals, to whom it is often more vital to know who is behind than before them, can see their own tails and trails without turning their heads. The much-hunted woodcock literally has his eyes in the back of his head (just abaft the center line) so that, when he is probing the ground in front of him for grubs with his long, flexible bill (where sight is hardly needed), he has excellent binocular rear vision in precisely the direction from which a hawk or a fox is most apt to approach. Something similar is true of the horse, who never has to turn his head to see behind him, as any Spanish *rejoneador* (bullfighter on horseback) can tell you. Each eye of a horse sees 215° (probably a wider arc than any other known single eye), which gives him better than a 70° binocular field forward without any real blind spot behind.

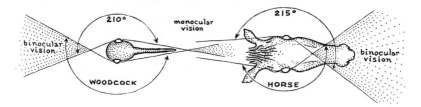

The pupil in a horse's eye, incidentally, is horizontal, nicely attuning his day vision to the shape of his preferred landscape in the same way a cat's vertical pupil aligns itself to the natural feline habitat: a tree. The shape of other pupils can be fantastic. The gecko's looks like a string of four diamonds, the skate's a fan-shaped Venetian blind, the fire-bellied toad an opening like a piece of pie, the armored catfish a horseshoe, the penguin a star that tightens into a square, the green whip snake (whose ancestors may have slithered in the Garden of Eden) an appropriate keyhole, and others resemble teardrops, bullets, buns, crescent moons, hearts, hourglasses, boomerangs.

The angles of pupils have a sort of polarizing effect on the light entering the eye. This brings up the subject of polarized light, or light allowed to vibrate in only one plane. I mention it because polarized vision (awareness of light's polarity) is now known to prevail among most animals who navigate by sunlight and who have the obvious need to steer their course at a precise angle to the sun, whether or not it is hidden behind clouds, something only perception of the angle of polarity of sunlight could enable them to do. It is not generally known that man too can see polarized light and that the average human may see it without help from filters or other instruments. But this is demonstrably true. In fact anyone with normal eyes who stares upward for several minutes into clear twilight should gradually become aware

185

of the shy and seemingly mystical retinal image known in optics as Haidinger's brush, a faint yellowish hourglass-shaped figure 4° long, squeezed at the waist between a pair of blue "clouds" and pointing exactly toward the sun.

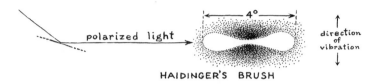

HAIDINGER'S BRUSH

Now going on to the very keenest natural vision among known creatures, we shall turn to the birds, particularly to such as hawks and kingfishers, who have evolved two foveas in each eye. The fovea is a special spot in the retina that's densely packed with cone-shaped cells (sensitive to daylight or its equivalent), where acuity is greatest, such as the image area of the two or three words on this page that you read with each little shift of your gaze. The reason hawks can see a mouse almost a mile away, or kingfishers a fish deep under the waves, is not that their eyes are bigger or more telescopic than yours or mine but because they have nearly eight times as many cells in their retinas, especially in the sensitive foveal areas which are packed with some 1,500,000 cone cells (as against your 200,000). The two foveas in each eye comprise a single lateral fovea for sharp monocular vision over a wide field to the side that's not seen by the other eye and a compound forward fovea for still sharper binocular vision straight ahead in the narrower stereoscopic field seen by both eyes. One eye or the other (not both) normally first notices the mouse or fish and, using its lateral fovea, concentrates on it during the downward swoop. Meanwhile the head turns and the other eye progressively converges forward until the compound fovea (compounded of both eyes) eventually takes over (binocularly) for closing in.

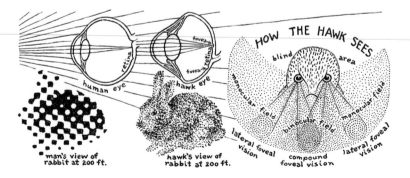

While daytime sight thus uses cone cells in the retina, night sight uses the 100,000 times more sensitive, long, rod-shaped receptors known as rod cells, which may respond to as little as one photon of light. These extraordinary cells are thought to have evolved in primordial fish during their prolonged struggles to go deeper, where it is safer and darker but where they could only see by gradually extending the detectable daylight with rod vision. Some 96 percent of all the kinds of fish who went down so deep that even the best rod cells could see no day, however, eventually evolved luminescence so they could at least see each other and recognize their fellow species by varied systems of colored lights and reflections, incidentally creating some fascinating problems in deception, camouflage, advertising, courtship and predation. But the other 4 percent of deep fish gave up vision altogether, including, surprisingly, a few species that evolved luminescence notwithstanding. Which would seem to confirm the discovery that it can be more important to be seen than to see, this being the reverse side of vision to which we will return presently.

Another aspect of vision is the functioning of the visual center in the brain, which learns to see through the actual visual experiences of the eye and retina. Evidence of this showed up in a classic experiment on kittens, in which they were allowed to see only vertical lines until they grew into cats. Then, exposed to horizontal lines for the first time, they could not see them at all and, since their fully grown brains had never developed pathways for horizontal vision, they were permanently horizontal blind!

You may recall that the eye, like other body parts, can be made out of any sort of flesh in the creature that grows it (page 153), but an eye can also be adaptable enough to migrate upon or through the body, even drift many times its own diameter on the initiative of its owner, without surgical or other outside help. I am thinking of nearly five hundred species of flat fish, most of which swim in a fish's normal vertical attitude when young with an eye looking out of each side of their heads. But fish of this order, in growing up, progressively take to resting on the sea bottom on one side, whereupon the down eye, finding itself blinded by mud and undoubtedly yearning for sight, begins a curious creeping migration up and around the ridge of the head (or, in some species, through it) until it arrives in a few weeks on the upper side, not only regaining but actually improving on its juvenile binocular conjunction. The mouth, incidentally, also twists upward. And in some species of flounders, ichthyologists report, the fish has a choice as to whether to lie on its right or left side, and whichever eye it finally decides to turn downward is perforce the eye that will migrate to the up side (with equal facility) almost, one might say, with the freedom of an independent organism!

If you can swallow such an eye on a fish, what about an eye that can help a frog to swallow? Strange as it may seem, the West Indian tree frog has one. When he stuffs his mouth full of anything, which can include his own offspring, he retracts his bulbous eyeballs into his throat to push it down. Before doing so, he naturally has a good look around to be sure the coast is clear, because of course this way of swallowing temporarily blinds him, a loss which ironically has been known to induce him to throw up something delicious just for the sake of a quick gander at what might be, but usually isn't, going on.

If you have always thought of an eye as a positive, material organ, you may be interested to know about a negative, abstract eye that is an evolutionary success of long standing. I am thinking of the eye of the chambered nautilus, which consists of a retina and optic nerve in the back of a hollow socket that opens upon the outside oceanic world by means of a small vacuous pupil through which seawater flows. As this mollusk has no solid eyeball, no cornea, no iris and no lens, the pupil hole admits both water and the light needed to cast an image on the retina by the abstract principle of the pinhole camera, the ocean itself in effect serving as the nautilus's eyeball.

Then there is the arrowworm, a kind of microscopic swimming eye in the ocean, who, being transparent, sees with the help of his entire body, which refracts light and serves as a living, breathing cornea and lens. The core of his vision though is the combination of his two localized eyes, each sectioned into five sub-eyes, each of which has its own lens and retina and its own direction for looking (forward, backward, sideways, up and down), the down sub-eyes being the ones that most use the rest of the body to look through.

While it is hard to visualize this specialized five-way outlook of the arrowworm, a good deal is known of the more general and shallower view of the common and familiar fish of ponds and streams. For, knowing the reflectivity and the refraction index of water, we can deduce that the fish-eye view of this world is something like looking upward through a round hole in a horizontal mirror, the mirror being the underside of the water's surface (where air meets water) which reflects the bottom everywhere except directly overhead. Thus the sun, moon and stars, seen through that round hole, appear to rise and set in fish heaven at an angle of only 49° from the zenith (straight up) instead of 90° as with us, their rays being bent as much as 41° on entering the water — so the fish can see the fisherman on the bank of a clear stream better and larger from the bottom than from near the surface. Indeed his porthole of upward sight is wider when looking

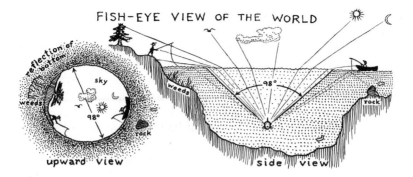

FISH-EYE VIEW OF THE WORLD

from there and it is said that the water's refraction has a magnifying effect — while all around the hole he sees only underwater objects and their upper reflections.

## OPTICAL ILLUSIONS

Fish probably share some of their illusions with humans, however, along with other creatures of vision. Consider the familiar bloated look of the rising or setting full moon, which affects any experienced eye viewing it from near the earth's surface. Although this remarkable illusion is known to have provoked theories throughout history, including one by Ptolemy (now accepted as true), Leonardo (off the mark) and others, most humans still seem puzzled by it. For this reason I am emboldened to explain that the true cause of the illusion is the natural capacity of the mind to compensate for the fact that all objects look smaller as you go away from them. Because you daily see familiar things like furniture, people, cars, birds, ships and airplanes diminishing as they go away or expanding as they come toward you, naturally the visual center of your brain makes allowance for your well-proven conviction that these things intrinsically stay the same size. So, in effect, they appear to you to maintain their constancy regardless of distance. This is true, among other things of course, of an object in the sky, like a balloon ten feet in diameter drifting a thousand feet above your head (subtending an angle of half a degree at your eye) ⊲———— ⬤. For although such a balloon will diminish to a dot (of less than a hundredth of a degree) as it moves toward the distant horizon, you feel little doubt that its diameter remains ten feet. Yet if, by some miracle, the balloon, instead of diminishing to a dot, surprised you by maintaining its visual fullness

189

(of half a degree) at the horizon, you would know it must have swollen enormously.

Now consider the moon, which after all seems not so different from the balloon when passing overhead for she subtends the same angle of half a degree and the average person (according to surveys) subconsciously thinks of her as only a few hundred feet up and moving across what appears to be the shallow but wide vault of the sky. Yet the moon does not behave like a normal balloon because she does not dwindle when approaching the far horizon. Instead she behaves just like the miracle balloon that somehow failed to diminish with distance, so your conditioned mind tends to assure you that the moon also must be swollen to look so big while yet so remote. In this case of course your mind, despite its habit of compensating, should not (under reason) do so since the moon's overhead distance (about 240,000 miles and far too much for the mind to grasp) remains practically the same at the horizon, holding its subtended angle within two percent of constancy. And you tend to forget that the moon is not just twice or ten times farther away than the hills on the horizon but actually 25,000 times farther away. So, out of habit and an eminently natural mis-

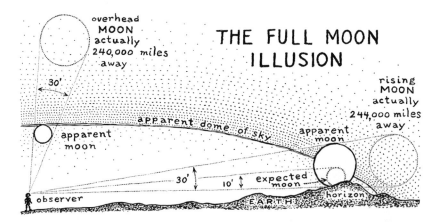

judgment of the apparent dome of the sky, the experienced mind (of man or animal) under such circumstances cannot help but compensate. In fact the mind clearly *over*compensates by bloating the moon to three times her actual size, somewhat as a drug addict's nerves (having learned to compensate for heroin) *over*compensate agonizingly when the heroin is withheld.

Other illusions are now understood to work for similar reasons, the mind unreasoningly compensating (often overcompensating) according to its long-accustomed perspectives, as you see in these drawings. Lines converging upward, for example, are apt to be interpreted by the conditioned brain as parallel and receding like railroad rails so that a line drawn crosswise between them looks longer when higher because the brain automatically accepts it (like the moon on the horizon) as farther off and reaching almost from "rail" to "rail" instead of less than halfway as in the case of a lower (apparently nearer) line. Indeed it is a cardinal rule of illusion that when two objects are optically equal in size (subtending equal angles at the observer's eye), the one that appears to be farther away will thereby seem proportionately larger.

Also, because we move mostly through horizontal landscapes, vertical lines make a greater visual impact on us and, in consequence, generally appear longer than equal horizontal lines — which explains this famous top hat illusion, where the crown seems to rise almost a third taller than the equal width of the brim.

Wavy lines resembling hair also give the illusion of movement, and a white lattice seen against a black background elicits little gray ghosts at the intersections where brightness (induced by dark contrast) is minimal — all such impressions presumably arising from what an engineer might call information-processing mechanisms in the brain that, under more usual circumstances, make the visible world easier to comprehend. The fact that one has a self and a local point of view of course accounts for many common illusions: the house rolling rearward as you gallop past it on your horse, the tree swinging into the road, the sun setting, the moon sailing through branches, the tower falling past the clouds . . .

Illusions occur in all the senses more or less, and, since all senses (like all things) are related, they often overlap or merge into one another. You may have noticed that, when you heft two objects equal in weight but different in size, the smaller one always seems the heavier because you expected it to be lighter. It is a case of errant subconscious compensation. Or, in homelier terms, when pints presume to speak in pounds, their language is liable to be misunderstood.

Another significant type of illusion was devised in 1832 by L. A. Necker, a Swiss naturalist, who drew a transparent cube on a piece of paper and noticed that its perspective had a strange way of reversing itself, as the face that seemed to be in front would suddenly flip to the back in exchange for the back face's seeming to flip to the front. Although a shift in eye fixation or a mental effort could induce this exchange in perception, Necker discovered it would also happen quite spontaneously every few seconds. And he designed a pair of weird "boxes" that, as you can see, flip even faster. Evidently the fact that

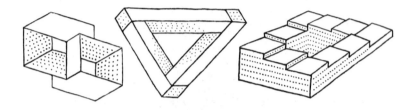

such a two-dimensional image representing a three-dimensional object is ambiguous and offers two possible interpretations as to what it represents makes the brain search for the true answer to the paradox by trying out the hypotheses alternately and never conclusively settling on either. And something similar occurs in the case of other so-called impossible objects, like the accompanying ambiguous triangle and endless stair on which you can mentally walk up clockwise or down counterclockwise, round and round, forever without getting anywhere.

## THE VISIBILITY-INVISIBILITY SENSE

Naturally illusions have been a factor in evolution by helping creatures deceive their prey or their predators, and so they introduce us to the passive side of the sense of sight, which takes two almost opposite forms: (1) the introverted sense of self-visibility with which one may *in*crease one's conspicuousness, like a peacock advertising his masculinity to woo a mate or a cat hunching his back to unnerve a dog. Or (2) the reciprocal sense of self-*in*visibility with which one may *de*crease one's conspicuousness through camouflage in uncounted ways, some of them so fantastic that few humans are yet aware of them.

Most obvious among forms of camouflage is the relatively simple principle of concealment by hiding that includes tactics such as the smokescreen of the squid who, when attacked, spews out either a black squid-sized cloud near the sea's surface or a blindingly luminous trail in the inky depths. A more subtle way is to make your body match its background by some such method as shifting skin pigments as the chameleon does. Or, in the case of some leaf-eating insects, having a transparent body, so that the eaten leaves inside you show through, automatically keeping you the same color as the similar uneaten leaves you are living on. And at least one ambitious caterpillar in South America, apparently unwilling to conform to his surroundings, contrives to make his surroundings conform to him — achieving this "miracle" by carving his own image in every leaf he eats, carefully forgoing just enough of certain parts of the leaf so that, after he has been feeding a few hours in a tree, it has begun to look like a tree full of caterpillars — a stratagem understandably frustrating to the hungry birds.

Slightly different is disruptive camouflage, in which concealment depends on breaking up an animal's outline with contrasting patches or stripes, like those on a zebra which could disguise him slightly if they ran lengthwise along his body but do a lot better by cutting across it. Or if you wonder how a tropical frog wearing a gaudy yellow stripe across his back from right cheek to left leg can afford to attract such attention to himself, you are overlooking the fact that the yellow stripe not only bears no resemblance to the real frog but looks from a little distance like a dry twig or a wilted blade of grass and, in the frog's natural setting, helps to conceal him.

Obliterative shading is another technique in which the normal contrast between sunlit upper surfaces and shadowy underparts of creatures is canceled by darkening their backs and lightening their bellies (in some deep-sea fish with luminescence) to camouflage them. And shadow elimination is an aspect of it by which a butterfly, say, rests with his wings together and aligned exactly edge-on to the sun so they cast no telltale shadow. These methods are also used by human camoufleurs to countershade big guns, rocket launchers, etc., eliminating their shadows with netting and even enticing bombers to drop their bombs off target by erecting huge sheets of composition board cut to cast shadows like those of important buildings designated for destruction but actually disguised and a safe distance away.

The eye is a conspicuous organ, not only inclined to be big, round and shiny, but it functions best when fairly exposed to view, so naturally it has provoked its own rather special camouflage adaptations

193

during evolution. Some vipers, for instance, hide the eye by having an iris that matches the rest of them, while the vital pupil remains (in daylight) only a tiny slit. Certain reptiles have transparent but lacy eyelids through which they can keep a semiwatch with restricted, inconspicuous vision until they are spotted. Many creatures disguise the eye with a dark stripe of nearly the iris color and a few, like the butterflies, divert attention with big scary fake eyes on their wings. There is also the stripe-eyed butterfly fish that, like a butterfly, has two perfect eye spots near his tail, giving the impression that that is his head end, an effect heightened by his habit of swimming slowly backward — the performance really paying off when some marauder tries to head him off by lunging at what seems a vital spot, only to be foiled when he unexpectedly darts away in the opposite direction.

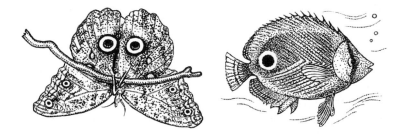

A good deal of animal camouflage is thus mixed with advertising, making use of the pickpocket's notorious distraction principle of having an accomplice bump provocatively into the victim's opposite side while he himself is gently lifting his wallet, a dodge that could have been suggested by the lizard's detachable tail that separates neatly when the animal is closely pursued, duping the voracious pursuer by continuing to writhe and slither in one direction while the rest of the lizard escapes in another to start growing a new tail. In some cases similar display-camouflage is used to lure prey to the business end of a predatory animal, say an angler fish with its luminous bait (pages 30-31).

In other cases the disguise may advertise the genuine inedibility of an animal known to predators as nasty tasting, or it may seek safety through pretended inedibility by mimicking such a creature. One caterpillar, to avoid being eaten, goes so far as to look as if he's already been eaten and disgorged — you know: that thrown-up look — while a certain Indian mantis pretends to be newly squashed and therefore not worth eating. The hunted animal may even flash a warning to his pursuer to look out for a really dangerous defense weapon, like the terrible-looking black dorsal fin of the tasty weever fish with its two poisonous spikes, the mere sight of which spares both him and most of his opponents a lot of unpleasantness. And of course there is no end to sham warnings like those of the bluffing South American fulgorid, a large insect with false teeth and fake eyes in a preposterous yet often successful attempt to palm himself off as the head of a ferocious young alligator.

Surely the most extraordinary of all such make-believery, however, is the construct of a Malaysian spider faced with the tantalizing double problem of avoiding being eaten by birds while attracting flies to his web — a defense-offense challenge he manages to meet with a single dramatic act that I can only appraise as exhibiting practical but divine imagination. The spider's black and white body, you see, is designed to look exactly like the solid lump of a bird's dropping, and he spins his whitish web with flat lobes extending in several directions upon some large leaf to resemble the splash so often associated with this kind of excrement, the effect being augmented by the faint ammoniac odor of the spider's own exudations. And the act is surprisingly successful, for the last thing a bird wants to eat is one of his own droppings, while it is hard to think of anything more attractive to the flies as they buzz toward the web in eager swarms.

Comparing visibilities correctly of course takes an artist's eye for color, and resistance to the common tendency to think of snow as white and grass as green, when both may actually be reflecting subtle shades of blue or lavender. And allowance is needed for the differing values of colors as seen by different eyes. The ancient Greeks, for instance, had no word for blue, while the Natchez Indians failed to distinguish between green and yellow, and the Choctaws between green and blue. Most fish and birds, however, can see red and some greens, but probably little if any blue, while bees see ultraviolet but not red, and green to them may look yellow. Hardly any mammals except man and his close relatives see color of any sort, yet the bee evidently notices contrasting hues even within the narrow waveband of what appears to us as white, and, as for night creatures, they naturally specialize in the dark-penetrating frequencies called infrared.

Colors also influence minds and moods, probably more than is generally recognized, and color engineers say their research shows that violet in clothes or home decoration induces melancholy, while yellow is an energizing color that stimulates thought, conviviality and optimism. Contrary to tradition, blue does not really give you "the blues" but rather relaxes you. In fact old people can become "blue thirsty," presumably because their yellowing eye fluids filter out blue light, 195 which seems to make some of them crave blue, like a nostalgia for carefree youth. Color produces its own illusions, making a red house appear some five percent closer than an equidistant blue house, or a pale yellow suitcase pounds (as well as shades) lighter than a dark blue one of the same weight. And while customers in a bar painted red seem to get thirstier and buy more drinks than people in any cooler-colored bar, an appeal for charity mailed in a light blue-green envelope has been found to bring a consistently more generous response than the same appeal in a white envelope. There is even a suggestion that this sort of color control may have been a factor in football history, for the famous Knute Rockne is reported to have regularly kept his Notre Dame team keyed up in a red locker room while consigning visiting rivals to one tinted a soporific blue.

## RADIATION SENSE

Now we must consider the parts of the radiation spectrum beyond visible light, which means radiation of wavelengths shorter than violet on one hand and longer than red on the other. The waves just beyond violet are called ultraviolet, and the first creature I can think of with senses in this range is the bee, who sees ultravioletly very well and is known to respond to the bright petals of the many flowers that flash rings of this radiation so alluringly they are virtually impossible for him to resist. If a bee were on the moon, which obviously could happen in this age of space travel, his view of the crescent Earth, in case you're interested, might look something like this ultraviolet picture of our planet taken by his human cousin on the moon in 1972.

the crescent Earth as she might look to the ultra-violet-sensitive eye of a bee on the moon

When it comes to the radiation of waves shorter than ultraviolet, like x- and gamma rays, or longer than infrared, like radar and the broadcasting bands of radio and television, few people seem to realize how naturally most earthly creatures sense them. Indeed there appears to be a general impression that these waves were invented by man and brought into the world only in the last hundred years. But the fact is that, far from being *invented* by man, these reaches of radiation were only *discovered* by him, for they are actually timeless parts of the great radiation spectrum that is a fundamental aspect of the universe. It is a spectrum, by the way, in which color (light wave frequency) corresponds to pitch (sound wave frequency) and all of its seventy known octaves (69 of them invisible) are beautifully analogous to music.

A case in point was the research done in the U.S. Naval Radiological Defense Laboratory in San Francisco in the 1960s in which sleeping rats were exposed to 250,000-volt x-rays. This sort of radiation had always been regarded as silent, invisible and unfeelable, but every time it was beamed on the rats, their heartbeats began to accelerate and within fifteen seconds they woke up, sometimes in a state of alarm. Again, when the animals were offered their favorite drink, sweetened water, while exposed to radiation, they couldn't swallow it. Even humans, it was discovered, can sense short-wave radiation when their eyes are adapted to darkness, for night vision somehow sensitizes their retinal rod cells to the point where they see both x-rays and gamma rays as a yellowish-green glow, accompanied sometimes by a tingling or burning sensation. And in the rare cases of runaway nuclear reactors when someone is subjected to massive lethal radiation, he usually tells of seeing a "vivid blue flash," even in sunlight, which unhappily amounts to an irreprievable sentence of death only hours away.

Long-wave radiation is much gentler, but has nevertheless yielded evidence of being naturally perceptible under certain circumstances. The New England Institute for Medical Research in Ridgefield, Connecticut, for instance, found in 1958 that pulsed radio waves (5 to 40 megacycles) could regiment many organic substances such as carbon, starch, red blood cells and even organisms like amebas, euglenas and paramecia which tend to align themselves and swim along the invisible lines of force of the radiating field, sometimes practically dancing to music. Although humans appear to need the aid of some form of metal or crystal in order to tune in to radio or TV waves, this doesn't always require much of a receiver, to judge by the innumerable reported cases of people who hear music or messages coming from tooth fillings, bridgework, bobby pins and steam radiators. A man in New Jersey with a mouthful of dental work wrote, "I've been getting Station WOR regularly." And inasmuch as bones, muscles and other parts of the body are crystalline in structure, to say nothing of trees and rocks, who knows what or who will be heard from next?

## Temperature Sense

Awareness of temperature is commonly considered part of the sense of feeling because one feels it. But it is really more logical to classify it as a radiation sense, akin to vision, because it is a tuning in on radiation — in this case infrared radiation, commonly called heat. That is why I turn to the temperature sense now, following most of the other radiation senses. It is a vital sense too, often protecting an organism against dangerous overheating or freezing. And it makes possible man's extraordinary range of heat tolerance, through which, I am told, naked airmen tested in dry air in 1960 set a world record by withstanding temperatures up to 400°F. and heavily clothed men to more than 500°F., while in the steamy atmosphere of a sauna bath some described 284°F. as "quite bearable" despite its being 72° above the boiling point of water.

The difference between sauna steam and boiling water of course is that the former is gaseous, the latter liquid. And, as any fish knows, a liquid is a very concentrated and penetrating medium to be in — certainly as compared with air or ordinary steam. It accounts for the fish's temperature sensitivity, especially since he is a "cold blooded" animal without the thermostat system of mammals and birds, who speed up their glandular and muscular activity (sometimes to the point of shivering) when their blood starts to cool, but sweat or pant to cool themselves by evaporation when their blood heats up. Lacking internal means of stabilizing body temperature, the poor fish must try to swim into water that has the temperature he wants, which means he usually heads down to cool down or up to warm up — often with only a few degrees of leeway between getting paralyzed with the cold or dying of the heat. The deadly seriousness of this problem to a fish was dramatically demonstrated at Harvard a few years ago when some goldfish were provided with a special and very delicate valve that could be pushed open by a fish, whereupon it would squirt cold water for a second into their bowl, lowering the temperature about half a degree. After a little training, most of the fish learned what the valve was for and how to use it. Then whenever their bowl got uncomfortably hot (above 96°F.), they would work the valve to cool it, and, when it was heated to 106° (which would kill a goldfish in a few minutes) they worked frantically, pressing the valve continuously and keeping at it until the water got back to their optimum 96°.

A few creatures living in air are about equally heat-sensitive in their individual ways. A dog's internal temperature is known, for example, to rise as much as 6° at the sight or smell of an approaching strange dog or human. And humans, using biofeedback and "thinking warm" (presumably to expand their outer capillaries), have raised the skin temperature of their hands as much as 15°F. If a dog (or human) sheds a few fleas and lice at such a time, it is presumably because fleas and lice are very sensitive to heat. In fact in some primitive societies lice living in one's hair are considered a prime proof of good health because such creatures have been observed to abandon anyone falling ill with fever. The great bubonic plagues of medieval times are now attributed to infected heat-sensitive fleas fleeing from feverish rats or humans, since, by doing so, they obviously advanced the spread of the disease.

The rattlesnake, in common with all pit vipers, seems to be a leading candidate for the championship in heat measurement (actually infrared vision) with his extremely precise thermal organ in each of two pits located between the nostril and eye on either side of his head, neatly focused to converge and overlap at striking distance a foot or so in front of him. Three thousandths of a degree Fahrenheit has been found to be the threshold acuity needed to alert the snake to any significant living presence in his neighborhood, to which he responds by turning immediately toward a warm-blooded potential prey or enemy, using his conical fields of heat vision to "see" the size, shape, motion and range of his adversary.

In humans the senses of heat and cold are less acute than in most animals and are usually described as twin senses using differentiated sets of specialized nerve cells: a set of about 150,000 in the outer fiftieth of an inch of the skin to register cold and another of 16,000 deeper in the skin to register heat. While human skin fluctuates a good deal in temperature, averaging around 81°F., and doesn't seem to be thermostatically controlled like blood, a mysterious skin area has been discovered in the center of the forehead that registers temperature differences as small as one sixtieth of a degree. I haven't heard any anatomist associate it with the "pineal eye," an eyelike structure in the forebrain believed by some to be a vestigial sense organ or "third eye." But I notice that, in the case of hagfish, lampreys and a few reptiles, the counterpart of this lobe evidently evolved what have been described as "visual structural adaptation," and there is growing evidence that it produces at least a little actual vision.

Probably the most interesting thing about this comparison of infrared and visual seeing is that these brother senses caught the world's attention only recently with the discovery that the hypothalamus is not only the body's thirst center and central "eye" of temperature awareness, but that, significantly, the temperature eye and the visual eye both evolved from the same matrix at the bottom of the third ventrical of the brain, out of which the visual eye moved forward to view the outer world, while the temperature eye turned inward to monitor the blood, their relation becoming one of the most elegant instances of sense complementarity in all the kingdoms of life.

## ELECTROMAGNETIC SENSE

Every creature on Earth, including plants and probably rocks, seems to react to electricity one way or another and thus may be said to possess an electrical sense. Man certainly possesses it and, though the occasion seldom arises, he can feel static electricity accumulating in his body just before a lightning strike when he still may have the seconds needed to reach a safer location. A young couple who barely escaped death from lightning on a 10,400-foot peak in British Columbia in 1948 remembered afterward (in the words of the girl, Ann Strong) that "there was a slow, inevitable rhythm about it. After each strike we moved in silence for a while, with only the tearing wind and slashing rain. Then the rocks would begin a shrill humming, each on a slightly different note. The humming grew louder and louder. You could feel a charge building up in your body. Our hair stood on end. The charge increased, and the humming swelled, until everything reached an unbearable climax. Then the lightning would strike again — with a crack like a gigantic rifle shot. The strike broke the tension. For a while we would grope forward in silence. Then the humming would begin again . . ."

Some animals, notably about 500 species of fish, not only feel electricity but generate it in large enough quantities to use it either as a weapon or a sort of electric eye for "seeing" their surroundings and navigating waters too murky for optical vision. Most spectacular of these is the electric eel whose fifty-pound, eight-foot body may be stacked with half a million wafer-thin plates called electroplaques, hooked up in series to compose a living 1000-volt battery that can kill a horse. This has been known to happen in shallow South American rivers where horses ford and may step on an eel burrowing in the mud. Men have been killed too, in several cases drowning after becoming dazed and paralyzed from repeated shocks.

The Nile catfish and the giant electric ray are two other dangerous electric fish, but most of the species generate too little current to mount an effective weapon, evidently using it almost exclusively as a sense system, of which they have evolved dozens of varieties. One variety, investigated in detail in an eel-like African fish called *Gymnarchus*, produces a spherical electric field around the animal with the negative pole in the tail and positive pole in the head. Such a dipole field clearly does not work on the radar principle of sending out radio waves to bounce off objects and return with information as to distance and shape — because radio waves don't penetrate water. But any objects near the fish do distort the paths of electrons circulating around the field, inevitably converging them toward those that are good conductors, diverging them from bad conductors, the distortions being "seen" by the fish through its many porelike electric "eyes" that are distributed about its body, especially near the head, and connected to internal nerve centers that lead to the brain. Like a bat broadcasting sonar (page 206), *Gymnarchus* thus puts out its field of electric current in pulses (about 300 a second), giving every sign that it visualizes the distortions as the shape of the surrounding world — with understandable supersensitivity to anything electromagnetic out there like, say, an approaching electric fish, who just might turn out to be an enemy or a mate. And, speaking of mates, many fish have recently been discovered to woo electrically, each of several kinds broadcasting its

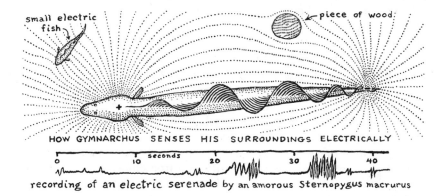

HOW GYMNARCHUS SENSES HIS SURROUNDINGS ELECTRICALLY

recording of an electric serenade by an amorous Sternopygus macrurus

own characteristic pattern of discharge, through which not only its species but its age and sex can be recognized. One researcher, Carl Hopkins in Guyana, using equipment that converts electric signals into sound, reported that when a typical female fish (*Sternopygus macrurus*) ready to breed entered the electrical field of a mature male, his "steady drone" changed abruptly to an electric serenade.

While the research is still far from conclusive, there are also indications that electric fish may navigate long distances by sensing the intensity and angles, if not the polarity, of the geomagnetic lines of force they almost continuously cross in the ocean. Here of course we are dealing with the magnetic aspect of the electromagnetic fields that enclose and permeate the earth, even the illusive emptiness between them, and which incidentally make it easy to accept the accumulating evidence that birds, fish, insects and other navigating animals must be influenced by magnetism, even as that lowly mineral called the lodestone. It may be enough in fact to cite the pioneering work of Frank Brown, zoologist at the Marine Biological Laboratory in Woods Hole, Massachusetts, who showed in 1960 that the mud snail ''could actually distinguish between different magnetic intensities and was also aware of the direction at which magnetic lines of force passed through its body.'' In any case, this breakthrough in biomagnetic research was followed by experiments on volvox, the planarian worm, paramecium, many species of insects from fruit flies to beetles, several kinds of birds, fish and man; and, in every case, the creatures tested showed reaction to magnetic forces. Planarian worms, for example, navigated along magnetic lines and repeatedly followed them within an angular tolerance of 15°. And human subjects in darkness could actually ''see'' magnetic fields, particularly a rapidly alternating field that would consistently register on the retina as a luminous glow, a type of nonoptical (closed eye) image known as a phosphene (page 237). There is still plenty of mystery about these magnetic senses, but at least a promising clue turned up recently when it was discovered that protons in hydrogen atoms all over the earth align themselves like compass needles parallel to the geomagnetic field. For hydrogen is abundant in living tissue and the solitary proton in the nucleus of each hydrogen atom, in effect a tiny spinning magnet a trillionth of a millimeter thick, just might somehow convey its bias to whatever organism it is in and so play the part of an infinitesimal compass.

Such a hypothesis, at any rate, is reasonably consistent with the facts that the electromagnetic polarity of a frog embryo (and presumably of other embryos) is irrevocably fixed before it starts to develop a skeleton or a shape, and that seedlings grow faster when their roots follow magnetic lines leading toward a natural or artificial south pole. All animals and vegetables are believed to generate electricity and put forth associated electromagnetic fields that are important in the workings of nerves, muscles, heart, brain and other organs. And Dr. Robert O. Becker, an orthopedic surgeon and researcher in New York's Upstate Medical Center, has said that ''subtle changes in the intensity of the geomagnetic field may affect the nervous system by altering the body's own electromagnetic field.''

In animals and man this field is found to be negatively charged at the forehead, positive at the back of the head and down the spine, gradually becoming negative again along the arms and legs. How important such body polarity can be is suggested by Becker's experiment in reversing it in a rat — which knocked the animal unconscious. He also found that electric fields can significantly help the healing of wounds and that the electrical potential of the human head, comparable to the electrical charge in a thundercloud, is directly related to the level of consciousness — so closely related in fact that anesthesiologists have begun to use the discovery in determining when their patients are ready for the surgeon's knife.

## HEARING

If the eye is the most notable organ for detecting radiation, many of whose sense variations we have been describing, the ear is hardly less important or complex as the outstanding organ of feeling — in this case feeling sound waves that mechanically vibrate the eardrum — and it may have an even longer evolution. At least its earliest discovered form seems to have been that of a simple balance indicator in primordial planktonic sea creatures who could not hear a sound. Known as a statocyst, it was a microscopic hollow cell containing an even smaller pebble of limestone, called a statolith, balanced on sensitive bristles so that, whenever the swimming creature got tilted, the pebble would roll toward the cell's down side, instantly triggering the down bristles' nerves and making them signal the animal how to shift back to an even keel. Such an organ hardly seems to have anything to do with hearing, yet, consistent with the interrelatedness of all senses, it evolved during hundreds of millions of years into the central gas bladder of modern fish which acts as a buoyancy balance or float to keep them at their accustomed depth in the sea. But it also vibrates when reached by sound waves, serving as an eardrum to convey hearable patterns through amplifying bones to the liquid channels of the inner ear, where they are transformed by hairs (more sensitive than the statocyst's bristles) into nerve impulses that go to the hearing center of the brain.

SIMPLE STATOCYST, ORGAN OF BALANCE, EVOLVED INTO FISH'S LATERAL LINE, ORGAN OF FEELING AND HEARING

This successful fish ear in some species can even be shifted into reverse, giving the fish a voice through muscular control of little drumstick bones that beat rhythmically on the bladder to "talk back" to whoever has addressed it. All fish and some amphibians in addition have an earlike organ along their sides known as the lateral line, through which they hear low-pitched sounds and feel water pressure fluctuations, including faint waves made by other fish — a sense indispensable to formation maneuvering in fish schools.

By the time birds evolved some 140 million years ago, hearing had taken another step forward with invention of the cochlea, an improved inner ear shaped like a spiral seashell and containing ducts of fluid separated by delicate membranes that vibrate and move around a kind of microscopic empyrean harp called the organ of Corti in which (as adapted to the human ear) some 23,500 hair cells somehow transform sound's mechanical motion into electric current that conveys it through about an equal number of fibers of the auditory nerve to the brain, where it is consciously heard. The organ of Corti presumably has thousands of resonators, each of which, like a harp string, responds only to one precisely pitched note, and there is evidence that it works on the piezoelectric principle (page 451). Certainly it is comparable to the retina, which transforms light waves from the eye into electrical optic nerve impulses and vision, but such organs are far too complex (not to say controversial) for detailed discussion in a book of this scope.

balance sense

OUTER EAR collects sound

ear drum

air

fluid

MIDDLE EAR amplifies it

INNER EAR transduces it

nerves

CROSS SECTION DETAIL of the ORGAN of CORTI

hair cells

shell-shaped COCHLEA is full of fluid

nerves to brain

HOW WE HEAR

Before the cochlea appeared, insects, living on land and in the air, evolved their own type of ear. Unlike the ears of fish, birds and man (all of which originated underwater), this bug ear had no fluid transmission of sound and therefore developed a simpler and more direct transition by running an auditory nerve from the eardrum directly to the brain. And it was this simple ear, I like to speculate, that heard the first messages ever uttered in the atmosphere of Earth. Three hundred and fifty million years ago it undoubtedly was much simpler than it is now, for it has evolved unceasingly, no doubt still increasing its sensitivity range in species after species, still trying out new body locations. The ears of moths and butterflies, for instance, are often in the base of their wings, mosquitoes hear with their antennae, and many kinds of insects have ears in their midsections, usually at the lowest point so they can detect the ground reverberating with the tread of terrestrial predators. Katydids, tree crickets and some grasshoppers, however, have slit-shaped ears just below their knees which are efficient for directional hearing because they can be widely separated and aimed in different directions, this being particularly important at mating time when males and females are out calling and sometimes work themselves up to the point of desperation to provoke positive and explicit responses.

Some ears, as you've probably noticed, serve more than one purpose like the gigantic ones of the African elephant and the dainty fennec, a fox of the Sahara (see chapter heading, page 177), which not only hear acutely but act as radiators to dissipate excessive body heat. This must be true also of the spotted bat of Mexico whose pink ears are as long as himself, not to mention the earlike petals and leaves of plants which may "hear" music (page 640), while they transpire moisture, refract and reflect light, attract pollinators and function in ways still unknown. Most aspects of vision have counterparts in hearing and other senses, you see, including aural camouflage and smell illusion.

The range of pitches heard by various kinds of ears varies widely, each animal family tending to evolve the range that will enable it to hear its predators or prey soon enough to escape or attack, and between times to converse and mate with its own species. Humans, incidentally, have a hearing range from about 20 to 20,000 vibrations a second, with maximum sensitivity at around 3000, which, significantly, is the pitch of a woman's scream! But hearing capacity is not constant, varying from man to man to woman to child and diminishing with age and masculinity for, although a young girl may hear bats twittering at an ultrasonic 25,000 cycles a second (as an acoustic engineer would put it), her mother can hear only a warbler singing at 15,000 and her aging father barely catch the top note of the piano at 4100. In fact, tests on men in their forties have shown that the upper limit of human hearing descends inexorably at an average 160 cps (cycles per second) for every year lived.

## THE SONAR OF BATS

The thought of inaudible, ultrasonic frequencies naturally brings to mind the echo-location technique called sonar that is used by so many high-pitched chirpers like bats, night birds and sea mammals. Bats, the most fantastic of these, have lived on Earth for 60 million years evolving some 1300 species from chickadee-sized ones to the "flying foxes" of Java with wingspreads exceeding five feet. Although they are mammals without feathers and cannot fly as fast as birds, they are champions of maneuverability — superior even to hovering chimney swifts and backward-flitting hummingbirds — for they can turn at right angles at full speed in little more than their own length.

Most of the time a hunting bat emits from ten to twelve chirps or beeps each second at a pitch averaging 50,000 cps, but when he detects a moth, say, five feet away, his output suddenly accelerates as he closes in on it, reaching what's called a buzz of some 200 beeps a second, which greatly increases his accuracy until he snags it — the whole pursuit and capture taking a scant third of a second with the bat's beeps (recorded on film) looking as you will see in the next illustration. Many bats thus catch gnats or moths when they're plentiful at a rate of one or two per second, which includes time for a reasonable percentage of misses.

Although the bat cannot possibly be conscious of it himself, his brain determines the moth's direction by comparing the echo's arrival time at one ear with that at the other — while the range is sensed in the lapse between each pulse out and its echo back, neither interval exceeding a thousandth of a second when the prey is close. It is probably not possible for a human to visualize accurately the kind of awareness a bat must have of the shape, size and motion of a flitting moth or of twigs, wires and other obstacles in his path, all sensed in darkness through his ears, but his sonar is certainly attuned to his specific needs, the distance between the sound waves about a tenth of an inch (just right for a small twig or a bug) and the aural images of these details all conveyed to his brain in electrical impulses of the auditory nerve (just as visual images are conveyed electrically by the optic nerve), combining into a sensation so complete it must be mentally equivalent to a visual impression.

Indeed the reason these remarkable creatures are not really "blind as a bat" (as tradition would have it) is not because they have eyes (which they all do) but because they have ears. For their ears give them what amounts to vision. In fact it has been proven experimentally that bats can fly with their eyes taped shut, but they cannot fly when their ears are plugged. So, in effect, they "see" with their ears. It is almost as if you had hundreds of ears, each one unbelievably sensitive to the exact direction of any sound you heard and that, while blindfolded, you listened to an orchestra until you could visualize the position and instrument each member was playing, all instantaneously, continuously and automatically in three dimensions, so that you could *see* the whole orchestra in action. Even if you can't imagine your *hearing* getting so intense that it could tune itself into an image as graphic as *seeing*, the evidence clearly shows that ultrasonic sonar accomplishes exactly that. And the bats, porpoises and other animals that use it discriminate between their own echoes and those of their companions, even where the frequency of the pulses is the same. This is hard to explain, but ultrasonic researchers reason that as the time dimension replaces some of the spatial dimensions when hearing replaces seeing, the creatures involved must begin to perceive both the amplitude and phase of the sound waves, discriminating among them in reference to a coherent background of ultrasound — which would mean that they have evolved a natural but unique bioholographic technique. This hypothesis in fact was tested in 1968 by Paul Greguss of the RSRI Ultrasonic Laboratory in Budapest when he made a model of a bat's brain which, operating at a frequency of one megacycle, produced actual ultrasonic holograms that could be recorded on sound-sensitive photographic plates, scanned with a microdensitometer and used to reconstruct three-dimensional images of objects "heard" by the "bat."

Although the ultrasonic voices of bats vary from jet engine intensity (in the swift ones) to a dainty whisper (in the hovering ones), some of the moths and smaller insects they prey on can hear them from more than a hundred feet away and take effective evasive action, sometimes diving to the ground where they may be completely out of sonar

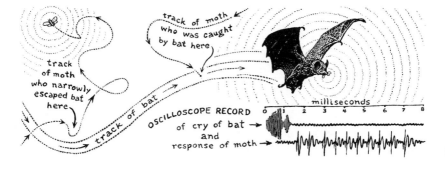

track of moth
who was caught
by bat here

track
of moth
who narrowly
escaped bat
here

track of bat

OSCILLOSCOPE RECORD
of cry of bat →
and
response of moth →

milliseconds
0   1   2   3   4   5   6   7   8

range. But, more amazing, is the fact that certain advanced moths not only dodge the bats but emit countersonar ultrasonic signals to confuse them when they get close. In one test, made with a high-speed movie camera synchronized with an ultrasonic tape recorder for playing back as audible slow motion, 85 percent of the bats hunting such moths actually abandoned the chase in the critical last second. And, as if jamming bat sonar weren't enough, parasitic mites have also been discovered living on the same moths, where they have quietly evolved a curious appetite for the soft flesh of one ear — only one ear, mind you, not both, because all the greedier mites that partook of the second ear have thereby long since doomed their breed into extinction by becoming suicidal passengers on a deaf moth who, through this very act, turned into easy bat fodder.

## PRESSURE SENSE

Although the pressure sense takes many forms, from warning a bird of a storm to informing a mole of a predator, the most observable of the various pressure organs is probably the fish's lateral line system that extends from his head along each side of his body to his tail. In many fish it is a clearly visible narrow dark line with thousands of sensilla (sense receptor cells) and their nerve connections that apprise the animal of low-frequency vibrations and pressure waves, coming through the water from anything solid nearby. Neurologists say it is a sense midway between hearing and touch, difficult to comprehend yet vital to the fish in locating prey, warning of enemies and enabling a school of fish to maneuver as a perfectly coordinated superorganism.

## SENSES OF TOUCH

Here we come to other senses of feeling that are more commonly understood as such or, more specifically, as senses of touch. For a little reflection will tell anyone that there are some half dozen different tactile sensations that not only have different conscious aspects but distinct sets of specialized nerves to convey their messages to the brain. When you touch something, for instance, you generally become aware not only of its size, shape and texture, of its hardness or softness and its roughness or smoothness, but also of its temperature (through molecular movement or radiation, as we have seen), its humidity and perhaps its weight, pressure or motion, on occasion including a degree of painfulness, itchiness, ticklishness, sexiness, etc., not to mention a possible clue as to its nature, needs, potentialities or intentions. And the relative importance of the sense of touch may be hinted at by the known fact that a blind cricket enjoys a normal status among his fellows while a cricket with his feelers missing is lucky to survive a day.

Because there has been so little serious research in the field of touch, I offer you a relatively frivolous experiment into its variegated facets that were dramatized in 1969 when a kind of "feel museum" was opened to the public at California State College. According to reports, the visitors groped in silent smell-less darkness through rubbery channels, oscillating fur muffs, swinging rods, erotic pillows, stone walls, bags of lukewarm liquid and scores of other sensations deemed needed by a touch-starved civilization. Reactions to this so-called Tactile Symposium ranged from "fearful" to "sexy." One young lady who had resurfaced in the buff, dress in one hand, brassiere in the other, murmured, "It's too much of an experience. I didn't understand why I was wearing these clothes." Another sighed, "It's like taking your bed to bed."

The nerves that feel shape and texture would seem to be more delicate than most of us realize, because tests reveal that the average person can detect an eminence on etched glass no higher than one twenty-five-thousandth of an inch, especially if a thin piece of paper is placed between fingers and glass and moved with the fingers to unmask the friction of direct contact. A barefoot housewife may have cleaner floors because she can better feel the dust than see it. And to someone deprived of other senses, touch alone can be exalting. Helen Keller, then still a girl, wrote in her diary: "I have just touched my dog. He was rolling on the grass with pleasure in every muscle and limb. I wanted to catch a picture of him in my fingers, and I touched him lightly as I would cobwebs. But lo, his body revolved, stiffened and solidified into an upright position, and his tongue gave my hand a lick. He pressed close to me as if he were fain to crowd himself into my hand. He loved it with his tail, with his paw, with his tongue. If he could speak, I believe he would say with me that paradise is attained by touch."

The main other locale of feeling is inside the body where our so-called proprioceptive or "muscle sense" keeps us informed (mostly subconsciously) of important bodily functions and especially of our posture, such as the precise positions of our limbs and fingers when they are in rapid coordinated motion.

## SENSE OF WEIGHT AND BALANCE

A different sort of discernment tells us where we are in relation to the force of gravity. A simple way to test it is by weight judging, which has shown that the average person can barely perceive the contrast between two objects whose weights differ by two percent. As to balance, it is known that we sense it by a loose, round, pebblelike bone called the otolith in the inner ear which, like the primitive statolith (page 203), continuously rolls to the bottom of a tiny cavity, touching hairs that tell us which way we are leaning and how to walk straight. The vegetable version of the same sense is called geotropism and works through hormones (page 61). Flies, on the other hand, keep balance in flight by means of vibrating rods known as halteres. And fish use buoyant internal bladders containing gas.

## SENSE OF SPACE AND PROXIMITY

The special awareness that enables animals to keep the same distance apart when moving in formation, and to know how near they may let a stranger or an enemy approach, is less well understood. But it works unfailingly and zoologists presume it is augmented by such other senses as sight, hearing, smell, sonar and the fish's lateral line.

## CORIOLIS SENSE

This is probably the least known among feeling senses if indeed it is properly so classified. For exactly what kind of inertial perception, if any, makes it possible for a migrating fish or bird to feel the turning of Earth remains unestablished. However the Foucault pendulum, which reveals earthly rotation by the relative change in its plane of swinging, is a manmade Coriolis instrument that could have an analogue in life.

This chapter has dealt mainly with the eyes and ears of the world. But it is time we moved on to Earth's more numerous and often less known senses, which involve invisible, silent molecules and the intangibilities of the psyche. These are hardly less important than sight and hearing, for all the senses yet discovered are part of the living Earth — indeed the very means by which she has begun to know herself and will soon know much more.

# CHAPTER 8

# Twenty-one Senses of Chemistry, Mind and Spirit

S O WE ARRIVE at what are often called the visceral or chemical senses, meaning those that enunciate the appetites for food, drink and, in some cases, physical love. And right off we encounter smell and taste, which work chemically and are thought to be the most experienced of all senses. If this is true, it is presumably because chemistry deals primarily with molecules, which are the material units composing the world, including all its organisms, and which therefore interact with creatures directly rather than indirectly through waves of radiation or compression, as in the cases of vision and hearing. Thus when organized life evolved on Earth several billion years ago, the first way it could sense anything almost inevitably had to be through direct contact, naturally at the molecular level, as earthly life had not yet organized larger units except such structures as crystals which, if they are definable as alive, may also be regarded as molecules or, more accurately, supermolecules.

So did smell and taste (originally one sense) come into our world. At first the smell-taste organ (if it could be considered such) must have occupied practically the whole body, making the viruslike creature in effect a living nose or tongue. Then, as macroscopic life appeared, the organ retained its central forward position at the business end of its owner, while the snout took shape and, doing so, started to create the face.

# SMELL

Life must have tried out a lot of questionable smellers during those primordial and invertebrate eons if we are to judge by the olfactory organs now to be found in the mycelia of fungi, in the palps of insects, on the heads of worms, in the feet of ticks, in the gills of mollusks . . . But by the time the backbone established itself as the prime feature of progressive earthly bodies, the nose had become more or less standard in its present dominant position in fish, reptiles, birds and mammals.

This is not to say that these animal classes are anywhere near equal in their sensitivity to smell. For birds are primarily creatures of vision, secondarily of hearing and, with the exception of ducks, petrels, shearwaters and albatrosses, relatively insensitive to smell. This, incidentally, may be advantageous to the great horned owl, he having excellent night vision as well as sonar to help him prey on nocturnal animals such as the skunk, because the skunk's fragrance is well known to repel practically everyone else and undoubtedly would repel the owl too if he were not so smuff. The word "smuff," I should explain, is the adjective I use instead of the sterile medical term "anosmic" to describe one lacking a sense of smell. I forged it out of "snuff" + "muff" because obviously smell-less creatures need an apt one-syllable word corresponding to blind and deaf, since, after all, the world contains not only those who are stone blind and stone deaf but, less conspicuously, an estimated two percent who are stone smuff!

If smuffness is prevalent among mercurial, clear-eyed creatures like birds, however, it is a lot rarer among the relatively plodding fish, almost all of whom are able to find their food, if not their way home, quicker through their nostrils than through their eyes. I'm thinking of the shark who smells blood two miles away and of the salmon who remembers the individual flavor of the brook he was spawned in from among hundreds of tributaries of a great river.

Reptiles, notably lizards and snakes, use their sense of smell pretty constantly, but augment it with a supplementary sense that seems to combine smell and taste in the manipulation of their forked tongue,

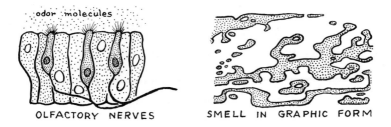

OLFACTORY NERVES          SMELL IN GRAPHIC FORM

which, as nearly everyone knows, they flick in and out at frequent intervals. What the forked tongue does is pick up molecules by the million, mostly out of the air, pulling them into the mouth and smearing thousands of them over two holes in the roof of the mouth which are entrances to a chemoreceptor known as Jacobson's organ, that instantly smell-tastes them. This is how a rattlesnake, for example, locates a rabbit he has bitten but which afterward scampers off in a panic for hundreds of feet before collapsing. The same technique helps reptiles find mates and to congregate when the season arrives for hibernation.

It is the mammals, however, warm-blooded and intricately social, who have developed the sense of smell the most of all, demarking their territories with odor fences and migrating long distances by smell navigation. A debate has been going on for centuries, for example, about how dogs that have been taken by strangers in closed vehicles to some strange town often manage to find their way home alone through a hundred miles of unfamiliar country. Alfred Russel Wallace theorized that such animals must automatically remember a "train of smells" when- or wherever they go and that they later somehow rerun the "train" in reverse, perhaps including many sidetrack trials and errors. Whatever truth there may be in the concept, however, hardly explains the dog's observed ability to take major shortcuts without getting lost, or his diverging sometimes scores of miles away from anywhere he had ever been. Could smell possibly carry that far? I wonder.

Even though odors can be wafted great distances on the wind and the keenest scented animals have proven they can detect a test aroma diluted to $10^{-13}$ or only one molecule in ten trillion of average air, it still is hard to believe the confirmed reports that bloodhounds and other trained dogs have found a lost wallet, a gun or a vial of heroin under tons of manure or concealed in a chemical factory reeking with fumes of sulfur or ammonia. The explanation seems to be that no smell can completely cancel or camouflage another smell because the          213 molecules both smells consist of are irretrievably diffused throughout the air they are in, and any two or more simultaneous odors, no matter how mixed, are smelled alternately in the olfactory cells and nerves, even though the alternations may be only milliseconds apart. Furthermore a dog who has sniffed, say, a man's cap can later recognize any other part of him and easily follow his trail because there are recognizable olfactory relationships between body parts as well as between species, races, sexes, ages, diets, diseases, neighborhoods, occupations or almost any other classifications of life.

If you want to avoid being tracked by a dog, then the first thing to do is wear brand-new shoes or cover your old ones with untouched plastic bags, so that the fewest possible molecules from your feet are left on the ground. But, in an actual case, even if no telltale molecules from your body get left behind (something manifestly impossible) an experienced dog may be able to follow you by smelling the freshly crushed grass or disturbed soil where you stepped, for this is the dog's specialty: he carries his nose close to the ground and the smelling part of his brain is not only disproportionately large but specialized to detect tiny traces of substances such as aliphatic acids in sweat that seep through shoes and diffuse steadily outward in air. In fact smell to him is a little like sound to a bat, giving him a degree of what we seeing-creatures call visualization. This was nicely demonstrated by a researcher who blindfolded a hound at a rabbit show and noted the "olfactory nystagmus" produced when a passing parade of rabbits caused the dog to swing his head back and forth as he smelled each in turn, something like an onlooker at a tennis match or a child scanning a book. Indeed one might reasonably have described the hound as "reading" rabbits with his nose.

Animals are usually more widely separated than this in their natural state, however, so it may be worth mentioning that auras of both sound and smell outdoors normally take ellipsoidal forms, extending mostly downwind from the animal or plant originating them. This is particularly true of smell whose dome-shaped diffusion is of course part of the total organism, its volume generally inversely proportional to the wind velocity, though a strong wind may waft a few thin odor streamers for great distances.

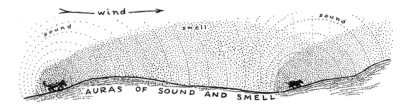

AURAS OF SOUND AND SMELL

While smell is a minor sense to most humans, a little training can develop it amazingly, and in former ages physicians seem to have relied on it for diagnosis, their medical treatises repeating such traditional olfactory observations as that the plague smells of honey, scarlet fever of hot bread, measles like fresh-plucked feathers, insanity like mice or deer. But a modern master specialist in odors, called a perfumer or flavorist depending on whether he concentrates on cosmetics or food, may be able (with an array of vials and sampling blotters) to identify something like 10,000 different odors.

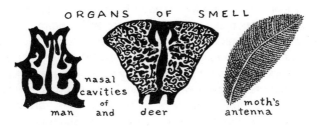

ORGANS OF SMELL

nasal cavities of man and deer

moth's antenna

It is in the realm of love, however, that the subtleties of smell really come into their own, particularly in the animal kingdom, where it is rarely love at first sight but much more generally love at first smell. A case in point is the seductive long-range scent of certain female flies and moths who may be smelled as far as seven miles downwind if one accepts the evidence of the upwind flight of marked male silkworm moths known to have arrived from that distance. Their type of perfume ($C_{16}H_{30}O$) in fact is believed to be the most potent substance known to physiology and one caged female pine sawfly exuded such a lecherous lure that more than 11,000 male sawflies turned up pining at her doorstep. But don't assume that male animals are unable or unwilling to put out love emanations of their own. For it is true also that the short-range perfumes some moth swains whiff out of glands under their wings to be flutter-fanned off special lacy hairs toward their sweethearts have such a universal appeal that even half-smuff humans notice them and have identified their flavors as similar to lemon, pineapple, chocolate, musk and several of the most fragrant flowers.

While most people do not seem to consider it very masculine nowadays for human males to entice females with perfume, this very tactic was widely approved as fair practice in ancient Greece, Rome, much of southern Asia, and it even caught on among the affluent in Europe and presumably America in the eighteenth and early nineteenth centuries, when, for example, Napoleon at Fontainebleau is said to have acquired the habit of garnishing his person with eau de cologne at the rate of two bottles a day. If some similar practice should regenerate itself in the West this century, even more significant may be the unpublicized tendency among modern and percipient young men to notice the natural body odors by which undeodorized and unperfumed persons of both sexes can ingratiate each other, by which ash blondes have been thought to exude the subtle aura of ambergris while brunettes have more often wafted to mind sandalwood, almonds or violets — all lovers of course being subject to the natural law that alkalis heighten sexual odors while acids diminish them.

Theories as to what an odor actually is, or how a nose knows a rose from a lily, have been speculated about for millenniums. They range from the idea that smelling is a natural biospectroscopic analysis of infrared heat rays that have passed through (and been patterned by) odorous vapors in the nose to the now widely accepted lock-and-key hypothesis that the smell sense is a biochemical process, specifically a recording of keylike molecules as they fit into various kinds of hospitable keyhole receptacles built into the olfactory center. The latter notion is actually very ancient, having been surmised by Empedokles, who said in 450 B.C. that "smell comes from very tiny particles drawn in along with the breath," and later by Lucretius, who wrote regarding the distinctive flavors of smells that there are "corresponding differences in the shapes of their component atoms. These in turn entail differences in the chinks and channels — the pores, as we call them — in all parts of the body . . . In some species these are naturally smaller, in others larger; in some triangular, in others square; while many are round, others are of various polygonal shapes . . ."

Lucretius of course had no way of proving his precocious ideas, nor did the dozens of other smell pioneers with rival theories until Linus Pauling announced the discovery in 1946 that "a molecule the same shape as a camphor molecule will smell like camphor even though it may be quite unrelated to camphor chemically." And three years later R. W. Moncrieff reformulated the whole lock-and-key concept in modern terms, followed by John Amoore, who called it the steric theory and, with the help of colleagues, pretty well established it by demonstrating that odors, like colors, can be sorted into a few primary ones of which all others are mixtures. The primary smells, it turned out, are seven in number: camphoric, musky, floral, minty, ethereal, pungent and putrid, each of them produced by various molecules approximating a distinctive shape or having a definite electric charge, and each smellable only when it is received in the right one of seven different kinds of complementary cavities distributed among seven corresponding areas in the molecular walls of the olfactory nerve cells.

It is quite a fantastic discovery in its way, this molecular revelation in osmics (the science of smell), and full of poetic analogies of love and the intimate courtships and matings of animals and flowers. The camphoric molecules (primary odor no. 1) smell something like mothballs, are shaped roughly round and fit into congenial basinlike hollows in the nerve cells. Musky molecules (no. 2) are styled to go into a microscopic lady's flat, oval hatbox, the kind in which she would keep her broad-brimmed spring hat. Floral molecules (3) carry the scent of a rose garden and the appropriate design of a gardener's wheelbarrow, deep in front and wide at the back. The minty molecules (4) are contoured for a romantic nook such as a cozy two-seated box at the opera where, between the acts, one might savor a mint with one's sweetheart. Ethereal molecules (5) carry the heady aroma of a more mature passion with soporific overtones and the lines of a long flat-bottomed barge like Cleopatra's on the Nile. The last two primary molecules (6 and 7), pungent and putrid, are small enough to fit into almost any olfactory aperture, the reception of the first, with the acrid tang of ants and vinegar, governed by its positive electrical charge, and of the other, with the stink of rotten eggs and dung, by its opposite negative charge.

SHAPES OF THE SEVEN PRIMARY MOLECULES OF SMELL

camphoric     musky     floral     minty     ethereal     pungent   putrid

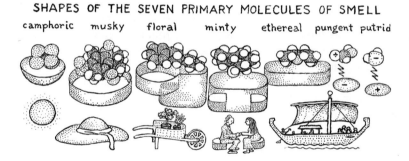

The stench of mercaptan, a garlicky sulfide (RSH), said to be "the worst odor ever compounded" and normally noticeable down to a dilution of a millionth of an ounce in sixty cubic feet of air, is presumably included in the putrid classification. In any case, any substance, to be smellable, must be both volatile and soluble — that is, warm enough so its molecules will actively zoom through the air around it in vast numbers and chemically adaptable enough so that, after they enter the nose, they can dissolve and penetrate the wet and lipid layers that coat the olfactory nerve cells. When the receptacles there are filled with enough odor molecules to block the acceptance of more, however, the sensation of smelling has to diminish while a kind of     217 smuffness often called "smell fatigue" sets in, at least for the scents involved, and this naturally confirms the lock-and-key hypothesis.

Another discovery was that certain alcohols and aldehydes have an extraordinary power to amplify smells. Presumably this is why they are present in so many successful perfumes and may be attributable to their molecules being composed of from four to eight carbon atoms in a chain with side branches that provoke extra contacts with receptor sites, thus intensifying and determining the specific odor. Basic mammalian sex lures, on the other hand, are apt to have heavy ring-shaped molecules: musk a 15-bead necklace of carbon atoms fringed with hydrogen and oxygen, civet a similar 17-bead necklace. But a smell molecule for general alarm purposes that other species can respond to is usually simpler, unspecific and small. All the evidence in fact suggests that every known smell is somehow spelled out by its molecule's shape and defined just precisely enough by it so that a pair of molecule twins, for example, identical in all respects except that one is the mirror image of the other, almost invariably have distinct and individual odors. Indeed the molecule called carvone with mirrored right- and left-handed configurations smells in its "right" form like spearmint and in its "left" like caraway, the amino acid leucine when right is sweet but when left bitter, and another handed molecule named limonene smells in its right version like an orange and in its left like a lemon. Which could mean that, gauche though it may seem, a lemon is in genetic essence only a left-handed orange.

Here we are touching on the symmetry and relativity of smells, which may relate to such curiosities as the fact that a derivative of indigo known as indole ($C_8H_7N$) smells like raw sewage yet, if you dilute it, it immediately gentles down to the genial scent of narcissus! On the other hand, if you sniff cedarwood for several seconds, its odor may tone down into that of violets, and bitter almonds should turn toward tar, while, if you add the odor of India rubber to cedarwood, it can cancel it so you smell nothing — even as balsam is canceled by beeswax.

This of course brings us back to the phenomenon of camouflage, which applies to smell about as much as to sight and sound, being found throughout nature, embodied in the chipping sparrow who exasperates the bird dogs of Texas by imitating the smell of a quail, or in the prowling rattlesnake who deludes his rodent prey by assuming the dank aroma of a cucumber. If these are olfactory lies, it is because smell too is a kind of language. The rattler, for one, has at least four "words" in his smell vocabulary. Besides his ambush scent of cucumber, he switches to a terrifying effluvium when in combat, exudes a socially somnific savor at hibernation time and wafts a love perfume when looking for a mate.

As organs of sense, man's nose and tongue are obviously less important to him than his eyes and ears, but the priority is reversed in the case of most animals, in whom the chemical senses are their main medium of communication. In social creatures, for instance, messages are commonly delivered in the "code" of pheromones or external hormones such as musk that circulate not inside the body but socially between the members of a family, a colony or a species. Pheromones are definitely a factor in population control for, when soldier termites secrete substances that inhibit the endocrine glands of the younger termite generation, these larvae are unable to grow up and reproduce. And, on the other hand, when the adult males of migratory locusts secrete a volatile liquid from their skins that is passed to the nymphs, these young females mature much faster, provoking in a few weeks the dramatic outbursts of huge foraging plagues of locusts.

Something similar may happen to lemmings, who also have catastrophic population explosions now and then and, with most species of the mouse family, the seductive odor of a friendly male mouse is known to initiate and synchronize the estrous cycles of female mice. Yet these same mice have a remarkably effective population regulator in that, if a strange male mouse approaches a newly impregnated female (as must occur when conditions get crowded), the merest whiff of him is usually enough to suppress the flow of her prolactin and end her pregnancy!

## TASTE

Taste is basically like smell except that it is less sensitive, requiring about 25,000 times as many molecules to elicit a sensation because it deals primarily with molecules in solid and liquid form rather than gaseous. And the primary tastes are but four in number: sweet, bitter, salty and sour; each key taste molecule having its own lock receptacle in what is called a taste bud on the human tongue, palate or throat. This means that each primary taste is tastable only in its own location: sweet at the tip of the tongue, bitter at the back, salty on the sides around the tip and in the throat, sour on the sides of the tongue farther back. Like smell, taste is very ancient and fundamental, and I feel no doubt that, when we know more about molecular structure, we will see geometric reasons why applesauce tastes good with pork, mint with lamb, cranberries with turkey, ketchup with baked beans and so on.

In the meantime the study of taste is increasingly bewildering. A couple of sample findings of research in the chemical senses indicate that the flavor of a common brand of coffee is a synthesis of about four hundred compounds (most of which are smelled more than tasted) while the formula for a different and more sophisticated artificial flavor "requires as many as 20,000 separate pieces of information." And this no doubt throws light on why man has been able to invent cameras, phonographs and associated techniques to record, amplify and transmit sights and sounds but has not yet devised any comparable method of recording, amplifying or transmitting a single smell or taste.

Relativity is another factor that makes smell and taste hard to comprehend. If, for example, you put on one side of your tongue a salt solution too dilute to taste noticeably salty but then add a little sugar on the other side, you will instantly begin to taste the salt along with the sugar. The opposite happens when you start with dilute, tasteless sugar on one side and add salt on the other. Furthermore, to most humans, a salt solution will begin to taste sweet when diluted down to .03 parts per million, particularly if it is cool, the amount of dilution needed to make this happen being roughly proportional to the temperature. On the other hand, although Epsom salts taste salty on the front of your tongue, they turn bitter when pushed back to the hind buds. And if you try many kinds of chemically graded salt, you will notice them tasting progressively bitterer as you get to salts of heavier molecular weight. The taste of salt of course is largely electrical (due to ionization of its constituent atoms) and every sort of electric current has its own flavor: a gentle direct current savoring subtly sour when the positive terminal touches the tongue but "like burnt soap" when the flow is reversed, while an alternating current that smacks of astringent sourness at 50 cycles turns steadily more and more bitter as it escalates toward 1000 cycles. Such relativity is not merely mental but rather an objective part of the natural and paradoxically complex simplicity of the chemical senses which, unlike senses that become electrical only in the final transmission of messages to the brain, may function electrically all the way from their first contact with whatever they perceive.

This is not to say that there aren't real differences of taste between individuals, for we all know such differences exist. Ordinary sugar, for example, is tasteless to a small percentage of children while saccharin tastes bitter to a few yet sweet to their brothers and sisters. And there is a synthetic chemical called PTC (for phenyl-thio-carbamide) that tastes intensely bitter to an average of two people out of three all over the world but utterly tasteless to the third person. And the common food preservative sodium benzoate tastes like almost anything or nothing, depending on who tries it. But there is such conclusive evidence that these phenomena are objective that a chemist named Arthur L. Fox has formulated a genetic theory of taste on his finding that 26 percent of people consider PTC bitter but sodium benzoate salty and like almost every kind of food, while 17 percent with different taste genes register both these chemicals as bitter and dislike most sorts of food.

Indeed, of the four primary tastes, bitterness turns out to be the easiest for a human to detect, perhaps because it signals danger in the form of poison. But there always seem to be a few unlucky people who just can't taste anything at all, and them I would call "smumb." They aren't necessarily the same ones as those who are smuff and their deficiency is more serious — but fortunately medical researchers have discovered that most of them, even those who've been stone smumb for years, can be cured within a few days by taking small doses of the trace metal zinc.

Animals naturally vary in sensitivity to taste, some insects going so far as to walk on their "tongues" and to taste with their feet. Other creatures, notably fish, may be trillions of times more sensitive than man with taste buds so densely distributed over their body surfaces that they literally swim in a sapid sea. There is the uncanny account of a coho salmon raised in a California hatchery who, when a year old, was dumped into a strange stream several miles away and allowed to migrate with his fellows down to the ocean. But at spawning time the next year he appeared back in his original tank, having followed the familiar flavor into his home stream, threaded a particular culvert under U.S. Highway 101 which enabled him to enter the hatchery's flume and storm sewer, from which he finally wriggled up a four-inch drainpipe past 90° elbows, climactically knocking off its wire cap and even leaping over the screen that surrounded the drain!

## THE SENSES OF HUNGER AND THIRST

Close cousins of taste of course are the twin senses of hunger and thirst. Hunger has long been known to be turned on by rhythmic contractions of the empty stomach and, more recently, by a decline in the sugar content of the blood. Indeed a transfusion of low-sugar blood from a starving dog into a well-fed one will make the latter hungry for the same reason that high-sugar blood from a satiated dog will ease the pangs of a hungry one. Yet hunger can hardly be as

221

simple as this. In fact certain researchers have recently found that the ratio of ions in the brain may regulate hunger and specifically that rats who have eaten to satiation can be induced to resume eating voraciously by injections of calcium ions in the cerebrum. Others predict that, when we fully understand it, hunger will turn out to involve a "hunger hormone," conveyable, if not normally from organism to organism, perhaps in some degree from organ to organ through lymph and blood.

Thirst is about equally mysterious but obviously different from hunger and much more compelling, at least to a water-dependent creature like man, as is proven by the fact that a man can live more than a year without food but, to the best of my knowledge, seventeen days is his world record without water. This record was made in 1821 when a prominent Frenchman named Antonio Viterbi committed suicide by refusing to drink, but of course he may have taken in a significant amount of moisture in whatever he ate. Doctors now say he probably would have survived if he had accepted water on the fourteenth or fifteenth day, but by the sixteenth it was almost certainly too late. There are cases on record of castaways deprived of fresh water for two weeks who were rescued just in time and managed to survive. Presumably all these sufferers were in humid environments, did not sweat and kept evaporation from their bodies to a minimum.

Something to consider also is that the thirst sensation is less influenced by the total amount of water in the body than by the amount of water relative to certain solids, particularly salts. And this accounts for the classic equation of thirst that assures the bartender he will sell in the end more than ten times as much in drinks as the cost of all the "free" salted pretzels, popcorn and potato chips he "gives" away to his customers to bolster their thirst. It also relates to the discovery in 1952 that a fraction of a drop of a salt solution injected into the hypothalamus at the base of a goat's brain will immediately make the animal thirsty, which, in combination with later evidence, seems to have pretty well proved the site of thirst to be the hypothalamus. Strange as it may seem, drinking beyond a certain quantity of water hour after hour *in*creases rather than *de*creases thirst because the body loses salt in urine and salt deficiency produces thirst, a kind of thirst, however, that can be relieved only by salt.

While on this subject, I should explain that the reason castaways are warned against drinking saltwater is not to deny that the watery 96.5 percent of it may immediately relieve them but only to avoid the ultimate problem of getting rid of their lethal excess of salt. Most mammals, including man, cannot eliminate salt in urine at the rate of more than about 2 percent so, unless one can be sure of soon receiving enough fresh water to dilute one's urine that low, to drink seawater (averaging 3.5 percent salt) is to take a deadly chance.

The most dramatic thing about thirst has to be the kind of death it brings when a man is lost, say, in a hot desert, a grim dissolution witnesses describe as unfolding in five fairly distinct stages. First comes the protesting phase of increasing discomfort and querulous disbelief (with 3 to 5 percent of body water lost). Second, the feeling of having a mouth "dry as cotton," tongue sticking to teeth, a lump in the throat that no amount of swallowing can dislodge, a face tight from shrinking skin (water loss 5 to 10 percent). Third is the burning agony of having the tongue shrivel into a knot, eyelids stiffen, eyes staring as the victim irrationally tears at his clothing or scalp, bites his arm for blood or even laps up his last drops of urine (10 to 20 percent). Fourth is the stage of the skin cracking apart (more than 20 percent of water gone or too much for any chance of recovery), lymph and blood oozing out, eyes weeping blood, arms digging aimlessly into the sand. And fifth, the final or living-death stage of gentle writhing on the ground and, often, calm acceptance awaiting the end.

The third stage is usually the worst, for nature seems to have evolved pain to save life before it is too late. But after the blood begins to thicken and dry and the damage becomes irreversible, the pain eases progressively away — which suggests why those who die of thirst are so often reported to be joyous in their final hour — even as one of them was heard to murmur, "I'm melting, I'm sinking . . . I'm drowning in an ocean of unexplainable peace."

## THE SENSE OF PAIN

When it comes to pain, it could be considered the final and most arresting of the feeling senses. However, I deem it more appropriate to place pain among the mental senses, the fourth sense group, which we now come to, partly because, like pain, the mental senses are in general abstract, if not nebulous, and closely involved with other senses. Any of the common senses in fact may become painful if experienced with sufficient intensity, like seeing a blindingly bright light, hearing a deafeningly loud noise, feeling an unbearably hot or cold temperature, or even smelling an aromatic "agony," which, I'm told, can torment such a creature as a tick whose odor-sensitive feet are olfactorily "burnt" if forcibly left too long in direct contact with even a normally pleasant-smelling substance.

Whether or not such evidence proves pain to be mental is controversial and continues to be debated among doctors and judges, few of whom can claim they really understand pain, although they daily deal with it. Indeed the specialists who know the most about this elusive subject are more likely the scientific experimenters such as a team of zoologists at McGill University who recently raised a batch of Scottish terriers from puppyhood in solitary, protective restriction so they could experience no pain. Then, to everyone's surprise, when the dogs were full grown, they were evidently incapable of feeling pain, for none of them reacted to such provocation as having a lighted match singe his nose or a pin jabbed an inch deep into his flesh. Pain had thus been revealed as something to be learned, a state of mind!

Such a concept should not be too hard to accept either, because it conforms to many people's experience of pain. I remember, for example, when I was a war correspondent in Dover, England, in September 1940, and got bombed out of a hotel, literally falling in two or three seconds from the fifth floor to ground level. Two British naval officers standing beside me were killed, but miraculously I was unscathed — or so I thought as I climbed out of the wreckage. Fifteen minutes later, however, I happened to notice a trickle of blood running down my ankle and discovered that a seven-inch hunk of flesh had been gouged out of my left calf. Perhaps it was the distraction of the fall, the expectation of being mangled or killed, that had served as an anesthetic. In any case I felt absolutely nothing of what would, under more normal circumstances, have been a painful injury. My mind at the moment was obviously not conditioned for pain. A neurologist, however, might say that, unknown to me, my mind had somehow influenced the polypeptide molecules that have recently been discovered to exist naturally in nerve cells and which, according to some researchers, can, under certain little understood circumstances, block pain as effectively as a shot of morphine.

Whatever the technical facts, there is a lot of difference between pain itself and someone's reaction to it, a reaction that encompasses an almost unlimited range of possibilities. Does the Eskimo calmly hacking off his gangrenous foot with a hatchet to save the rest of him feel less pain than a similarly afflicted European who requires a delicate and expensive operation? A researcher who conducted a heat test on several dozen Eskimos, Indians and Caucasians to find out, discovered that they all began to feel pain when their skins got to 113° F., but they varied widely as to how hot and how long they could stand it, the more "primitive" ones usually (but not always) being the more relaxed. Pain, in other words, appears to be not a thing but a concept and therefore, in essence, abstract.

In every investigation so far, to my knowledge, culture and training have fairly demonstrated themselves to be the critical factors in any sensation describable as pain. The famous Russian physiologist Ivan Pavlov, experimenting with dogs many decades ago, found that they reacted violently to the pain of a strong electric shock applied to a paw. But he conditioned the dogs to the electricity and the "pain" by consistently feeding them right after a similar shock every day. In consequence, a few weeks later, he noticed that the dogs were beginning to salivate as soon as he shocked them, because they were hungry and the daily shock was turning into a dinner call. Eventually the dogs became so receptive to the shock, even after Pavlov increased it to the point where it was burning and injuring them, that they would wag their tails in anticipation of something that had blossomed from a painful into a joyful experience.

Another researcher went farther, this time with cats, not only training them to associate a shock with food but providing them with special switches so they could administer shocks to themselves whenever they got hungry, thereby proving that the shocks were their own free choice. Some of the cats in fact got so greedy they even took to fighting each other for possession of the switches.

Comparable to this may be the curious fact that, if a certain kind of dragonfly is held firmly by the wings and its tail is put into its mouth, it will bite the tail and keep on eating it until the stump gets out of reach while the dragonfly slowly dies. Do you suppose such an insect would commit such an apparently pointless "suicide" if doing so were painful? Would an angleworm, cut in two right through its main nerve cords, its blood vessels, intestines, peritoneum and circular muscle system, continue on its way as two "carefree" angleworms if it, or they, were suffering in any comprehensible sense? Could the shriek of stripped gears in an overloaded old automobile be called a mechanical cry of "pain"? What is pain really anyhow? And is it an essential part of our world?

We have characterized pain as abstract, but there are countless and conflicting opinions as to how necessary it is to the world. Hundreds of men and women of all the major races are known to have been born congenitally incapable of feeling pain. Perhaps they happened to have abnormal quantities of polypeptides in their neurons. In any case, they are not numb for they have fairly normal senses of touch, temperature, humidity, etc. They are like Edward Gibson who appeared in vaudeville at the turn of the century billed as "the human pincushion" because he encouraged customers to stick pins into him by the dozen and, at the climax of his career, smilingly submitted to crucifixion on the stage with real nails through his hands and feet until the show had to be stopped because "too many members of the audience fainted." Actually people like that are more to be pitied than envied, for nature and evolution are probably weeding them out. Indeed they often fail to survive childhood because they overlook the normal danger signals of pain, and neglect their unfelt hidden wounds and symptoms, which consequently get infected or deteriorate unnoticed until it is too late. The philosophical case for pain in the world and its polarity aspect under the paradox of good and evil are important subjects that we will return to in Chapter 18, but I can tell you now the good news that at least the physical aspect of pain has been found to have finite dimensions. For researchers recently discovered that the pain of heat reaches its "excruciating maximum" at a skin temperature of 152°F., beyond which, no matter how hot a fire gets, it feels no worse. I don't know how this maximum was established — it must have been agonizing for someone — but it is interesting to think that it might have saved the medieval torturers a lot of fuel and trouble had they known.

Even so, there can be a mysterious benediction following torture which must be attributable to the mysterious potential for joy inherent in pain, a paradox that is a prime example of the relativity of the opposing aspects of life's qualities, and one of the Seven Mysteries presented in this book. Certainly a benediction came to a seaman named Kim Malthe-Bruun who suffered horrible torture in a Nazi prison but was able to write the next day: "Suddenly I realized how incredibly strong I am. When the soul returned once more to the body, it was as if the jubilation of the whole world had been gathered together here."

## THE SENSE OF FEAR

A hunted animal is commonly assumed to be very frightened when his pursuers close in on him. Yet his understandable urge to escape is not necessarily based on actual fear — at least not on "fear" in the human sense. Take a running antelope overtaken and killed by a lion. There can be no doubt that he suffers a severe shock, but it is a natural shock which may involve little or no pain or fear. The normal kind of pain that forces an animal to reduce activity until his wounds heal has obvious "survival value" in evolution, and so does his reasonable fear, but what could be the survival value of pain or fear in a mortally wounded creature who is soon to be out of the living world? I can't think of any value beyond the warning of danger in his despairing last cries, cries which, however, could be just as effective if produced by painless shock.

Studies of predation suggest that the prey usually behaves as if stunned, once the predator has seized him. He rarely struggles significantly and often does not even protest the *fait accompli* that has overtaken him. He may indeed be anesthetized by the shock if not the inevitability of being devoured by his natural superior. A case in point is the story of Major Redside, a British hunter in the Bengal jungle some fifty years ago, who had stumbled when crossing a swift stream, dropping his cartridge belt into the water. His companions happened to be beyond earshot and, now out of ammunition, he advanced in their general direction until he noticed a large tigress stalking him. Turning pale and sweating with fright, he began retreating toward the stream. But it was already too late. The tigress charged, seized him by the shoulder and dragged him a quarter of a mile to where her three cubs were playing. As he recalled it afterward, Redside was amazed that his fear vanished as soon as the tigress caught him and he hardly noticed any pain while being dragged and intermittently mauled when the tigress played "cat and mouse" with him for perhaps an hour. He vividly remembered the sunshine and the trees and the look in the tigress's eyes as well as the intense "mental effort" and suspense whenever he managed to crawl away, only to be caught and dragged back each time while the cubs looked on and playfully tried to copy mama. He said that, even though he fully realized his extreme danger, his mind somehow remained "comparatively calm" and "without dread." He even told his rescuers, who shot the tigress just in time, that he regarded his ordeal as less fearful than "half an hour in a dentist's chair."

Something of the kind also occurs during battle and on other occasions of severe stress and danger. And it indicates that the more active the role one plays the less one feels afraid. Soldiers in World War I, for example, often said that staying still in a trench under bombardment was harder on the nerves than going "over the top" to fight in the open; that, although it took guts to expose yourself, once you did so, the excitement of action almost always made you forget the danger. Or, as a physiologist might say, the adrenalin flow that triggers fear is greatly influenced by the state of mind.

Another aspect of fear is the interrelation between alternative, sometimes conflicting, anxieties. I remember walking along crowded Oxford Street in London one morning in 1941 when, without benefit of the usual air-raid warning, a German bomber streaked overhead and bombs began to fall. It must have been tempting to some of the pedestrians in that moment to shelter themselves by ducking into doorways or perhaps even diving into the gutter. But what self-respecting Londoner could bear the thought of being the first to show the white feather? No one did. For, in fact, the courage it would have taken to duck in full view of that crowd of shoppers clearly exceeded the courage it actually took to keep on walking as if there were no danger. And some Londoners were killed that day because they lacked the nerve to take reasonable but "shameful" precautions.

A different example is the British captain who, perhaps a year later, drank a cup of tea on the bridge of his destroyer under dive-bombing attack. When the lookout shouted, "Aircraft on the starboard bow, sir," he didn't even look up. When he heard "Aircraft diving, sir," he sipped once more and took a casual glance skyward. And when the lookout said, "Bomb released, sir," he ordered, "Hard a-starboard," and finished his tea as the bomb hit the water nearby. Although such restraint made the captain a hero to his men, the very real conflict between fears hidden inside his mind was revealed to the ship's surgeon, who later chanced to glimpse him sitting alone in his cabin — weeping.

Still another side of fear is its slow development over a period of time as the mind is enabled to assess the actual risk. An air force general in Vietnam once remarked that "after six months' flying, many pilots have aged ten years." And I have noticed the same thing over a much briefer span on the several occasions when my life was in sudden jeopardy, most memorably the time I was bombed in Dover. For I realized, as the floor dropped out from under me, that I might well get crushed to death by the collapsing building. Yet I distinctly recall my amazement that at that moment I felt no fear — for it did not occur to me until days afterward that the two or three seconds in which I heard the bomb falling and exploding, followed immediately by my own fall, simply weren't a long enough time to permit my adrenal glands to generate the sensation of fear. And notwithstanding the fact that the naval officer beside me was instantly killed by a bomb fragment and the other officer had his skull broken in the fall, fear had not arrived. Because fear is truly a conditioned state of mind.

## The Procreative Sense

Continuing with our roster, the mating urge and capacity for sex arousal is still another important member of the mood and mental senses, although of course it often also involves seeing, hearing, smelling, feeling, etc. And its essence is expressed in the orgasm that comes at its climax and is related to such other climaxes of bodily and mental function as "blowing one's top" in uncontrolled anger, the "flash" of creative inspiration or a paroxysm of anguish. The pioneer or prophet of the orgasm is widely considered to be Wilhelm Reich, a pupil of Freud's, who proclaimed it a vital expression of "Cosmic Life Energy." And there are other modern thinkers who agree that this climactic procreative sensation includes a spiritual counterpart to its physical aspect. Also, although the orgasm undoubtedly represents the most intense natural physical joy one can sense while in one's earthly body, I see no reason to deny that even greater ecstasies may await us, at least potentially, in transcendent states beyond.

## The Sense of Play and Laughter

Although laughter is widely accepted as the supreme outward expression of human pleasure, a little thoughtful research shows it can just as often be the surface manifestation of escape from a hidden conflict. Or, as Freud explained it, the humor that sets us laughing is a benign device for mastering our forbidden urges. And a modern psychologist might nod and say, "Indeed: like a cork on a bottle of potent brew."

But if laughter, like weeping, is a surface phenomenon, it too is prone to be delusory, and sometimes on a mass and commercial scale. French theatrical producers long ago started paying *rieurs* (laughers) as well as *pleureurs* (criers) to provoke audience response, and any competent TV comedy director today knows how to "lay in a laugh track" with "canned laughter" or, increasingly often, to use a tape console with keyboard for fading in desired shades in the spectrum of laughter, shrieks and applause, from male belly guffaws to feminine hysterics or from a roaring ovation to scattered titters or sobs. And no less delusory, if a lot grimmer, is the weird disease called *kuru* or "laughing sickness," whose obvious symptom is anything but a laughing matter, for it has been found only among members of the obscure Fore tribe in eastern New Guinea where, to those who catch it and laugh, it is 100 percent fatal.

The sense of pleasure (as distinct from laughter) has been studied less than that of fear, but it appears to be simpler and presumably easier to understand with its known centers in the brain, some associated with the appetites for food, drink and sex. There is even a spot in the septal area of the brain called the pleasure center, where, if delicate electrodes are implanted and just the right electric current applied, the delightful satisfaction imparted (according to one estimate) is "greater than the satiation potentials of all other known appetites combined." Although most of the research so far has been on animals because of the obvious human emotional, ethical and philosophical questions involved and the dangers of misuse if unscrupulous operators ever usurped these means of controlling humans, research is now progressing in man's own cortical pleasure centers and, in a few cases, techniques of therapy explored.

## THE SENSE OF TIME

Awareness of rhythm and chronological sequence is as much a sense as sight or touch, I believe, and properly called the sense of time. It derives from sensitivity to a basic dimension of our finite world and is involved in the principle of transcendence (as we shall see in Chapter 19), one of the Seven Mysteries of Life this book is named for. Although closely associated with such feeling senses as awareness of space and vibrations (page 203), which have subtle organic manifestations, it is much more predominantly a mental sense and so fits in here among the other impalpable and hard-to-measure senses that are more and more indisputably classified as abstract.

Of course time can be measured with the aid of such a modern invention as the stopwatch. That is how the traditional test of manhood in the old West, quickness on the draw, was finally accurately defined when a champion gunman took exactly three fifths of a second from touch to target. And a crack modern typist does about equally well in hitting 11 keys, or a piano virtuoso 15 musical notes, in the same fraction of a second. The latter feat, if you can believe it, was performed by Simon Barère playing Schumann's C Major Toccata with its 6266 consecutive notes in 4 minutes 20 seconds at the astonishing average pace of better than 24 notes per second. Which is slightly faster than the standard speed of movie frames that flash on the screen to create a nonflickering illusion of continuous motion.

The movie illusion, however, can be provided for some creatures at only one thousandth the speed of others — because their time sense is geared that much slower. I am thinking of the difference between, say, a horsefly, who can easily see the gaps between frames in a standard movie, and a garden snail, who, researchers have found, cannot react to any visual event in less than four seconds and therefore theoretically could see a continuous movie when the frames were changing at about the pace of a slide show.

A more important use of the time sense is for sex and species recognition, as in courting. The common firefly *Photinus pyralis* of course courts at night, flashing his light as he flies in a U-shaped dip. The female, sitting on a tall grass stem with a wide view, is excited by his U but waits almost exactly two seconds before flashing back. He gets excited in turn by her timely response. Then, as he turns to fly toward her, repeating his message, she answers again in her precise two-second timing and he "homes in" on the light of his life. The maximum error the male firefly has been known to tolerate in this exchange is a fifth of a second, which gives the approximate range of accuracy of his time sense — and, by it, if she is more than a fifth of a second too eagerly fast or more than a fifth of a second too indifferently slow, he just knows she can't be his true flame. His skepticism is well advised too, for, it has just been discovered, several species of fireflies have evolved preying *femmes fatales*, who mimic the flash sequences of females in other smaller species through which fakery they manage to lure a percentage of overanxious males close enough so they can catch and devour them.

Man's own time sense is seldom nearly so precise, and its range has obvious limits. When you are told, for example, that heavy subatomic particles are created in high-energy collisions lasting only a one hundred-sextillionth of a second, but that these same particles "decay" much more slowly, taking a ten billionth of a second, you probably have trouble realizing the distinction between such seemingly instantaneous events. Yet actually the time ratio between them ($10^{-23}$: $10^{-10}$ sec.) is the same as that between a second and a million years!

In the north woods where I live, the time sense is commonly regulated by the sun, as when the lengthening daylight of February triggers the mating of the great horned owl, ensuring that its eggs will hatch in March when red-winged blackbirds arrive in flocks, followed by the main spring bird migration and the myriads of young, awkward mice, squirrels, chipmunks and rabbits appearing exactly when the owlets must daily be fed their own weight in food. It is a precise but complex phenomenon, the stimulus coming from an exact mental measure of each day, the result unfolding into a neat biological schedule of developments.

Animals in general and almost certainly vegetables, not to mention minerals, have something known as a biological clock that, in many tested circumstances, has been found to run accurately without temporal clues from the sun, moon, stars or earthly environment. It seems to be mental, having no proven location in the body, despite a theory that its timing could be regulated by the pineal gland, and its existence is known almost entirely through its behavior. Slugs — the kind that live on the undersides of stones in a pasture — lay their eggs about August first every year, even when confined in a seasonless, weatherless laboratory under artificial light (eleven hours a day), darkness (thirteen hours a day) and unchanging temperature and humidity. Ground squirrels kept in "total" darkness at near-freezing temperature for their life span of three or four years hibernate about a third of the time in a cycle that approximates 365 days. One squirrel, unable to hibernate because of being kept in a room as hot as 95°F. for nearly a year, nevertheless lost weight at hibernating time, when his biological clock forced his appetite and digestion into low gear.

Humans feel their body clocks most unmistakably when they fly east or west through several time zones, because this temporal shift makes their sleep-wake, digestive, adrenal and other body cycles reset themselves proportionately, accompanied by a sudden metabolic loss of nitrogen and sulfur. It seriously handicaps the likes of actors, tycoons, diplomats, athletes, chess players and race horses whose internal "clocks" are forcibly advanced or retarded six hours for every 90° of longitude they travel through, with an inevitable corresponding quarter-circle shift in the angle at which they stand, walk or sleep.

The solution to such problems has so far escaped us because we know so little about the elusive biological clock — indeed not even whether the body *has* a clock or *is* a clock. There is no shortage of theories though — such, for instance, as that of A. T. Winfree of Purdue University, who surmises the mainspring driving the bio-clock must be spiral-shaped, something like the gene. He developed the idea as a geometric generalization of periodic biological processes, which, being recurrent, can be presumed to be circling around axes of some sort, if only abstract ones. But each turn of such a bio-spiral would have to be geared to one sort or another of temporal influences, from cyclic enzymes to seasonal day-length changes. And one of the better of many possibilities for such a timer is even included in an independent theory proposing a mysterious regulator in the brain with characteristics of both an hourglass and an oscillator. Yet no one to date has really found the bio-clock or convincingly explained what

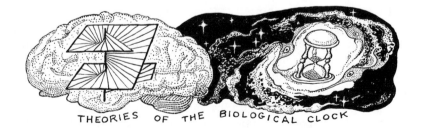

THEORIES OF THE BIOLOGICAL CLOCK

might cause it to be stopped or reset (as many have been) by a whiff of oxygen, a flash of light or a dip into a freezer.

To my mind, the greatest value of the biological clock, as it is revealed in both the microcosm of the body and the macrocosm of world travel, is its positive demonstration of the intimate regulation of earthly life by Earth herself — proof, you might say, that life, in at least one sense, is actually celestial.

## The Other Mental Senses

The remaining mental senses don't need detailed individual discussion here because all of them will be covered as occasion allows in the chapters that follow. Let me say only that every mature and normally active human knows them, if not from personal experience, through contact with others, including the animals.

The navigation sense that we will take up in Chapter 12 is one of several that is far more widely distributed in animals than in man. To a lesser degree that's true also of the domineering and territorial sense (Chapter 18) and the colonizing sense (Chapter 19) but probably not of the horticultural sense. On the other hand, the language sense (Chapters 9 and 11) is obviously most highly developed in man, and likewise the sense of reason (Chapter 11). But the other five mental senses (all in Chapter 11) of intuition, esthetics, psychic capacity, hypnotic power and sleep seem to broaden mysteriously out toward elemental qualities, the last being universal enough to include all creatures.

## The Spiritual Sense

I don't know how to subdivide the sense (or senses) of spirit, so I simply call it one sense. It is far too mysterious and immeasurable to do otherwise. It also seems to be the ultimate of all senses, something we can't help but return to at the end of this chapter as well as in Chapter 22 and, most completely of all, in the final and all-embracing Seventh Mystery of Life.

## Intersense Relations

Miguel de Unamuno once wondered whether eyes and ears might somehow become aware of each other's worlds. He could have broadened the question of course to include smell, taste and other senses. And he might eventually have even explored the curious, if little known, relation between them. It is the sort of subject at any rate that one would hardly expect to hear talked about except in some such fabulous place as the flavor laboratory of Arthur D. Little Inc., consulting engineers in Cambridge, Massachusetts, where the rooms are not only immaculate to reduce the number of extraneous molecules and silent because "noise dulls the perception of odors" but constantly maintained at 65 percent humidity, which, I understand, permits the keenest smelling — all this exhaustively determined by the head sniffer whose awed assistant once said, "He can smell to three decimal places."

I suppose it was in a laboratory of this stamp that nearly a century ago something called "colored hearing" was discovered. At least that was the accepted name for what William James described in his *Principles of Psychology* as "a strange idiosyncrasy, found in certain persons, consisting in the fact that visual impressions on the eye, touch on the skin, etc. are accompanied by distinct sensations of sound, these being only extreme cases of a very general law which says that all our sense organs influence each other's sensations." James went on to cite cases in which persons, who could not read because the letters they were looking at were too far away, suddenly found themselves able to read them when a tuning fork was sounded close to the ear. He also reported opposite cases of persons unable to hear sounds too faint or far, who suddenly heard them when colors were flashed before their eyes. And similar uncanny amplifications applied, he noted, to smell, taste, temperature sense and touch when these were combined with exactly the right colors and tones, suggesting that some mysterious sort of resonance must interjoin one sense with another, despite the fact that the vibrational frequency of the one (say, light) could exceed that of the other (say, sound) by a factor of a trillion.

The first measured interaction between light and sound I ever knew about was produced at Massachusetts Institute of Technology in 1964 when a laser beam was made to generate sound waves in a sapphire at an unheard-of radarlike acoustical frequency of 60 billion vibrations a second. It exploded the sapphire and spurred thinking among intersense resonance engineers, to judge by the subsequent development of the acoustical holograph, which "sees" with sound waves like a bat, though not through "ears." Instead it sees through eyes by transforming its sound waves into light waves for reassembling the three-dimensional image on a screen so it will be visually looked at.

Before we leave the subject of intersense resonance, if that indeed

234

it be, I should mention that I've heard of music lovers who claim they habitually see colors when they hear certain melodies, indeed that they enjoy the music the more because of it: like an aria that "spreads as northern lights in ice pale blue" or a fugue flashing red and orange flames. Is this purely subjective? Or could there be some objective explanation, as is suggested by the fact that a surprisingly high percentage of nearly a hundred different listeners, questioned separately, have observed the same colors with the same tunes, and too often for it to be explained as coincidence? Historical research has even unearthed old records to show that Bach called E flat a grayish tone but B flat yellow-green while Schubert saw E flat as reddish-gold and B flat simply green, and Rimski-Korsakov viewed C as white and E flat a gloomy bluish-gray. Psychologists term such correlations chromatic phrenopsis (possibly an emotional cross-leakage of current between optic and auditory nerves), which, under various names, continues a tradition going back at least into the Middle Ages, possibly as far as Pythagoras, to whom is attributed the original concept of an analogy between the seven colors of the rainbow and the seven notes of the musical scale: C = red, D = orange, E = yellow, F = green, G = blue, A = indigo, B = violet.

An even more controversial cross-sense is the repeatedly observed finger vision known as DOP (dermo-optical perception) through which certain extraordinarily sensitive persons, mostly young women in both America and Russia, have demonstrated, while blindfolded under stringent laboratory conditions, that they could read newspapers and recognize colors, some with opaque shields between their hands and the material they "saw," a few reading fluently even with their toes, feet, shoulders and elbows. One Russian girl also proved able to taste with her forearms. There was general agreement that light colors felt smooth and thin, while dark ones were rough and heavy. Yellow was slippery, soft and light in weight; blue smooth and cool; red warm, sticky and coarse; green stickier; indigo like glue; black like thick tar. Theories to explain the phenomenon ranged from that of Professor Richard P. Youtz of Barnard College in New York who thought it might be an infrared sense to Russian psychologist Abram Novomeysky's electromagnetic theory that it is akin to the "seeing" of electric eels.

## ABSENCE OF SENSE

Space research seems to be the latest discipline to contribute to sense science, mainly by conducting experiments into what the ab-

sence of any sense stimulation might do to a man — say, a man lost alone in space. And it found, to nobody's great surprise, that floating limply in a tepid bath in the silent dark hour after empty hour is very trying for most of them, normally leading to hallucinations within a day. Various subjects have reported seeing "prehistoric demons," rows of little yellow gnomes, squirrels marching with knapsacks through futuristic beehive cities, five-dimensional teeth and the ultimate "gone" feeling of being swallowed down an "astral throat" into a "stomach outside the universe" — some such effects persisting for days after returning to normal living.

The world record for enduring "total" sense deprivation — staying alive, conscious and sane without appreciably seeing, hearing or feeling anything — is 3 days and 20 hours, recorded in 1962 at Lancaster Moor Hospital in England. This ordeal of course did not include motionlessness, the world record for which is only 4½ hours, and I doubt if the darkness, silence and feellessness were anywhere near total.

In actuality there is a large, sometimes fearful, amount of background noise to be heard in one's ears if one rests quietly, listening for it and tuning on it. Even on a still night in subfreezing weather I find I can hear something like crickets chirping, wind whistling, machinery grinding and of course my own heart thumping. And something comparable occurs with vision for I see what appears to be the Brownian movement of molecules in the air and other mysterious moving forms and colors, especially when my eyes are shut.

## PHOSPHENES

This brings up the subject of a kind of inner sight that is hard to categorize because it is not yet well understood but seems too important to omit from our discussion of senses. It is the phenomenon of images known as phosphenes, the scientific word for the "stars" you see when your head gets banged and for the scenes that appear when you're half asleep or when you meditate with your eyes closed. Derived from the Greek *phos* (light) and *phainein* (to show), phosphenes may appear whenever visual input from outside fails to penetrate your eyes. They are believed to originate primarily inside the retina and brain, "reflecting the neural organization of the visual pathway," and may be the nearest thing to a scientific explanation for the visions of religious mystics. Pilots flying alone in empty skies at very high altitudes habitually experience phosphenes, and presumably astronauts on long interplanetary voyages will be familiar with them, al-

though at least some of the flashes already seen by astronauts going to the moon are deduced to have been caused by the heavy particles known as cosmic rays.

Phosphenes are also seen, probably inevitably, by all normal young children (not to mention animals), to whom they may be as real as the external world — that is, until, little by little, the unfolding years of growing up shed light on how to tell the difference. Between the ages of two and four, when the child can hold a crayon but knows little of how to draw objectively, he is most apt to draw things with a distinctly phosphene character. And this is about equally true of primitive humans who lived during mankind's childhood, to judge by the phosphenelike figures in some of the prehistoric cave paintings, in folk art and Indian blanket designs. Drugs likewise bring on phosphenes, particularly hallucinogenic drugs like mescaline, psilocybin or LSD. So does alcohol, as anyone who has been through delirium tremens can tell you, and diseases of high temperature, particularly scarlet fever.

migraine          psychedelic          eyeball pressure
PHOSPHENES

Probably the simplest way to see phosphenes though is to shut your eyes and rub your eyeballs hard. This is pretty certain to ignite for you an array of lights like a city viewed at night from an airliner, a dramatic crystalline checkerboard or moiré pattern featuring many colors and flashing rubies, diamonds, sapphires and emeralds. That these patterns are not just random is now well accepted, especially as they were recently classified into fifteen categories by a researcher named Max Knoll on the basis of reports from more than a thousand volunteers. While psychologists seem reluctant to conclude more than that "certain forms are characteristic of each pulse frequency for each individual," to my mind these fifteen characters are rather otherworldly and exciting and I let myself imagine they just might be the alphabet of some still undiscovered interworld code or script — or maybe even the signs of a mental zodiac of the universe.

237

THE FIFTEEN BASIC PHOSPHENE FORMS IN ORDER OF COMMONNESS

## UNSENSES

If phosphenes are a manifestation of an inner sight that blossoms into being when one's normal outer vision is cut off (by eyelids, injury, drugs, etc.), they are a living testament to the abstract nature of the world, the relativity of its qualities and to the paradox that we must *unsense* some things in order to *sense* other things. The fact that only when it is dark can you see the stars is thus reconciled to the fact that only when you lose some senses do you become aware of others. Helen Keller became the classic and eloquent advocate of this principle when she exclaimed "I sense the rush of ethereal rains . . . I possess the light which shall give me vision a thousandfold when death sets me free."

Sometimes it is easier to adjust to losing sight, as John Howard Griffin did when blinded by a bomb in World War II, than to the return of sight, which shocked him when his vision unexpectedly came back more than ten years later. The sudden intensity of the light then seemed cruel. When an understanding attendant turned off a lamp, he said, "I felt as though a burden were lifted from me — safe, at home in the dark . . . Certainly this adjustment [to vision] is more difficult than the one to blindness . . . Then I was alive to all stimuli. Now I am blurred to all of them except sight."

The aspects of relativity may be even more striking in the case of hearing. At least that is suggested by the story of the aging English earl who liked to entertain diplomats and, when he grew hard of hearing, trained his servant to beat a drum in a certain rhythm whenever one of the guests spoke. The earl wasn't really perverse, as you might think, but had made a practical discovery as to how to hear better. The noise of the drum, hardly noticeable to his own half-deaf ears, nonetheless forced the speaker to raise his voice to the earl's hearing amplitude. At the same time, for other listeners, it effectively drowned out what was being said, so the earl could enjoy his most sensitive conversations in exclusive privacy.

A lot more important of course were the deafnesses of such composers as Beethoven, Bruckner and Smetana, whose aural failures may well have helped them and the world by intensifying their concentration on listening and by blocking extraneous sounds. Beethoven

238

is the most dramatic case of all. When his First Symphony was first performed in 1800, young Ludwig, already famous at thirty, was also hard of hearing; two years later when his beautiful Second Symphony was completed, he could barely hear a full orchestra through his ear trumpet. Yet, as his deafness reached totality, that work was surpassed by his Eroica, Moonlight Sonata, Fifth Symphony and, after a quarter century of deafness, his glorious Ninth Symphony. If he hadn't gone deaf, one may presume that he would have continued his brilliant career as a Viennese piano virtuoso, as well as the teaching and social activity that traditionally went with it. But he was an extremely sensitive man and so deeply embarrassed by his deafness that he became a recluse, retiring shamefacedly to the little suburb of Heiligenstadt at the age of thirty-two, largely shutting himself off from the world while moodily, broodily withdrawing inwardly, dreaming, composing and recomposing, hearing music only in his imagination. Indeed without the loss of his hearing, his unequaled profundity of creation might never have flowered. And a paradoxical truth we may conclude from this is that sense in general is not only life's bridge to the world but also its inexorable cloud that veils and distorts reality.

## THE MEANING OF SENSE

Surely the world adds up to something greater than just what our senses tell us. But what? And if the real world is not the material one we sense, what is matter for? And is the stuff of Earth and the universe in any way intrinsically base or unworthy as some spiritual leaders seem to imply?

I submit that the essence of matter may be that it is the means (God's means, if you can accept God) of acquainting us with facts during this elementary finite phase of our existence. For, if there were no material world of sense, would not some other kind of world have to replace it, assuming we are to learn anything or grow or evolve? And would not that substitute world have to be sensed in some way also if it were to serve its purpose? And, in doing so, would it not demonstrate that it too must be a material, palpable world — indeed a world just about like the one we live in (and presumably *must* live in) at this stage of our development?

If this line of reasoning be close to the mark, perhaps the material world is nothing but the stuff of consciousness, composed of things that can be tuned in on by senses. For even an odorless gas like carbon monoxide can be sensed deductively or indirectly with instru-    239 ments while, if something cannot be sensed, it hardly can be termed matter. This, then, would make the material universe subjective in essence, something each of us tunes into a part of with his own individual organs of sense, leaving the so-called objective aspect of the world a mere mosaic of the innumerable subjective facets we behold!

What is really heard in a sound? Not merely the pattern of air compression and rarefaction but also the time-space relationship we call pitch in a musical note and, beyond that, the harmony and timbre of multi-notes blended together. And, beyond that, a melody of sequential changes. And beyond that, nothing less than the SPIRIT of the composer!

Thoreau must have understood this intuitively, even as a boy, for years later he wrote, "In youth, before I lost any of my senses, I can remember that I was all alive, and inhabited my body with inexpressible satisfaction; both its weariness and its refreshment were sweet to me. This earth was the most glorious musical instrument, and I was audience to its strains . . ."

But he became aware also of forces beyond any known senses, profound and mystic forces that whispered to him of the abstract nature of the world. Naturally he wasn't thinking of artificial senses or bionics, a science interrelating animals and machines that hadn't yet been imagined, nor of future sensuous inventions for the handicapped like aural readers that "see" printed words and "utter" them with sound, of audible ink, visible voice prints, electronic muscles, nor of technological sense aids like metal that, when it weakens from fatigue, will scream ultrasonically for help, of chemical diagnoses by body effluents or even of the density waves in galaxies that travel 120 miles a second and may somehow someday be heard as a kind of celestial melody. No, the music of Thoreau's spheres obviously carried beyond the material realm and into what is generally regarded as the divine, of which he wrote in humble simplicity, "I perceive that I am dealt with by superior powers. This is a pleasure, a joy, an existence which I have not procured myself. I speak as a witness on the stand, and tell what I have perceived . . ."

What he perceived must naturally remain partly veiled in mystery, but it is not alone Thoreau's mystery nor that of any man. For it is the mystery of the sculptor of mountains and the molder of seas, of the inspirer of every wind ever seen or heard or breathed or felt. It is the mystery that out of stillness engendered motion and out of silence created sound and out of sound, music. It is the mystery of all the uncounted, unseen worlds that ever await unveiling for the eyes that are capable of beholding them.

# CHAPTER 9

# Emergence of Mind

O NE DOESN'T HEAR much talk about the noosphere these days — not even among astronomers or astronauts despite the fact that the noosphere is a vital layer of any mature planet and so something a person dealing in celestial bodies might expect to come face to face with often.

The noosphere is invisible and intangible. It is not made of matter. It is describable best perhaps as a boundless mental aura surrounding a living planet. It is a concept Teilhard de Chardin introduced in *The Phenomenon of Man* as the "thinking layer" around Earth, located more or less outside her metallic core, the barysphere, and her rocky crust, the lithosphere, enveloping her watery hydrosphere and airy atmosphere which together comprise the biosphere of life. It is a rather new abstraction in this part of the firmament, one that adds mind to what once seemed a mindless world: that pale globe I see down there rolling ponderously across its counterpane of stars, rolling, as we seem to have presumed, for billions of years, unconscious, blind and deaf to its own existence as a corpuscle of the universe.

Whence, one must ask oneself, arises thought from such a body? Could thinking have germinated somehow in Earth's rocks or molten interior, in her shifting magma, which is known to create electromagnetic fields? Such a hypothesis would appear a logical maturation of Niels Bohr's profound surmise that the wave aspect of the atom is the mind aspect of matter (page 408) and that the electron may possess in significant degree a germinal will of its own. Or should one think of mind's emergence as happening at the higher level of the atom where the waves and forces seem to have settled into a more stable and balanced state? The answer, I believe, depends very largely on one's definition of mind although, in general, the more developed any kind of matter is, the more easily we can accept it as including some degree of potential mentality, perhaps even a hint of inherent purpose behind the direction of its evolvement.

When it comes to larger and relatively more complex units of the material world such as molecules, particularly the DNA molecule and certain viruses, evidence of motivation and purpose becomes definitely easier to imagine. In plants it may manifest itself as interaction between bacteria in a root, as consultation among branches (page 62) or possibly of trees in a forest. Darwin seemed convinced that vegetables not only had strong feelings but also a rudimentary power to think. In writing about the radicle, the part of a seed that grows downward and becomes a root, he said: "It is hardly an exaggeration to say that the tip of the radicle . . . acts like the brain of one of the lower animals; the brain being seated within the anterior end of the body, receiving impressions from the sense organs, and directing the several movements."

## THE FIRST MEASURE OF MIND

Probably the earliest well-accepted mind in evolution is that of the simplest of animals, the one-celled ameba, after whom the mind seems to diffuse outward and complexify. But even in this elementary creature one can scarcely fail to notice the evidence of mind, as the following ooze-by-ooze account of an ameba learning by experience will show. At eight o'clock one morning a large ameba surrounded and swallowed a small piece of a medium-sized fellow ameba but did not completely engulf it, leaving it still attached by a protoplasmic thread to the main body of the medium ameba, the thread passing through a tiny opening in the large ameba and by means of which the swallowed piece made strenuous efforts to ream the opening and escape. The large ameba, however, foiled these attempts by outflanking the ameba morsel and finally "biting" off the thread while backing up, which left the now-independent small ameba almost completely

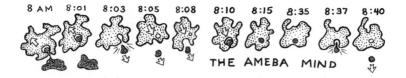

enclosed. But the small ameba continued its efforts to get away by flinging appendages out the opening (now in the rear of the retreating large ameba) and, after two minutes of struggle, it managed to break out and get entirely free. At this point the large ameba, presumably hungry and remembering it had just lost a meal out its backside, quickly "reversed its course" until it overtook the quarry again and surrounded and swallowed it, leaving no opening. This time the embattled little ameba, seeming to realize a shift in tactics was called for, contracted into a tight ball and remained quiet, biding its time while stoically resisting all digestive attempts by its captor. It must also have been aware (probably through some sort of pressure sense) of how thick were the walls of protoplasm enclosing it, because when, in the course of the motions of the large ameba, there chanced to be a momentary thin place in the wall next to it, it lost no time in poking its way through and escaping so swiftly the large ameba seemed surprised, confused and discouraged — at least enough so that it made no further attempt at pursuit.

If such responsive behavior by microscopic one-celled animals demonstrates some sort of stream of consciousness, intermittent or continuous, how much more evident is conscious thinking in hydra and jellyfish, which have fixed body polarity and coordinating appendages that deftly pass food from tentacle to tentacle until it is put into a mouth? Do such animals have a real choice as to where to turn or what to grasp? Or are they simply taking nature's orders, following the line of least resistance, obeying the dictates of vim and whim?

These sorts of questions have been pondered by philosophers for thousands of years, as the famous conversation in Chinese between Chuang-tzu and Hui-tzu recorded in the fourth century B.C. suggests. The two men, perhaps brothers, had strolled onto the little bridge over the Hao River when Chuang observed, "Look how the fish are swimming in the water! How happy they are."

"But you aren't a fish," said Hui. "How can you know the fish are happy?"

"And you are not I," retorted Chuang. "How can you know what I know of the fish's happiness?"

"True," said Hui, "I'm not you. So I cannot know what you know    243

of the fish's happiness. But the fact remains that you're not a fish. So, by your own reasoning, you cannot know the fish are happy."

"Ah," said Chuang. "Though I am not a fish, I know how happy the fish must be — by my own happiness when swimming in the water."

Today this sort of knowledge is called anthropomorphic intuition, and biologists reject it as unscientific, preferring to study the animal mind by testing, measuring and analyzing animal behavior.

## Mind of the Ant

The ant, for instance, is known to be guided in most of its activity by instinct, yet in half an hour the average ant can learn its way around a square yard of ground it never saw before and will usually remember it well enough to return to a nest entrance in a direct line from any point. Ants have been tested in mazes too and proved they could not only correct their mistakes but, after a few dozen tries, avoid making them — that is, if they weren't too old. For youthfulness in ants, as in most animals, is more flexible than age and young ants will adjust to being suddenly put into an artificial nest, an experience so traumatic to their elders that few of them survive it. Ants, after all, have only a tiny, simple brain and almost no comprehension of what we call cause and effect. When researchers hang a piece of food just out of reach above them, they make desperate but futile efforts to reach it, never realizing they could easily do so by piling up half an inch of dirt under it, or a few twigs to erect themselves a ladder.

Ants also seem absurdly gullible when it comes to hospitality toward some of the dubious creatures they live with, a matter that inevitably reflects their mental level. For although they will fight to the death to keep an obvious stranger from entering their nest, ants have often been observed taking in, and being taken in by, some of the most thinly disguised villains in the whole creepy, crawly world, who do things like eat ant larvae and eventually wipe out the entire tribe, and among whom entomologists have actually recorded more than three thousand different species of beetles, worms, flies, spiders and other arthropods.

But ants do act with logic in many situations. When an orchardist paints a band of sticky material around a tree trunk to block invaders, ants have been known to carry mud from the ground and use it to build themselves a causeway over the barrier. When table legs are put into buckets of kerosene so ants can't climb up them to get food on the table, the resourceful creatures sometimes go to the ceiling directly overhead and simply drop onto the food. Harvester ants store seeds for food and seem to realize that, when their earthen granaries get wet, the seeds will either rot or sprout and spoil, so they take them one by one out to dry when the weather is sunny or, if it remains wet, they bite out each seed's germ to degerminate it so it won't sprout, thus saving a whole winter's food supply by the only means they know, a means whose evolution seems to have required some degree of thought.

When slave-keeping ants attack a strange ant city seeking more slaves, they usually first send out scouts to reconnoiter and ascertain the exact location of the city approaches, the strength of its garrison, the entrances and how best to reach them. And they seem to study their chances of success with military objectivity, sending for reinforcements when they find they are dangerously outnumbered. Even after launching an attack, if the enemy forces turn out to be stronger than expected and too many of the attackers are getting wounded or killed, a strategic retreat may be called and the assault postponed until more reinforcements arrive or, in extreme cases, the whole plan is summarily canceled and the army marches prudently home.

In some observed cases, on the other hand, a kind of cold war of nerves is waged, featuring provocative skirmishes and menacing gestures around the perimeter of the enemy position, a stratagem that, often as not, wins a bloodless victory. Just how this works and how the ants communicate strategic ideas is not easy to understand but it is known that when an ant city is threatened on one side, its defending citizens commonly move their most precious possessions (eggs, pupae, aphids, etc.) out of their deep chambers and to the upper opposite side, obviously in anticipation of a forced retreat and rapid evacuation should such moves become necessary.

Under more peaceful circumstances, when an ant colony begins to

245

feel dissatisfied with its nest location and is contemplating moving to another place, its citizens may hold something like a referendum to settle the issue: ants in favor of the move picking up their eggs and other possessions and carrying them over to the new place while ants opposed tag along "empty-handed," but, on arrival, repossess the baggage and carry it back the same way they came. Thus appear two parallel columns of ants with baggage moving continuously both ways. At first such a procedure would seem inefficient to the point of absurdity, but, if you reflect on it, you soon realize it is like a dance marathon of the polls and that the "dancing" column with the most endurance and determination (usually the majority) is bound to last longer and win the election. In any event it is a recognized phenomenon in the ant world, a voting system that is both democratic and well suited to social insects.

An even more remarkable function of insect government is the matter of police work and executions, which often involve close (and perhaps controversial) decisions as well as very determined action. The mental processes that lead to the action are not yet known but entomologists have noted that ant colonies that tolerate beggars and exploiters in prosperous periods tend to change their attitude in times of famine, when they will suddenly turn on these objectionable ones and kill them. Termites have a comparable practice in their system for reducing the number of their soldiers after a war. When defending themselves against besieging ants at the onset of war they rapidly breed extra guards but, not having any veterans' organization and the specialized guards being nonproductive and unable even to feed themselves, when the siege has been lifted the termitary evidently feels an urgent need to get back to its normal ratio of one guard to a hundred other termites. So the order goes out to destroy certain of these idle and hungry soldiers, which is done by starving them, then eating them. A startling but significant fact in this dire performance, however, is that the victims are individually selected. Then, in the presence of their fellow guardsmen, they are refused food. This is known through many delicate experiments which show that the victims are not picked for being tasty or nutritious, not at random nor, so far as we know, by criterion of age, health or length or quality of service, but simply as surplus troops. Yet why certain individuals, presumably as innocent and worthy as their fellows, are sentenced while others are reprieved seems so far to have eluded all investigators.

The eventual answer may turn out to have something to do with the mysterious group mind of social insects, which has been speculated about by both scientists and philosophers for generations. Who runs an ant or termite city anyhow? Or a school of fish? We will return to this subject later when we take up transcendence in Chapter 19.

## GROUP MIND OF BEES

Meantime let us consider the beehive as an organism with its own mind. The fact that an individual bee cannot live apart from his fellow bees longer than two or three days, that two bees together can survive scarcely a week while no number of them less than forty has ever been able to maintain a hive to my knowledge, should establish their colonial nature if not their collective mind. Like other societies of social insects, the hive lends itself to appropriate and serious consideration as a volatile yet integral organism. It weighs about ten pounds in the form of 50,000 cells (bees) ranging over a square mile or more of territory. It breathes, circulates, metabolizes, regulates its temperature, considers, decides, defends itself, eats, drinks, evacuates, mates and procreates. Most remarkable of all, it intracommunicates with itself in a formal coded dance (with sound and smell effects) that amounts to the most abstract and specific animal language yet discovered by man.

This bee language was pieced together little by little in the 1920s and 1930s by Karl von Frisch, the famous Austrian zoologist who has gradually become so intimate with his subjects that, as he says, he literally senses a hive from the inside and "feels himself a bee." It is a language used primarily by bee scouts who have discovered a source of nectar too far away to be readily findable by smell or sight and who want to share it with their fellow workers. They tell about the new nectar by doing a very precise dance on the honeycomb inside the hive, which consists of quivering the abdomen while going through a curious figure eight maneuver. But the meaning of this dance, so

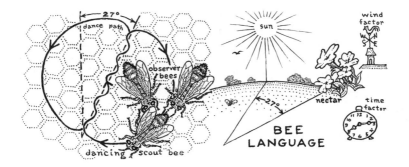

247

painstakingly translated by von Frisch, is that the middle axis of the figure eight gives the direction of the nectar in relation to the direction of the sun, which (replacing light with gravity for the purpose of the dance) is assumed to be straight up or exactly above the top of the comb. Thus if the dancing bee moves up the comb at an angle 27° to the right of vertical, she is telling her sisters to fly 27° to the right of the sun to find something good to make honey out of and, after they have "read" the message several times and understood it, they actually go out and fly the designated course. Even in cloudy weather a bee can usually see the sun's position through her awareness of ultraviolet and polarized light. As to how far bees are to follow the course, that instruction is imparted by the tempo of the dance, a near destination being indicated by faster dancing and a far one by slower dancing, always in direct ratio. And the kind of nectar is also revealed in the dance, this time by smell since the scout inevitably has picked up a sample of it on her legs and body. Besides that, she hums quietly, and not just to herself, the tone intermittently coming from her wing beats at 250 cycles a second which, with mystic appropriateness, turns out to be the key of B. I said "intermittently" because the bee hums audibly only during the middle or "straight run" portion of her figure eight dance which, being a fixed percentage of the whole dance, has a duration in direct proportion to the distance to the new nectar source and therefore tips off all the bees who hear her as to how far away it is.

Sight being scarcely possible in the dark interior of the hive, the bee scout's message is not read visually like seeing black ink upon a white page but more as a blind person "reads" Braille: by the sense of touch. Usually several curious bees cluster eagerly around the dancing scout, feeling her with their palpitating palps, noticing everything: how fast she moves, what she tastes like and, above all, the exact angle of her path across the comb. Incredible as it seems, the observing bees take in every detail and remember them so well they have been seen to fly minutes later within ten or fifteen degrees of the designated course, an acceptable margin of accuracy at the ranges involved. And, like all competent aerial navigators, they steadily shift their flight directions to match the movement of the sun through the sky, this obviously by instinct with the aid of their version of the biological clock. And they even make corrections for the wind by heading just far enough to the right or left of their assigned course so that the wind blows them back to where they would have been if they had headed exactly on course in a dead calm. But, as if wind correcting were not remarkable enough, they always uncorrect their headings when doing the dance, so the witnesses receive the simple sun angles, leaving it to each bee to apply her own wind adjustment according to the weather conditions of the moment.

It would seem only natural for a society with such a highly developed language to use it in governing itself. And that is what beehives actually do. When the bees decide to move to a new location, an operation called swarming, they begin by sending scout bees out looking for building sites, inspecting hollow trees, holes in the ground, boxes and other possibilities. Then, instead of choosing one of these locations by some sort of "marathon" voting system as do ants, the bees hold what might be called a nominating convention in the hive, each scout describing where her favorite spot is by her waggle dance language, and expressing the strength of her feeling by the number of times she repeats it. Then, having narrowed the choice down to two or three places, the scouts go out again for a final look at the most popular sites. After that they meet in a conclusive plenary session where, by conversation and genuine persuasion, they almost invariably reach a unanimous agreement and fly to the new home. I say "almost" because I've heard of one case in which a swarm conferred for two whole weeks yet still couldn't make up its mind about two equally attractive sites. Then on the fifteenth day the bees finally gave up — and gave up so utterly that their strange solution amounted to hive suicide. They actually turned erratic and built a new hive in an obviously unsheltered spot in the nearest bush, where they froze to death the following winter.

If one of the marks of a sophisticated language is its propensity to diversify into dialects, the bees' language must qualify. For von Frisch found many dialects of it in use by the various kinds of bees. Primitive dwarf bees, for example, dance on a horizontal comb outdoors, aiming the "straight run" of their figure eight directly at the destination, which obviates any need for transposing the angles of light and gravity. Most honeybees substitute a round dance for the figure eight when the nectar is near, and Italian bees do a rather sickle-shaped figure eight dance to indicate middle distances. But different races of bees use these "scripts" differently, Italian bees doing the round dance to indicate nectar within 30 feet, a sickle dance for ranges between 30 and 120 feet, and the figure eight dance beyond, while Austrian bees do the round dance all the way to about 500 feet and dance all their dances noticeably faster. Indian bees, on the other hand, do the figure eight to within 10 or 12 feet of the hive and much more slowly than the European bees. Which has the consequence that bees of different species, when mixed, often misunderstand each 249 other, an Austrian bee, for example, taking an Italian bee's figure eight dance (completed every two seconds) to indicate the nectar is 1000 feet away when the Italian bee meant to say only 700 feet.

## SOME OTHER ANIMAL LANGUAGES

Other examples of animal language are hardly less extraordinary, and a few even depend on writing. Best known is probably that produced by the spider who spins a circular web that "writes" a silken script through which he converses with his neighbors, his mates, victims and, in some degree, with whoever passes by. He does it not with eyesight but by feel and instinct, usually at speed and in the darkness before dawn, measuring and adding a new rung to the orb every second and making two complete revolutions around it every minute, the whole taking less than half an hour.

Of course there are many kinds of orb webs (not to mention webs shaped like funnels, bowls, domes, tubes and purses), each of which represents but one of the thousand cobweb designs or dialects these little creatures have evolved in the last hundred million years. And though indelibly fixed in their genes, the designs outwardly express the spiders' inward moods and minds, being somehow influenced by

SPIDER WEBS SPUN UNDER THE INFLUENCE OF

marijuana        benzedrine        caffeine        chloral hydrate

every environmental factor from weather to diet. In fact a surprisingly potent influence on webs is any drug the spinner may recently have absorbed. If he was fed coffee, for instance, his next web will be a loose, ragged array of crooked, unfinished spokes. A Benzedrine web is more organized but rather one-sided. A marijuana orb lacks outer threads. A scopolamine or mescaline one is smooth but disoriented, while one spun under chloral hydrate, the barman's "Mickey Finn," is barely begun before the spider passes out. And the remarkable thing about all drug-affected webs is that they so clearly characterize each causative ingredient that research laboratories have started domesticating spiders for the sole purpose of using their webs to test drugs.

Grasshoppers are not known to write, nor does one usually think of them as linguists, but entomologists have discovered they have a vocabulary of twelve calls appropriate to twelve situations, notably courting, in which subtle variations of dialect are the main difference between one species and another, a difference implemented by prevention of mating through inability to comprehend and respond to another's love notes. Some male butterflies have a clicking language to which not only other butterflies but occasionally birds respond. And some can speak both when they are caterpillars and while pupating inside a cocoon. Even earthworms talk to each other in faint crackly voices, and some ants emit dainty squeaks, while tree ants drum with their heads to converse tree to tree, although the ten-word ant vocabulary, as presently known, contains more words of smell (such as "I'm one of yours" or "Beware!") than of sound. All in all there are estimated to be at least ten thousand species of singing insects, each with its own language that is identifiable to almost no one but closely related insects, each kind tuned to its unique individual cadence of "words" and intervals.

Fish have a surprising range of voices, though not all of them hear well. The horse mackerel grunts like a pig, the doras growls like a wolf and, more appropriately, the electric catfish hisses like a cat. The maigre, a large Mediterranean fish, has the most varied vocabulary of all, being known on different occasions to purr, buzz, whistle and bellow. And evidence that his finny audience really grasps the meanings of such utterances of the deep comes from experiments in recording and replaying them, like the study of damselfish language by Arthur Myrberg in Bahamian waters in 1969. Having learned the meaning of damselfish words, Myrberg found he could play a chirp and make one of these fish go into a spawning dance involving a 45° twist followed by a U-shaped dip. A slightly different call would persuade the fish to change color. Myrberg also discovered a low-pitch tone that is a potent lure for sharks, bringing dozens of the big marauders out of "nowhere" in less than a minute. And now the U.S. Navy is testing the idea of mobilizing a shark patrol to protect ships against enemy frogmen and perhaps developing a technique for talking them away from sinking ships in the interest of human safety.

Such examples of explicit animal language are obviously but a few out of the planetary proliferation in every meadow, barnyard, desert, sea or forest on Earth. As Carl Sandburg once put it, the horse utters 251 "the whicker of expectancy, the whinny of excitement, the neigh of love." Comparably the lion coughs, grunts, sniffs and roars, while his lioness moans to control her whining cubs. And the tiger may have a bigger vocabulary: a gentle moan of disappointment and a louder groan of warning of his approach, a barklike "pooking" to a fellow tiger, a cough to frighten scavengers from a kill, a whistled mating call, a deep confident roar, a higher, wilder roar of anger that announces his charge, and an imitation of the mating cry of the sambar stag by which the tiger lures a hind within range, the only instance I can think of where a nonhuman predator uses aural camouflage to influence his prey.

Some animal talk is revealed in their social games too, which are often similar to those of human children. Lambs and gibbons are fond of playing "follow the leader" with various bleatings and chortles. Young deer and otters prefer "hide and seek." Badgers like "king of the castle," and almost all young animals enjoy "tag." Solitary games are common also like the favorite of young hippos who love to blow a floating leaf high into the air from under water accompanied by gleeful gurgles, sometimes repeating the trick a dozen times just for the fun of it.

## Bird Language

As for birds, they have probably evolved more languages than any other class of animals despite their bird brains that are often smaller than their eyes. Even chickens, not noted for wit among birds, enjoy a vocabulary with several explicit clucks and calls for their chicks: to summon them, admonish them, alert them to danger, tell them about things to eat. And, in adult relations, they utter words of menace, of suspicion, challenge, triumph, fear, submission, affection — about thirty expressions in all.

Bird language in fact ranges all the way from conversations between eggs (page 143) to the ultrasonic subtleties of the antiphonal love duet (page 636) and is usually used quite judiciously. If a sea gull finds a small piece of food, for example, he may eat it without a word, but if he discovers a large supply, he generally utters the food call that brings dozens of other gulls to share his good fortune. Also birds evolve local dialects in the different places where they live and ones that migrate need to learn these variations if they expect to communicate fluently on their travels. Crows in North America have very definite alarm calls that tell other crows to fly away from danger, a cry for help when caught and an assembly call for when they sight an owl or large hawk or perhaps a prowling cat they would like to mob. If such language is played on a high-fidelity recording in American woods, any crows present have been found to recognize and respond to it. But if played in France they obviously misunderstand it. In fact French "crows" (including rooks and jackdaws) assemble rather than flee when they hear the "scram" alarm of Pennsylvania crows. Pennsylvania crows, in turn, do not respond normally to the cawings of Maine crows if they have been confined and deprived of contact with strangers, but crows free to migrate between the two areas soon get to understand both dialects, and some Pennsylvania crows, shipped to France, have even learned to converse with French jackdaws.

The same goes for gulls, French herring gulls making neither head nor tail of the conversations of American herring gulls. Yet ocean birds that visit both continents understand not only the dialects of all closely related species but many of the languages of quite alien ones. And there are countless known instances of one order of animal standing sentinel for another or telling another where to find food. The classic case is that of the black-throated honey guide (*Indicator indicator*) of Africa and southern Asia who loves honey but fears getting stung by bees. This alert little bird is adept at finding wild bees' nests, presumably by following the bees. But once he has discovered such a nest, instead of going after the honey directly he looks for someone to do it for him. The someone usually turns out to be a man and the bird flutters around him uttering such desperate cries reinforced by such unbirdy behavior that the man cannot help realizing the creature must be purposely attracting his attention. When the bird in turn recognizes that the man realizes what he is up to, he immediately flies toward the bees' nest still crying out his version of "If you want honey, follow me," and returning again and again to make sure he is being followed, his excitement steadily rising as the man gets nearer the honey. Of course the man is usually a native who understands the bird's intention, honey guides being well known in Africa, and naturally he has been taught how to open the nest and help himself to honey, gratefully leaving a well-earned commission for the bird.

In cases where the honey guide cannot locate a man, he usually seeks out a species of badger called a ratel instead and tells him about the honey in the same fluttery language of which the ratel also has an inborn understanding and an appetite that makes him more than glad to cooperate, including his inevitably if inadvertently leaving behind more honey than the bird can eat. To sum up, the honey-guide language is so well known in Africa today that, if a man or any articulate    253 animal imitated the bird in front of a ratel, the ratel would almost surely be reminded of the honey guide and would presumably follow him, visualizing a hive full of honey.

If you wonder where birdsong comes from, you would be interested in the recent research using apparatus for accurate sound analysis which shows that a nestling songbird, if reared out of hearing of any birdsong except his own, will develop singing ability as he grows up only in a rudimentary way and without the intervals and flourishes of his wild relatives. His limited song in fact may be taken to represent the part of his singing potential that he inherited through the egg.

If several young birds are confined together in a room, however, but prevented from hearing any outside birdsong, their singing efforts will stimulate each other enough to enrich their average song with a distinctive "group melody," even though this remains much simpler than the full natural melody (or melodies) learned and developed by similar birds allowed to hear their elders. And it turns out that there is a short and critical period (in some cases lasting barely a month) late in a bird's first year of life during which his song (whether learned or only inherited) is imprinted in his mind and so fixed that it remains essentially the same thereafter, only minor changes being possible for the rest of his days.

Birdsong of course can become very exuberant in the wild, including both antiphonal melody shared by several birds and four-note chordal harmony sung through the Y-shaped syrinx of a single bird. And it is a more specific language than you might suppose, including as it does the courting and mating songs that not only attract females, warn off competing males and delineate territory but that stimulate or regulate nest building. The Reverend James A. Mulligan, S.J., professor of biology at St. Louis University, indeed demonstrated the latter in 1968 by surgically deafening canary hens, which resulted in their taking an average of thirteen days to build their nests instead of the normal five days and, in a few cases, stopped them from building any nests at all.

Other research about the same time indicated that the male songbird, say a chaffinch, sings a staccato soprano note called a "chink" when courting a female, which both attracts her and is a sound of short enough wavelength to be deflected by small objects, enabling her to cock her head in different directions and locate him by the pattern of sound "shadows" cast by her head. The abrupt beginning and ending of the staccato note also reinforces her directional sense through the micro-interval between when each "chink" reaches one ear as opposed to the other.

On the other hand, when a hawk swoops overhead, the chaffinch has a distinctly different language problem. It then suddenly becomes vital for him and chaffinchery not only to warn his companions of the danger but to do so without revealing his own whereabouts to the hawk. This he accomplishes with one very special call word, which is pronounced something like "seees." The essence of this call is that, although it is enunciated loud and clear, it begins and ends so gradually the hawk cannot time it with his ears to ascertain its direction. Also the pitch of "seees" is precisely attuned by evolutionary feedback to the hawk's head, its wavelength so nearly equal to the distance between his ears that there is no "shadow" to betray the source.

## HUMAN-ANIMAL COMMUNICATION

For millenniums man has been trying to communicate with animals. And he succeeded fairly well a long time ago with wolves, dogs and later draft animals, a good example of human receptivity being the account I heard recently of an Eskimo who remarked to a visiting anthropologist that a man with a dog sledge had been approaching his village for three days and would arrive that night.

Asked how he knew, the Eskimo said, "The wolves told us." For it is true that the tone and quality of wolf howling varies subtly but explicitly with their mood, which naturally responds more to a man with a sledge and dogs than to a man on foot alone. The wolves speak a little more gently also about a woman or a child, and differently about more than one person. And the tempo and excitement of their voices inevitably rise (as in the language of the bees) in ratio to the nearness of whatever stimulates it.

Modern studies of the languages of animals, including the partly ultrasonic utterances of dolphins and whales, suggest further that most of them have recognizable vocabularies of from roughly a dozen to a score of "words" that, on land, are almost as likely to be expressed by behavior or smell as by "speech." The dog, man's most sociable animal companion, is exceptional in his variability of expressions that include barking, baying, growling, courting, mating, sniffing, tail wagging, baring of teeth, fighting, baiting, leg lifting, panting, scratching, tracking, hounding, killing, mothering, playing, etc., which, with their nuances, add up to the largest known natural animal vocabulary of some 50 words. Indeed the only larger animal vocabularies appear to be those of a few man-trained primates, particularly chimpanzees. 255

The chimp might possibly acquire this larger vocabulary *without* human training if his larynx were like a man's, for he seems mentally capable of it but evidently doesn't have the voice to express more than a few babyish sounds. Which explains why psychological and lingual researchers in recent years have been laboriously teaching chimps to express themselves by gesticulating like deaf and dumb people or, in a few cases, by pushing buttons in special communication computers. And I think it worth mentioning that one of the most remarkable cases so far, to my knowledge, is that of a female chimp named Sarah who, by the age of seven, had a working vocabulary of 128 "words," which enabled her to converse with her trainer, psychologist David Premack of the University of California, in a large number of simple four-word sentences, not only comprehending his ideas but responding with original sentences of her own.

The "words" used took the form of symbols cut out of plastic and mounted on metal bases so Sarah could easily "write" them on a magnetized board. After months of painstaking practice, she thus learned that a blue triangle meant an apple, a red square a banana, while other symbols stood for various people, colors and familiar objects. Then she began to learn syntax: verbs, prepositions and eventually how they could be combined into sentences she could understand, as proven by her response, at an accuracy rate of about 85 percent. In one test she was handed an apple and asked to choose symbols, red or green, round or square, to describe it. She immediately picked the correct ones: red and round. Even when the apple was replaced next day with its symbol, the blue triangle, she confidently "described" it as red and round, demonstrating that her animal mind was capable of at least this much abstract thought.

On another occasion Sarah invented a sentence-completion game and taught it to her human pupil. She began a sentence: "Apple is on . . ." and arranged several possible completions, only one of which she regarded correct: "Apple is on banana." Then she induced her pupil to try her completions one after the other until, by elimination accompanied by "deep" suspense (which delighted her), he discovered the right one.

Meanwhile, at Yerkes Regional Primate Research Center in Atlanta, researchers have been teaching another female chimp named Lana a somewhat different language called Yerkish, using nine simple geometric figures that can be superimposed on each other to form hundreds of "words," each of which (unlike those in most languages) has only one meaning. Lana is already fluent enough in Yerkish to ask for things she wants, to express thanks when she gets them, apologize for misbehavior, talk to herself and, as soon as another chimp learns Yerkish, to confer with one of her own kind, which may go down in history as the first time two animals ever had a conversation in a man-made language! When that happens, perhaps we will learn, direct from nonhuman minds, something new about the potentiality of the mind in general.

The next step in mental evolution, I suppose, is man's communication with himself, which undoubtedly amounts to the most sophisticated expression of mind on planet Earth to date. We will get into that in Chapter 11, but first let us take a look at the brain, how it evolves and even probe as best we can into its mysterious interconnection with thought.

# CHAPTER 10

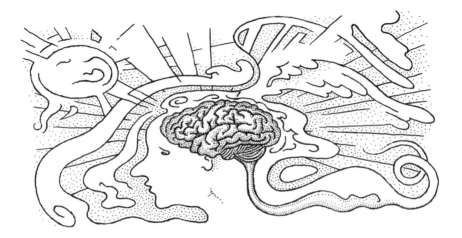

# The Body-Mind Relation

IF THERE REALLY IS such a thing as a noosphere, it seems to me
that, as it is composed of many minds, its relation to the Earth with
her many bodies must be closely analogous to the general relation
between mind and body. And this being so, the body-mind relation is
a key question for any student of life and the universe — something
he must deal with in reconciling the worlds of the concrete and the
abstract, the seen and the unseen.

With this thought then we come to take a close look at the brain,
the mind's most accepted tool, which, being the body end of the body-
mind connection, seems the obvious place to begin this chapter. The
brain of course is the exchange center of the nervous system, the place
where sensations generate ideas and ideas are expressed in action. Its
importance to the body is suggested by the fact that, while it has only
2 percent of the body's weight, its operation uses 20 percent of the
body's oxygen and blood. Its form, curiously enough, is like that of
a resting bird with folded wings, for its hundreds of billions of cells
correspond to the barbs, barbules and microscopic barbicels of feathers which, like neurons, fit together exactly, compactly, and so inti-

mately coordinated as to make the whole thing workable in shape and size. Perhaps that is why brain tissue is wrinkled and dense like a walnut or a bowl of spaghetti. And also why one must visualize the wings outspread if one is even to begin to understand the complexity and potentiality involved, a complexity that in a single human brain is comparable to all the telephone switchboards, exchanges and wiring patterns of computers, radio, TV and other electric equipment on Earth.

It used to be thought that the intelligence of animals was proportional to the size or weight of their brains. This would make the sperm whale with his twenty-pound brain the most intelligent of earthly creatures, with no real rival on land but the elephant with his thirteen-pound brain. So someone proposed a new criterion: not brain weight alone but brain weight as a percentage of total body weight, which had the effect of promoting the marmoset and other small monkeys to the top in intelligence with the genus *Homo* only a bright second.

However, further research, you may be relieved to know, has indicated that neither the relative weight nor size of the brain is a reliable clue to intelligence, nor even the size of the cells, but rather it turns out to be the number of cells that counts. A critical experiment in this field was performed at Princeton University in 1955 in which normal salamanders (called diploid because they have two sets of chromosomes in each cell) were pitted against triploid salamanders (with three sets per cell) to see which could learn to thread a maze the faster. The two types of salamander are the same size despite the fact that triploid cells are proportionately larger, because triploid salamanders for some reason compensate by producing fewer cells (including brain cells) in the same ratio. So evidently it must have been their advantage in cell numbers that enabled the diploid salamanders to master the maze in less than a third the average number of trials required by the triploid ones. Not that this should surprise anyone though, since the same principle clearly holds for telephone exchanges and computers, whose potency or value derives not from their size but almost solely from the number and interconnectability of their units.

## EVOLUTION OF THE BRAIN

A good way to get to know the brain's intricacies is to review its evolution. That means starting with an invisible nerve junction in whatever tiny primeval worm or arthropod may be presumed to be the

259

common ancestor of more complex creatures, and it includes some sort of acknowledgment of feeling and thought from there. Since brains, like hearts, lungs, livers and other organs, are not vital to all life, the earliest nerve systems did not have them, and even some human nerves still work independently of brains. If you touch a sea anemone, it will instantly collapse like a pricked balloon because its sense cells connect directly with its muscles in a simple reflex action similar to the human knee jerk and involving neither brain nor thought. But such an automatic nervous response, while fast, is also crude and, in situations requiring discrimination, apt to be detrimental to the animal involved. That, evidently, is why the internuncial cell (from "internuncio," an envoy of the Pope) evolved between the sense cell and the muscle cell to add a little leeway or flexibility to the system, a beginning of subtlety, the look before the leap — eventually a choice between right and wrong.

And from there, by trial and error in a long series of additions, the brain evolved. Significantly, however, the primordial worm's nerve junction still exists at the base of our skull, but with internuncial cells and other complexities all around it, something like the oldest growth ring of a tree, which, in the tree's old age, remains hidden in its inmost heart. And each subsequent major evolutionary improvement has annexed a layer or appendage to it, creating a kind of irregular, lumpy onion that houses the brain's main centers.

The medulla oblongata is the most ancient and primitive of these, presumably evolved almost directly out of the original nerve junction, which itself seems to have been accidental or a mutation although, to

EVOLUTION OF THE BRAIN

worm    fish    bird    rat    gorilla    man

students of deeper meaning, it must inevitably have had some mystic relation to the overall plan of Earth, to say nothing of the universe. The medulla oblongata is specifically the oblong terminus of the spinal cord and it regulates the venerably vital automatic functions like breathing and heart pumping. The cerebellum or "little brain," next oldest brain center, looks like a ball of yarn just above it and is well developed in most of the larger animals, coordinating voluntary muscles and enabling one to keep one's posture and balance. And the cerebrum is the big new brain center that fills man's whole upper skull, actually the most sensitive and advanced part of the most complex organ yet evolved on Earth, and the place generally regarded as the seat of consciousness and mind.

The cerebrum has evolved to its present maturity only in the human species but its specialized response to the conscious use of body parts is extraordinary. The brain area devoted to control of fingers and hands, for example, is much more discriminating and therefore obviously larger than that assigned to toes and feet. And the tongue and mouth section exceeds by a hundred times that of the stomach and intestines. In a hound almost half the brain may be geared to the perception of smells. In a hawk the eyes and visual cortex can actually be bigger than all the rest of the brain. And a giant dinosaur dug up in Wyoming a century ago was found to have a brain the size of a hen's egg in its modest skull, but another "nerve center" ten times bigger near its gargantuan hips which required elaborate coordination and prompted my old paper, the Chicago *Tribune*, to comment that:

> *You will observe by these remains*
> *The creature had two sets of brains . . .*
> *If something slipped his forward mind,*
> *'Twas rescued by the one behind.*

It would be a mistake to think of the human brain, or even an animal cerebrum, as an organ fully created at birth, for, like life itself, the brain is a growing, flowing thing. It first becomes discernible (by microscope) in the human embryo about two weeks after conception from when, it is calculated, every second of its early rapid growth in the womb retraces more than a century of its ancestors' long-drawn-out evolution. Starting as a tiny flat plate on the flea-sized creature, a groove appears across the center line which steadily deepens, cleaving it into two hemisectors, while the plate's edges curl upward like petals until they touch and fuse into a closed tube that is the beginning of a central nerve. A few days later a thin, gray film of cortex starts to spread across its upper surface, crinkling as it grows into the cerebrum, luxuriating like a tree until, at birth, the baby is about one third head and literally top-heavy with brains.

At that point the infant brain is still only a kind of overgrown seed, however, weighing but 25 percent of what it will attain after its long ripening as its nerves lengthen, strengthen, branch and insulate themselves with fatty glial cells. Both neurons and glial cells, by the way, are amazingly active, especially the neurons whose nuclei have been observed (for unknown reasons) to rotate, sometimes as fast as 30 revolutions a minute, which rather justifies the common saying (of       261 thought) that "the wheels are grinding." Another observation repeatedly made is that the growing nerve cells behave like colonies of bees, vibrating and buzzing as they spin, the while shooting out jets of protoplasm and probing their companions, which seems to help them learn how to transmit messages and convey meanings. For just as muscles grow with exercise, so does the brain with its uncountable parts and pathways. And its structure is created in elaborate detail by the mental experiences being lived by the growing child, mainly in preschool years, yet continuing in lesser and lesser degree the rest of his life. In this way something like half of anyone's brain is genetically inherited while much of the other half is culturally inherited (by teaching and example) and only the relatively small remainder is what the possessor of the brain creates individually by living his own unique life. Thus does the living brain day by day shape itself biologically as a physical organ through all the language, images and ideas that reach it. Thus is something obviously concrete and finite created continuously by something so abstract that it may be infinite.

## BRAIN FUNCTION

If you wonder now what factor determines which images and ideas will reach the brain and which will be screened out and kept away, the answer is that it is a very sensitive and complex matter involving both conscious and subconscious censorship as well as physiological filtering in sense organs. Tests have shown, for instance, that a dog who learns to expect 10 flashes of light per second may still see 10 flashes after the experimenter has slowed the light to 5 flashes. And a man made to wear glasses that make everything look upside down will suddenly a few days later see things right side up through the same glasses, because the pattern of images coded and conveyed by his optic nerves, then decoded in his brain, has somehow been redecoded for reality in the visual center of his cortex.

It is well known that the cerebral cortex or its underlying cerebrum is composed of two hemispheres connected by an isthmus of nerve tissue, but less known that each hemisphere specializes in its own function: the left one in expression and the right one in perception. A dramatic demonstration of this was made a few years ago in which the word HEART was flashed before several subjects but partitioned so that their left eyes could see HE and their right eyes could see ART. Asked what word they saw, all of them replied ART, presumably because this portion of HEART was read by the right eye, which

connects directly with the left hemisphere which controls speech. But when asked not to speak but instead point with their left hand to one of two cards, HE or ART, to identify the word they had seen, all of them pointed to HE, which had been read by the left eye connected to the right or perceptive hemisphere. This showed that each hemisphere had read and expressed itself according to its capacity. And, to go a step further, if you can take the word of a research psychologist named Robert Ornstein, the left side's rational, objective and analytical modes of thought generally tend to represent the scientific outlook in contrast with the right side's intuitive, subjective and holistic modes that lean more to the mystic view of life, each of which, for good reason as it now appears, regards the other as one-sided and narrow.

There is a case on record in which a man's right brain began a violent hatred of his wife. His left hand would repeatedly make obscene gestures at her and once tried to strangle her. On that occasion in fact it was only by using his reasonable right hand to pull off his emotional left that the man saved her. More recently, cerebral polarity was tested by cutting a brain's isthmus so as to completely block all nerve messages between the hemispheres, thereby demonstrating that the separated halves of the brain may take on two distinct consciousnesses, each independently capable of sophisticated thought. And when a similar operation was performed on a monkey's brain and the brain's halves deliberately given contradictory conditioning that would produce one kind of response in the left hemisphere, another in the right, the result was severe hesitancy followed by reluctant performance of both responses, one after the other.

Brains can't be expected to take kindly to being thus divided, however, for, in many ways still not understood, they depend on their wholeness and, when naturally interconnected as in a normal human cerebrum, the hemispheres are known to hold a kind of consultation in which the left one is given to articulation and talk and the right one to listening and perceiving things, including melodies, which it conveys mainly by nonverbal signs. This goes more than a little way to explain the difficulty one sometimes has in making a decision, when debate in the congress of one's mind, like that in a beehive, seems to lean 55 percent toward approval of an impending move only to encounter an unexpected 15 percent opinion shift to a 60 percent majority in the opposite direction.

When a hemisphere has to be removed from a human brain, as when it is racked with cancer, the patient does not usually survive long enough to enable the other half to learn how to take over the missing functions. But occasionally the brain somehow works it out and I

know of a strongly right-handed man whose dominant left hemisphere was removed in 1965 and who made a seemingly miraculous recovery despite his forty-seven years, not only becoming very dexterous with his left hand and arm (controlled by his right hemisphere) but having his remaining half brain compensate by learning to move his right limbs enough so he could walk with a cane, not to mention whistle a tune and even speak intelligible short sentences. With children it is easier, and the most successful therapy so far for the young, after removal of brain tissue, is to use a regimen of precise muscle manipulation to pattern-train the brain remnants with the motions and functions of the missing brain parts they will be substituting for, for the rest of a lifetime.

As this suggests, brain tissue is more adaptable than has been generally realized. In fact it will grow in a test tube, sprouting its intercellular connections and spontaneously generating its normal bioelectrical current. And researchers have taken brain matter from an unborn mouse, coaxed the cells apart with the help of enzymes and mixed them separately into a growth-sustaining solution, whereupon, like a sponge whose cells were separated by forcing it through a sieve (page 509), the brain neurons almost magically rearranged themselves, reorganized and reassembled into a reconstituted brain that was indistinguishable from the original one. It is as if an unseen director were telling each cell where to go, or an abstract mind in some infinitude beyond space-time materializing a concrete, finite brain to enable it to think here and now.

Although brain transplants must still be a long way off for humans, they are already beginning to be experimented with in animals and most successfully in very small embryos. At Yale University as long ago as 1958, for example, parts of the brains of chicks were transplanted inside the egg after 35 hours of incubation, in a few cases producing birds that hatched and lived for more than two months with normal responses, one of them even learning to answer a dinner whistle, which many birds never learn.

Soviet researchers meanwhile have kept grown dogs' heads and brains alive for days by attaching them to the necks and bloodstreams of other dogs, although without the dogs' nerves being joined. For splicing the living nerves of two animals together so they will mesh and function has not yet been achieved, nor even the regeneration of a single spinal cord in the same individual — at least not to the satisfaction of the medical profession. Yet work in smaller nerve generation and regeneration goes on advancing, so that completely severed human hands and even whole arms have recently been re-

joined to their owners and made functional. And most nerves have at least the theoretical potential of adapting to new functions such as "hearing" through eyes and the optic nerve, as Beethoven obviously did when he read musical notation after he turned deaf, or "seeing" through ears and the aural nerve, as bats and certain other creatures have been doing for millions of years. One authority in this field, Dr. Wilson P. Tanner, director of the sensory intelligence laboratory at the University of Michigan, went so far as to speculate that a baby born with crossed nerves — his optic nerve connected with his cochlea and his aural nerve attached to his retina — might hear and see as well as a normal child because it could be presumed from present knowledge that the parts of his cortex receiving the sense signals would naturally develop the appropriate interpretations of them.

Returning to brain transplantation, however, the difficulty of course is that the brain is only half the nervous system, to the rest of which it is attached principally by the spinal cord. Until the art of splicing the spinal cord has been learned (including splicing the cords of two individuals into one), transplanting a separate mature brain would be roughly equivalent to transplanting a tree after cutting it through its main trunk and leaving all its roots behind. And no one seems to have seriously faced the intriguing question (in the event that brains become transplantable) of who is the real survivor of the operation: the donor or the recipient?

## NERVE MESSAGES

It has been said that nerve messages are analogous to telephone conversations transmitted in one or two directions through wires while hormone signals are more like radio broadcasts transmitted in all directions through space. This view is of course a great advance over the ancient Greek concept of the nervous and glandular systems as a network of tubes conducting "animal spirits" or even of the nineteenth-century notion that nerves transmit hormones. For now we know there are not only two distinct systems, one sending electrochemical waves through the neurons and the other sending hormone secretions through the blood and lymph channels, but that they are interrelated and interdependent in marvelously intricate psychosomatic ways that influence (and are influenced by) everything from the growth of the body to the peace of the mind.

In this connection I can't resist mentioning a small-scale but significant case that was reported in the *New England Journal of Medicine* 265 in 1967 concerning thirteen children whose growth had been abnormally slow. When investigators found that all of them were emotionally upset by their respective parents, who in every instance were notoriously quarrelsome, often to the point of violence, the children were placed in a convalescent hospital, where within a few weeks their growth rates accelerated spectacularly. But several months later, after they were discharged from the hospital as normal and returned home, their growth rates all slowed down again to about where they had been originally — which, follow-up tests revealed, was accompanied in every case by diminished activity of their pituitary glands, a symptom almost certainly attributable to the renewed emotional strain of living in a chaotic household.

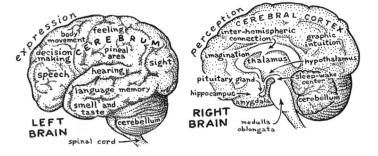

As the pituitary or growth gland is only one among many suborgans of the brain, it may be appropriate here to mention some of the others, like the thalamus (pain center and relay station), hypothalamus (vital sign monitor), hippocampus (memory implanter), amygdala (regulator of instinctive behavior) and the reticular activating system (of sleep, wakefulness, etc.), all of which, plus the lesser ones, are constantly influencing the still unfathomed workings of the mind. And it is just these mysterious doings of the mind that, in turn, lead us to such age-old fundamental questions as what is the essential nature of thought, of consciousness, of memory? Exactly how do these functions relate to the brain or other parts of the body? Is there any sort of boundary or dimension between body and mind? Do nerve cells individually think or remember? Or is each one something like a cog in a machine or a note in a tune, of minor significance in its individual self but of major meaning when combined with others to compose a chord or, when extended through time, a melody of thought and action?

These are tough questions that for millenniums have engaged the greatest philosophers of East and West. Anyone may of course com-

266

pare a thought to a bird that alights in the tree of the mind, but it takes more than a poetic analogy to make a useful measure of the relation. For the body-mind interface, if any, is far from simple. Certainly the mind is not the only factor that organizes, unites or directs the body. The body manifests perfection at its conception and in the days thereafter when it is still a microscopic embryo that could hardly be said to possess a mind. Its developing blood and immune systems also react directly and independently to throw off poisons or diseases without any knowledge or known mental influence. Even blood from a stranger, freshly transfused, will serve the body perfectly without direction from the nervous system. And unknown food automatically digests itself without the slightest thought.

If the body then does not depend for many of its vital functions upon the mind, why should the mind depend absolutely and forever upon the body? I suspect that the two are capable of somewhat separate existences and I have found no evidence under science or philosophy to persuade me otherwise. By separate existences I mean of course separate in the way the earth's north pole is separate from her south pole. This is the polarity principle (Chapter 18) that says the two are different only to the degree of being opposite aspects of the same thing.

## CONSCIOUSNESS

Increasing numbers of researchers in recent decades have investigated the mind and its relation to the body, some measuring brain waves through electroencephalography, others doing things like implanting hair-fine wires into muscles all over the body and interconnecting them with supersensitive galvanometers, after which it was found impossible for the subject (no matter how relaxed) to *think* of moving any part of his body without the *thought* registering in measurable electric current. J. B. S. Haldane, the English biochemist, experimented far enough to conclude that the average thought or mental "event" interrelates with about seven square centimeters of brain surface, lasts almost a second and consumes an unimaginably tiny amount of energy: $6 \times 10^{-27}$ ergs. Someone else discovered that a man may see as few as one or two photons of light under favorable conditions, a threshold now believed to be the minimal visual "event." But naturally such and larger "events," including nonvisual ones, continually overlap and interrelate in the mind, a fact that enjoyed a rare confirmation recently when a researcher, measuring the electrical response to a flashing light in the visual area of a hungry cat's cortex, noticed that the response decreased when the cat smelled fish.

Can a single cell think? Can a subdivision of a cell experience a mental event? No one has proved it can or cannot. But there is a good deal of advanced opinion holding to the concept of the nonexistence of any lower (or upper) limit to consciousness, which, incidentally, seems to have as many definitions as life itself, a few of them going so far as to suggest that consciousness and life are synonymous.

Some pioneers in the field speak of the consciousness of crystals and molecules, whose pristine awareness, they deduce, exists only in individual pulses of less than a thousandth of a second's duration. Human consciousness in turn may arise through the integration of the pulses of trillions of molecules into a patterned weave of memory, a synthesis of innumerable threads into a mother rug of ultradimensional apperception. Even the atom may know a beginning of primeval consciousness with its electrons or its undiscovered quarks choosing their own paths under the law of indeterminism. Indeed the very universe may integrate its component consciousnesses into the sublimely all-inclusive megaconsciousness we commonly attribute to God.

One definition has it that consciousness is merely awareness of oneself. Another, that it is awareness of awareness. Still another, that it is awareness of the relations between oneself and one's environment, a concept that may be extended to families, tribes and nations as they gradually become aware of their dealings with other creatures and peoples and consciously draft these relations into history. If there is truth in this idea, perhaps it would throw some light on consciousness to suggest that the siege of Troy marked the primal awakening of the ancient Greeks, who had lately begun the process of rousing themselves up from tribal oblivion into a great people via an imaginative movement that succeeded largely through Homer's work in dramatizing their story and inspiring them with a collective spiritual consciousness.

This idea of course supports the related notion that the first essential for consciousness is change, particularly noncyclic, irregular or startling change. And such a concept explains why mere knowledge, which is generally static even when not instinctive or inherited, is seldom very conscious (depending on your definition) while learning, which always involves change, is much more likely to be conscious (by any definition).

The factor of relativity likewise comes into considerations of consciousness because there are many types, degrees and states of consciousness which may be symmetrical, if not complementary. And physiologists have recently determined that, whether the brain entertains rational or irrational thoughts, it consumes oxygen from the blood at the same rate. In other words sanity is not physically measurable. Indeed there seems to be a mystic polarity between states of mind, which was beautifully expressed by old Chuang-tzu in the fourth century B.C. when he dreamed he was a butterfly. "I was conscious only of fluttering hither and thither," he wrote, "following my fancies as a butterfly. I was unconscious of my individuality as a man. Then suddenly I awoke. So here I lie, myself again. But now how can I know for sure whether I have just wakened from being a real man asleep, dreaming I was a butterfly, or whether I am instead really a butterfly who has just fallen asleep, dreaming I am a man?"

The ghosty, illusory quality of the mind must surely be attributable in some degree to the indirectness with which material things impinge on it, for apparently they can only reach it through sensory nerves which in turn send coded messages to the brain, where they undergo a mysterious decoding process describable as mental interpretation. Thus, as Bertrand Russell once wrote, it is really prejudice that tells you you see the same table in your neighbor's dining room every time you go there. For actually the table may have been replaced by a duplicate more than once. Indeed you have not *experienced* the table so much as *prejudiced* it.

The problem, however, goes deeper than prejudice. Conscious thinking, you see, takes in information bit by bit, one thing at a time. But this method of linear scanning is not very efficient for learning in our nonlinear world where everything happens everywhere at once. Besides, the appearance, sound and feel of a thing, particularly when it is moving, are all profoundly deceptive — because the successve "orbits" of subatomic particles in the atom, of atoms in the molecule and of molecules in larger structures have a way of blending like movie frames into multisensuous continuity.

This realization of course does not bridge the gap between matter and mind but rather exposes the abstraction and mystery of matter. Yet the exposure seems to be at least a step into the matter-mind    269 problem which many thinkers think of as primarily an energy question, energy being probably the most widely postulated medium of liaison between mind and mind if not between matter and mind.

## THE BODY-MIND QUESTION

A way of considering the matter-mind relation is to observe its influence on the intense competition between human beings that is revealed in sports, both physical and mental. For almost all top athletes recognize the factor of mood or state of mind in winning a race or game. A good example is Parry O'Brien, former world champion shot-putter, who used "positive thought" to tap the last reserves of his inner strength by playing tape recordings of his own voice exhorting himself to victory: "Keep low, keep back, keep your movement fast across the circle. Fast now! Fast! Fast! And beat them! Beat them all!!"

Olympic weight lifters depend even more on a similar kind of self-hypnotism, concentrating for many seconds before each lift in a kind of prayer for faith, each in himself and his body — in the ascendancy of mind over matter — ascendancy to the supreme level of victory over his unrelenting opponent, the weight. And the reverse is probably as true for mental competitors like Bobby Fischer and Boris Spassky, who, in their famous chess match in 1972, made such a point of playing tennis or badminton, swimming and doing calisthenics to keep in top physical condition for their nerve-racking bouts at the board.

INTERRELATION
OF
MIND AND
MUSCLE

Although recognition of the polar relationship between body and mind has been slowly diffusing for millenniums in the consciousness of man, it seems to have been only recently that anyone realized that such a polarity logically implies that these opposites are diametric aspects of the same thing. For William James appears to have been the first philosopher to deny that there is any real line or interface between body and mind, and I wonder if it could be said that he intuitively anticipated Einstein and the mass-energy equation $E = mc^2$ when he theorized that thoughts are made of the same fundamental stuff as material things. In any case Sir Charles Sherrington went on from there to declare of body and mind that "it is artificial to separate them. To separate the one as 'action,' the other as 'thought,' the one as physical the other as mental, is artificial because they both are of one integrated individual, which is psycho-physical throughout." He thought of the mind as a natural outgrowth of Earth and inherent in it. "If the vertebrates be a product of the planet, our mind is a product of the planet. It senses each and all gear into the ways and means of our planet, which is its planet . . . The dry land created the feet that walk it. Our situation has created the mind that deals with it. It is an earthly situation. If the agent is terrestrial and the reaction is terrestrial, is not the medium of the reaction terrestrial? The medium is the mind."

If it is wrong to ask where body and mind meet, as scientists and philosophers have been doing since philosophy began, what is it right to ask? The earliest realistic answer to this question I can think of is that of Henri Bergson who pointed out the absence of any boundary between the growth of the embryo bird inside the egg and the ensuing development of its behavior as it begins to peck its way out of the shell, to walk, eat, fly, sing, mate and raise its young, all these functions (physical and mental, concrete and abstract) being parts of the same continuous emergent life force. This of course is what Ashley Montagu meant when he defined mind as "an abstraction from behavior" for, although what we call "life" is generally very different from what we consider "nonlife," the difference is not absolute but essentially a difference in degree — perhaps a development in dimension with no real border or clear shift in between. Memory, in this context, is but a record of past behavior and events. Intention is but pressure developing for future behavior. Emotion emerges out of sense awareness, will springs up from microcosmic uncertainty, thought out of atomic vibrations, of matter waves, of nodes and resonances.

Even a down-to-earth, working scientist, J. M. J. Kooy, professor of theoretical physics and mechanics at the Royal Military Academy in Breda, Holland, must have had something of the sort in mind when he wrote in the *Journal of Parapsychology* for December 1957 that "it is of great importance to realize that a 'body in itself' is a product of mere mind-spinning, not belonging to the real world of events, which is extended in space-time. What we call a 'body' is only a kind of permanence perceived by us in the 'course of events.' In a similar way, in quantum mechanics, an atom in a definite state of energy, or an electron by itself, must be considered only as a metaphor . . . All these incongruities disappear as soon as 'matter' is replaced by 'action,' the basic material of becoming, of events . . .''

In the same vein, Bertrand Russell in *The Outline of Philosophy* declared categorically that "everything in the world is composed of 'events' . . . An 'event' is something occupying a small finite amount of space-time . . . Events are not impenetrable, as matter is supposed to be: on the contrary every event in space-time is overlapped by other events."

Turning then to the mind, Russell added: "An important group of events, namely percepts, may be called 'mental'. . ." Mind he defined as "a group of mental events . . ." whose "constitution corresponds . . . to the unity of one 'experience.'" This perceptual experience of mind, he went on, must be closely associated with a brain because "the events that make a living brain are actually identical with those that make the corresponding mind." In fact the only generally acceptable difference between mind and brain is "not a difference of quality, but a difference of arrangement."

All of which may fairly be said to represent the forefront, if not the consensus, of current thinking about the body-mind relation, and which suggests that by natural law there must be a gradual evolution of matter into mind in any viable world where the various mechanical, chemical and electromagnetic responses of mineral elements progressively become sensuous, where crystal accretion develops into protoplasmic replication and the *growth* (by division) of cells sprouts into the *behavior* of larger, more complex groups of cells, and where finally their emergent sense response (a vital ingredient of behavior) blossoms into perception, consciousness and mind.

Evidently no one has demonstrated the existence of any clear seam or shifting of gears in this progressive emergence of mind through the kingdoms, although evolution includes all known changes of molecular combination, crystallization, fermentation, germination and flowering. So, to my way of thinking, it is almost certainly an evolutionary unfolding of dimensions: a process mathematicians think of as moving into higher derivatives, such as the shift from position to movement (at constant velocity) to acceleration (at constantly increasing velocity) to hyperacceleration (at constantly increasing acceleration) to hyper-hyperacceleration (at acceleratedly increasing acceleration), etc. Something similar must also occur whenever a single (phase) wave becomes part of a group wave (as I explained in my *Music of the Spheres*), when a circle takes on motion relative to its surroundings and becomes a spiral, when a musical note is blended simultaneously with others to become a chord and when the chord is combined

sequentially with others to create a melody. So does matter become more than matter through addition of the time dimension, which brings on atomic movement or metabolism, turning the particle into a wave, the material structure into a flow (inevitably a changing flow) of energy — which again is like existence awaking into motion, motion spreading into growth, growth fermenting into awareness, awareness germinating into consciousness and consciousness flowering into mind and spirit.

It is as natural a progression as if you were measuring an angle with a protractor and you saw that it read 71° and then looked closer to see if you could get the reading any finer. But, to dramatize my point, let us now suppose that an angel (a very special "angle angel") appeared and leaned over your shoulder to help you, making your protractor grow magically larger so it reached out farther from the angle while more and more details of its scale appeared in your range of vision. By this means you could see the angle was not just 71° but 71° 12'. And then it might sharpen to 71° 12' 26". Then to 71° 12' 25.982" — and so on in a mysterious intensifying flow of divisions and relationships of steadily increasing precision.

But think now. Need there be any end to this? Perhaps you have begun to recognize it as the ancient enigma of Leukippos and Democritos who postulated the atom in the fifth century B.C., a momentous surmise that only in our time seems finally to be resolved by the quantum theory. But is the problem really solved? Can it be solved? What are quarks made of? Which are the ultimate monads of creation, if any — or are they all relative and illusory in this seemingly commonplace but really very drastic world?

J. B. S. Haldane once made the ingenious suggestion that mind might be "a resonance phenomenon." I suppose he was consciously agreeing with Niels Bohr's inspired idea that the wave function of matter represents its mental aspect (page 408). Or he may have meant that the electromagnetic rhythms of the nerve cells could be produced by tuning in from some unknown source, perhaps in a dimension still unknowable to this phase of existence. In any case, the narrowing of the "circle of confusion" as it resolves itself into a sharp point in the focusing process may be as applicable to consciousness and mind as

273

to optics or harmonics — and just possibly it could turn out to be an interdimensional noumenon at the ideal center of this seam of the worlds.  But, to me, it now seems a lot simpler just to realize that the essential difference between body and mind is that the body is bound by space and time while the mind is not, and therefore that the body is finite while the mind is free to transcend toward the infinite.

# CHAPTER 11

# Memory, Intelligence and States of Mind

E VEN IF ORBITING out here is in essence a mental exercise, it does at least serve to acquaint our thoughts with space and time. And if we ever have to choose between the two, recognizing our dimensions of space as cyclic and constant, while only our time dimension is cumulative and changing, may persuade us to give the time dimension its due as a percept. To that end, in this chapter we will look first at memory, the interaction between different times (specifically past and present), and then to the intelligence that depends on it, as well as the states of mind that seem somehow to emerge out of it.

## MEMORY

I know memory is hard to visualize or relate to the material universe, but it really shouldn't be because it is basic to mental continuity and consciousness. And it is built into rocks and stars as well as vegetables and animals, being intimately involved with everything right down to such laws of physics as the conservation of energy. In fact energy conservation itself *is* a form of memory. And this will be clear to you if you consider that the brick laid at the top of the chimney "remembers" the energy exerted by the hod carrier who lifted it up there and that many years later when the brick falls to the ground, it expresses that memory kinetically by hitting the earth with exactly equivalent energy.

An elastic band likewise "remembers" the energy put into stretching it and voices that memory in its insistence on returning precisely to its original state of relaxation. Some kinds of plastic, however, have a subtler memory. Irradiated polyethylene may be heated, molded and cooled, after which it obediently keeps its new shape but without "forgetting" its original form, the memory of which is retained in the distorted crosslinking of its chain molecules. As a result, if this kind of polyethylene is heated again outside a mold, the aggrieved molecules will eagerly pull it back to its former shape — which, in a sense, is like reversing the flow of time.

Liquids seem much too disorganized to have as much memory as do solids but, surprisingly, water can "remember" its invisible patterns of circulation for days, even inside small containers. Laboratory experiments have shown that, if the faucet is angled enough to give the water even a slight rotary circulation as the tub fills, it cannot be drained a day or two later without the water swirling out in the same direction. This of course presumes the water was left completely uninfluenced the whole time, during which its silent memory currents must have kept coasting round and round their orbits like asteroids circling the sun. There is a limit of about four days to this sort of memory, however, the imparted movement having by then been slowed down by friction to the point where the Coriolis effect (of the earth's turning) takes over, after which undisturbed water invariably drains counterclockwise north of the equator or clockwise south of it.

Understandably an ocean has a much longer memory than does a tub. And it gets it naturally not only from its currents but also through other activity, from storing up heat in summer and releasing it in winter to transporting it from the equatorial regions toward the poles, in either case retaining the effects of the weather of bygone years as a measurable record built deep into its dynamic body. But there is no end to such memories, from the salt in the Dead Sea that can never forget it was once part of the oceans to the mountain chains whose very shape is a pressing reminder of how they got there. Not to forget the Earth itself, whose body (including the atmosphere) is just one fat volume of the history of everything it has ever done — and so on to other worlds and systems of worlds and systems of systems without end.

Memory is everywhere. Yet it is as elusive as the answer to the question of whether a river is a flow of water or a channel cut by water. Is it, in other words, positive or negative? Abstract or concrete? If one could make an analogy that consciousness is like a needle moving forward through space-time followed by a thread representing the memory of that consciousness, the act of remembering it could logically be pictured as a reversal of the same motion whenever the thread shifts and guides the needle back through space-time to where and when it "remembered" it was.

The trouble with such efforts of course is that they don't really make memory understandable or visualizable. But there still may be a way of getting a down-to-earth grip on the "stuff" of memory. If you stare hard for about ten seconds at some bright object like a tree in sunshine, then shut your eyes, it is possible to "see" the tree still in a kind of under-eyelid vision or afterimage which presumably uses the same cones of your retina that saw it originally. If this isn't memory, it is at least close to it, for visual memory may well use the retina as well as the optic nerve to imprint its graphic images. Something similar probably occurs in aural memory, where repetition of words or sounds helps one to retain them through the auditory nerve, as has long been practiced in learning by rote all over the world. And no doubt there are comparable reinforcements in smell, taste, touch and other senses, memory of which is known to involve nerve cells in the spinal cord, the solar plexus and, in conditioned reflexes, even below the waist.

The exact role of the cell in memory is not yet known completely, but recently animal experiments at the University of California have revealed nerve cells that can actually count, doing so by holding back the discharge of their electrochemical signals until an exact number of clicks has been sounded. By such means one type of monkey brain cells, in repeated tests, proved it could reliably count as high as nine. And in other research at the same university it was discovered that so-called Golgi bodies inside brain cells sprout out branchlike fibers as learning proceeds (see illustration, page 104), particularly in the parts of the brain most closely related to the senses used — this apparently explaining why the visual areas of the brains of artists have been found to contain a higher than average proportion of densely branched Golgis, and in musicians similarly developed cells in their auditory areas.

Even more strikingly specific has been the evidence of memory registration discovered by Dr. Wilder G. Penfield at the Montreal  277

Neurological Institute about 1948 when, during a brain operation on a twenty-six-year-old woman with epilepsy, he touched a spot on her hearing center with his electrode and she exclaimed, "I hear music!" The music was a familiar tune, "Marching Along Together," and it kept playing as long as the electrode touched that exact spot but, when he removed the instrument, it stopped abruptly like turning off a tape recorder. Dr. Penfield knew he had happened on something important and repeated it some twenty times in the next hour, noting that the same melody was played every time and, significantly, that after each silent interval, the tune began again at its beginning as if the "tape recorder of memory" was somehow automatically able to rewind itself.

Experimenting further with other brain surgery patients, Dr. Penfield soon found that the "tape" was at least an audio-video "tape," because most patients could be made to see things as well as hear them, and none of their senses seemed excluded. These were not hallucinations either, for careful checking showed them to be reruns of real life experiences, and all the patients insisted they were much too vivid to be mere memories. Nor could more than one recording be played at the same time no matter how many electrodes Dr. Penfield touched to the various responsive points in each brain.

Another development that tended to confirm Penfield's brain-tape discovery occurred when hypnotism showed it could elicit similar remembrances in certain subjects. Indeed an elderly bricklayer under hypnosis was able to describe minute cracks, chips and other surface details on several ordinary bricks he had laid forty years earlier, just as if he were looking at close-up photographs of them.

But the pianist Arthur Rubenstein, who in his late eighties could learn a new sonata in one hour, hardly bothered even to hypnotize himself in guiding an automatic memory through extraordinary performances all his life. "When I play, I turn the pages in my mind," he explained to an interviewer in 1966, "and I know that in the bottom right-hand corner of this page is a little coffee stain, and on that page I have written *molto vivace* . . . At breakfast I might pass a Brahms symphony in my head. Then I am called to the phone, and half an hour later I find it's been going on all the time and I'm in the third movement."

## RESEARCH IN MEMORY

Work on the structure of memory storage has led to all sorts of fascinating experiments, many of them involving animals. To test the

278

theory that memory necessitates continuous neural activity, rats were taught to go through a maze without an error, then they were frozen until their rate of metabolism dropped virtually to zero, then, revived, they still were found able to negotiate the maze faultlessly, demonstrating that uninterrupted neural function is not required.

An experiment with goldfish at the University of Michigan helped to prove that if a lesson just learned is not to be forgotten within minutes, it must somehow immediately deposit itself in a kind of memory bank that seems to be in the brain and presumably made of protein. The fish in this case were taught that when a light went on in their specially built training tank it would be followed 20 seconds later by an unpleasant electric shock, but that the shock could be avoided if the fish promptly swam over a barrier into another and unlighted compartment. Thirty exposures to the light-and-shock threat in the training tank were generally found to be enough to teach a fish to swim away from the light every time it went on and, most important, the fish would remember to make this evasive maneuver on seeing the light a day or a week later or even after an absence of several months. But if the experimenters injected a drug called puromycin (which inhibits the growth of protein) into the skull of a fish immediately after his training, the fish promptly and invariably forgot everything he had just learned. However, if the same ten microliters of puromycin were injected in the same way an hour after the training ended, by then the lessons had become so consolidated in the fish's memory that the drug no longer could block his normal power to retain his training and he would go on evading shocks for the rest of his life — indicating that the manufacture of protein must somehow enable the short-term memory of learning to be inscribed into the long-term memory of permanent knowledge.

If lasting memory is thus stored in protein, with the molecules imprinted by thoughts the way tin cans are indented by kicks, the question arose as to how many molecules or cells are needed for memory. It wasn't an easy question but had the interesting consequence that, in seeking the answer, researchers in the nineteen fifties turned seriously to the worm for enlightenment.

They picked the worm because it is not only cosmopolitan enough to be the average in complexity of all animals but also because it is ideally suited to experimentation into basic memory. The particular worm chosen was a half-inch flatworm called a planarian, unique because he is both the lowest creature in the evolutionary hierarchy who has a central nervous system and the highest creature who reproduces by dividing. He also offers a good deal of fascination with his

279

disconcertingly humanoid, cross-eyed look plus a nonchalant capacity to be chopped into as many as six pieces, each of which will grow into a whole new worm. Besides he is probably the simplest creature

possessed of a nerve junction that can reasonably be called a brain.

In the laboratory a batch of these worms were first subjected to a bright light, which normally makes them stretch out, then two seconds later they were given a mild electric shock lasting one second, which made them curl up. By the time this light-and-shock sequence had been repeated a couple of hundred times, they had begun to regard the light as a warning and would curl up before the shock as if deliberately bracing themselves against something they had come to fear. Next many of the worms were cut in two and the remainders allowed to grow back into new worms, which were then tested with the light-and-shock sequence. So far it had been presumed that the head ends of the half worms, which contained the original brains and only needed to grow back their tails, would remember their previous training better than would the brainless tail ends that had to grow new heads and brains. But, in fact, the tail ends did about equally well. In some cases the recovered tail ends actually remembered and responded to the light signals faster and better than did the recovered head ends, strongly suggesting that memory does not depend upon a brain.

In a later experiment trained worms were cut not just in two but chopped into hundreds of tiny pieces and fed to worms who had not been trained at all. And the untrained worms who ate the trained pieces began to curl up whenever exposed to training lights, indicating that memory of the training of the chopped-up worms had almost certainly been transmitted to the untrained ones through what they had eaten.

This of course, even more than the previous finding about brainless memory, dramatically supported the chemical theory of memory — including the ancient superstitions of cannibal tribesmen that, if they could only contrive to eat certain parts of a brave enemy, they would become as brave themselves. And there was a noticeable flurry of speculation about the future of education, which could be expected to become more liberalized, with increasing probability that one day an ambitious young man might make himself into an intellectual, if not by literally partaking of a professor, at least more and more likely by regularly taking protein pills containing scientifically filtered RNA or mutated enzymes from tissue purportedly sophisticated or even wise.

As for the question of how many cells are needed for memory, we already know that one cell will do, because we observed in Chapter 9 the one-celled ameba who remembered the prey he had lost several seconds earlier. And we don't need to wait for what we are going to learn in Chapter 12, that a bird's egg, which (like other organisms) was initially one cell, may remember the patterns of stars, the melodies of songs and a great deal more that was sensed, learned and bequeathed by its ancestors — undoubtedly going back for thousands of generations.

## ABSTRACT ASPECTS OF MIND

Among other discoveries is the fact that memory can be duplicated and stored in numerous areas of the brain and that sense memories (visual, auditory, etc.) are recordable not only in their respective specialized zones but each item is filed in several distinct (sometimes widely separated) places, a system that may have evolved as a safeguard against local damage. Indeed, although a severe electric shock or a blow on the skull may erase memory, particularly recent memory (the span erased being proportional to the severity of the shock or blow), as much as half the human brain has been surgically removed without noticeable memory loss.

Now a researcher named George Ungar, at Baylor College of Medicine, and his colleagues have advanced from worms through fish to mammals and are claiming to have isolated the first known "mammalian substance involved in the chemical transfer of learned behavior." Specifically this means they have discovered a memory protein named scotophobin (Greek for "fear of the dark"), which was obtained from the brains of rats trained to avoid the shadowy places they normally feel safe in and which, when it is injected into untrained rats, makes them wary of the dark too.

The same team says they have also isolated a compound from the brains of fish trained to avoid two different colors, a completely artificial behavioral trait that can be passed on to other fish by injection. And rather charmingly macabre is their recent experiment in training some 6000 rats to ignore the ringing of a loud bell, then collecting and purifying the rats' brains by repeatedly testing fragments of them on mice and keeping only those portions that made the mice ignore the     281 bell. It was a process not only of distilling the animals' musical memories and pouring them into test tubes but of establishing that the very faintest recollection of a ringing bell has a chemical structure at once abstract and tangible, a chain of six amino acids that can be visualized both as coded sheet music and as the most exquisite of microgeometries.

When I encounter such tenuous abstractions in the seamlike interplace between body and mind, I can't help wondering where ideas come from. If we allow that it really has taken something like five billion years for Earth to evolve mind to its present state, and that mind is now speeding up its development by learning to leap forward in time as well as outward and inward in space, to grapple a little with the unknown, who can gainsay any rate of acceleration it may yet attain toward things still undreamed?

Memory is also an abstraction of abstractions, rather like the group wave, because it is conserved in the chemical patterns inside cells such as brain cells, patterns that survive metabolic changes among the atoms of the molecules composing them, probably even changes in whole molecules. Yet the actual materials that make up these abstract relationships are themselves, in latest analysis, also abstract, being made of "waves of probability" of no determinable concreteness. Space-time-matter itself indeed has, and is, a kind of memory, for anything done by it — any movement or acceleration or physical change of any sort — is recorded in its material position, velocity and orbit — in its exact physical state. This state also is always the sum of its history up to that moment, its present the equal of its total past. And presumably human memory works similarly, what one potentially can recall at any instant being the accumulation of all one's past nerve impulses or chemicomagnetic traces that were strong enough to have left impressions in the "tape" of one's brain cells or in the genes of any kind of cells that could accept such impressions.

Having thus concluded that a single cell (including a sperm or an ovum) is definitely endowed with the possibility of memory, it is easy to see how any offspring might remember the experiences of parents or earlier ancestors. And this brings up the question of the inherited memory that may reside, as we shall see, in an egg, that can enable an animal to navigate across oceans by star patterns and is generally classified as instinct. And another example of instinctive animal orientation is the case of termites inside their houses, building columns in which two teams have been observed a foot apart, each constructing a vertical pillar in the darkness, seemingly without interpillar communication, yet curving the structures toward each other at the same height so that they eventually met to form a perfect arch, which would later support a gallery floor.

The more socially sophisticated animals seem generally to inherit some such maps of the mind, maps that not only enable them to visualize their migratory way but somehow to objectify whatever nest they are assembling or even the pattern of a courtship dance. And senses other than the visual ones are included, so a bird may "hear" in his mind's "ear" the song he wants to sing as well as "feel" in his throat the muscular and other sensations of how to sing it. Such inherited awarenesses give strong evidence also of life *before* conception, to say nothing of life *after* death, and demonstrate (as we will see in Chapter 19) that our present earthly consciousness can well be just the finitude phase of our continuing transcendence toward Infinitude.

## Intelligence

The kind of intelligence acquired by animals is naturally different from that of humans and has a far greater component of instinct, which is usually specialized for survival value. An example is the case of double foster parenthood which, despite the indifference wild creatures generally show toward young not their own, was revealed recently in an ornithologist's report of a nest of young Savannah sparrows being raised by parents who happened to be clearly identifiable by their color bands. For just after the little birds were hatched, their mother was killed and eaten by a snake. But the father quickly took on a new wife to help him with the hungry youngsters. The following week the father was also lost, evidently caught by a hawk, whereupon the foster mother attracted a new husband to her nest and the lucky orphans were raised by these two strangers until fledged and flown.

Another instance of specialized, instinctive sensitivity, probably augmented by experience, was that of a pair of foxes with a den in the bank of a shallow stream. One bright April afternoon the mother fox was seen from a nearby house carrying one of her babies in her mouth across the stream to a sheltered spot under a high rock ledge on the other side. And then she appeared with a second baby, and a third, one trip for each. That same night there came a lashing rainstorm which turned the stream into a raging torrent that certainly would have drowned the fox cubs, had they remained in their den. Four days later the vixen carried her family back home to the damp but drying den. Later that spring on two more occasions the mother moved her

283

young ones to high ground and each time within twelve hours the den was flooded. Such natural prescience of stormy weather, possibly through acute perception of dropping barometric pressure, rising humidity, wind direction or other clues is not known among humans to my knowledge.

As for the perennial question of which is the smartest of animals, there is a wide difference of opinion, depending on one's definition of smartness. But at least among animal trainers, who tend to judge intelligence by sensitivity and quickness in learning what will (or won't) bring a reward, something of a consensus (admittedly controversial) seems to be developing. The majority appears to rate the ape, particularly the chimpanzee, as first, then the various monkeys, followed by the raccoon, the porpoise and the pig. The average dog, they claim, although very responsive to man after his long domestication, is generally outwitted by the pig. Then there is the shy fox, followed by the mysteriously independent cat. The crow, next in line, seems to be the brightest of birds, seconded by the parakeet and parrot. Then come the cow, the sheep, the goat, the nervous horse and, in succession, the beaver, the rabbit, squirrel, hamster and other rodents, various barnyard fowl, and so on . . .

An animal who uses a tool is not necessarily highly intelligent, if he was given it and taught its use by a human. But there are known instances of animals taking to tools on their own initiative, even to the point of shaping them in advance. The best documented case I know is Jane van Lawick-Goodall's account of a chimpanzee who broke off a long, straight twig from a tree, trimmed away its side sprouts and leaves, then poked it deep into a termite nest, after which he pulled it out with a few clinging termites and licked them off as from a spoon. Several young chimps were later seen learning to make and use such termite spoons by observing their elders, suggesting this was likely an example of early cultural evolution handed down from generation to generation, not genetically but by education. The best toolmaker among birds may well be the great spotted woodpecker of England, who cuts clefts in trees, in this case mostly by instinct, and uses them as vises to hold pine cones in place while extracting their seeds.

As for "tools" used without modification, all the apes and many varieties of mammals, birds and insects avail themselves of them, particularly of sticks and stones. Baboons have been known to break open hard-shelled fruits with rocks and use smaller stones to crush scorpions before eating them, not to mention adopting anything from a stick to a corncob as a napkin. Gorillas have been seen using leaves for toilet paper. And elephants to pick up long sticks to scratch themselves where their trunks could not reach, or a branch to switch away flies. A few animals even make tools out of the stolen parts of other animals, as the octopus's habit of using a jellyfish's stinging tentacle to shock and immobilize shrimp for food or, on one known occasion, to halt the attack of a moray eel. And the ubiquitous thorn, sharp and tough, is a natural chisel, needle and nail, being used to dig out tree insects by a Galápagos finch who thus fills the role of a woodpecker, as well as by a shrike who adopts it for a meat hook to hold his prey.

Although a man-made mirror does not qualify as an animal tool, its occasional use by an animal produces not only a reflection of its body but, more interestingly, of its mind as well. This was shown in the case of the lion in California who was brought up in a normal house and evidently regarded himself, if anything, as just another pussycat until the appalling truth was finally forced on him by a full-length looking glass. But a baby giraffe was offered no such help when separated from his sick mother at birth and raised in a zoo pen all alone — with the result that, when he was eventually introduced to her several months later without having ever seen or heard of giraffes, he was terrified. To him, who hadn't the slightest idea there could be any such animal, it must have seemed incredible that this unimaginable

creature could somehow imagine she owned him. Even the human mind depends much more than you would think on the kind of revelation only its reflected image can provide, for it became dramatically evident in several recent cases of abandoned ghetto orphans who reached school age without ever having seen themselves that, without a mirror, not one of them could really be said to have a realistic comprehension of his own existence!

The difference between animal and human intelligence is seen thus as primarily a difference of degree, for even though few animals are capable of realizing that their mirror image represents themselves, most of the great apes learn about reflections from experience almost as readily as does a human child. Indeed learning by experience is observed throughout animal life, even in lowly one-celled protozoa, who often actually try several different responses to their environment 285 before settling on the one that works best. And when any animal repeatedly finds the same behavior successful, he almost inevitably becomes conditioned to continuing it. Such a stable response of course depends on memory too, which appears to evolve in rough proportion to the complexity and longevity of the animal. That is why planarian worms need about a hundred trials to learn how to get through a simple two-fork maze while rats easily learn and remember a single right-left alternation pattern in a more difficult maze, although they rarely can master double alternation such as two right turns followed by two lefts. Yet rats and more advanced mammals usually can solve relatively complex problems involving unfamiliar principles if they have time to learn by experiment, even if the experiment is more accidental than intentional. A case in point is that of an exceptionally imaginative female snow monkey in Japan who discovered, while playing with a potato at the beach, that dipping it in sea water not only cleaned it but gave it a delicious salty taste. She also discovered she could separate wheat grains from sand by dropping the mixture into water so that the wheat floated free, and both discoveries were quickly recognized and adopted by her troop.

When animals are guided by humans they often learn even faster. In fact the rat's capacity to learn was measured recently at a research "rat school," where a class of furry, young "scholars" who had been given several weeks of intensive instruction in the laws of nature were found to have developed brains 5 percent bigger than uneducated rats, their nerve cells having increased slightly in number but more definitely in size and complexity. And probably the most important single advance they made (which goes for the education of young humans too) was that they learned how to learn!

Obviously there are limits to animal intelligence, assuming a reasonable definition of the animal as it evolves through future eons, presumably under increasing human control. But individual animals, like individual humans, show occasional mysterious flashes of genius, particularly if the animal's intelligence is greatly enhanced by human training, as in the case of the border collie of Scotland, a smallish, black and white breed, now considered the best sheep dog in the world. I heard a story about a border collie named Commodore owned by a sheepherder in Arizona. One evening a nearby rancher saw a column of sheep being driven over a rise by Commodore, who was carrying a man's hat in his mouth. When the dog had neatly penned the sheep, he took the hat to the rancher. Then he turned back to the sheep trail, looking over his shoulder to make sure he was being followed. By heeding the dog, the rancher found the sheepherder pinned under a fallen boulder, in shock but alive. Commodore had known his master was in serious trouble but, being also responsible for bringing his flock home for the night, he took the hat, saw his sheep safely put away, then used the hat to get help. This is the kind of intelligence found in a first-rate sheep dog, who controls his charges not by running and barking but mainly by getting the attention of the flock leader by fixing his eye upon him and holding him mentally with a kind of hypnotic concentration, augmented by subtle anticipatory movements that make him respond in precisely the desired direction.

Intelligence in humans certainly goes far beyond the intelligence of any known animal and, in the perspective of evolution, has no known limit. Nor has any end been found to human genius or creativity, which is not the same as human intelligence and seems to depend as much, if not more, on such other traits as persistence and caring, not to mention the mysteries of abstraction and divinity which we shall soon encounter in Part Three, where we meet the main theme of this book.

## HUMAN LANGUAGE

The fact that untrained animals know nothing of verbal symbols is probably the principal reason why man so completely dominates them. It also makes a fundamental distinction between animal and human languages, for, unlike the animal who can voice little more than his mood, a human using words may utter any abstract concept from *abatement* to *zygosis* and use mathematical or other symbols to solve problems all the way from the composition of an atom's nucleus to the correction of a rocket's orbit in space.

The most historic benchmark for man's final commitment to language may have been Confucius' immortal saying that "the beginning of wisdom is calling things by their right names." But even he could hardly yet have realized that language would not only clarify human thinking but also greatly accelerate the advance of knowledge by enabling each generation to record and pass on what it had learned so the youth of the next generation could start farther ahead. This is not the same as saying that language is as new as what we call civilization, for human language is thought to be about as ancient as man himself.

Moreover, curiously enough, unlike the rest of proliferating evolution, languages and dialects seem to be getting fewer. Evidently they developed very gradually over millions of years, presumably starting in the animal phase with mothers and babies cooing to each other, lovers murmuring, workers grunting, hunters yelling with excitement. Each family or clan developed its own dialect of words — words that were likely to be misunderstood by any outsider — with the result that, even into historic times, there were about as many languages as there were clans or tribes, most estimates running into the thousands. But as tribes combined into city-states and eventually nations and empires, the best-established tongues tended to absorb the local and minor ones, particularly if the latter had no literature or script. And today there are only an estimated 130 significant languages ("significant" meaning spoken by at least a million people), which include many you may never have heard of like Wu in China, Tadzhik in the Soviet Union, Bagri in India, Xhosa in South Africa, Pashto in Afghanistan, Quechua in Peru. And the vast majority of people in the world speak one or more of the top dozen, which, in the order of the numbers using them as their native tongue, are:

| | |
|---|---|
| Mandarin — 450 million | Japanese — 95 million |
| English — 350 million | Arabic — 90 million |
| Hindi — 180 million | Bengali — 90 million |
| Spanish — 160 million | Portuguese — 80 million |
| Russian — 160 million | French — 75 million |
| German — 110 million | Italian — 55 million |

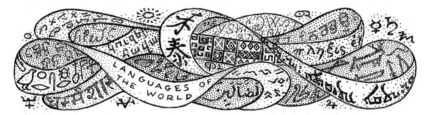

Even with the continuing reduction in the number of surviving languages, however, an immeasurable burden of confusion and disaster continues between people who do not have a common speech. An outstanding example was the reply of Japan's Premier Tojo to President Truman's ultimatum of July 26, 1945. When Tojo said Japan would "*makusatsu*" the ultimatum, he meant that his government would "consider" it. But the translators at Domei quoted him in English as saying the Japanese would "take no notice of" it. So atomic bombs destroyed the cities of Hiroshima and Nagasaki — perhaps for nothing! And the same sort of oriental misunderstanding continued through the Korean and Vietnam wars, where the much publicized "peace talks" bogged down for years over the occidentals'

assumption that "to negotiate" was "to compromise," while the orientals interpreted it as "to get something by talking."

Individual injustice through language of course must be even commoner than the more conspicuous bungles in international diplomacy, and I've read that before the Russian Revolution an Assyriologist named Netomeff was exiled to Siberia for life on a charge of blasphemy and treason because he wasn't given a chance to explain that his book about Nebuchadnezzar did not mean "*Ne boch ad ne tzar*" (Russian for "no God and no tzar"). And such irrationality of language interpretation continues to plague world understanding in the United Nations Assembly, where a translator on one memorable occasion translated "out of sight, out of mind" into an expression the Russians understood as "invisible insanity."

Even when no translation is involved, most languages have ambiguities that can cause serious misunderstanding. And this made the history books in 1851 during Napoleon III's coup d'état when one of his officers, Count de Saint-Arnaud, on being informed that a mob was approaching the Imperial Guard, coughed and exclaimed, with his hand across his throat, "*Ma sacrée toux!* (my damned cough)." But his lieutenant, understanding him to say "*Massacrez tous!* (massacre them all!),'' gave the order to fire, killing thousands — needlessly.

The English language, now beginning to be considered a leading candidate for a universal tongue, is still not only seriously unphonetic but full of illogical idioms. A London house on fire may not only "burn up" and "burn down" at the same time but it itself can "put out" the same flames and smoke that the firemen are simultaneously "putting out" with their hoses. And, speaking of smoke, a Chinese student of English rang a fire alarm at Fort Bragg, California, emptying a big building to which fire engines dashed with sirens screaming, all because he needed a light for his cigarette and had carefully followed the directions printed on a red box: PULL FOR FIRE.

Up to now the established languages have evolved naturally without conscious guidance or design or with anybody seeming to mind that the English phrase "I assume" translates into French "I deduce" and into Russian "I consider." Yet there is an unobvious mystique about language associated with the fact that it grows by itself, both in individuals and throughout the world, like a sentient being. For, as linguist Noam Chomsky of Massachusetts Institute of Technology has pointed out, all normal children at birth possess an innate capacity and compulsion to acquire speech within their next three or four years, not just by imitating their elders but, more significantly, by comprehending and creating a constant flow of new combinations of words and phrases never expressed in exactly the same way before. Actually this is a very subtle two-way flow, with the young mind both shaping the language and being shaped by it in return, depending greatly on the characteristics of his particular language as well as on how it is used — on whether cautious noun thinking eventually "achieves success" or instead more aggressive verb thinking simply "succeeds."

The best-known pioneer in this field was probably Benjamin Lee Whorf, who, in the 1930s, began the first scientific investigation into the way thought is molded by language, noting that the *content* of thinking directly influences the *process* of thinking, indeed so much so that all our mental images of the universe literally shift from tongue to tongue. He noticed that people visualize size in the ways language has segmented it: that a few battleships may be 3 while a few molecules may be 300, and that a sentence is to a language what an equation is to mathematics. He pointed out that bare words are essentially individual numbers which acquire their true significance only in combination because no word alone can have an absolute meaning.

Whorf spent years studying primitive tongues and found the Algonquin languages "marvels of analysis and synthesis." Their pronouns, he discovered, have four "persons." From the European viewpoint, this means two kinds of third person, their difference indicating whether the subject or the object is referred to. Thus, using (s) to denote subject and (o) for object in the sentence "The chief called his(s) son and told him(o) to bring him(s) his(s) bow," the different pronouns distinguish "him(s)" from "him(o)," etc. As a result Algonquin legal documents should never need to adopt cumbersome phrases like "the aforesaid person," "party of the first part," etc.

Another primitive language is Chichewa, related to Zulu and spoken by a tribe of unlettered blacks in East Africa. It contains an extraordinary perspective on time through its two past tenses: one for the real or objective past (o) that influences the present, the other for the mental or subjective past (s) that does not influence the present. In it one can say "I came(o) here while I went(s) there," meaning "I came here while thinking about there." If a man says in Chichewa, "I ate(o)," he is presumably full and satisfied, but when he says, "I ate(s)," he implies he is hungry because he merely thought about eating. Such syntactic devices might conceivably in time lead their users to rare insights into relativistic physics and philosophy.

The Hopi Indians of Arizona have a still more remarkable metaphysical language which is intimately related to their view of creation. Instead of a noun for "wave," they have only the participle *"walalata"* (waving). Instead of a past and future, their temporal concepts embrace only the ideas of things objective or manifested on the one hand and things subjective or unmanifested on the other without specific time comparison. Matters "objective" include everything accessible to the senses, all history and known present factors. Matters "subjective" include the future, the unseen, the mental, the emotional, what's in the hearts of creatures and man, even the spirit of the universe: God. As in the theory of general relativity, there is no universal simultaneity here. What happens beyond the mountain is not of us now, and cannot be discovered until later — maybe. If it doesn't happen *here*, it doesn't happen *now*. Things farther away in space must also be farther away in time.

As remoteness in space-time stretches away to the unfathomable in distance and to the unrememberable in history, the objective and the subjective eventually merge into one all-encircling beginning and ending — the abysm of antiquity, mythland, dreamworld, the nethersphere, the antipodes . . . indeed, as Ouspensky, the Russian philosopher, once expressed it, "a noumenal world, a world of hyperspace, of higher dimensions, awaits discovery by all the sciences, which it will unite and unify — awaits discovery under its first aspect of a realm of *patterned relations*, inconceivably manifold, yet bearing a recognizable affinity to the rich and systematic organization of *language*, including *au fond* mathematics and music, which are ultimately of the same kind as language."

## Consciousness Divided

The objective-subjective polarity we have seen expressed in human language is evidence that man, even in his primitive cultures, recognizes that consciousness has multiple aspects. And this, I notice, is often manifest in the not-rare experience of hearing oneself saying something that surprises one. On occasions when I am angry at my wife — bless her — I am apt to hear myself saying things I don't really mean and which the more reasonable part of me immediately realizes I will be sorry for. In fact sometimes the fairer-minded faction of me is already sorry for the behavior of the other faction at the moment that faction acts. It's as if my adrenalin were speaking independently of my mind. Or would it be truer to say my adrenalin commandeers a minor segment of my mind to its ends, seeking thus a kind of release from imperative tension?

In any case, a different and less natural division of consciousness occurs under the influence of drugs, which may give an animal or a    291 man a life utterly different from that of his normal, undrugged life. Thus a dog trained to respond to a bell while drugged with curare responded a week later when drugged again, but did not respond during intervening periods of being undrugged. Which suggests that both learning and memory must be blocked by some kind of a barrier between the drugged and undrugged states. This hypothesis at least has been supported by some alcoholics who insist that drinking improves their memory of things that happened during earlier periods of drunkenness — something that, I hear, has led a few pioneering therapists to give therapy to alcoholics while they are drunk so it will have more effect the next time they start drinking.

Amnesia is probably the simplest and most specific possible division of consciousness, and cases are known, though rare, in which loss of memory is so complete a person is mentally reborn, in effect experiencing a second life (sequentially) in the same body, neither life knowing directly of the other. Obviously I am not thinking here of hallucination, although hallucination also can produce a sort of second life, albeit of a different and more morbid nature. As a matter of fact I've heard of a retired English schoolteacher who, shortly after her husband's death, noticed a woman walking beside her on the sidewalk one day and, without thinking, tried to shake hands with her. But although the woman also put out her hand, the two hands went through each other without touching, the "woman" being only an image without flesh or bone. Then the schoolteacher noticed that the "woman" resembled herself. "It's exactly like looking in the mirror," she told a psychiatrist, "except there is no mirror."

Such phantoms are anything but common but, when they do occur, they can be impossible to shake. When the teacher closed her eyes, she could still see her double with *its* eyes closed. "In a detached, intellectual way," she explained, "I am fully aware that my double is a hallucination, yet I see it and I hear it. And I always know it is there."

The doctors admit that so far they know no acceptable therapy for treating such a disorder, which may be brought on by a brain injury but is more often considered to be a natural form of clairvoyance, whatever that turns out to be. There is, however, an equally rare kind of divided consciousness, exemplified in such wide-selling books as *The Three Faces of Eve* and *Sybil*, in which a person (usually a young woman) suddenly flips into a new personality as if her body and brain were somehow suddenly taken over and totally possessed by someone else's mind and soul. Then, after years of painstaking psychoanalysis, it turns out in each fully treated case that the new personalities (some-

PLURAL CONSCIOUSNESS

times numerous and hard to identify) are actually suppressed aspects of the original personality mysteriously created by, yet completely unknown to, that person. Although much remains to be learned about this phenomenon, which seems to require a scientific alternative to the ancient and classic exorcism, it is found to develop out of some sort or another of terrifying experience so unbearable to the small child (who may not yet know how to talk) that she blacks out into amnesia, but then, while temporarily blocked from returning to the same agony, resurfaces in a different personality. If the second personality also suffers beyond endurance, a third consciousness may appear, or more. In the case of Sybil, originally a battered infant, an amazing total of seventeen personalities turned up during 2354 sessions in eleven years conducted by psychiatrist Cornelia B. Wilbur, some of them resembling babies, some adolescent boys, some grown women. But the final personality is described by Dr. Wilbur as a cured, whole and basic woman who alone of them remembers all the feelings and tortures of her former selves. In fact she is now so well balanced that she can get along as a mature, respected artist and teacher, living normally under a different name at a midwestern university, her extraordinary past still unknown (at the present writing) to her neighbors and associates.

Multiple consciousness is almost certainly more common than is generally realized. This is partly because those who experience it rarely get sufficient professional help to integrate the different consciousnesses, and notice only some of their periods of forgetfulness, which, if long, are regarded simply as amnesia. Dr. Wilder Penfield has said that multiple consciousness occurs unpredictably when the cerebral cortex is stimulated by a mild electric current — which, it may be postulated, can happen naturally whenever the electrochemical nerve impulses are short-circuited under conditions of traumatic shock. This also must be nature's way of diluting agony through subdivision of consciousness, presumably one of the unrecognized manifestations of the Polarity Principle (Chapter 18) under which good and evil and other double-sided qualities form the grit and grain of the Soul School of Earth.

Surely all of us are plural in a sense. Why therefore need all of you or me be confined to one body or one lifetime? Is not the mind more than the sum of the body's nerve interactions and sensations — just as life is more than the carbon, hydrogen, oxygen and nitrogen in common clay?

When a fertilized ovum divides into identical twins, its budding consciousness must divide too — a process that seems simple, as when the proverbial fisherman digging worms asked one of them, "Which way d'you think you're going?" while raising his shovel to cut him in two. To which the worm most appropriately replied, "I think I'll go both ways" — slice — "didn't we?"

Something similar but opposite happens to people at conception, most of us, however, never becoming aware of anything particularly remarkable in the merging of the potential consciousnesses of a sperm and an ovum or in their ensuing genetic reorganization into an embryo that can twin, both the multiplication and the division being natural functions in life.

## SLEEP

Sleep is of course another way in which consciousness is divided, in this case temporally and into cycles that are normally (but not necessarily) geared to the motions of Earth. Sleep also seems to be vital, particularly to the functioning of a highly evolved mind. However, animals, especially primitive ones, may live out their entire lives with little or no sleep. Ants, for instance, seem not to sleep although they may do something equivalent in short intervals still unnoticed by man. On the other hand, animals that are preyed upon on the plains cannot afford to relax their vigilance for more than a few seconds, so they take turns sleeping, usually with their eyes open, and some solitary antelopes go to places where they can see without being seen and, after a careful look around, sleep deeply for perhaps a minute or two at a time.

Creatures with safe refuges, like the mouse in its hole, the hippopotamus in its pool and the bird in its tree, sleep very soundly. As do the large predators, who almost never have anything to fear even when exposed to full view. A few, if they have eyelids, not only close their eyes but cover them with paws or tail or, if they have very keen ears, like the bush baby and the big-eared bat, sleep with their outer ears folded. Seals are known to sleep underwater while holding their breath. The deepest sleeper of all, not counting hibernators or the sperm whale, is reputedly the sloth bear of India who has such a fierce temper that even a hungry tiger has never been known to risk attacking him.

Most deep sleepers, however, including humans, have a sort of filter in the central nervous system that discriminates to some degree between harmless and dangerous sounds, or scents or other stimuli. A circus elephant will sleep through having another elephant sit on his head yet will rouse instantly on hearing a faint metallic clink. Or a mother will sleep through her husband's snores or coughs but wake the moment her baby whimpers.

Sleep provides a vital rest for most muscles, although a few work rhythmically or intermittently during it, resting only fractions of a second at a time in the case of heart, lung and digestive muscles, or irregularly for those in limbs, neck and torso, which have been found to shift the average sleeper to a new position about five times every hour. But in many known cases, muscles can function in sleep almost as when awake, particularly if their movement is rhythmic. This has been known not only in the familiar manner of people who get out of bed to walk in their sleep but also among dog-tired soldiers in an all-night forced march, in dance marathons, in soaring albatrosses that almost certainly must doze on the wing and in the proverbial punkah wallah (fan servant) of India who can work his hanging punkah (fan) with one foot quite satisfactorily while sound asleep all night long.

In modern sleep laboratories, it has been found that sleep often begins physically with a relaxation of the tongue and jaw, progressing down the throat, through the head and out the limbs. Sleep's mental side also involves a progression (not yet well understood) of states of mind leading into deeper and deeper slumber, the stages of which can be distinguished to increasing extents by the kinds of brain waves they produce in an encephalograph, by eye movement, by phosphenes (page 236) and by dreams or dreamlike experiences.

An example of one of the less recognized states of mind might be an experience I first became aware of one morning in Spain in 1956, just after awaking and while still half asleep with my eyes shut. It wasn't a daydream but rather a kind of shut-eyed consciousness beyond the self — something imposed on me and out of my control, something akin to dreaming but definitely conscious and not really     295 part of sleep. It was entirely visual. With my eyes shut I could see a kind of phosphene consisting of moving cloudlike forms. Large, hazy, whitish clouds, dimly outlined against surrounding darkness, seemed to be condensing or drifting inward toward the center of my vision, continuously getting smaller and smaller until they disappeared into a point as down a drain. And while each diminished, taking three or four seconds, it was replaced by new and different clouds appearing in the outer circle. It made me think of starbirth, the presumed condensation of dust clouds in space that continues (although extremely slowly) until the steadily increasing pressure of gravity ultimately ignites the new system of worlds.

I began to experiment with this mind state, making a strong effort to control the clouds either in form (by willing certain shapes) or in speed (by slowing them down), but found I had no more influence on them than if they were clouds in a real sky. For they appeared independent and a law unto themselves. They seemed more than imaginary, moreover, for imaginary clouds are controllable to some degree by any artist who imagines them. Each of these seemed independently perfect in itself: beautiful, individual, willful. I felt that by opening my eyes I risked losing the clouds for good yet, if I opened them for only a fraction of a second, I found I could regain the cloud phosphene as soon as I closed them again. Then there came a time when I opened my eyes for several seconds and, on closing them, I was suddenly "blind" and saw nothing but black, which gradually became dark brown. It was a shock, and I tried in vain to will the clouds to reappear. Then, about five minutes after I gave up, while still in a drowsy state, the clouds returned as before — just as if they had a will of their own. I have scant doubt that there must be many such little-known states of mind, particularly on the sleepy side of consciousness.

The modern, objective way of measuring or identifying a state of mind or the depth of sleep is by the sleeper's brain waves. So-called alpha waves (oscillating about ten times a second) are what the oscillograph shows during the relaxed, shut-eyed wakefulness of "transcendental meditation." If the subject then opens his eyes or moves around, his alpha waves speed up and intensify until (at about thirteen cycles a second) they become beta waves and perhaps soon sort of hash out. If he dozes, they slow down and gradually laze into looser forms. When someone consciously generates a particular kind of brain wave, say alpha, when meditating, the process is known as biofeedback, and developments of it may be used in everything from psychotherapy to self-hypnosis. The waves of full wakefulness, one no-

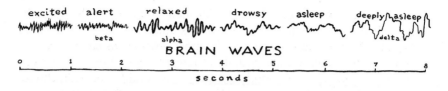

tices, display a relatively high frequency, although their amplitude, denoting voltage, is inclined to be low. The waves of sleep, on the other hand, have a low frequency (generally less than half that of wakefulness), but with a correspondingly high amplitude and voltage. And the so-called delta waves of deepest sleep have the lowest frequency and highest voltage of all — a fact that has recently led more and more sleep researchers to begin thinking of sleep as not just a nightly submergence below consciousness but as perhaps the mind's basic (high voltage) condition, out of which wakefulness arises as a daily (high frequency) flight of excitement.

Meanwhile a discovery in physiology supports the same concept of sleep as more active than passive by revealing that certain cells in the brain "turn on" to inhibit physical activity and thereby promote subconsciousness. This finding has been credited to Dr. Carmine D. Clemente, who has been exploring it since 1959 at the University of California's Brain Research Institute in Los Angeles. He reports that a class of specialized neurons he calls "sleep cells," located in the basal forebrain, begin to fire as one gets sleepy, sending impulses to muscles, glands and other active parts of the body, ordering them to hold still and, where possible, relax. And a further step attributed to Dr. Dominick Purpura, anatomist at the Albert Einstein College of Medicine, is the realization that this natural process can be promoted by artificially passing electric currents through the brain. A mild current in fact brings on unconsciousness by jamming brain circuits and disorganizing neuron activity to the point where information transfer is stopped. This is now known as electroanesthesia and is beginning to be used in surgery where other anesthetics are dangerous or impractical.

## DREAMS

If sleep thus may be the basic state of mind, the dream in turn may be the basic state of sleep. At least the "dream lab" researchers' consensus to date definitely indicates no human can survive long 297 without dreaming — a conclusion evidenced by the nervous system's steady deterioration when dreams are systematically interfered with. It has also been demonstrated that the most obvious, or easily remembered, dreams come during periods of lighter sleep, when one's eyeballs move rapidly (their motion readily measurable through closed lids), and that these periods of REM (rapid eye movement) occur cyclically, about every ninety minutes. This doesn't mean the intervening periods of deeper sleep are dreamless, for evidence has long been accumulating of significant mental activity during delta sleep. Edgar Allan Poe, for example, had a strong intuition that dreamless sleep is nothing but a myth and he declared with authority that whenever we wake "we break the thread of some dream."

My own experience confirms this, for nearly every time I awake, even when I can't recall the subject of my dream, I am aware that I have been dreaming (or thinking) about something. In fact I can almost always "guess" what time it is within half an hour, which I do by reflecting on the duration (if rarely the content) of my less-than-conscious thoughts since I fell asleep. And, to test the universality of continuous dreaming, I have made a point of rousing innumerable other sleepers to ask them as their eyes opened what they had been dreaming (or thinking) about, and I have yet to find one completely dreamless.

Ouspensky goes even farther in corroboration of the perpetuality of dreams, writing: "I became convinced that we have dreams all the time, from the moment we fall asleep to the moment we awake, but *remember* only the dreams near awakening. And still later I realized that we have dreams continuously, *both in sleep and in a waking state* . . ." In reality, he continues, "dreams never stop. We don't notice them in our waking state amidst the continuous flow of visual, auditory and other sensations, for the same reason we don't see stars in the light of the sun. But just as we can see the stars from the bottom of a deep well, so we can see the dreams that go on in us if, even for a short time, we isolate ourselves . . . and achieve 'consciousness without thought.'"

"Consciousness without thought" presumably means a meditative state, but whether the phosphenes and "daydreams" accompanying it qualify as true dreams of course must depend on one's definition of a dream. In any case, the dream is a phenomenon that offers us prime evidence, if not understanding, of basic mental processes — processes used by animals, who also seem to dream, if not by plants and lower life forms. To me, a striking thing about the dream is that it is not subject to the will, but rather seems imposed from beyond consciousness. Like a law of nature, we accept it because we must. It swallows us and digests us. And the same may be true, if less obviously, of all mental processes.

Where, in fact, do thoughts come from? Historically it has turned out to be a lot easier to deduce the source of comets, meteorites and cosmic rays than to discover where thoughts and dreams come from. We certainly have no assurance that dreams just originate in ourselves. Nor can we rule out the possibility that some unsuspected amalgam of heredity and environment might somehow someday reveal the hidden nature of the thought and the dream.

In a dream the deep sea becomes a hospitable world for breathing, talking men and women, the clouds a drifting staircase, the sky a banquet hall. You drive a car and it takes wing. You float through a house and it changes. You meet a strange girl and you know her thoughts.

What is the purpose of the great traffic of dreams, the extravagant scenery that shows no regard for economy or efficiency, the vast crowds and fantastic cities, the struggles and adventures that so easily slip out of memory before we awake? Are they hints of a world to come, of the struggle for transcendence from our present finitude of space-time to the dimension-free Infinitudes beyond?

The preponderance of current authoritative opinion is in surprising accord with many such ideas. It considers dreams the sleeping mind's arena of confrontation with the stresses of wakefulness, with the daily emotional problems that need nightly therapy, not through reason but through feeling — feeling that often reaches down to the level of a child or an animal. For dreams have a way of simplifying things, of boiling their messages down into pictures and parables. And they avoid making moral judgments. They express symbols, not thoughts, omitting all the qualifying conjunctions like *if*, *so* and *but*. And there is some recent evidence that they even help inscribe short-term memory into long-term knowledge.

Sigmund Freud, the great pioneer of dream analysis, surmised that dreams in general seek to fulfill wishes, particularly sexual wishes. His reasoning was that, because sex wishes are the ones most apt to be unfulfilled in most people, sex has to be the biggest common denominator of dreams. Other psychologists believe, however, that although almost all children's and many adults' dreams deal with wish fulfillment (including wishes for sex, power, popularity, etc.), other dreams may be reactions to such physical or psychic stimuli as noise, hunger, anxiety, bladder pressure and indigestion, which, incidentally, the dreams tend to disguise so as not to wake up the sleeper. Still other dreams deal with anticipation, with going to a party or off on an exciting journey . . . And, as I've suggested, dreams of all these kinds seem to express some sort of conflict and, in one way or another, to work at resolving it.

Carl Jung, originally Freud's close colleague, accepted all this, and broadened dream interpretation by adding to it interpretations of myths and religions, explaining that all three symbolize unconscious human thoughts and feelings, which together use "libido" or psychic energy to promote mental well-being. Specifically he theorized that the symbols featured in these dreams, myths and religions include archetypes or primordial images that enable the mind to resolve its neurotic dilemmas superconsciously, if not mystically, in the sense that such visualizations are not limited to one individual alone but are "historical strata" woven through the "collective unconscious" of mankind (just as star patterns are imprinted in the minds of birds) and thereby presumably imbued with some teleological purpose. In fact it is said that Jung himself was more than once inspired by dreams to change the course of his career, indicating his mystic sense of the dream's oracular portent and his acceptance of its being part of some universal mind.

## Psychic Dreams

The dream's power to foretell the future is of course a very ancient concept, going far back into the forgotten ages before history began. Pharaoh's dream of the fat and lean kine and Joseph's dreams of the sheaves and stars were widely accepted as of this sort. In modern times there was Mark Twain's dream that his brother had been killed and was lying in a metal coffin, a bouquet of white flowers on his chest with a red blossom in the center. Although the brother on that date was still alive and well, it was only a month later that Mark actually witnessed the very scene of the dream, including the red blossom, when he went to view the remains after his brother was killed in a boiler explosion on a Mississippi steamboat.

More remarkable still — in fact perhaps the most astonishing of all prophetic dreams ever published — must be that of Robert Morris, Sr., father of "the financier of the American Revolution," who dreamed he himself was destined to be shot to death by one of the big guns on a naval vessel he was scheduled soon to visit. Because of the dramatic vividness of his dream, he felt apprehensive about the ship and tried to cancel his visit, but naval officers told him he was being absurd, if not cowardly, and eventually the ship's captain convinced him that it was absolutely safe to go aboard as no guns would be fired until after he left. At the end of his visit, the captain tried to make good on this promise by ordering that a salute (without a cannonball) be fired only after he raised his hand when the party, including Morris, had safely reached shore. But while the boat was still within gun range, a fly alighted on the captain's nose, causing him to raise his hand to brush it off. And the ship's gunner, taking this as the signal to fire, fired with the astounding result that a tiny fragment of the discharge hit Morris and wounded him fatally!

In this century J. W. Dunne's book *An Experiment with Time* offered a theory that prophetic dreams result from time's little-known multiple dimensions which actually extend perpendicular to one another, so that anyone's consciousness may impinge on the future as easily as on the past, a fact rarely noticed since, by the time the events foretold in the dream take place in real life (sometimes many years later), the dream more than likely has been forgotten.

Another possibility, it seems to me, is that the illusion of dreaming the future in *time* may come from an interrelation between differently dreamed perspectives in *space*. If a boy views a horse race through a knothole in a fence, for example, he may see a white horse about to overtake a front-running brown horse in the stretch and conclude the white horse will win. But his father standing beside him and looking over the fence may take in a much wider view that includes a very fast black horse overtaking both the others so rapidly he can safely predict the black one must win. If the father tells the son he can foresee a black victory and it proves true, it could seem to the boy

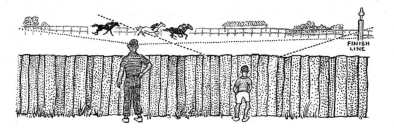

that his father has prophetic vision into the future. Yet in fact the only advantage the father had was his larger view of space which gave him more information with which to make a more accurate extrapolation of the same future already extrapolated by the boy. Such at least is the function of one of the causal seams joining space to time.

Some dreams do more than foretell the future. Some dreams create the future. Robert Louis Stevenson attributed all his best plots to his dreams, including the famous story of Dr. Jekyll and Mr. Hyde. And Professor Hermann Volrath Hilprecht, the great German Assyriologist, dreamed that a Babylonian priest gave him the key to two ancient and cryptic inscriptions he had spent years trying to decipher — whereupon he woke up to find the dream key was also a "real life" key that soon enabled him to translate both texts. Kekulé, the German chemist, had been baffled in his struggle to learn the true shape of the benzene molecule when, in a brief daydream while riding a London bus, he saw a snake holding its tail in its mouth and instantly realized that the solution was a loop, now known to chemists everywhere as the benzene ring.

Dunne's hypothesis of multidimensional time may be supported by the fact that dreams can enfold each other relativistically, a phenomenon I noticed when I "awoke" one morning from a dream and started writing down what happened in it — whereupon, to my utter surprise, I found myself waking again, which instantly apprised me that my previous awakening and writing was all part of an outer dream enclosing the inner dream I'd been writing about. How many such folds or shells of dreams might encase one another like Chinese boxes, I know not — nor when I will awake again from my present conscious life, revealing it too to be a dream relative to some still larger consciousness — presumably with more dimensions and beyond the grave.

Still another confirmation of Dunne is the fact that one often knows, while dreaming, that a particular scene or episode, apparently freshly encountered, is yet already familiar in detail from previous dreaming whose scope must be much vaster than our waking memory can normally recall. I noted this most strikingly when dreaming of getting a film developed and printed, which produced lots of wonderful photographs I had taken of past adventures, which, having nothing to do with my conscious life, were yet recognizable as part of dream life. And this dream life pretty clearly comprised fields outside time, space and self-consciousness (even possibly outside other "tools" of finite awareness) unfolding continuously outward in unknown dimensions.

## Dream Analysis

The therapeutic function of the dream as the great resolver of mental problems has become widely accepted since psychology joined the sciences. When a young English mountaineer suffered a dangerous slip that left him breathless, though uninjured, he dreamed the next night he had a terrifying fall that almost killed him. And the night after that, he dreamed the same dream again, except that this time, in falling, he tried to clutch a jutting ledge but missed and crashed as before upon the rocks. On successive nights he kept repeating the nightmare, never quite catching the ledge until the sixth time, when he finally got a good hold on it and saved himself. And this succession of dreams, according to analysts, had the important therapeutic effect of giving the mountaineer not only insight into what was at stake should he slip again but repeated experience in coping with it. Indeed repetitive dreams of this sort are not uncommon among pilots, stunt men, fire fighters, soldiers, divers and in fact among all the many occupations that invite harrowing experiences.

Often dream characters seem to represent aspects of the dreamer's own character and thus dramatize a truth he or she couldn't realize unassisted. A young woman under psychiatric treatment for a sexual phobia had a nightmare about a hungry tiger loose in the next room and trying to break through the door to kill and eat her. The tiger of course personified her repressed eroticism. A week later she dreamed she was swimming in a pool with an ugly dog, her sex feelings no longer dangerous but still unpleasant. Next she had a dream of bicycling along a country road until she found a cow lying in the way and had to alight to walk around it, her sex having become not so much unpleasant as a nuisance. The following month she dreamed she was awakened by burglars in her garden and was frightened enough to scream for help. But the "burglars" wasted no time explaining they were really only baggage men who had responded to her request to have her luggage taken to the station, thereby indicating that her sex feelings, though alarming at moments, were no longer a serious bother and were actually beginning to serve her in a useful way. By the end of her series of dreams, the progression of symbolic figures from tiger to cow to man had signaled complete recovery from her fears, closely followed by her full acceptance of her sexuality — even on occasion, to her surprise, to the heights of delight!

That is the way dream symbols work. Often they seem trivial or as crassly simple as a punning advertisement that makes its point at some basic level below full consciousness. I heard of a man who dreamed he went through the ordeal of having the sole of his foot operated on to test his toughness: to see, according to his mental associations, if his soul (sole) was strong. And there was the character I met in 1957 in my own dream about a mental institution, who portrayed his irrationality by going into a "Julius Seizure"!

There is also the levitation dream that many people (including me) have repeatedly enjoyed, and which may express not only a distant memory, such as ancestral tree leaping, but perhaps a premonition of our future in zero-G space. Dreams surely stem from a deeper well than ordinary consciousness and I am convinced that the dreamworld is no mere shadow of the waking world but rather something far more fundamental, universal and continuous, something creatures in all worlds must experience in their various ways, even though right now it remains unestablished whether any of them, except man, enjoys a waking consciousness as full and memory-packed as the kind we are daily familiar with.

Helen Keller's dreamworld naturally endowed her with senses she knew nowhere else, so it became supernormal for her. "I dream of sensations, colors, odors, ideas, and things I cannot remember," she wrote at the age of seventy-two. "Sometimes a wonderful . . . light reaches me in sleep — and what a flash of glory it is! In sleep I never grope but walk a crowded street freely. I see all the things that are in the subconscious mind of the race. When I awake I remember what I have dreamed."

William James too sensed the scope of this hidden realm and declared: "The whole drift of my education goes to persuade me that the world of our present consciousness is only one out of many worlds of consciousness that exist, and that those other worlds must contain experiences which have a meaning for our life also."

## HYPNOSIS

A state of mind not easily distinguishable from light sleeping is the hypnotic trance, which, as is becoming well known, can be induced in most people above the age of four by placing them in a relaxing environment and voicing some sort of a soothing monotonous patter such as "Relax. Close your eyes. Make yourself comfortable. You're feeling more rested than ever in your life. You're floating on a warm, soft cloud — sinking gently into a feathery bed — so soft, so drowsy, so comfortable . . . You're getting sleepier all the time — drowsy, sleepy, comfy — deeper asleep, deeper . . . deeper . . ."

The easiest hypnotic subjects are obliging, suggestible people who feel secure enough not to mind entrusting their minds to someone else. Such ones can be hypnotized even from a phonograph record of the patter, and it is not rare for certain types among them to learn to hypnotize themselves. Self-hypnosis in fact (despite its difficulties and limitations) is said to be a common accomplishment among the people of Bali and so widespread in many oriental lands that I have heard of people there who had apparently lived through a whole waking lifetime in something described as "a light hypnotic trance." In America, on the other hand, self-hypnosis is well enough established among psychologists that it is not only being researched in psychotherapy but some doctors are seriously trying it out as a technique for easing chronic pain.

Once asleep or in trance, of course a hypnotized subject of any sort may readily be guided into any reasonable present or future activity by positive suggestion, such as "Your right hand feels light and ready to rise. Now it's starting to come up — and up — and up . . . " Or "Your lips are dry and you're getting thirsty. In five minutes you'll be so thirsty you'll wake up and go to the basin for a drink — a drink of cool, refreshing water . . ."

In a recent series of tests at American University in Washington, D.C., with dozens of men and women, it was found further that when they were allowed to fall asleep naturally, they responded to these verbal suggestions almost exactly as much as when they were put into a hypnotic trance. And even when tested while fully awake, most of them responded to the same suggestions at least halfway. Although it is hard to know why some people (and animals) are so much more suggestible than others, it turns out that suggestibility is virtually universal. Which probably explains why a yawn, a titter or a whisper in a crowded room is so generally contagious.

Nevertheless the kind of suggestions that sink deep into the subconsciousness have proved to be the most potent ones of all. I mean that hypnotic suggestion has definitely been credited with having eliminated pain in major operations, removed warts, created blisters (when the subject was told he had touched a hot stove), raised and lowered skin temperature by as much as 7°F. and turned memory off and on like magic. An example of the last came when a German author wrote a novel about his concentration camp experiences but lost the manuscript several years later, then he discovered he had forgotten almost everything he had written. He was fortunate, however, in meeting a hypnotist for, under hypnosis, the novel came back to him so quickly and vividly he was able to dictate it all over again in a few weeks and completely recover his loss.

If hypnosis is a way of reaching into the subconscious mind, the free and spontaneous expression of ideas is another. And this was accidentally discovered as a therapeutic technique about 1880, when an Austrian neurologist named Josef Breuer, who had been unable to

305

help a mentally ill woman by talking to her when she was under hypnosis, suddenly realized that, when she herself was permitted to talk, she "experienced an emotional catharsis" which progressively improved her. When young Sigmund Freud got wind of this development, he joined Breuer and, through further research, soon discovered that this new "free association" therapy not only did not require hypnotism but probably worked better without it — a key step in the launching of the new science of psychology and its precocious offshoot: psychiatry.

## EXTRASENSORY PERCEPTION

The advent at the same time of still another branch of psychology called parapsychology or psychic research, supported by such scientific pioneers as William James and Sir Oliver Lodge, the physicist, only added fascinating new evidence, not to say mystery, to the evolving knowledge of man's mind. Which wasn't made any more digestible by one researcher's designation of ESP (Extra Sensory Perception) as more accurately "Error Some Place."

We haven't space here to go into the vast and shadowy realms of occultism adequately, but I must mention that the last few years have evoked a considerably more respectful attitude toward them from established science — like the sober American Association for the Advancement of Science granting the Parapsychology Association an affiliate membership in 1969, the National Institute of Mental Health awarding grants for psychic research in 1973 and the *New Scientist* of Britain recently polling its readership to find that only 3 percent of 1500 science-minded respondents think ESP an impossibility. And even that 3 percent, if they concluded ESP is impossible because it remains unproven, took the untenable position that the nonexistence of proof amounts to the proof of nonexistence, a logical absurdity.

Among the more astonishing of supernatural phenomena investigated in modern history are the feats of Uri Geller, the young Israeli who apparently bends and breaks steel without touching it, makes the mercury in thermometers rise and fall at will and reads other people's minds with almost infallible accuracy. In 1972 he underwent a series of "cheatproof" experiments at Stanford Research Institute in California, amazing the scientists with such performances as making a balance inside a bell jar respond as though a physical force were applied to it and later correctly naming, eight times in eight tries, the numbers shown on a die shaken inside a closed metal box. It is reported that only scientists handled the box, none of whom knew what number the die showed until Geller each time made his prediction and the box was opened. If the scientists had been trained in detection of fraud, this might have been a convincing, if not a conclusive, demonstration (one way or another), but scientists rarely get interested in the details of deception and seem to have a significant skepticism as to the motives of magicians, presumably because the latter are masters of deception. So, even though magicians (honest ones of course) would logically have been the best judges of genuineness in such a critical performance, no qualified magician was even invited to be present, still less allowed to pass judgment on the reality of what happened. So Geller's standing as a phenomenal subject, to my present knowledge, remains unproven though he may actually be tapping (as some think) a kind of bioenergy (or even cosmic energy) that neither science nor he himself has yet learned to assess.

The case for the lie detector expert Cleve Backster is obviously somewhat different. His results do not depend on his own personal psychic powers so much as on sensitive galvanometers and polygraphs that can be worked by others, including any magicians who suspect him of trickery. At least that appears to be the reason Backster was taken with what may be called skeptical seriousness by many botanists and other scientists when he casually announced in 1969 that he had been testing plants and other cellular tissues for three years with his galvanometer and had discovered a mysterious electrical responsiveness in them apparently akin to the feelings and emotions of animals and humans, even including in some cases an evident telepathic sense that seemed undiminished by any barrier from a lead plate to a thousand miles of atmosphere around the convex solidity of Earth. The most obvious emotion revealed by plants, according to Backster, is "fear," which can be so severe they "pass out" and become totally unresponsive, as if dead. This "fear" seems to be provoked mainly by threats (such as the approach of a dog) or injuries, including those inflicted on other organisms (like a shrimp being boiled alive in a saucepan) within the "territory" they consider theirs. Backster calls this amazing perception "primary" because, as he says, "we have found this same phenomenon in the ameba, the paramecium and . . . in every kind of cell we have tested: fresh fruits and vegetables, mold cultures, yeasts, scrapings from the roof of the mouth of a human, blood samples, even spermatozoa."

While Backster's findings in plant sensitivity remain largely unconfirmed and therefore unacceptable to science as a whole, there is another kind of awareness, on the part of primitive animals as well as    307 plants, that has been studied longer and more widely. It is the creature awareness of various large-scale conditions on Earth, most notably of future weather. Although still far from understood, it is dramatically portrayed in the accompanying graph showing the metabolic rates of potatoes and salamanders (in sealed airtight chambers) during a typical month in conjunction with outside barometric pressure. And it testifies that even the most ordinary organisms of two kingdoms can consistently predict the weather for their localities two days ahead of time!

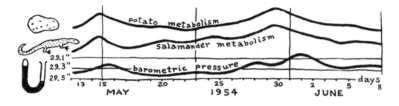

Basic sensitivities at this level are not generally considered occult, but probably they should be, for similar phenomena have been remarked by many empathic biologists from Aristotle to Linnaeus to Darwin, and perhaps most of all by Luther Burbank, who became world famous as the plant wizard of California and who created more than 800 new strains and species of plants, ranging from the Shasta daisy to the spineless cactus, using a combination of unorthodox science and mystic flower communication he never fully explained. Significantly Burbank was brought up by a telepathic mother and admitted in 1923 in a magazine article: "I inherited my mother's ability and so did one of my sisters." So much so, in fact, that he seldom used the telephone to talk to sister Emma, generally relying on telepathy alone.

As for his plants, he not only hand-pollinated them with intuitive skill but would commune with them daily and intimately like a lover, alternately praising, entreating and tenderly soothing their apprehensions. To persuade one stubbornly prickly species of cactus to quit sprouting its needles, he would gaze on it every morning with loving eyes, whispering mentally, "You have nothing to fear. Relax and trust me. No longer will you need your defensive thorns. I am your friend for always. I love you and I will protect you . . ."

He believed that the vegetable kingdom has so far evolved "more than twenty" senses, most of them naturally having to do with each plant's vital relations with things like the wind, rain, soil, and sunlight, rival neighbors and pollinating animals. And he is known to have confided to his close yogi friend, Paramahansa Yogananda, shortly before dying at the age of seventy-seven, that he felt sure all plants possessed minds of some sort, probably simpler than the human mind but in essence the same. He also increasingly saw "men as trees, walking" and seemed to consider this a sign of the evolving superorganism of Earth. Indeed Yogananda specifically remembered Burbank's unique philosophical summation: "I see humanity now as one vast plant, needing for its highest fulfillment only love."

## Universality of Mind

One could sum up mind as a universal aspect of life and energy, an aspect with a relationship to the body mystically similar to the wave's relationship to the particle. And, according to modern religious sources, mind also has a resonance relation to brain cells, which vibrate in response to spiritual energy under laws far beyond the scope of science. This doesn't mean that a few far-sighted scientists are not trying to bridge the gap — such as physicist David Finkelstein of Yeshiva University who for years has been seeking a measurable connection between particle physics, relativity and human consciousness, and recently said, "The way has been prepared to turn over the structure of present physics to consider space, time and mass as illusions in the same way temperature is . . . a sensory illusion."

And there are pioneers like Jung who wrote in his *Psychology and Religion: West and East*, "My psychological experience has shown time and again that certain contents issue from a psyche more complete than consciousness" — presumably meaning intuition, dreams and the universal mind. Which he elucidated in *Modern Man in Search of a Soul* by adding that "spirit is the living body seen from within and the body the outer manifestation of the living spirit — the two being really one . . ."

All this is a way of saying that science offers us only one of the basic paths to truth and even that path includes such factors as the uncertainty principle, abstraction, polarity and transcendence. Our here-present finite world, in other words, with all its limitations, is profoundly mysterious, confirming Goethe's observation that "remarkable discoveries and great thoughts that bear fruit, do not lie within anyone's power. For they are exalted above all earthly influence." They even could be, in fact, emanations of a cosmic mindscape that permeates the universe — a soul hypostasis that dreams and dreams and somewhen dreams not just our planet but, were we capable of knowing it, all life in all worlds.

309

PART THREE

# THE SEVEN MYSTERIES OF LIFE

# CHAPTER 12

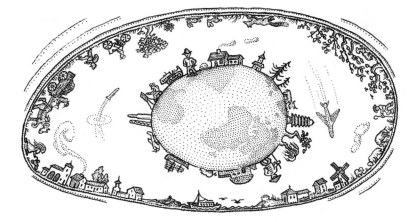

# First Mystery: The Abstract Nature
# of the Universe

A THOUGHT that comes to me as I arc above our world, watching it pensively out of my mind's eye, is that, although Earth appears stationary, she is actually moving in many ways — swiftly, subtly, abstractly.  Not only is her blue-flecked surface spinning around its axis at a quarter-mile a second but, as a whole, she is orbiting around the sun at 18½ miles a second and the sun's entire system of planets is drifting through curved space toward the star Vega at 12 miles a second, while virtually all the stars we can see (including Vega and the sun) are swinging at 150 miles a second around the Milky Way. And even the Milky Way, a wheel of stars an unimaginable 100,000 lightyears in diameter, is speeding away from other galaxies at thousands of miles a second, depending on which one you compare it to, in what has been aptly described as the exploding universe.

All these relative motions of Earth of course convey man right along with them, but in an abstract way — without his feeling the effects in    313 the slightest. For this is the world where objects, without much plausible reason, shrink with distance, where thrushes pull up worms to turn them into songs, where an acorn becomes a giant oak in a century because it was forgotten by a squirrel. In other words there is something otherworldly about our existence here — something more than matter, more than the body and mind we have been discussing — in short, something fundamentally and profoundly abstract. And I mention this aspect because it is not at all obvious, indeed scarcely noticed by the great majority of us as we go about our daily lives.

Yet I find abstraction so importantly mysterious that it almost unavoidably falls into place as the first of the Seven Mysteries of Life this book is about. The usual definitions term it unreal, visionary, theoretical, intangible and abstruse despite its being the very stuff our world is made of, the ubiquitous mortar that keeps it together, the vital spark that makes it alive. The evidence is seemingly everywhere — in the sky, in the sea, in the incredible numbers of our fellow creatures living unseen, unintroduced around us. Did you know that a shovelful of ordinary soil contains a microbe population greater than that of mankind? And that, if the Milky Way were reduced to the size of Earth, our planet would vanish into a mote of dust too small to see. It makes one wonder whether there is any inherent realness to anything. Whether seemingly dissimilar things like a flight of stairs, a flight of birds and a flight of fancy are basically much different.

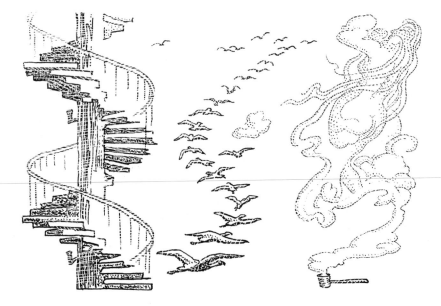

Although the world must always have been abstract, it was probably impossible for its inhabitants to realize this until language had matured to the point of articulating abstract concepts. For this reason it seems unlikely that even such a gifted people as the ancient Egyptians could have really understood abstraction because their speech had not evolved beyond visual terms. This is shown by their writing which consists only of pictograms representing the material things they saw around them. But the Greeks, with their more subtle language, really broke through to the hidden profundities of existence. Have you read of the famous discourse between Leucippus and Democritos as they walked the beach of Abdera in Thrace in the fifth century B.C., debating the nature of matter? Logically, they concluded, there has to be a limit somewhere to the subdivision of grains of sand or drops of water, because, if any matter could be divided and divided forever, the various sizes would be so utterly relative that no particle would be distinguishable from a world. By that abstract reasoning, without ever seeing the microcosm, they conceived the "partless part": the atom.

Yet the questions about matter still go on. And on. Does any grain or atom really need to be distinguished from a world? From a universe? Whose voice in this earthly node of flesh can declare with authority that our universe is not an atom of some unknowably larger megaverse outside it? And what atom anywhere has been prohibited from having a microverse inside it? Furthermore is there any evidence that relativity does not pervade all dimensions, even transcending finitude so that ultimately space and time and self unravel into some sort of Infinitude . . . for Ever . . . ?

## WHAT'S IN AN EGG

"In the beginning was the Word." Something abstract. "And the Word was made flesh, and dwelt among us." So said the apostle John as abstraction turned into matter and matter into life — as all three were one and of the essence of the world, and of its rocks and seas, its dreams and seeds and eggs.

If you've never thought of an egg as particularly abstract, may I say there are depths to the egg far beyond the nest, the brooder or the omelet? Would you believe that a common bird's egg contains a song neatly tucked away in chemical notation and packed into its genes? That it also has detailed instructions for nest building, a menu, a compass and a clock, a map of stars?

If such an egg cargo seems fanciful, please know that although long, 315

painstaking, expensive research went into establishment of the facts, the list is far from complete. In the 1950s in Freiburg, Germany, for one example, a research ornithologist named E. G. F. Sauer took about a hundred warblers' eggs of several species that fly at night and are known to migrate in autumn from northern Europe to southern Africa and experimented to find out if the birds might be learning celestial navigation from their parents or other environmental sources. He hatched the eggs one spring in separate, soundproof boxes and raised the chicks in complete isolation so they never knew the existence of other birds, the sun, moon or the stars until fully grown. Then, when it came time for the fall migration, he noticed the birds grew restless and he initiated them to the outside world by putting them into cages through which they could see the clear night sky, the while observing them very closely. Although they could not escape, they obviously wanted to fly and Sauer was excited when he noticed that they consistently aligned their heads and bodies in the correct direction of their migration routes, neglecting to do so only when clouds or some other cover made the stars invisible.

Next he placed them, one by one, in a special planetarium so that each bird could see only artificial stars while being monitored by a trained observer from below who recorded the exact direction in which the bird headed while fluttering its wings and trying to fly. And Sauer was delighted to find that the birds responded to the planetarium's tiny lights exactly as if they were real stars, the garden warblers instinctively heading "southwest" toward "Spain, Morocco and West Africa" and the lesser whitethroats "southeast" toward "Greece, Egypt and East Africa" just as their respective ancestors had done for millions of years. He checked all his birds in this way and even gave them course problems to test their navigation senses. For example, he confronted some of them with a starry sky that looked exactly as it would have looked at that moment if they had been in Siberia, to which they reacted by heading westward to get back on course. Then he arranged the sky to look as if they were in America and again saw them turn to get back on course, this time eastward, both responses confirming the birds' extraordinary navigation instinct.

Of course it was an instinct almost certainly inherited through some sort of species memory of the patterns of the "fixed" stars imprinted in the cells of the visual areas of the birds' brains and handed down for millions of generations through the genes in the egg. You may wonder, since the stars are not really fixed but rather drifting continuously about on their remote and inscrutable orbits, how the birds' genes could keep up with their motions. But just remember that the stars are so remote their apparent motions are extremely slow and that genes, like other body parts, metabolize and change, and they would only need to adjust slightly in a thousand generations to enable their host birds to recognize the star constellations, even as men today still see the belt and sword of Orion essentially as did Aristarchos in the third century B.C. Thus when the migrating warbler sees Cassiopeia's

Chair overhead of a September evening, he feels not only reassured by the familiar pattern but impelled to head southward toward the beckoning Square of Pegasus. And his happens to be a dynamic memory, incorporating and compensating for the clocklike rotation of the sky (relative to the turning earth) as he feels the instinctive urge to veer more and more to the left of Pegasus (at the rate of 15° every hour) as the night wears on — an inner genetic drive that somehow pervades his mind and body straight out of the egg, a drive geared to the rolling planet of which he is in very truth a part.

Eggs also talk to each other and to their mother (page 143), hatching plans to hatch and, if you consider the question "Which came first: the hen or the egg?" as classic and unanswerable, you should be interested in the egg's answer, proclaiming itself the winner by half a billion years! Yes, the egg has existed on Earth at least several hundred million years while the hen has been here only fifty million. And the difference between them is also the difference between mortality and immortality, as we will see (page 510) in our chapter on transcendence.

As the egg's single cell divides, multiplies and grows into a multi-celled organism, consciousness is presumed to develop along with it.

Consciousness implies an appreciable awareness (and control) of matter, an interaction involving both the developing body and the emerging mind that is at once abstract and close to the quick of life. Indeed the fact that you can move your legs and walk, or your tongue and talk, makes you alive. And so does the fact that you can control the engine and wings and tail of your airplane when you fly. You may object that the airplane is not really alive because it is not a natural organism but only man-made and artificial. But I reply that so is a bird's nest artificial for it is bird-made and not strictly a part of the bird's body (although the know-how of its making is inherited through the genes in the egg). And so too is coral artificial in the sense that it is made (or excreted) by the coral polyps. And so is the oyster's shell built of calcareous substances out of the sea. And so also are the shells of birds' eggs and a bird's feathers made of things the bird eats. And so are even your teeth and bones and your fingernails and hair, in fact your whole body. There is no definite line, you see, where artificiality begins. And there is no absolute boundary between life and the world. Your shoes are about as much a part of you as a horse's hoofs are of a horse, even though you can change them easier. After all, hoofs are trimmed and changed too when the horse is reshod, just as hair can be cut or dyed and nearly any organ in your body replaced with either an organic or inorganic substitute. So your consciousness makes the automobile alive in essence when you drive it and thus extends your will and senses out into its steering system, engine, instruments and devices. Just as your house is your shell and your coat your pelt, in effect, so does your consciousness form your aura of personal life, and whether your mind expresses itself through natural or artificial muscles and limbs doesn't make much difference. Even a robot would be alive if it had some sort of consciousness in the computer that controlled it. Of course that hardly seems possible in this early stage of earthly culture when we do not yet have any real understanding of the nature of consciousness.

## THE ABSTRACT BODY

When it comes to the nature of the physical body, on the other hand, it appears the very opposite of abstract at first. But I think anyone would find ample reason to change his mind when he had had time to ponder the deeper significance of the very common elements that are there.

318    I remember reading half a century ago about some chemist who

added up the chemical value of the material in an average man to a sum total of 98 cents. In 1963, probably due to research and inflation, the value had risen to $34.54 and today, I suppose, it would be well over $100. Still the ingredients seem ordinary enough when listed (as one investigator did) in the form of an inventory for a country store: water sufficient to fill a ten-gallon keg, enough fat for seven bars of soap, carbon for 9000 "lead" pencils, phosphorus for 220 matches, magnesium for one dose of salts, iron enough to forge a nail, lime to whitewash a chicken coop, sulfur to purge a dog of fleas . . .

The reason a living body can be made of such everyday stuff of course is that it is complex and flowing and the stuff is not really the body but only what passes through it, borrowed in the same sense that an ocean wave borrows the water it sweeps over. If one could ignore the dimension of time, a body or a wave would naturally be much more material than it is, less abstract and more concrete, for at any instant it is obviously made of the atoms and molecules that then compose it. But as time takes effect and the wave moves on, leaving bubbles, foam and water behind it, it proves itself less and less material. In fact science knows a wave to be made not of matter at all but purely of energy, which is an abstraction. That is why a wave (made of moving air and wheat) will move at the speed of the wind across a wheat field without carrying with it a single blade of wheat. That is why a church bell will appear unchanged in the belfry after centuries of ringing out waves of music. And that is also why a living body will remain about the same in weight as waves of metabolism flow through it year in and year out.

What then, I ask, is a body made of? At any given moment it is made of the world, for there is no fixed borderline between you and your surround — yet, reflecting on it at length and in the full context of time, the body progressively becomes as abstract as a melody — a melody one may with reason call the melody of life. Does such an answer surprise you? A surprise it certainly was to me when the idea first entered my head. For, although I had intuitively assumed life itself abstract, the physical body had always seemed simply material and I did not see how it could be otherwise. Then I tried to define the physical boundaries of the body and began to realize they are virtually indefinable, for the air around any air-breathing creature from a weed to a whale is obviously a vital part of it even while it is also part of other creatures. The atmosphere in fact binds together all life on Earth, including life in the deep sea, which "breathes" oxygen (and some air) constantly. And the water of the sea is another of life's common denominators noticeable in the salty flavor of our blood,     319 sweat and tears, as are the solid Earth and its molecules present in our protoplasm compounded of carbon, hydrogen, oxygen, nitrogen and a dozen lesser elements.

Yes, life as a whole breathes and owns the common sky and drinks the mutual rain and we are all embodied in the sea and the clouds and in fire and forest and earth alike. As the God of Egypt was quoted as saying about the year 2000 B.C., "I made the four winds that every man might breathe thereof like his fellow in his time . . ." And thus, He might have added, I made sure that all life mingles and shares the most vital elements. Indeed oxygen (then unknowable to man) is the leading substance of life as it is presently known, making up some 60 percent of the weight of the human body, surging and blowing through it in the rhythmic torrent that Sir Charles Sherrington called "a draft of something invisible" to fuel life's flame. It is plain to see that we all breathe the same sky and we are becoming aware that it pours through our lungs and blood in a few minutes, then out again to mix and refresh itself in the world. But it is still easy to overlook the completeness of airy suffusion throughout the planet, so easy in fact that I would like to offer a few quantitative statistics to point up some of the significances.

Did you know the average breath you breathe contains about 10 sextillion atoms, a number which, as you may remember, can be written in modern notation as $10^{22}$? And, since the entire atmosphere of Earth is voluminous enough to hold about the same number of breaths, each breath turns out, like man himself, to be about midway in size between an atom and the world — mathematically speaking, $10^{22}$ atoms in each of $10^{22}$ breaths multiplying to a total of $10^{44}$ atoms of air blowing around the planet. This means of course that each time you inhale you are drawing into yourself an average of about one atom from each of the breaths contained in the whole sky. Also every time you exhale you are sending back the same average of an atom to each of these breaths, as is every other living person, and this exchange, repeated twenty thousand times a day by some four billion people, has the surprising consequence that each breath you breathe must contain a quadrillion ($10^{15}$) atoms breathed by the rest of mankind within the past few weeks and more than a million atoms breathed personally sometime by each and any person on Earth.

The rate of molecular mixing and diffusion in the sky is an obvious factor in such a calculation and so is the net speed of the wind under normal atmospheric turbulence, which is dramatically revealed by the accompanying world map plotting the tracks of six random parcels of air followed by free-floating radio balloons in 1964, the air circulating around the earth in from about a week to a month, the period no doubt varying with latitude, altitude, season and other factors. With such information you can more easily accept the fact that your next breath will include a million odd atoms of oxygen and nitrogen once breathed by Pythagoras, Socrates, Confucius, Moses, Columbus, Einstein or anyone you can think of, including a lot from the Chinese in China within a fortnight, from bushmen in South Africa, Eskimos in Greenland . . . And, going on to animals, you may add a few million molecules from the mighty blowings of the whale that swallowed Jonah, from the snorts of Muhammad's white mare, from the restive raven that Noah sent forth from the ark. Then to the vegetable kingdom, including exhalations from the bo tree under which Buddha heard the Word of God, from the ancient cycads bent by wallowing dinosaurs in 150 million B.C. And don't forget the swamps themselves and the ancient seas where atoms are liquid and more numerous, and the solid Earth where they are more numerous still, the gases, liquids and solids in these mediums all circulating their atoms and molecules at their natural rates, interchanging, evaporating, condensing and diffusing them in a complex global metabolism.

While the sky thus breezes through our bodies in a few minutes and the rain filters through us in a day or two, our solid parts, such as bones, change more slowly, taking a couple of months to renew themselves, as I noticed a few years ago when recuperating from a broken leg. Nerve cells are slower still to metabolize, certainly a lot slower than bone. Yet practically all of our material selves is replaced within a year — actually excreted, barbered, manicured or just washed, evaporated and sloughed off while new living protoplasm unobtrusively replaces it.

To back up such a statement, I will mention that Dr. Paul C. Aebersold of the Oak Ridge Atomic Research Center has reported that his radioisotope tracings of the numerous chemicals continuously entering and leaving the body have convinced him that about 98 percent of all the $10^{28}$ atoms in the average human are replaced annually. "Bones are quite dynamic," he declared, their crystals continually dissolving and reforming. The stomach's lining replaces itself every five days, skin wear and tear is completely retreaded in about a month, and you get a new liver every six weeks. As for how long it takes to replace every last neuron, the toughest sinew of collagen and the most stubborn atom of iron in hemoglobin, all of which are notoriously reluctant to yield their places to substitutes, it may well take years. But there ought to be some limit to this stalling of the final few holdouts and my late friend, Donald Hatch Andrews, professor of chemistry at Johns Hopkins University, who seems to have given the matter long consideration, put it at about five years, after which one can presumably consider one's physical body completely new down to the very last atom.

Assuming this approximately so, then of what does the body really consist? Where resides the continuity of consciousness and memory which may last a hundred years? For a while I thought the body's essence might somehow lurk in the nucleus of each cell where the genes physically direct growth and development. But suddenly I realized that the double helix molecules of DNA, the material genes, are made of the same old carbon, hydrogen, oxygen, nitrogen and a few other atoms that constitute ordinary matter, and they are not excused from replacement any more than any other of the body's atoms, all of which metabolize in life's flamelike flow, partly coordinated by enzymes, and which in the context of time are something abstract that is really beyond mere matter. In a man, to be sure, atoms cannot renew themselves nearly as fast as in a hydra, where they do the whole interchange in two weeks, but the process is clearly the same in principle. Essentially no single atom or molecule or combination of them can be indispensable to a body for they are all dispensed by it. It is only the pattern with its message that proves really vital to life. On the ocean one could make the analogy that it is not the saltwater but the abstract energy that shapes and powers the wave. Likewise it is not the atoms in DNA but their geometric relation that makes the gene. And it is not the paper and ink but the words and meanings that compose the book.

In the same sort of way a body, I think, may quite appropriately be compared to a corporation of cells and elements, for the word "corporation" literally means "body" and my dictionary defines it as "an association of individuals . . . having a continuous existence independent of the existences of its members . . ." This in fact is the means by which a business corporation may exist for centuries while one by one all its members die off and are replaced every generation by younger ones and younger ones again who keep it going like any other living organism with a kind of animative momentum that has hardly begun to be understood.

If you can stand another analogy, the genes are like blueprints, each full of detailed plans of a building, which is the body. If the building should catch fire and be destroyed, it is not an irreparable loss as long as the blueprints exist, for, by following their guidance, a new building can soon be built to replace the lost one. Even if all such plans are destroyed, it is not a vital loss as long as the architect exists who made them, for he can probably remember them more or less and, with effort, redraw them and thus recover the building. Or even if the architect be dead, perhaps a photograph or sketch may turn up somewhere. The point is that it is the pattern of design itself that is the indispensable thing, and not just its representation on paper or in bricks and mortar. Of course the design is not really a thing in the material sense for it is abstract. Indeed it is a kind of intangible essence, something like Lao-tzu's best knot which, as he explained, was tied without rope.

Thus our very bodies that we always thought were material, because they are formed with atoms, fade away into immaterial abstraction, turning out to be essentially only waves of energy, graphs of probability, nodes of melody being mysteriously played in our time. And this is true of life generally: of birds singing in the sky as their feathers molt in bilateral progression, of worms hunching through the soil, of forests and tundras and plankton in the sea. And, there being no clear line defining where life begins, it includes the sea itself and the volatile sky, and mountains and glaciers and deserts, in fact the whole earth and all planets and stars and systems of stars. In other words: all matter everywhere, as we shall see in Chapter 14 about omnipresence.

## ESSENCE OF LIFE

The universality of life is strongly suggested again by the growing realization among scientists that every sort of matter is found to move and metabolize if it is only observed for a long enough time. In Chapter 15 we will speak of the metabolism of pebbles and dunes and beaches, of rivers, lakes, glaciers, storms, islands, volcanoes and even of fires, all of them composed of atoms that sooner or later disintegrate to be replaced by new ones (new in their positions), showing in many such ways the basic attributes of life.

Naturally all this action is relative: the positions, the movement, the rate of living. For no kind of matter remains the same forever, and matter in essence cannot be a mere *substance* but must be an *event*, an *abstraction* or even (if you can countenance the pedantry) a *fleeting pattern compounded by space-time into a definable entity*. It is almost like saying that, to the degree that the function of substance is matter, the function of living is life. At least this becomes evident when one accepts life as matter's highest development, analogous to a river or a flame where cells and molecules not only flow through the body from day to day but diffuse through evolution from generation to generation. If you've ever been seriously ill you may have noticed that when you don't use your body it begins to disappear, for nature has in effect commanded you: "Use it or lose it!" A calf muscle, prevented from functioning by a plaster cast, can dwindle to half its normal size in two weeks as the protein and fat are released back into your blood and lymph streams. When you stop eating, your organism keeps itself going by digesting your flesh. It is a case of your body borrowing from itself, of life lifting life by life's own bootstraps until, in better times, it may hope to restock with fresh substance from its world. This is the way any limb or brain or family or field or town not in use is bound to wither, turn fallow and, in a surprisingly short time, cease to be what it was.

But the law of use has a constructive side too, for, where disuse destroys, use creates! The classic experiment establishing creative body use took place in the 1860s when C. E. Sédillot of France performed a series of operations on puppies, excising the larger bone or tibia from the lower hind legs of each animal so half its body weight would have to be supported only by the remaining tiny fibula bones, each with less than a fifth of the tibia's diameter. And as the puppies recovered and began to walk falteringly around on their feeble fibulas, these reedlike little bones in every case were so stimulated by their inordinate burdens that within a few months they all grew into bones that averaged bigger than the ones removed. It was as though each fibula, somehow sensing the emergency, had loyally resolved to save its host organism by becoming a tibia. Then, with appropriately dogged determination, it had set about building itself up to the seemingly unreachable goal. Yea, like the salamander in Spemann's famous experiment that proved a tail can become an eye (page 153), the fibula demonstrated anew that intrinsically any material containing the common organic elements is good enough to grow a living organism, or any clay a man.

324     This is a concept virtually everyone seems to share when it comes

to disposing of a dead body or even a worn-out part of a living body, for, I notice, the man whose leg was amputated yesterday does not ask to keep it around today. It is no longer of him nor he of it. Neither does he consider himself less alive without it, especially after he has been fitted with a good artificial leg which, in a few months, may feel almost as intimately his as the original. Curiously, the very deadness of an artificial limb may become a vital factor in its embodiment of the afterlife of the limb it replaces, and in the same basic way that the dead limb of a tree sometimes lives on a higher evolutionary level when it has assumed the role of the afterlife of a formerly green branch. I refer to the fact that the favored twigs and boughs that birds most often alight upon (when they're available) are the bare dead ones which liberate the birds' movement and vision from the blinding obstructions imposed by the leafier branches — this being a nice example of how death in one kingdom (the vegetable) can so often enhance life in another (the animal), which quite literally serves as its Kingdom Come. Indeed what better heaven could a dead tree hope for than to harbor a beautiful bird or later, in moldering senescence, to serve as fuel for the hearth of a philosopher?

A different kind of afterlife is that of the amputated human limb, which medical science long ago discovered capable of surviving indefinitely as a phantom, not only in the mind of the amputee but almost inevitably in certain nerves, tendons and adjoining muscles that have sensations ranging from gentle tingling to excruciating pain and all seeming to originate in the limb that isn't there. In treating such a nonlimb, doctors have found that it can often be exorcised by exercise on the pragmatic hypothesis that it is real and alive even if only an abstraction. To cite a case: a young electrician lost his left arm a few years ago in a severe electrical burn that left him too miserable to accept an artificial arm, mostly because, as he said, he felt his missing arm rigidly doubled behind his back with the hand numb and electric shocks intermittently shooting through it so violently they sent "sparks snapping off his fingertips." However he was persuaded to stand before a blackboard with his eyes shut and practice writing with his imaginary left hand. And after months of this phantom exercise, with nothing physical happening except possibly in the nerve stumps and adjoining muscle fibers, he became assured that he had swung his phantom arm around and forward to where he could raise it over his head. His shriveled scar tissue meantime softened and stopped hurting while the exercise proved to have loosened it so much that "an extensive grafting operation became unnecessary," and the unexpected removal of this burden gave him such blessed relief that he eagerly accepted an artificial arm, which in turn so satisfyingly incarnated the phantom that he gratefully began training for his new profession of bookkeeper. Thus was a seemingly hopeless case cured, one might say, by faith in its own ghost.

Yet where are the fingers or toes a man feels after his arm or leg is

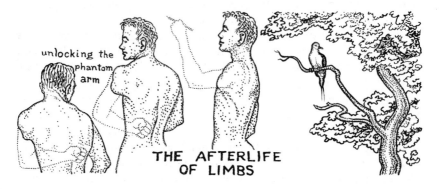

unlocking the phantom arm

THE AFTERLIFE
OF LIMBS

removed? And where are the scenes the blind man "sees" after losing his sight, which in a few cases, I am told, have been so vivid he did not even realize he was blind? No doubt these presumed illusions are outside of space and time — at least enough outside that one might consider them the stuff of dreams, abstract and not normally definable. But for some reason their capacity to confuse human minds makes me think of the woman who painted the inside of her garage pale green, and, when a neighbor encouraged her by remarking that the light color would make the place seem bigger, responded with airy enthusiasm, "I sure hope so. We really could use the extra space."

A rather different illusion was experienced by a polliwog I heard of living in a farmer's pond. One November he was seen at the moment of getting frozen in the ice and looked, and probably felt, gone or "dead" if he felt anything. In any case he remained in "rigor mortis" all winter. Then during an April rain his icy tomb melted and he was suddenly free and alive. But it was as if his five months' sleep had been an enchantment for, like the sleeping beauty, he awoke apparently the same age as when he had dropped off and then recovered his active life from precisely where he had left it. He even finished biting the weed that had frozen in his mouth in November. The magic spell of winter was over. Time had stood still for a third of a year. The last breath inhaled in autumn had turned into the first breath exhaled in spring. Life and mind were simply flicked out like a light . . . . . .

. . . . . . . . . then flicked on again unchanged!

## UNIT OF BEING

In returning now to our pursuit of life's essence it seems important to define the living unit of being. Is it an atom, a molecule, a cell, an organism, a family, a species, a world population . . . ? Just what best constitutes a living individual? If reproduction is vital (and who can deny it?), a mating pair of rabbits is alive but one rabbit is not. In this context a pregnant rabbit is, strictly speaking, not alone. And perhaps (in special cases) even an unpregnant rabbit or single individual of another species capable of virgin birth (page 149) should be considered more than singular. Also a mating pair at the sophistication level of Adam and Eve probably could not qualify as a living continuum if too few others of their kind existed to enable them to avoid imminent extinction. It is relevant in this connection to recall that the passenger pigeon was doomed last century when its numbers got so low it could not roost in its customary, congenial multitudes. But that doesn't make it noticeably easier to predict how small a human community, even after thorough survival training in body, mind and spirit, could be expected to sustain and revive a stricken mankind. And such a key community, as much as any in any species, might qualify as the model unit of life.

## NORMALCY

We come then to the question of what is the range of traits an individual might have without becoming a different species or a lethal mutant. What is normalcy? The *Atlas of Human Anatomy* by Barry Anson shows the variety of stomachs in ordinary adult humans, some

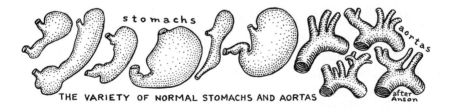

THE VARIETY OF NORMAL STOMACHS AND AORTAS

of which hold fifteen times as much food as others and are as different in shape and function as in size. The aorta has three main branches near the heart in about 65 percent of humans, says Anson, two branches in 27 percent, and any number of branches up to six is considered normal. Nerves are even freer to steer or shape themselves and, like veins, hair, guts and many muscles, so long as they do the job they may express themselves as individually as vines and roots. A throat, for example, seems a pretty standardized affair, but some men have them so big they can swallow swords with professional aplomb while others have such narrow ones they have choked to death on an apple seed. So-called normal testicles vary from 10 to 45 grams in weight while abnormal ones range from a half to more than 300 grams, and Alfred Kinsey reported an even greater latitude in sexual performance, listing one "sound" man who ejaculated but once in thirty years while another, a "scholarly lawyer," did so 45,000 times in the same period, averaging more than four times daily.

Individuals differ chemically perhaps more than physically, and a recent medical study of 5000 "normal" people shows that gastric juices vary a thousandfold in their content of both pepsin and hydrochloric acid. Vision, hearing and other senses have an astonishing spectrum too, one hungry baby in a sensitivity test refusing to suck milk except when it was kept between the temperatures of 73° and 122°F. while another guzzled it contentedly all the way from freezing (32°) to steaming (149°F.) And similar findings appear in inborn capacities to reset one's biological clock after flying across longitudinal time zones shifting night and day, in mental traits like intuition or spiritual ones such as immersing oneself in a mystic Cause. Clearly every individual from a germ to a man (and indefinitely beyond in both directions) is discrete unto itself and freer in important ways than has been generally realized — and I doubt not that this realization will in time give a fresher and more spiritual meaning to the concept of brotherhood whose oneness is certainly profounder than uniformity, whose harmony is indescribably more beautiful than conformance.

Some anthropologists still seem to assume that being normal means being average, one I've read of going so far as to declare that only one person in 6500 could be credited with achieving true averageness in every measurable respect, that person presumably having molded himself into the nicest, dullest and most proper "freak of normalcy" for miles around. So this may be a good place to recall that the specter of averageness was effectively dispatched by Walter Heller a few years ago when he introduced the man with one foot in a freezer and the other on a stove, who felt on the average perfectly comfortable.

## THE ABSTRACTION OF IDENTITY

If it is characteristic of an individual then to be unique, though in some cases less than extraordinary, does he absolutely *have* to be unique in order to exist? Does he have to be identifiable to anyone besides himself or his twin? Indeed is identity a life principle?

The classic case in point broke into the news in Detroit in January 1969 when a pair of identical twin sisters named Terry and Tracy got mixed up by their mother, who, despite having known them intimately for the six months since their birth, could not for the life of her figure out which was which, being driven ultimately, as she shamefacedly confessed, to "the eenie, meenie, miny, mo method."

This unusual event was fortunately no tragedy, for the babies themselves fairly reveled in interchangeability, neither knowing nor caring whether Mama could tell them apart. Yet a philosopher might well ponder, aside from the fact that the twins had started life as one fertilized ovum with a single potential consciousness, whether their two individual consciousnesses, be they recognized or unrecognized by the world, really required separate identity. It is an abstract question of course, yet deeply germane to the abstraction we know as life. After all, what an individual is as a conscious entity may be definable as a continuing pattern of memory, an abstraction analogous to music and which might conceivably be duplicated (or nearly so) from cognate genes in another (similar) body at the same (or an equivalent) time. This other body too, unlike an identical twin, could as well belong to the same consciousness or, putting it another way, a mind could have more than one body (as occurs naturally when a mind survives a drastic body change after a crippling injury, when one mind appropriates another in temporary hypnosis or some longer lasting form of psychic "possession") and vice versa (page 292). In other words, bodies and minds are not (necessarily) indelibly associated with each other on a one-to-one basis. Indeed body-mind disengagement has been known to occur, sometimes followed by separate existences of both, not to mention shifts in alignment and reassociation in new combinations — something that will not surprise anyone who can accept the possibility of life after death or the transcendence of consciousness (as in dreams) beyond space-time.

## IMPROBABILITY

What makes YOU as an individual genetically distinct from others (assuming you are not an identical twin) is the extreme improbability that any parents but yours could have achieved the exact combination of some hundred thousand genes that directed your development, a combination considered largely random as it was drawn generation by generation from the much larger gene pool of all mankind and one among incalculably more potential combinations than the number of atoms in the entire known universe, certainly one so high that the chance that two different sets of parents might have happened to produce exactly the same gene combination in the same megamillennium of time or the same supergalaxy of space is astronomically so small as to be not worth consideration as a realistic possibility.

The fact that every individual of us is so fantastically improbable then, so negentropic if you like, is of life's essence as surely as life is the mysterious force that resists entropy and increases order in the worlds. And improbability seems to be the only factor preventing many, if not most, people from having genetic duplicates of themselves here and there among the world's population, something more than theoretically possible, as is demonstrated both by the occasional phenomenon of identical twins (where improbability is naturally circumvented) and by cloning, a similar asexual reproductive process (where improbability is artificially circumvented) and which some prescient researchers say will someday enable anyone of means to keep a deep-frozen identical twin called a clone on hand in a "clone bank" for organ transplants, a rather nightmarish service that conceivably will ultimately include the transplanting of the entire body or mind (two faces of the same coin) through some sort of transference of memory cells, perhaps involving temporarily induced amnesia, hypnotism, astral projection or a form of psychic anesthetic today not even hypothesized. When (or if) such a day comes, the study of death will presumably have become a promising branch of science closely allied to the already growing researches of memory and consciousness with psychologists inducing losses of memory so complete the patient can be mentally reborn with a second mind, later a third, fourth, etc., amounting to a multiple consciousness, not to mention a multimortality and heaven knows what other states of mind that in turn might lead all the way to colonial or group consciousness or even some sort of unprecedented macrocosmic immortality.

## DIMENSIONS

Unfortunately it is more than possible that a discussion like this will turn less than comprehensible because, when we talk about consciousness, death, time and other dimensions as criteria of life, we are in the same hard-to-visualize field Einstein explored when working out his relativity theory along with its contingent concept that every individual's personal orbit through life is representable as a "world line" framed in the common four-dimensional crystalline coordinates of space-time. And in case you didn't notice it, a prime philosophical deduction from world lines is that, if relativity be true, an independent "I" bounded by birth and death is an absurdity, since, as we will explain in Chapter 17, the field concept now so well established as a foundation for relativity implies continuums in virtually everything, including space-time and most certainly its best-known derivative: life. In my view, furthermore, the key to comprehending space-time is the obvious (to me) fact that space is the relationship between things and other things while time is the relationship between things and themselves. The time relation thus requires some establishment of identity (between things and themselves) seeing as identity is indispensable in temporal continuity. But if identity is of the essence of time, it follows that when a human being gives himself to a cause, letting his individual identity be absorbed in something larger than himself, he is proportionately liberating himself from the field of time. Which tells one something about the relation between mortality and immortality and between life and death, for it presumes that, as one's self is swallowed by universality, to a comparable degree one becomes immortal.

How this relates to dimensions in the universe is not very apparent, but if you've ever had the feeling P. D. Ouspensky once hinted he'd had, that your "piece of universe" is not as big as it should be, it just could turn out that one of your dimensions is slipping. Two-dimensional people from Flatland, for example, who have length and breadth but no depth, can see one-dimensional worms and two-dimensional flatfish, and they naturally accept the flat plane they are in as the whole universe, but any evidence of a third dimension that shows up is likely to be regarded as a curious phenomenon in that "universal" plane, something presumably supernatural. Should their world plane be wrapped around a sphere, let us say, so that they dwelt on its surface as humans do on Earth, they would be able to draw circles only as big as a certain size (a great circle like the equator) but no bigger. For no matter how hard they tried, if they were dimensionally confined to the surface, they would be unable to draw a bigger circle because there would be no room for it in their finite world. A virus swimming through the bloodstreams of a flatfish or a tapeworm likewise might be considered to be in a 2-D world (assuming the tapeworm

331

had approximately zero thickness) even though the tapeworm be tied into a knot. For the knot would be a knot only from the 3-D viewpoint of an outside observer (say a 3-D bear) and the virus could swim on through all the tapeworm's convolutions and twists without knowing they existed. If he ever got outside 2-D he would presumably think he were "dying" until he discovered he was being "born."

Now there is ponderable evidence that the world beyond our present mortal life on Earth is far greater than what we can know here. I mean greater in dimension and, from the earthly view, it probably should be considered infinite. For if we are limited here-now to the familiar three dimensions of space and the one of time, which together construct our finite 4-D world line, the fifth dimension looming up after death might be a line perpendicular to and intersecting all world lines — I mean a line passing through some moment of all life, an encompassing orbit of vital eternity, a symmetric circuit of infinite simultaneity simplifiable in mathematical notation to NOW$\infty$. The sixth dimension could be an evolutionary spiral composed of all such 5-D orbits, amounting to the totality of all the events that happen on all world lines, the unending expanse of which weaves the texture of a world surface. The seventh could be an evolutionary lattice crystal made of these 6-D woven expanses in the form of endlessly layered surfaces that together compose a world volume. The eighth could be a sequence of world volumes, the ninth a succession of world sequences, and so on . . .

## PROGRESSIVE NUMBERED DIMENSIONS

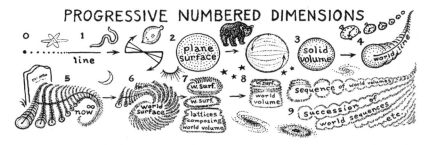

The new mathematical concept of *fractals* adds the complication of fractional dimensions: partway between whole numbers, produced by curves or surfaces that wiggle enough to partially fill the gap between one dimension and the next. And Ouspensky speculated that there must be negative dimensions also — that is, immeasurable ones such as the lengths and breadths of streets in cities that are but dots on a map. And equally plausible negative dimensions come to mind like those of seeds harboring potential massive trees and forests (ultimately planets), of galaxies so distant they are but sizeless points of light in the black sky, of atoms enclosing mystic worlds too tiny to measure. The curvature of space is parcel of this of course, not to mention the curvature of time or higher dimensions, and it raises such questions as to whether the space-time in the known universe curves positively like a ball or negatively like a saddle. Although the preponderance of evidence so far points to negative curvature, this may be illusory because of the nearer, more visible, space flowing continuously through the waist of the hourglass of time, a saddlelike region, while the more distant space progressively yields to positive spherical curvature in every direction beyond our powers of detection.

Some physicists have tried out mass-energy and temperature as dimensions, I am told, also wavelength-frequency, spin and presumably some of the less-known attributes of subatomic particles. And there always seem to be hypothetical entities in the offing, like the new tachyons that "travel faster than light" and conceivably could compose the "bodies" of intelligent "beings" with whom, fortunately or unfortunately, no one has yet figured out a way to communicate. This all seems pretty fantastic, but is not out of line with the way many hypotheses have established themselves. The fact that temperature is the rate of molecular motion, which is a quotient of space ÷ time, evidently denies temperature a separate dimensional status. And the same with frequency and spin. But it is hard to know what other aspect of matter may yet turn out to qualify as a dimension, if any. There seems to me a significant and perhaps dimensional analogy between space-time, mass-energy and particle-wave, the first member of each pair being quite concrete as compared to the second one, which is so obviously abstract. On page 324 I pointed out the dimensional difference between a concrete body at any moment and its abstract essence over a period of time, and on page 446 a dimensional compromise between order and movement. Later (page 491) I will try to show how the classic paradox of free will vs. fate may be resolved by dimensional perspective.

And could it be that life itself is a dimension? Life seems to be an aspect of everything (both concrete and abstract) and in some ways remarkably independent of space-time, not to mention mass-energy, particle-wave or even body-mind. Dreams and thoughts certainly overleap the accepted spacial and temporal dimensions of mass and materiality and who can say where or when or to what extent life may exist outside our familiar physical span of a few score years? Should I deny then that whatever the factor is, if any, that distinguishes life from nonlife, its nature just might be dimensional?

## MATHEMATICS

As dimensions are magnitudes measured in numbers, they naturally and logically lead us to the venerable science of numbers known as mathematics, which is also called the·language of abstraction because it is made up not of words but of symbols of quantity. Mathematics indeed is the only language based on pure logic which can therefore be understood equally well by trained people in every country.

But how mathematics originated remains a mystery. One version has it that two cavemen got to quarreling over a woman, but, being friends, decided to try to settle the issue by peaceful contest rather than just belaboring each other with clubs. Then, after agreeing that the one who could think of the highest number could have her, one said, "You go first." The other retorted, "No, you go." Finally, after deep thought, one of them said, "Three." The other scratched his head a long time but eventually gave up. So the one who had thought of three became our ancestor and taught all his many children to count.

As the millenniums passed, however, people thought of new numbers with which to count higher and, little by little, discovered ways to compare one thing with another through measurement and eventually by counting the numbers of standard units. "Things are numbers," taught Pythagoras. And in his later years he added, "There are also numbers beyond things."

Thus mathematics advanced into abstraction and took a major stride forward when the Hindus introduced the zero. At first this seemed nothing but witchcraft, and it provoked fierce opposition throughout Europe when the Moors started promoting it in ninth-century Spain. Then slowly it took hold after a learned Arab explained that "the zero is not nothing. It is more than nothing, for it is something that has to be there in order to show that nothing is there. Also, like magic, it turns 1 into 10 because, of no value itself, it gives value to others."

Negative numbers like $-2$ or $-7$ logically followed the zero which, serving as a mirror, reflected them as images of the positive numbers. Next came irrational numbers like $\pi$, the circumference of a circle divided by its diameter, ten decimals of which give the circumference of Earth to the fraction of an inch, thirty decimals the circumference of the known universe to a precision imperceptible to any microscope, yet whose decimals, for no known reason, go on and on — apparently forever. Then, like a revelation, appeared impossible numbers like $\sqrt{-1}$ which, meaningless in itself, mystically led to the discovery that three-dimensional space could be combined with time into the four dimensions of general relativity.

334

And when the distances discovered by astronomers became too big to visualize, mathematics helped mankind see the relation between actual and possible worlds by providing a quicker, simpler way of saying difficult things, such as $1/r^2$ for Newton's law of gravitation or $E = mc^2$ for Einstein's equation of the atom. Even an extreme comparison like that between the masses of the observable universe and of the unobservable electron became easy when you could list the former as $10^{55}$ grams and the latter as $10^{-27}$ grams. Or if you wanted to know, say, how thick a piece of fine tissue paper would become if you could fold it over upon itself fifty times, a straightforward exponential calculation would tell you that anything one thousandth of an inch thick folded double 50 times (if that were possible) would become 17 million miles thick, or fat enough to reach to the moon and back 35 times! The secret of this calculation of course is that 50 doublings multiplies to the fiftieth power of 2, which happens to be 1,125,899,906,842,624. And if you just harness the power into modern notation as $2^{50}$, it is almost mystically simple without being any less powerful.

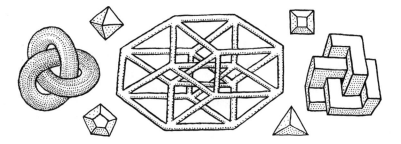

Mathematics is obviously a vast, as well as mysterious, science — and keeps on telling us such things as that four colors are enough to color any flat map unambiguously, though a map on the surface of a doughnut takes seven; that $3^3+4^3+5^3=6^3$ which, represented as a cube 6 inches on each edge, contains exactly the same number (216) of square inches on its surface as of cubic inches inside it; that a skillful high jumper lets his center of gravity pass just *under* the bar abstractly while his body, bit by bit, goes *over* it; and that, if $n$ represents any number, the sum of the first $n$ odd numbers is always $n^2$. But, logical as mathematics is, it can't solve everything, for it contains both unsolvable problems and undecidable propositions — like, for instance, the sentence: "This sentence is not true," which cannot logically be either true or false since, if true, it confesses its falseness and, if false, its confession is true.

Also confirmations of a hypothesis sometimes, instead of supporting it, will paradoxically disprove it. Like, for example, when three cards are marked (on their faces) 1, 2 and 3, then shuffled and laid face down on a table to demonstrate the hypothesis that no card will be found in the same order as its number. For if the first card turned over is the 2 and the second the 1, the hypothesis has been confirmed in detail so far — and yet the revealed certainty (by elimination) that the third card must be the 3 has totally disproved it.

Perhaps the most abstractly beautiful revelation in the history of mathematics came in the Pythagorean discovery of the ideal and elegant five regular convex solids: the tetrahedron, cube, octahedron, dodecahedron and icosahedron, all known ever since to be harmonically related in concentric order, the simple 4-faced tetrahedron in the center, the cube with 6 faces and 8 corners next to its reciprocal octahedron with 8 faces and 6 corners, followed by the dodecahedron with 12 faces and 20 corners next to its reciprocal icosahedron with 20 faces and 12 corners — all five complying with the mystic rule that the numbers of their faces plus their corners equals the number of their edges plus 2.

Neither Pythagoras nor Plato, who adopted these famous figures after him, nor Kepler, who looked for them in the heavens, had any idea that they might be alive or actually living on Earth. Yet recently in fact all five of them have been found alive among the crystalline plankton in the sea, rather gaily betassled yet minding their own business in benign indifference as to whether Pythagoras, Plato or mankind ever existed. The tetrahedron, somewhat rounded as if from internal pressure, is embodied in a protozoan called *Callimitra agnesae,* the cube is *Lithocubus geometricus,* the octahedron *Circoporus octahedrus,* the dodecahedron *Circorrhegma dodecahedrus* and the icosahedron *Circogonia icosahedrus.*

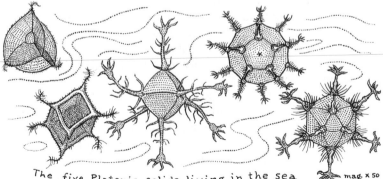

336    The five Platonic solids living in the sea    mag. x 50

Of course these living Platonic solids express an extraordinary synergy between abstract geometry and life, as do the patterns of leaves and florets in trees and flowers in relation to the Fibonacci series of numbers (page 58), but most wonderful of all is the fact that these various synergies are part of one abstract whole that amounts to far more than all of them individually added together. There is also a particularly mystic and irrational number or ratio known to mathematicians and architects as the Golden Section and sometimes represented by the Greek letter $\tau$, which could, for aught I know, signify Tree or Topology or Tao. Certainly it is one of the world's magically beautiful relations that directly affiliates protozoa and trees with geometry. Heaven only knows where it ends, or if.

I might define $\tau$ by saying that it is the ratio between two incommensurable quantities of which the lesser is to the greater as the greater is to the sum of them both. Or, taking a specific example, $\tau$ is the length of the diagonal of a regular pentagon or of the radius of a regular decagon expressed in units equal to any side of either of these equal-sided figures. Such units of course could extend any distance: an inch, a mile, an angstrom, a light year. It makes no difference, for $\tau$ always comes out 1.61803 . . . which is the length of the diagonal or the radius in whatever units we use. $\tau$, in other words, is the ratio between two lengths (expressed in numbers) in the same way that $\pi$ or 3.14159 . . . is the ratio between the diameter and circumference of a circle. Like $\pi$, $\tau$ is an irrational number because it discriminates distances that lack a common denominator. And so there is no known end to its decimal places and therefore no way it can be expressed exactly, a discovery credited, by the way, to Pythagoras himself and which shocked science in the sixth century B.C. Neither does it help much to turn trigonometrical and say that $\tau = 2 \cos \pi/5$. Better, if you have the philosophy: just to realize that this Golden Section is exemplified in a rectangle whose width (1) times $\tau$ equals its length (1.61803. . .). This sublime figure (despite its irrationality) remained sacred to the Pythagoreans, who first drew it out in rows of dots after probably gleaning its approximate proportions from Egyptian priests. There is no doubt that they knew how to erect a square ($\tau^2$) on the rectangle's longer side ($\tau$) so that the two figures combined would create a new and larger rectangle of exactly the same proportions as the original. Such a square was one form of the Greek gnomon, which Hero of Alexandria later generalized as "any figure which, added to any other figure, creates a resultant similar to the original."

Gnomons of course apply to triangles among everything else and any triangle may be divided so that one section is a gnomon to the rest. But the point here is that almost any shape of figure, including

337

solid ones in three dimensions, can be built up, gnomon upon gnomon. And this, I take it, is the prime geometric and genetic secret of the chambered nautilus (whose chambers are gnomons) and of many sensitively balanced spiral shells and horns, and it's the key to the self-congruence that enables them to grow year after year without appreciably shifting their centers of gravity.

How the Golden Section relates to Platonic plankton in the sea, however, is deeper than can be casually understood. Indeed it may seem downright mystical when you discover that three identical Pythagorean rectangles with sides in the ratio $1:\tau$ will exactly fit together at right angles (as shown in the illustration) so their twelve corners become the twelve vertices of a perfect icosahedron with twenty equilateral triangular faces, and that $\tau$ turns up as a vital ingredient in the formulas for several other polyhedral functions, including those for the surface and volume of the pentagonal dodecahedron (this last an abstruse $4\sqrt{5}\tau^4$ counting the edge length as one).

Among still more affiliations of the Golden ratio, the final one I must mention is its almost uncannily intimate identity with phyllotactic leaf and seed patterns. To understand this relation, you need to study the ratios between alternating pairs of successive numbers in the Fibonacci series. If these values are added on both sides of the progression (as I have done in the illustration), it soon becomes clear that the

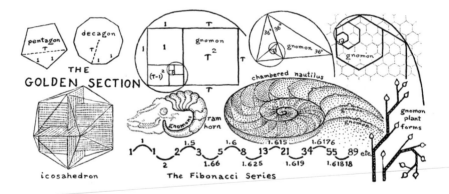

solid ones in three dimensions, can be built up, gnomon upon gnomon.

Fibonacci series is basically a succession of gnomons whose alternating intergnomon ratios are both (from different directions) approaching the mystic $\tau$ value of 1.61803 . . . and that the Golden Section truly lives in flowers and trees as well as swimming in the sea, climbing high mountains, and that it even sires human designs from the playing card (of China) to the Parthenon (of Greece).

Obviously it is a revelation fit to open the kind of fresh, incisive insight into the hidden heart of our universe that thrilled, and would again thrill, Pythagoras and Plato and Copernicus and Kepler and Newton and Einstein. For it is part of the real music of the spheres, parcel of the inaudible, the irrepressible melody of life that pervades everything while known only to the rare few. Awareness of it can reveal the seams of things and the nodes of being that almost all of us miss in this contingent phase of existence.

## OTHER ABSTRACTIONS

Of course there are so many worlds and worlds of worlds and sons and grandsons of worlds, all living, spawning and evolving at the same time, some inside others, that it is practically hopeless to discover them, let alone keep track of them or find out what they're made of. And, to some extent, this goes even for Earth. For, if you feel any doubt about this planet's having otherworldly aspects, you may be interested to know that physicists have already found in meteorites (born in other worlds) evidence of elements 111 and 115, which have not yet been really discovered here or named by man but only deduced and dreamed about. And besides man's cities, which, as I've suggested, have various crystalline grains like the rectangular grid of New York or Chicago, the circles of Paris, Washington and Moscow, there are the latticed patterns of country fields and roads, of herringbone clouds and squall lines, of stellar and planetary orbits, of rainbows and sand dunes and animal territories and all manner of invisible and abstract structures, like business cycles, spheres of political influence, lobes of language, even the games of children which have a life of their own sometimes lasting for centuries on several continents. There are visual patterns like the palm-prints of newborn babies too that correlate with vocal graphs of their first cries, both these having been successfully used to diagnose serious abnormalities, from mongolism to rheumatic fever. The subtle, crystalline derivatives of life we call spider webs serve also as a kind of genetic script, revealing each spider's inheritance in the wheel he spins when young (perhaps the very shape of his spinning genes), then showing his developing mind in changes to a more modern design as he matures, to say nothing of the already-mentioned aberrant forms (page 250) when he has been drugged. Regimentation of individual organisms into armylike ranks creates crystal textures in schools of swimming fish, birds flying in formation, viruses and certain rod-shaped bacteria who (for a still

339

unknown reason) have been seen to array themselves in parallel rows in three dimensions without actually touching one other.

It is well known that small creatures from insects to amphibians can adapt to the loss of some of their wings or legs and still fly or run quite well on what is left. But recent laboratory experiments in physiological rearrangement have gone a step deeper into abstraction by transplanting, for example, the front half of one lizard onto where the head of another had been amputated, thereby assembling a six-legged composite lizard that soon learned to run and coordinate all six of its limbs in the natural gait of an insect. And another experiment reconstituted a complete ameba out of parts from three (the membrane of one, cytoplasm of another, nucleus of a third) integrating them so skillfully that the new synthetic ameba lived and propagated normally, apparently without ever dreaming it was basically just a patchwork of spare parts.

Then, shifting from biology to music, I must say that music is so abstract I constantly have to remind myself it is a wave phenomenon lacking not only tangible but even intangible existence until at least one vibration has sounded. For any briefer sound can hardly have a definite pitch or be a note. And when it comes to fitting notes together into a melody, any piece of music can be taken apart and reassembled, indeed a lot more freely than can an animal. Someone has even invented a synthetic symphony that can be played and integrated entirely by one versatile musician, who performs solo on all the orchestral instruments, one after the other, recording each part of the symphony separately before acoustically blending all the parts into one complete performance that, however beautiful, is no less artificially abstract than the reconstituted ameba.

## ABSTRACTION IN REVIEW

We will now wind up this chapter on abstraction by trying to enumerate the main evidences of it in a dozen categories:

1. Matter, although it is commonly regarded as concrete because (at any given instant) it is composed of a particular system of atoms, inevitably becomes abstract with the passage of time because metabolism, erosion and other forces, second by second, year by year, millennium by millennium, replace its atoms and molecules with other atoms and molecules, leaving only an abstract pattern to persist indefinitely. In a similar way the spokes of a wheel start to "dematerialize" when the wheel begins to turn, fading farther and farther out of sight as its rotation increases in speed. Billiard balls appear to

340

touch when they bounce off each other but modern molecular science gives us irrefutable evidence that they really interact more like comets rebounding from the solar system with lots of "emptiness" all around and in between. And a hologram image, which can be electronically turned over and looked at from all sides, is obviously an "object" in the abstract, yet in essence it may be no more abstract than a chair or table made of atoms whose real substance is still completely unknown.

2. Subatomic "particles" are being discovered by the hundreds but the most tangible thing they offer physicists is a mysterious new world of ultrahigh energy full of unseen, unproved entities named quarks, antiquarks, unitons, etc. These "things" somehow resonate like musical notes, behave like waves, obey strange laws of symmetry and can be visualized only with the help of such analogies as the "quantum string," whose single dimension gives it zero mass while, under 13 tons of tension, its ends (tipped with quarks) move perpetually at the speed of light. As far as can be determined, neither space nor time reaches much below the magnitude of the atom where a pervasive uncertainty prevails (governed by Heisenberg's Uncertainty Principle) — so there may be no distance anywhere as short as a quadrillionth of an inch, nor any instant as brief as an octillionth of a second.

3. The universe, as far as we can tell, is just as abstract as the atom or its nucleus. Because remote galaxies and supergalaxies of stars are calculated to be receding at speeds proportional to their distances from us, those moving away at the speed of light and at a range of twelve billion lightyears are practically undetectable because their radiation (including light) is carried off as fast as it is generated. And remoter ones, if they exist, would presumably be receding faster than light and therefore not only unknowable but in effect nonexistent as matter in our finite world. Something similar applies to the phenomenon of the black hole, which is so massive its gravitation implodes all its matter plus all its radiation (including light), making it invisible and almost unknowable. Some cosmologists think the entire universe is a kind of super black hole and that matter draining out of it must therefore appear in a "white hole" elsewhere, perhaps in an anti-universe made of anti-matter where time flows in reverse, these two universes forming what can be represented as mirror-twin cones joined into an hourglass-shaped figure. And at least one of them (page 401) even postulates a third universe shaped like a funnel surrounding the hourglass and inhabited by tachyons, intangible (almost unimaginable) particles that move faster than light — than which I don't think it's possible to be more abstract.

4. Light is the visible octave of a vast abstract spectrum of radiation that includes everything from gamma rays to radio waves and much

more, the most important fact about it being that its speed (186,282 miles a second) is the absolute maximum for the propagation of material influence anywhere. It is interesting to note that light is material in the sense that it responds to gravity, for a beam of light will fall like a stone (16 feet the first second, 48 feet the next second, etc.), although the fall is not likely to be noticed since 16 feet is so negligible compared with the 186,282 miles it may go in some other direction. Light can also be twisted like a cable or pumped (through valves) like a gas, and it has been calculated (by computers) that if one could drive along a road lined with telephone poles at speeds approaching that of light, the poles would appear to lean toward you, their tops curling closer and closer as your speed increased, until they seemed to touch you at 186,282 mps. Also that, on a space voyage at comparable speeds, the stars ahead would not only crowd toward your line of sight but also turn bluer and change in brightness (depending on their color), and the stars behind would get redder and most of them disappear — all this because of the Doppler effect of squeezing the approaching light waves ahead and spreading the departing ones behind. The velocity of light or radiation, to posit a generalization, may be a kind of seam in the universe where space meets time, where body meets mind or even, in a sense, where finitude transcends into Infinitude.

5. Mathematics is probably the most obvious abstraction of all, comprising everything from the Pythagorean theorem to Einstein's $E = mc^2$, Bode's Law of planetary distances, Balmer's Ladder in the spectrum of hydrogen, the seven octaves of the Periodic Table of the Elements, the seven shells of the atom, the Fibonacci series, the Golden Section $\tau$, the rings of planets like Saturn and their harmonic relation, the Platonic solids, $\sqrt{-1}$ and a lot more (some of it discussed in my *Music of the Spheres*).

6. Music is mathematical in structure of course and brings abstraction to the earth as a gong, to the atom as a harp, to life as a melody, while the Pythagorean inspiration that celestial bodies have musical relations is virtually a key to the universe.

7. The gene is also abstract as a pattern, independent of the atoms that implement it in any given moment, a meeting point between matter and energy, a message that moves in a wave of meaning through life, an acorn harboring an oak, an egg containing feathers, menu, songs and a map of stars.

8. The interrelation between all creatures and things is abstract too, something we will go deeply into next chapter. And it includes the ecologies of everything from rivers, glaciers, islands and fires to those of wars and worlds.

342

9. There is also a pervasive polarity between such opposites as good and evil, mass and energy, male and female, predator and prey, cause and effect, body and mind, yin and yang, concrete and abstract — relationships that themselves are abstract, which we will take up in Chapter 18.

10. The concept of mind (Part Two) is likewise abstract, elusive, unlimited by either space or time. For where do dreams reside? Does consciousness begin in the atom? Does it merge or divide in procreation? Does memory depend on a brain? Can a tree yearn? Can the impossible ever become possible?

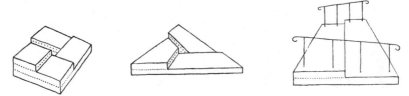

11. Life and death are both very abstract besides being aspects of a profound transcendence, as we shall see in Chapter 19. This concept derives from an understanding of life as a river or flame with currents and eddies where organisms evolve into superorganisms, mortality is a phase of immortality and finitude steadily unfolds toward Infinitude.

12. Evolution is another abstraction of life (Chapter 21) involving time, place, tributaries, multiple trunks, branches and the continuous development of matter, mind and spirit, particularly as exemplified in creatures, beings, machines, ideas, virtues, worlds . . . Evolution in the cosmos progresses from the circle to the helix in a new dimension, creating genes, muscles, rivers, galaxies — also cycles of fermentation and germination from cells to worlds (Chapter 22) to metaworlds — culminating in the spiral grain of the universe.

13. The divinity aspect of the world may be its most profound abstraction (Chapter 23), especially if, as I surmise, the world turns out to be in essence a Soul School. There is ample evidence, for example, that some unfathomable influence has somehow permitted the atmospheric flow of oxygen to evolve breathing, solar radiation to initiate vision, nerves to induce thought, thought to ignite spirit. And whatever the whole of this mysterious creative force may be, as with other wholes, it shows signs of being greater than the sum of its innumerable parts. Could any creator possibly design and bring into being a world better than Earth for teaching life and spawning spirit? Whatever the answer, there seems little reason to think the universe will soon, if ever, yield any concrete solution to its abstract Mystery.     343

# CHAPTER 13

# Second Mystery: The Interrelatedness of All Creatures

W HEN YOU LOOK at a planet whole, from outside, for the first time, certain things come clear that you never could be quite sure of before. One is a persistent feeling that all the inhabitants of that world must be related. Surely this is an intuitive inference from the obvious oneness before you, bestowed by the spatial perspective of the astronaut, which is, in the case of Earth, a crucial step in her dawning consciousness of herself.

But there is a profound mystery in Earth's inner relatedness, for relatedness implies common ancestry, just as first cousins share grandparents, and all life here must ultimately have come out of the womb of Earth. Yet if in fact this happened, how did it happen? How could the steaming, volcanic planet, or even her later, salty seas, spawn the vast and vital proliferation we know today?

## COUSINHOOD OF MAN

I will begin our exploration of this Second Mystery of Life with the most familiar of species: Man. I know man is something of an upstart on busy Earth, having existed for only the most recent one tenth of one percent of her history. And it is almost certain that he is now entering a critical period which may show him up as a catastrophic bungler. For his position is precarious, he being but a single species, only one of millions of species, most of whom have not only been around a lot longer but are much more solidly established. His total numbers who have lived their lives within the couple of million years of his existence add up to a few dozen billion individuals — which, in planetary perspective, isn't much. Specifically the anthropological estimates of those numbers range from about twenty to fifty billion, depending on how far back he can be said to have attained his station of being human, on the duration of his average generation and on how widely and densely he may have populated the world.

Although few of his members seem to realize it, man's relation to himself is fairly easy to measure and is surprisingly close. In fact, no human being (of any race) can be less closely related to any other human than approximately fiftieth cousin, and most of us (no matter what color our neighbors) are a lot closer. Indeed this low magnitude for the lineal compass of mankind is accepted by the leading geneticists I have consulted (from J. B. S. Haldane to Theodosius Dobzhansky to Sir Julian Huxley), and it means simply that the family trees of all of us, of whatever origin or trait, must meet and merge into one genetic tree of all humanity by the time they have spread into our ancestries for about fifty generations. This is not a particularly abstruse fact, for simple arithmetic demonstrates that, if we double the number of our ancestors for each generation as we reckon backward (consistently multiplying them by two: 2 parents, 4 grandparents, 8 great-grandparents, 16 gr. gr. grandparents, etc.), our personal pedigree would cover mankind before the thirtieth generation. Mathematics is quite explosive in this regard, you see, for the thirtieth power of two (1,073,741,824) turns out to be much larger than was the earth's population thirty generations ago — that is, in the thirteenth century if we assume 25 years to a generation.

But you cannot reasonably go on doubling your ancestors for more than a very few generations into the past because inevitably the same ancestor will appear on both your father's and your mother's sides of your family tree, reducing the total number (since you can't count the same person twice), and this must happen more and more as you go back in time. The basic reason it happens is that spouses are not just spouses but they are also cousins (although the relationship is usually too distant to be noticed), which means they are related to each other not only by marriage but also by "blood" because somewhere in the past they share ancestors. Another way of saying it is that your father's family tree and your mother's inevitably overlap, intertwine and become one tree as their generations branch out backward, a

345

process forced by geometry until eventually all your ancestors are playing double, triple, quadruple and higher-multiple roles as both sides, all quarters, eighths, etc., of your tree finally merge with others into one common whole and the broad tops of the family trees of everyone alive and his family become identical with their ancestral world populations. These populations of course are the fertile portions of past societies and naturally cannot include "old maids," cautious bachelors or anyone who fails to beget at least one continuous line of fecund descendants to disseminate his genes into mankind's future.

If people all over the earth always married at random, choosing their mates by lot regardless of what country, race, class or religion they came from, it would take something like 35 to 40 generations for all their family trees thus to merge completely in common ancestors, the time beyond 30 generations needed because of the aforementioned slowdown in backward ancestral multiplication, a rate which progressively decreases from two to one and ultimately even below one when the family trees begin to encompass all humankind, whose numbers of course were smaller in earlier centuries. But obviously people do not mate at random, for there have always been barriers to boys and girls getting together: oceans, high mountains and vast distances between them; rigid marriage laws, religious sex taboos and innumerable racial and cultural prejudices. On the other hand there have been almost as many compensating influences tending to overcome these barriers and increase the scope of mating: long hunting trips for vital food, nomadic herding, expeditions for business or war with its spoils in slaves and women; also strict laws against incest and inbreeding, all naturally abetted by man's primordial appetite for novel amorous adventure.

Probably most significant of these factors, geneticists agree, are the nearly universal rules of endogamy (inbreeding) and exogamy (outbreeding) that respectively permit marrying persons to pick each other only from *inside* the membership of a specified larger, *end*ogamous social group (such as a tribe or caste) yet also only from *outside* a specified smaller *ex*ogamous group (like a clan). In many sophisticated countries today of course the endogamous group is nothing less than mankind and the exogamous group is one's own immediate family, which gives a man latitude to propose to practically any uncommitted female in the wide spectrum between his niece and a chimpanzee. But virtually all the more primitive of human societies have traditions that greatly restrict marriage. In rural tropical Africa, for example, the average tribesman is expected to marry inside his tribe of perhaps ten thousand people yet outside his family clan of one or two thousand, giving him a total population of some eight thousand to pick from.

346

In a few rare cases tribal members have been even more restricted endogamically, as in the inbred clan called Foldi in the Hyabites tribe of Arabia, all of whose people have six toes on each foot and five fingers plus thumb on each hand and who have come to value their 24 digits so highly they will kill any 20-digit baby born to them as the illegitimate issue of an adulterous mother. The same goes for the monkey-tailed clan of Kali who traditionally refuses to keep any infant who doesn't have a tail.

Laws of exogamy, on the other hand, can be at least as extreme in the opposite (outward) direction and the marriage law in much of western Europe during the time of the Crusades defined incest as extending out to fifth cousins. In some other parts of the world the outbreeding groups were even larger, as in imperial China, where one was forbidden to wed anyone with the same family name, a rule that diverted the average Chinese bridegroom away from about a million of his closer relatives and undoubtedly contributed much to the homogeneity of that most populous nation.

Weighing the marriage customs of Earth, then, as annotated in reports from Westermarck's classic *History of Human Marriage* to the latest sexological research, it appears that the rules of endogamy have generally been enforced less rigorously than the rules of exogamy. This is presumably because the natural suspicion a few people may feel toward someone else's exotic consort from an alien land is less apt to arouse public disapproval than the natural revulsion those same people feel toward a liaison that can be called incestuous. Also if a man in ancient or medieval times married his sister, he would forgo his opportunity for aid in the form of brothers-in-law to hunt or fight with or ultimately to avenge his death. And so these various regulations of outbreeding and inbreeding have probably not, in net effect, greatly inhibited man's genetic circulation within his species. In sum, it would seem a reasonable estimate that, while geographical factors in historic times must have held back the gene flow between a man and mankind by perhaps ten generations, other factors have more or less canceled out to keep human kinship confined to about the fiftieth cousinhood range.

It is practically impossible to set limits on where any man's family connections may reach upon the habitable earth. And the very proliferation and complexity of his relations is incomprehensible to most of us for, even if you assume only two children to a couple, as I have done in the accompanying simplified diagram (a quota actually too low to permit humanity to survive), everyone on the average must have 4 first cousins (the children of his uncles and aunts), 16 second cousins 347

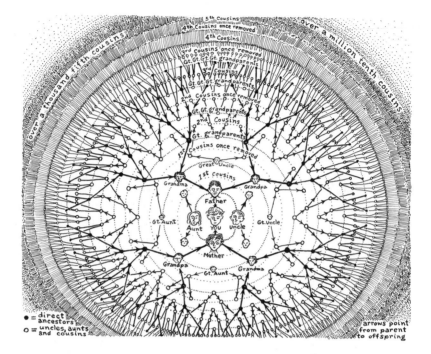

(the grandchildren of his great-uncles and great-aunts), probably 64 third cousins, about 250 fourth cousins, roughly 1000 fifth cousins and some million relatives as close as tenth cousin. Then, if you extrapolate on into the billions, which means extending your relatives to all humanity, it is evident that, even after allowing liberally for the shared ancestry of spouses and all the barriers to intertribal and intercontinental marriages, the range of fiftieth cousins will still easily cover the planet.

This is not to claim that there has been more than a trickle of intercontinental travel in bygone millenniums in which Celts, Phoenicians, Vikings and Polynesians are now known to have sailed to America, and Papuans, Melanesians, Malays and Negritos to have trafficked in Australia — but, in the nature of things, a trickle is all it really takes to establish close cousinhoods. It is scarcely possible, I admit, for the most perspicacious of minds to visualize cause and effect reliably over thousands of years and tens of thousands of miles. One could as reasonably expect a mayfly who lives a day to understand the love life of a turtle who lives a century. But it is demonstrable nonetheless that a single indirect genetic contact between Africa and Asia in a thousand years can make every African

348

closer than fiftieth cousin to every Chinese. Surprisingly, this may happen without any natives of either continent doing any particular traveling at all, but simply in consequence of the wanderings of nomads in intermediate territory. No single nomad would have to travel more than two or three hundred miles in his lifetime either, a meager enough total for a carefree tent-dwelling herdsman. Yet inevitably every generation produces at least a few would-be Marco Polos, mettlesome young bloods who keep going for thousands of miles, zestfully overcoming every hazard, swimming rivers, running borders, fighting brigands, making off with girls as they find them — and obviously it needs only a couple of brothers of this stamp to make *first* cousins of their own children on opposite sides of the earth — first cousins who, like as not, do not even know of each other's existence. If you can accept this fact, that an African family in the Sudan, for example, can be *first* cousins of a Chinese family in Hangchow without any Africans leaving Africa or any Chinese leaving China, then it shouldn't be too much for you to understand that there must be thousands of relationships at least as close as fifth cousin at comparable distances and that the tremendous genetic circumfusion of fiftieth cousinhood cannot help but take in the bulk of mankind.

To see how such interrelationships work, look at the illustration (next page), in which an old nomad living in Persia has eight sons who, one after the other, set off to seek their fortunes, each in a different direction. The son who goes to China predictably marries a Chinese girl and has half-Chinese children who are first cousins of the half-black offspring of the son who went to Africa and married a black girl on the upper Nile, all of this new generation of course being grandchildren of the old nomad of Persia. Later the African son's descendants naturally increase in number with succeeding generations until they include practically everyone in his tribe along about the thirteenth generation (even assuming only two children per family), by when his genes must inevitably have spread (through raids, wars, migrations and resulting infractions of endogamic law) to various other tribes, whose members in turn all become descendants of the old nomad by about the twenty-fifth generation. And after that his spreading waves of progeny must irregularly continue to advance tribe by tribe all over Africa and beyond, relentlessly filling up each endogamous pocket until by the fiftieth generation it can hardly help but include everyone.

"How do you know," I hear you ask, "that there were actually enough raids, wars and migrations in the past twelve hundred and fifty years to carry the old nomad's genes from tribe to tribe at the rate you claim?"

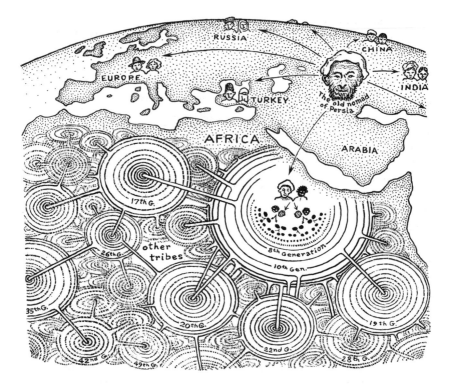

"How do you know," I reply, "that flowers in the garden will be cross-fertilized by the bees as the bees collect their food?" The two questions are analogous, both having to do with probability in genetic combinations on a broad statistical scale. And they are close to the well-known mathematical laws of rolling dice and the roulette tables of Las Vegas. For as the number of bees multiplied by their feeding rate and divided by the number of flowers tells you how often the average flower will be pollinated, so, by more complex equations, you can estimate the minimal cognate circulation of man, which, it turns out, affirms the extreme unlikelihood of any viable segment of human population being able to avoid all outside contact for anything like a thousand years.

Of course a few snobbish people will protest, "But my family is an exception. We traced our ancestry back eleven hundred years and they all came from an isolated community on the Isle of Man where they spoke Manx and never had any marriage contact with the outside world."

Statements of that sort cannot possibly stand up to close scrutiny. Genealogists, I admit, will often trace a prominent family name back for hundreds of years — which involves only the relatively simple procedure of following one male line of descent. But the average person has approximately a million ancestors with some thousand names just in the last 500 years and it is patently impossible to trace any large portion of them. Even such a well-publicized lineage as the *Mayflower* descendants from Plymouth, Massachusetts, can hardly begin to track down their relatives of 350 years, and a little knowledge of early Yankee seamanship and fecundity in the tea and slave ports of Asia and Africa, plus mathematics, will show that their ranks probably now include more than a million Chinese in China, a comparable number of Hindus in India and blacks in Africa — not to mention several million Americans and Europeans.

Even if at some period there really was an isolated, highly inbred clan on the Isle of Man, could anyone possibly prove it had maintained complete sexual segregation for even a hundred years? Any boy who escaped to the mainland or adventurer who arrived from anywhere might have established a relationship between it and the rest of the world. Is it not likely that some such incidents took place in every generation? To be sure, the traditionally proud family archives may show one or two honorable descent lines of reputed purity, but they have a way of neglecting to include the black sheep and their illicit or unorthodox doings, if indeed such ever got beyond being just suspected in the first place.

There is probably more than enough interracial gene flow right inside your own country to make you a close cousin of practically anybody imaginable. Dr. N. C. Botha, the immunologist who worked with Dr. Christiaan Barnard's heart transplant team in South Africa, calculated that the "white" people of Dutch extraction there have an average of 7 percent "nonwhite" genes, while the two million so-called colored people have 34 percent "white" genes. And the *Ohio Journal of Science* has estimated that 155,000 "Negroes" passed over the color line to become "whites" in the United States during the decade 1941–1950, indicating the dimensions of the accelerating flow of genes between the black minority and the white majority that is largely fueled by the waxing tide of black-white espousals now known to exceed one percent of all American marriages.

## What Is a Race?

It should hardly be necessary now for me to point out that there is no such thing as a pure race, nor any race of men on Earth that is unrelated to other races. Fact is: if such a race existed, it would be a species. But races are not species: they are (by definition) only subspecies. And all the peoples and races now existing on Earth are demonstrably of one single species, that being the largest group classification of organisms that regularly interbreed and reproduce their kind. This biological fact actually tells a lot about the homogeneity of the genes of *Homo* (man) whose species name is *Homo sapiens*, rather immodestly bestowed on him by Linnaeus in the eighteenth century. In a sense it is also science's own measure of the brotherhood of man.

It may be instructive here to contrast man with the mule, who enjoys no comparable brotherhood because he is not a species. The reason there is no species of mules is that the mule is not an interbreeding organism, for, with rare exceptions, a mule is sterile and cannot have little mules. So if you want to raise a mule, you must find a jackass and breed him to a mare (as has been well known at least since the days of Aristotle), the mule being a donkey-horse hybrid, an explicit kind of cross between two species. Normal humans of course do not have the mule's sterility, mankind being a species, so any mulatto or cross-breed between any two (or more) races of man may marry and have normal children. That is one of the important proofs of close human kinship, demonstrating that intermarriage between races must have always been going on, and obviously it has never stopped long enough to make interbreeding sterile in its results. Strong evidence for such a persistent prehistoric interracial mixing of genes comes also from numerous archeological excavations of the remains of divergent races in the same areas, such as bones from Aurignacian (white), Grimaldi (black) and Solutrean (yellow) peoples all evidently living as neighbors in southern France about 25,000 years ago.

Man's propensity to diversify into races, please note, seems to be characteristic of all the earth's most successful species because, say evolutionists, it gives species additional ways of adapting to environmental challenges such as an ice age, a warm age or a wet age (when all continents sink into the oceans) and, even in the most stable of times, it lets them live in more extensive and much more varied territories than any one single race possibly could. An example is the fact that polar bears and European brown bears interbreed and produce fertile cubs, showing they are races of a single species. Another is the recent discovery that herring gulls have diversified their species into 13 races living in various regions of the northern hemisphere, two of which share the same territory of northwestern Europe although they are unable to interbreed directly with each other. These two races, however, are known to interbreed with fellow races in adjoining areas so presumably their genes are constantly commingling through the indirect route of the half dozen other herring gull races ringing the Arctic Ocean. Of course races occasionally get separated by seas or mutations or fateful quirks of courtship, whence over the millenniums they diverge so far they evolve from subspecies into species. By such means presumably the hundreds of thousands of species of insects, reptiles, birds and mammals arose on all the continents of Earth. But man, outdoing the most pliant of his competitors (perhaps the rat, cat or housefly), became so adaptable he learned to move freely over vast distances (even more so than the birds) and to live in any country in any season, hot, cold, thin-aired, arid or wet, while still avidly inter-marrying back and forth among all his races, remaining thereby, for better or worse, a single species.

A mental hurdle to acceptance of man's close kinship may come from the resemblance usually noticeable among brothers and sisters, a family similarity which may understandably (if illogically) give one the notion that any two persons who do *not* look alike (say an Arab and a Swede) cannot be related. This common error is due to failure to realize that diversity persists in the most inbred of families, often between twins, in every race without exception and in humanity as a whole, which today obviously receives no "new blood" from chim-panzees, gorillas, orangutans or other of its nearer outside relatives. I do not, mind you, discount the statement of geneticist Geoffrey H. Bourne of the Yerkes Primate Center that there is "very little phys-iological reason" why man-ape hybridization via artificial insemination might not succeed — something that predictably will become the subject of an actual experiment one of these days.

But, regardless of any such development, diversity is a profound attribute of life and basic to Darwin's theory of evolution through natural selection among the ever-new, ever-different variations in all life forms, a diversity inexorably bestowed upon the species man by complex but fairly regular mixtures that produce gradients between all the so-called black, white and yellow races on Earth, as suggested in the accompanying illustration. And no amount of cross-breeding will eliminate it for, as Theodosius Dobzhansky categorically states, "if all peoples on earth were to intermarry at random, the resulting hu-manity would not, as is frequently but mistakenly supposed, be some kind of a compromise between all the now existing races. It would rather be an extremely variable lot: some persons resembling each of the now existing races would continue to be born, but other individuals would have combinations of traits that are rare or non-existent at present." His genetic explanation is that "'pure' races could exist    353

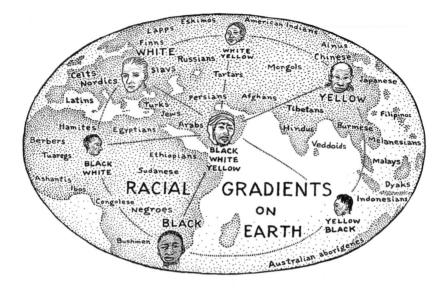

only if heredity were like a fluid transmitted through 'blood,' and if the hereditary 'bloods' of the parents would mix and fuse in their offspring." But, as you will recall (page 161), Gregor Mendel discovered last century that the old "blood" theory of heredity is plainly wrong. And in this century the microscope of the modern biologist has clinched it by proving that red blood cells have neither a nucleus nor any kind of heredity, can therefore be safely transfused from person to person, and that blood actually has less to do with heredity than any other part of the body.

To use the classic analogy, heredity is like a hand dealt from a huge deck of cards. Each child in effect (page 164) is dealt half the combined parental deck. If your parents together possess a total of 20,000 genes, which may well be true, the hand dealt you by heredity would contain 10,000 genes. Each of your brothers and sisters must also receive 10,000. Except in the case of identical twins (who originate from the same sperm-egg union), each brother or sister receives a different combination of 10,000 genes. Almost all your own genes may be duplicates of the 10,000 received by one of your brothers or sisters, or almost all 10,000 may be different. Any combination may chance to happen, but the number of genes is so large that, under the aforementioned law of probability, it is considered virtually impossible that the same exact combination of genes could happen twice in all the past or future existence of humanity.

354    The wonder of this secure, almost sacred, individuality of each of

us in the ocean of life has been dramatized by an eminent American biologist, Hermann Joseph Muller. He calculated that if we could collect together all the nuclei from all the sperm and egg cells from which the present human population of the earth has sprung, this whole material controlling mankind's heredity could be compressed into the size of an aspirin tablet. Of such extreme tenuousness is the architect's blueprint of human growth, of our body shape, our quality — while the great mass of building material that is physically ourselves has come quite literally from the food market.

The way in which all men are thus linked through humanity's pool of genes is a discovery that inevitably leads students of genetics to believe in human brotherhood. It is a great absurdity of the so-called race problem in the United States, for instance, that anyone who admits having any African or Hebrew ancestry is classed as a black or a Jew regardless of his or her appearance. This idea is sheer snobbery. A person may in consequence have to have only 10 percent of "Negro blood" to be called a Negro or a black, while he would be required to have more than 90 percent of "white blood" to be called a white. When it gets to be realized someday that there is no absolute criterion of race, that all of us literally have some white, some black, some yellow and some other kinds of heredity, the race issue may well fade away into the notebooks of anthropologists where it belongs.

I am not naïve enough to suppose that an Arab guerrilla sneaking a bomb into a Jewish market in Jerusalem or a white policeman loading a rifle to defy blacks in the heat of a riot in South Africa could easily be persuaded that he is engaged in a family squabble, but surely the significance of "race" is exaggerated in his mind and is in sore need of reasonable definition. Actually the problem of how to define the word "race" has been tantalizing anthropologists, ethnologists and geneticists for centuries, while the number of "races" identified by competent scholars has ranged from two to two hundred. Some of these authorities have classified humanity according to skin color, some according to head shapes, nose shapes, hair, ear wax, finger-prints, blood types and combinations of these and in other ways. But when it gets down to cases, such attempts have not worked well, for all Chinese do not have yellow skin nor long heads or round heads, and all people with straight black hair are not Chinese. Blood types, taste blindness, susceptibility to disease, innate intelligence, physical aptitude and most other human traits are distributed fairly uniformly throughout humanity. It is true that the people of some continents show a few traits more conspicuously on the average than the people of other continents, but it is unusual for any trait to be so infallibly

355

associated with any one group of people that it can identify them as belonging to one race. The nearest such traits would probably be the black skin pigmentation of the Negro, which evidently evolved in the tropics because it is vital in filtering out excess ultraviolet light and controlling the synthesis of vitamin D, and the Mongoloid eyelid of the oriental, which seems to have great survival value as the best protector of the eye's inner corner from the frequent dust storms of central Asia. But there are numerous exceptions even to these.

Reviewing all these worthy attempts, I can find no more generally accepted, if vague, definition than Dobzhansky's that "races are populations which differ in the relative commonness of their genes." Reasoning from this definitive statement, it is clearly possible, though rare, that two brothers can have relatively few genes in common and be as unlike as a Negro and a Chinese. In fact if the definition of a Negro or a Chinese is an individual with a certain blend of genes (as Dobzhansky suggests), then obviously it is possible for two brothers to be actually a Negro and a Chinese and conversely for a member of one race to have a brother in another.

Before you toss this idea aside as far-fetched or implausible, please consider that brothers and sisters sometimes differ as much as their parents and that, since parents can be of different races, it is logical that brothers or sisters may in such cases be of different races also. Come to think of it, isn't this an uncommonly elegant example of the brotherhood of man?

Another significant thought occurs to me. Although your brother or sister is more apt to have a set of genes like yours than is any single one of your cousins chosen at random, the fact that you have so many more cousins than siblings builds up the chance that someone among their multitudes may very well have genes closer to yours than has either your brother or sister. Indeed a friend of mine who is a probability researcher at M.I.T. calculates that this is almost certainly true, which is a way of saying that each of us is so much a kin of humankind that all the world's children are your children and mine, and not only spiritually but genetically as well.

Summing up the genetic compass of man as including the entire species within the scope of fiftieth cousinhood, I must admit that I have been unable to find any geneticist who has seriously tried to determine just how many generations ago all the family trees of today's population were merged into one, so the figure fifty is only a rough limit. But those I consulted who worked the closest to this question are agreed on the general order of the fact. And the fact means that your own ancestors, whoever you are, include not only some blacks,

356

some Chinese and some Arabs, but all the blacks, Chinese, Arabs, Malays, Latins, Eskimos and every other possible ancestor who lived on Earth around A.D. 700.

It is virtually certain therefore that you are a direct descendant of Muhammad and every fertile predecessor of his, including Krishna, Confucius, Abraham, Buddha, Caesar, Ishmael and Judas Iscariot. Of course you also must be descended from millions who have lived since Muhammad, inevitably including kings and criminals, but the earlier they lived the more surely you are their descendant.

And a thought-provoking parallel to this earthwide family tree is contained inside your own body, for it has been calculated that a mere fifty generations of cells multiplying their population by repeated fission could create the average human with his approximately $2^{47}$ cells. This being roughly right, the relationship between all men on Earth is comparable to the relationship between all cells in a man. And as cells metabolize and circulate in the body, so do bodies and their genes circulate throughout mankind, joining everyone to everyone at least once in fifty generations, so that not only does the ancestry of each of us include all fertile humanity of fifty generations ago, but our descendants fifty generations hence in turn will include every living being. If therefore your appetite disdains any kind of man, shake not your family tree. For its fruits appear in every color, in every stage of ripeness or rot, and its branches encompass the earth.

## COUSINHOOD BEYOND MAN

This is kinship awareness long overdue. Even more interesting in the large view, it is part of a relationship that reaches beyond man — that extends in fact everywhere, could we just become aware of it. For the family tree was not pruned or polled by Muhammad. Neither is it absolutely bounded by the dimensions of our species. While there is a kind of circulation (a metabolism if you like) among its spiraling boughs that quietly but insistently propels all human genes through humanity within fifty generations — roughly once in a thousand years — this genetic surge can be traced back millions of years to where it connects with kindred currents and ancestral rivers and ultimately the seas and clouds and streamways (visible and invisible) of all life.

Putting it another wise, the collective heartbeat of the colonial superorganism we call man throbs once each generation to pump its vital gene pool through all its arteries and capillaries at the rate of a circuit

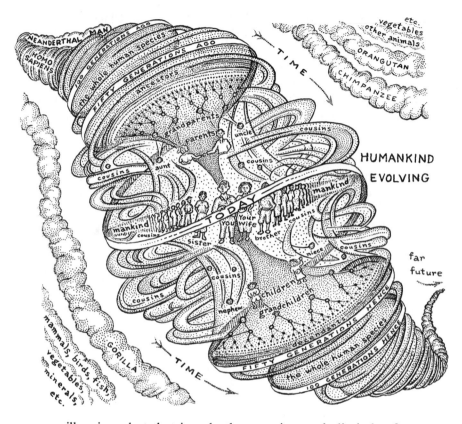

HUMANKIND
EVOLVING

per millennium, but that is only the genetic metabolic index for one species, *Homo sapiens* — and inescapably there is a larger flow and a more majestic circulation rate for the whole order of primates (including not only man but apes and monkeys), for the broad class of *mammalia* (from shrews to whales), the subphylum of *Vertebrata* (all back-boned creatures from eels to ostriches), the phylum of *Chordata* (everything with spinal cords from worms to elephants), then the whole kingdom of animals everywhere, followed by life in all its kingdoms, which, as we will see, may include not only the mineral but other kingdoms not yet designated, possibly some neither imagined nor (to us) imaginable.

Muhammad must have comprehended some of this propinquity of life when he led his believers in a prayer credited to him, in which he enjoined them to "Honor your aunt, the palm, who is made of the same clay as Adam." I wonder, had he been asked, whether he would have been able to say what magnitude of cousinhood joins man to beast or Earth to Moon? In any case, with our present swift-growing knowledge of evolution, it is not hard to approximate the date of a common ancestor when, if you can somehow establish a generation length, all you need do further is to divide the number of years since the common ancestor was alive by the number of years between succeeding generations to find the ordinal magnitude of the cousinhood. Thus the great apes are approximately man's millionth cousins, because the presumed common ancestor (Proconsul) lived about as many million years ago as the number of years in a generation averaged between those of man and the apes. By the same reckoning, most of the larger mammals are within the range of ten-millionth cousins of man, and all but the tiniest of animals roughly within that of billionth cousins, while the rest of life, including microbes, plankton, bacteria and viruses, comes generally within the trillionth-cousin magnitude.

Naturally creatures like bacteria and viruses, with generation spans measurable in minutes, must have thousands of times more generations between them and their common ancestors than do trees, which measure their generations in years, so the larger plants are usually less removed from us (generationally speaking) than some of the insects (say fruit flies) that buzz among them, even though the insects, with heads, legs, eyes, mouths, muscles, nervous systems and social organizations, are obviously closer physiologically and psychologically to us, and the ancestors we have in common with them 800 million years ago lived more recently than did our common ancestors with the vegetable kingdom some 1 billion, 500 million years ago. So cousinhood's range is more logically proportional to the time elapsed since the common ancestor than to the numbers of generations generated by the disparate branches of his descendants. And by this criterion, our quadrillionth cousinhood surely must include the mineral kingdom and even the superorganism of Earth herself with all her elements. Further, by applying celestial genetics from the time Earth's closest ancestor (the sun) spawned his family of planets and moons, we discover close sidereal cousins among the Milky Way's stars and more distant ones in remoter galaxies and supergalaxies — all these being relatives of estimable propinquity, which, if you can stomach specificity to its ultimate, bring every last one of them within an ordinal compass of cousinhood delineable within about twenty figures.

Reverting for a moment to the traditional comfort of Scripture, if you have read Jesus' announcement that "other sheep I have, which are not of this fold . . . and there shall be one fold and one shepherd" (John 10:16), you perhaps did not imagine he could have been thinking of the far ranges of the universe. Yet the profound fact of oneness not only on Earth but in all creation is becoming more and more evident as knowledge expands outward, and nowhere are to be found any absolute boundary lines or uncrossable barriers between any kingdoms of life. Not even between life and nonlife, nor between body and mind and spirit.

These realizations naturally bring to memory many a fascinating surmise, such as Ernst Mach's unexpectedly affirmative reply to the question: does a falling stone obey the stars? — which he explained to science with his now generally accepted theory of inertia, holding that inertia is the natural response of any material object to "the masses of the universe." The celestial navigation instincts of many creatures from walruses to bees to birds (who, as we have seen, follow actual star patterns on their night migrations) is about equally fantastic, as is awareness of the sun's direction by the blind termite who builds his castle from inside yet aligns it for optimum midday heat control. And in countless other ways the sky mingles closely with life and Earth, its oxygen circulating continuously through air, sea, land and blood, completely replenishing itself through photosynthesis (page 52) every 2000 years while its carbon dioxide component is renewed in a mere three centuries.

The volatile stuff of the world, you see, is ever in bubbling ferment, flowing and eddying through the turbulence of the biosphere, the soil surging forward to slake the root, the rock radiating abroad its crystal signature, the wind spiriting invisible spores and plankton across continents and seas, each molecule squirming for equipoise among its fellows, each atom singing in its orbit: the carbon cycle adrift through trees, leaves, soil and sky, the nitrogen cycle hitching every conveyance from digestion to lightning. The sheep grazing dreamily in the meadow is thus revealed to be a vital part of the grass and earth and air, a key segment of the meadow itself — even as the meadow in turn must be a living integrant of the total sheep.

cracks in a mud flat
MINERAL

veins in a grape leaf
VEGETABLE

veins in a grasshopper wing
ANIMAL

In countless ways all the organisms extend across the supposed boundaries between the kingdoms, being in fact animal, vegetable and mineral at the same time, the degree of each varying from case to

case. Did you know that many animals actually have meristems of growth similar to those of plants, the tiny hydra a prime example with not only a specialized growth zone in its flowerlike stalk (illustration, page 528) but branches that sprout in a regular Fibonacci spiral, delicately apportioned like a beech tree at intervals of about 120° from one to the next in what has been described as "a perfect phyllotaxis" or vegetal melody of growth. Jellyfish too have meristems and vegetate rhythmically in the sea, among the smallest of them being the amazing little polyps who not only build coral continents but are considered by evolutionists to approximate the common ancestors of most of the advanced animals on Earth, indeed probably all creatures who use legs, wings, fins or eyes and many who haven't yet evolved that far.

Empedokles of Agrigentum in Sicily intuitively guessed the kinship in various biological tissues when he wrote in the fifth century B.C. that "hair and leaves, and thick feathers of birds, and the scales that grow on mighty limbs, are the same thing" — a divination that has generally been verified by increasing human knowledge, specifically in the fact that all keratin (the stuff of hair, feather, horn, scale, etc.) is composed of microfibrils packed into crystalline arrays and, more generally, in that the sperms of animals and plants (such as ferns) are similar and embryos almost indistinguishable. Sometimes viruses cause mutations by actually transporting genes from a chromosome in one species to a similar chromosome in another, and it was recently discovered that a certain traceable primate gene was transferred that way from an ape to a cat thousands of years ago, retying those two diverging species into a slightly tighter bond. And now the much closer cousinhood of monkeys and men has been precisely measured in the discovery that human DNA will combine chemically (to the extent of about 40 percent) with monkey DNA, demonstrating that the two DNAs have at least that percentage of similar sections, suggesting that something approaching a half of the genes in monkeys and men are mutual to them both. Mice are obviously less closely related to men, sharing only about a quarter of their genes, yet even ten-millionth cousins like mice and men, who obviously cannot interbreed as whole organisms, were shown in an experiment at New York University School of Medicine in 1966 to be microcosmically congenial when normal cells of each were hybridized with the other's cells in a culture medium and the offspring "colonies of man-mouse cells grew successfully through more than 100 generations" during a period of six months.

As for our more distant relatives, a protein known as histone-4, isolated from the thymus gland of a calf, has turned out to be almost identical with histone-4 protein isolated from pea seedlings, despite

the fact that these animal and vegetable organisms could hardly have had a common ancestor more recent than several hundred million years ago. It is enough to convince me that all things are really involved in all things. Then there is the fact that certain kinds of bacteria in the oceans consume molecules of dissolved iron, concentrating them in their bodies over millions of years until they eventually settle into vast beds of iron deposit on the ocean floor that during a hundred million more years may be slowly pushed upward into mountains for men to dig into. And in the working of such mines a significant portion of the iron molecules inevitably diffuses into the air so that we unwittingly breathe and eat iron as invisible dust and our marrow forms it into the iron-hungry nuclei of hemoglobin for our red blood cells. Thus, when you walk down the avenue and see a woman blush, you can be certain the red of her cheek is closely related to the steel girders that are the bones of the city around her. Or when you glimpse the planet Mars shining red in the night sky, you can be assured it is akin to that blush because the deserts of Mars are made of iron-bearing sands. And, even more basic, the red light of the planet, like the red light of Betelgeuse, Arcturos, Antares and other red stars, is essentially an abstract frequency, a low pitch of colored light (about 5000 angstroms in wavelength), a kind of visual music of the spheres that can exist anywhere in the material universe — and that, like iron, rust, blood, moon, planet, star and galaxy, it is just one small part of all the things that are involved in all things.

## INTERKINGDOM RELATIONS

What I am saying of course is that all the kingdoms interact upon one another continuously. When a farmer plows his field, in effect the human kingdom (a man) is persuading the animal kingdom (a horse) to induce the mineral kingdom (a plow) to influence the vegetable kingdom (corn) to grow food. The kingdoms don't always harmonize, however, and bitter wars are fought between them, as any forester can tell you, with the so-called lower kingdoms winning at least as often as the higher or more intelligent kingdoms. I mentioned earlier the South African grapple plant whose seed case sprouts a dozen sharp claws which savagely penetrate and cling to any foot that steps on them, and which have been known to kill lions who carelessly got them into their mouths, which made eating so excruciating that the lions had no choice but to starve — a case of the "king of beasts" being literally vanquished by a mere vegetable.

More commonly, I am glad to say, the animals and plants make "love" to each other, using organs evidently evolved for just such a purpose. Indeed all flowers and blossoms, apparently without exception, are reproductive organs! Something like 10,000 species of plants actually manage to get themselves pollinated by insects, most commonly by bees and butterflies, who, in terms of the business world, correspond to middlemen merchants shipping high priority goods from anther to stigma with commissions collectible at both ends. And research into this animal-vegetable commerce has shown, interestingly, that it could hardly have been initiated by either kingdom alone, but rather must have evolved symbiotically through gradual adaptive changes by both parties, the basic deal being a direct barter of food for fertilization that, during tens of millions of years, has developed countless specializations and elaborations, not only to guarantee the exchange but also to exclude outsiders from fooling around, cutting in or hijacking any particular channel of traffic. Thus flowers are shaped to accommodate a certain class or family of animals, deep bell-shaped ones welcome those with long beaks or proboscises, curved or odd-shaped ones take animals fitted especially for them . . . So we hear of bumblebee flowers (blue and yellow), butterfly flowers (bright), sunbird and hummingbird flowers (mostly red). We discover moth flowers (pale), rat flowers (orange-red) and bat plants (the calabash, candle tree, areca palm . . .) that are open for business only at night. We hear of the saguaro cactus that is often pollinated by doves. And blossoms that specialize in particular kinds of flies, in mosquitoes, beetles (the wild rose, pond lily, magnolia . . .), ants, spiders, snails, frogs, possibly fish and surely man who is domesticating and engendering new varieties of plants at an alarming rate.

The flowers, for their part, act as if they wanted to make sure that their long-standing contracts will be honored by their cosigners, showing it both by their display of colors and perfumes that invite said pollinators to serve them before anybody else gets there and by advertising themselves with all sorts of floral scripts in the form of dots, lines, arrows and other labels that identify them as specific species and pointedly guide unfamiliar visitors to the right spot to tank up. Expediting the efficient loading and unloading of the pollen to a degree that is virtually foolproof, many flowers have stamens and stigmas hinged in such a way that when any intruding insect pushes hungrily toward the nectar, he forces the male flowers to dab sticky pollen on his back and the female ones to lift it gently off again by adhesion.

Most flowers seem naïve enough to accept this sort of automation as sufficient, but an insidious minority of them, presumably after eons

363

of bitter frustration from perpetually uncontrolled bees in their bonnets, have resorted to sterner and more fail-safe methods of making the bugs stop bugging them, and really keeping their part of the bargain. They actually have evolved traps baited with osmophores effusing scents that have shown themselves to be "irresistible" to certain passing insects, perfumes with unfloral aromas ranging from cidery to uric to fecal to musky, which lure the susceptible creatures into a type of funnel-mouthed blossom botanists call "the caldron," where they are literally overpowered and imprisoned. Darwin knew that flowers were somehow able to enslave insects, but details of their subtle techniques have only recently come to light, such as the lady's-slipper's bossy way of making a bee beehave by tricking him into sliding down a slippery chute into her jail-like interior, where one-way hairs permit him no movement except toward her sticky stigma that mops the pollen (from other flowers) off his back for cross-fertilization on his way out a special exit. And apparently the shimmering of tiny hairs around the rim of many cup-shaped flowers looks to flies like the fluttering of wings beckoning them to join a feast for that very cup (or funnel) is a favorite landing place even though coated with a substance that neutralizes the suction pads on their feet so that they slip, despite their desperate struggles, steadily and inexorably down into the caldron's pit.

Once an insect has thus been taken into custody in a blossom brig, the flower warden wastes no time in exacting his due in pollination, the hundreds of different species of rapacious plants wielding an impressive arsenal of more ingenious than scrupulous means to achieve their vital ends. Indeed some of the loveliest blossoms are really deadly dungeons from which there is no hope of reprieve, a few of them ruthlessly digesting their prisoners in a matter of hours. Certain subdolous lilies offer both of these fates, presumably on a semisporting basis to reconcile the harsh contradictions of sex. Thus we find the jack-in-the-pulpit, whose male jail accommodatingly flings open a hatch to parole the mosquito convict freshly powdered with pollen, yet, if the flushed mosquito is gullible enough then to let himself be enticed by pungent perfume into the identical-looking female quarters of the same flower, he learns too late that Jill is a grimmer jailer than Jack and quite capable, thank you, of relieving him not only of his pollen but of his life.

That a few plants may be evolving animal-like sexual behavior is suggested by the male members of certain ones like the barberry whose stamens erect themselves when stimulated by touching or stroking, almost as if they were penises. Sensitive plants like the mimosa (page

368) even have a motor organ, the pulvinus, that stiffens them when its cells become gorged with a fluid corresponding to blood. In a few cases all a flower's stamens will rise to the occasion together when excited by the bustling arrival of a bee, thus ensuring that the animal will be smeared with their pollen for transport to the female organ of another flower. This is a nice example of sex relativity too, for it shows the bee to be in effect female in relation to the male stamens, yet, a moment later, male in relation to the female stigmas it fertilizes.

Hundreds of kinds of flowers, moreover, not content with emulating animal behavior in their interkingdom dealings, actually have evolved their whole outer forms in candid imitation of animals. This is particularly true of the orchid, which is now widely recognized as zoomorphic because of its resemblance to spiders, wasps, birds and other animals with a hundred disguises enshrined in such common names as adder's tongue, tenderwort, marbled crane fly, green fly and butterfly orchids, some of whom hold sway over their animal namesakes not only by looking like them but by smelling like them, provoking their belligerence as "rivals" and, in some amazing cases, even mating with them! The genus *Ophrys* is an orchid famous for this interkingdom "marriage," which evolved perhaps thirty million years ago in a strange symbiosis with solitary male bees of the genus *Andrena*. The exotic flower masquerades as a voluptuous female Andrena during the weeks just before the real female Andrena has matured, enhancing the illusion by exuding a balmy bee perfume, which, in combination, exerts such a seductive allure that the male bee seems utterly bewitched by his blooming "maiden" and, before he knows it, falls head over heels in love — or at least into such a state of infatuation that he impetuously descends to embrace and deflower her — and has repeatedly been observed actually copulating with her (insofar as that is possible with a vegetable). This is far from a unique phenomenon, for it has spread to other orchids and to flies and wasps, and somehow

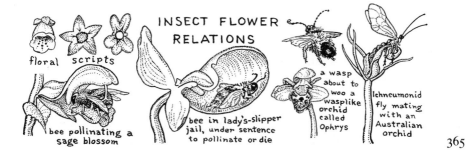

INSECT FLOWER RELATIONS

floral scripts

bee pollinating a sage blossom

bee in lady's-slipper jail, under sentence to pollinate or die

a wasp about to woo a wasplike orchid called Ophrys

Ichneumonid fly mating with an Australian orchid

365

must have proved its evolutionary survival value for these insects over the long intervening period, not to mention its absolute indispensability to the orchids, each of which in effect is thus "impregnated" through the inadvertent cross-fertilization performed by its enraptured mate!

Some flowers, believe it or not, literally get "in heat" in order to achieve procreation — even more so than do animals — and the large, lush voodoo lily, which grows in the tropics and is pollinated by flies and scavenger beetles, generates such rapid metabolism when respiring its reek of carrion and dung to lure them that the temperature of its reproductive organ, called the spadix, rises near its tip as much as 27° F. in a couple of hours. Another lily, *Cryptocoryne ciliata*, which takes root on the bottoms of ponds and streams, raises a "flag" above the surface that exhales such hot, pungent breath that insects are drawn to it and cannot resist landing and crawling all the way down into its foot-long submarine chamber, where the flower sees that they remain until they have paid the expected forfeit in full.

Among the countless interkingdom cause-and-effect relationships are what ecologists call positive and negative feedback between such cyclical factors as the warmth and coolness of days and nights and the pace of swarming ants, the pitch of chirping crickets and the bustle of human crowds. Did you know that the number of bees born each summer day is precisely regulated by the number of flowers that blossom in the nearby countryside? That forests humidify the sky for miles above them, sometimes saturating it enough to form rain clouds capable of returning a vital mist even in the dry seasons? That stag antlers are known to the French as *bois de cerf* because they

represent plant life in an animal and are sometimes more asymmetric than trees, the comparison being substantiated by modern biological investigation into antlers' meristems, abscission zones and the similar ways in which hormones control the growth in both kingdoms? That hydrodictyon cells colonizing one of the tiny round vesicles in your body mysteriously crowd together in one dimension while spreading apart in the others, with the curious result that they form not a ball but a disk shaped like the assemblage of stars of the Milky Way?

The field of geology has always been full of its own special inter-kingdom relationships, many of them known to modern prospectors who have taken to making chemical tests of plants to find out what sort of metal traces their roots have absorbed out of the ground. If you are looking for gold, for instance, it would behoove you to note where horsetails grow because these ancient plants have a fondness for it, and one botanist recently reported a case of horsetails absorbing gold in the proportion of four ounces per ton of plant material. And a kind of mold was recently discovered in Russia that can extract up to 98 percent of gold in liquid solution. Wild buckwheat, native to the western United States, has a similar affinity for silver and is known to be abundant near silver mines. Likewise the wild poppy and the dandelion are clues to copper in the American Southwest, locoweed often marks uranium deposits there and tumbleweed and milk vetch are the tip-off for selenium.

Animals share in this attraction for minerals too, because all protein molecules seem to welcome a few metallic atoms in their complex spiral latticeworks. Australian sheep have been known to die for lack of a tiny amount of cobalt in the soil, and other animals have succumbed because they needed a tiny trace of copper or manganese. The latter metal in fact is so vital to ants that miners in the mountains of New Mexico have been reported to use anthills to plot the courses of manganese veins. What we call organic molecules, it seems, almost certainly evolved out of simpler inorganic ones, so the swapping around of atoms is presumably a basic characteristic of all matter, dead and alive. Which may well be why life has gotten so involved with metals, metallic atoms being the most swappable because of their footloose outer electrons, which, having few ties, tend to be chemically venturesome — venturesome in this context meaning not merely flirtatious but actually promiscuous.

Applying such factors to the elusive question of where are the borderlines between kingdoms, it becomes more and more obvious that none of the kingdoms has really unequivocal boundaries anywhere. Not only do a number of animals, such as the barnacle and the sponge, live passively like vegetables, spending their adult lives rooted in one spot, but many vegetables are as active as animals, even mobile (page 55) and, in numerous cases, carnivorous to the extent of catching and eating insects. I am told there are a good 450 known species of meat-eating plants from microscopic worm-snaring fungi (page 97) to the jack-in-the-pulpit and foot-deep leafy urns in Borneo which often capture amphibians and reptiles. The Venus's-flytrap, best known of these predators, has animal-like senses in the form of three hairs on each half leaf which enable it to distinguish between a living and a dead object. In fact its ingeniously spiked trap does not close unless at least two hairs are touched in succession or the same hair twice within a few seconds. Even more extraordinary as an animal's pitfall is the beautiful water weed called bladderwort, whose submerged threadlike stems grow dozens of bladders with trap doors that snap open when a swimming insect touches their sensory hairs, letting the water rush in, sweeping the insect with it, whereupon the door closes again and the plant promptly pumps out the water to be ready for the next victim.

Some pitcher plants, which normally trap flies and moths, are so large they have been observed and photographed catching mice, digesting them in their usual period of about ten days, after which they spit out the bones and reset themselves. There are also the sundews and butterworts that work on the flypaper principle, the latter exuding a sticky mucilage whenever an insect alights on a leaf, followed by acid with a digestive enzyme that quickly paralyzes the victim while the edges of the leaf gradually curl up and wrap it in a tight roll for thorough digestion. It is significant that the "stomachs" of all these preying plants secrete virtually the same juices as animal stomachs, although the plants are inevitably not only slower but gastronomically much less durable than the animals, rarely managing to tuck away more than three meals (per stomach) in their lifetimes.

Perhaps closer to being an animal is the sensitive mimosa plant, native in warm countries like India (where it is called "coy maiden" because of its shyness), which has contractile hydraulic tissues (pulvini) similar to muscles at the junction of leaf and stem and a nervelike control system that includes something comparable to a reflex arc. These animal-like developments enable it to dodge a browsing cow by furling and withdrawing its frond from a wide open to a closed position in what Sir J. C. Bose measured as less than a tenth of a second after the first touch, with rapid successive closings of other fronds down the length of the branch. This startling responsiveness of a mere vegetable is not as fast as that of a frog, who reacts to a touch or a mild shock in about a hundredth of a second, but it compares well with that of a turtle and is much faster than that of a snail. This mimosa is animal-like too in the way it gets tired, for it normally takes about twelve minutes to recover from furling its leaves, but, if touched several times in succession, responds more slowly both in closing and reopening and, if driven by repeated blows (as in a hailstorm) to complete exhaustion, it may become too paralyzed to respond at all. However, after half an hour or more of rest, it will usually recuperate and return to normal.

Being acutely sensitive to temperature, light and radiation as well as touch, this plant's capability naturally varies with the weather, and tests show it to be most active at 91° F. in sunshine, becoming a bit sluggish when clouds cover the sky, literally turning numb with cold at 50° F. and suffering a kind of heat stroke at about 150° F. Like a cat that purrs when fondled but flees from a kick, the sensitive mimosa turns its leaves toward the friendly sun but pulls them cringingly away from any light much brighter. Besides being a lot more sensitive than man to light and (some botanists think) to radio waves, plants in general are known to feel very faint electric currents, one called bio-phytum responding to a shock only one sixth as strong as can be detected by a human. The nervelike impulse in the average plant has been measured to travel about an inch a second, but in some mimosas it has been clocked at 18 inches a second. As in animals the impulse can move in either direction but "prefers" one direction and will go faster and easier that way. Also as in animals, there are both sensory and motor nerve impulses in plants, according to Bose, the motor impulses traveling generally faster by six times.

The idea that plants can become dissipated is by no means just an anthropomorphism, for Bose and more recent experimenters have offered them alcoholic drinks and repeatedly succeeded in getting them drunk. When Professor Wilder D. Bancroft gave a sensitive mimosa a drink in his laboratory at Cornell, presumably submerging its roots, its initial reaction was a kind of glow of confidence, as from one's first cocktail. The little plant held its leaves straighter, closed and opened them faster and looked positively eager to perform. The stimulus had given it a mood that one observer described as "cocky." After the next drink, however, it began to lose control. Its movements turned haphazard like those of a sot with wobbly knees until only its stiff lower stem held it from staggering. Next it grew dopey and its leaves folded and dropped and it figuratively slid "under the table." But after a quiet four hours of "sleeping it off" it "opened one eye" in effect, then, despite its "hangover," placidly climbed back "on deck" as if nothing had happened.

Perhaps most suggestive (not to mention controversial) of all Bose's findings of animal-like traits in plants is his assessment of their specific, if not individual, sensibilities. Thus he not only particularized the pea as suffering an actual death agony in the pot, but found the undemonstrative carrot one of the most excitable and nervous of all vegetables, concealing feelings even keener than those of the sensitive mimosa. Celery, on the contrary, he rated as low-strung, tiring easily and caring little whether it be eaten or left to rot.

Although few biologists anywhere seem to have accepted such interpretations as justified, at least they have served to make us aware of the distinct possibility that to some degree a tree, any tree, is capable of feeling the woodsman's ax. And, beyond that awareness, a new and different sort of insight into vegetable senses may come from a look at a branch of life where the genetic union between vegetables and animals is most obvious: that of plankton in the sea. The sea of course is where life has done most of its evolving from the vegetable into the animal during the last billion years, a process believed to be still going on, particularly among the swimming flagellates, which are continually diversifying into myriads of very dissimilar varieties with anything from one to scores of ciliated propellers. As you may have suspected, both botanists and zoologists study these creatures for the reason that they behave like vegetables and animals at the same time, some of them (such as the flagellate *Euglena*) actually glean the advantages of the two kingdoms simultaneously, using green chlorophyll to get energy from sunlight by photosynthesis like a tree, while spending that same energy to pursue and capture their prey like a cat, then eating it for more energy with which to grow more chlorophyll, etc.

Still another example of an animal-like vegetable, this time rather sinister in its form and behavior, is the deadly dodder, a parasite upon other plants, which advances toward its host-prey with a kind of malicious deliberation like a very slow snake with fangs in its "stomach" instead of its "head." When the dodder sprouts from the ground, its tendril-like stem seems to sniff and listen for a few hours, then, somehow sensing the presence of its victim (probably through the light-sensitive cells in its leaves which direct it to relative shade), it turns its tip and grows rapidly toward it — and often, by the end of the second day, has coiled itself firmly around and started sinking its rootlike fangs (called haustoria) straight through the cambium layer into the sap channels. Then, having "tasted blood" so to speak, it pulls its roots completely out of the ground and becomes a full-fledged parasite, as shown in the illustration, well able to advance like a

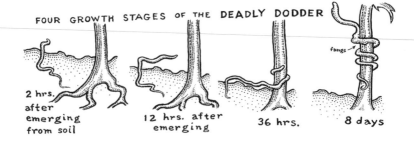

FOUR GROWTH STAGES OF THE DEADLY DODDER

fangs

2 hrs. after emerging from soil

12 hrs. after emerging

36 hrs.

8 days

serpent toward any prey that appeals to it. One further snaky feature of the dodder is that it is bare of twigs or leaves, having abandoned them millions of years ago when it evolved, in their place, scales!

Although plants in general, being relatively conservative in evolution, move and respond much more slowly than do most of the complex animals, the plants' behavior as seen in a time-lapse movie becomes surprisingly animal-like. I saw a month's growth of a climbing vine thus shortened into a minute in which the vine's tip would rise upward under the stimulus of light and warmth each morning (one second long), then relax and settle back as the sun went down each evening (another second). But with the coming of every new day the vine reached upward with a fresh effort and always in a slightly different direction, usually a little higher than the day before. With the time dimension thus speeded up by a factor of 40,000, it was obvious that the plant was not only eager to climb but also, through the combination of progressively augmenting its strength and varying its aim from day to day apparently in a random manner, along with its presumed hormone guidance from light and gravity, it was groping upward with a kind of vegetal purpose, methodically and rhythmically reaching to and fro, up and up, mostly in empty air, but nevertheless rising higher and higher and still higher with a deliberate attitude of expectancy — even of confidence. And every now and then the vine's groping tendril or finger would touch and discover a new support somewhere above it, whereupon, with a seeming sigh of gratitude, it would feel and fathom its way around this handhold, embrace it like a trusted friend, then continue upward with renewed faith. Could this be, I thought, the way animality arose out of vegetality — out of the patient and constant struggles of vegetation to achieve a position, a condition, a station in life almost but not quite beyond reach? Could this be what Bose meant when he wrote, "There is no life-reaction in even the highest animal which has not been foreshadowed in the life of the plant?" Could this be what God was saying when He told Isaiah (40:6) that "all flesh is grass, and all the goodliness thereof is as the flower of the field . . ."?

## DOMESTICATION

Right here might be a good place to raise the fundamental question as to whether there can possibly ever be two objects or factors in the universe that are not in some sense related. It is not a new question, for the fact of universal relatedness — assuming it is a fact — has been

371

curiously anticipated by philosophers for millenniums. Anaximander of Miletos in ancient Greece, for one, is said to have taught that "the primary substance of the world is infinite, eternal and all encompassing." Two thousand years later Bruno went further in writing that "all reality is one in substance, one in cause, one in origin . . . and every particle of reality is composed inseparably of the physical and the psychic. The object of philosophy, therefore, is to preserve unity in diversity, mind in matter and matter in mind . . . to rise to that highest knowledge of the universal unity which is the intellectual equivalent of the love of God."

Expectably Bruno got burned at the stake for this and other heresies, but his thought lives on and has been reinforced both by modern science and modern religion, a reinforcement probably most specifically evident in the remarkable recent growth of the Baha'i Faith, of which we will hear more in Chapter 23. No educated person should be surprised, therefore, that Earth's minerals, vegetables and animals have been found to converse and interrelate in almost every conceivable way. And one of the most significant and fascinating forms of interspecific, and often interkingdom, relation is the phenomenon called domestication, which also increasingly affects all life. I used to think that man many eons ago must have deliberately set out to tame the dog, the goat, the cat, etc., and that his eventual domination of the animal signaled victory in a one-sided campaign of pursuit, capture and control. But the evidence strongly suggests that domestication was largely unplanned and in most cases either accidental or entered into as willingly by the animal as by man.

The dog seems to have been the earliest creature to join man in close partnership, perhaps very tentatively hundreds of thousands of years ago, but pretty finally somewhen around 20,000 years ago. And this presumably happened as much because the dog was attracted by scraps and bones left around the campsite as by man's appreciation in return of barked announcements whenever a stranger approached. Indeed it likely took form as a true symbiosis or working partnership for mutual benefit — a deal in which four-footed sentinels freely served for their keep on a day-to-day basis with no contract and no questions asked. Only much later, probably without realizing it, did man gradually enslave the dog by controlling his food and his breeding, little by little turning a partner into a parasite.

And perhaps it was only a short while after the dog joined man that the reindeer, a salt-loving ruminant, likewise made a compact of sorts with him, in effect offering his freedom for the chance to lick up any salty substance such as human urine. But this time it was man who accepted the major role of parasite, as he took to following the migrating herds of the large beasts like a flea following a dog. This happened in many northern countries from Lapland to Siberia and, from all accounts, it persisted intermittently for millenniums before man learned to corral the reindeer, to breed them, decoy them in hunting, milk them, ride them and eventually harness them to pack loads and haul sledges.

Again, in the case of goats, sheep, cattle and probably asses of the Nile valley, these animals gravitated toward easier grazing conditions near human habitations beginning some 8000 years ago and most certainly where and when agriculture had started to produce tempting orchards, gardens and fields of grain. So eventually fencing in the creatures along with their fodder was, in a sense, only a formalizing of an already-established relationship.

In similar ways such scavengers as the pig and the duck entered man's domain as willing servants with a natural capacity to convert stale vegetable scraps into fresh whole meat. Then came the buffalo, the elephant, the rabbit and the goose, all opportunistically raiding man's crops, gaining confidence with each success, in due time bringing along mothers with their young which could readily be captured and raised as tame dependents.

A few creatures also ventured tardily into the human dominion as eager, often welcome, pest-destroyers, among them the ferret, the mongoose and, most notably, the cat which, although held sacred in Egypt, was eventually shipped from there in significant numbers to Sybaris and lesser European ports during the last millennium B.C. Still other kinds of animals may have been purposefully tamed by people already familiar with domestication, like the inhabitants of the Indus valley who captured and bred the shy jungle fowl in the third millennium B.C., probably initially for cock fighting but soon for egg production. There were also the Egyptians who long experimented with captive gazelles, ibex, oryx, addax and other antelopes, with monkeys, cheetahs and even hyenas (as early as 3000 B.C.) which they force-fed for the table — this being part of the unending succession of animals enslaved all over the world: the horse of Turkestan broken to chariots in the third millennium, the camel of Arabia and the Near East saddled for caravans about the same time.

Birds subjugated before the Christian era included the falcon in much of Asia and eastern Europe, the peacock from India and Burma, the pheasant in Persia and the Caucasus, the dove and quail of the Near East, the pelican and crane in Egypt and the ostrich and guinea fowl from several parts of Africa. Silk moths likewise had been made 373 domestic by the Chinese, reputedly about 3000 B.C., and the honey bee possibly as early as 5000 B.C. in Egypt, whence beekeeping spread to Asia and Europe. And there were four kinds of domesticated fish bred by man in antiquity: the Roman eel, which was not only fattened for eating but often kept as a pet and on occasion dressed up with jewels, including earrings; the carp in both ancient China and later Europe; then the goldfish under the medieval Sung Dynasty, followed by the colorful little paradise fish in modern China.

In America about the only domesticated animals in sight when the Spaniards arrived were the turkey, guinea pig, llama, alpaca and a few kinds of dog, even including (among the Incas) a bulldog and a dachshund. The load-carrying llama and wool-bearing alpaca are of course the domestic cousins of the wild guanaco and vicuña, these four interbreeding Andean animals being closely related to the camels of Asia and Africa, which, like the early horses, are known to have migrated westward from America while there was still a land-bridge in the present region of the Bering Sea.

But although the so-called Indians of America evidently domesticated few animals and had no tame goats, sheep, bison or even caribou, surprisingly, they were ahead of the Europeans in the number of vegetables they had domesticated, which included potatoes (white and sweet), corn, squash, pumpkin, lima beans, manioc, coca, chocolate, vanilla, chili, chive, amaranth, sunflower, quinine, tobacco, sarsaparilla, peyote, such fruits as tomato, avocado, pineapple, papaya and guava, and scores of berries, herbs and drugs. Domestication in the vegetable kingdom, however, may have been no more of a one-way conquest than it was in the animal kingdom, for lowly weeds have shown that they are just as able to choose civilizations as are dogs or swans. In fact this vegetal capability is about all we need to explain how the barley plant, rejected by the earliest human farmers as a "weed" in wheat, nevertheless managed in time to become a valued "grain" through its patient persistence in "adopting" man as the likeliest provider of enough space to grow in. Something similar happened in the case of corn in America, which the Indians domesticated but which in turn domesticated them as their economy, settlement pattern and even social organization followed the life cycle of this indispensable plant and its many races which now exceed 300.

The roster of domestication of course increases by several species yearly and it already amounts to hundreds, including the names of new animals being added at an accelerating rate as every creature from crickets (in China and Japan) to the remora (a fish trained to catch sea turtles by suction in tropical seas) to musk oxen (in Vermont) is drawn into man's fold. And the many kinds of experimental animals in research laboratories seem to be almost erupting into what could be called their own special new subkingdom of regimented, computerized members, already totaling hundreds of millions of mice, rats, hamsters, guinea pigs, rabbits, dogs, cats, monkeys, pigeons, fruit flies, flatworms, etc., some of them being species tailor-made for research, like a synthetic breed of fast-growing miniature pigs that has become the cheapest, handiest "pseudo-human physiology" yet available. Come to think of it, it is doubtful whether any large land mammals remain on Earth that are not at least nominally under the purview of some nation's game laws, if not restricted directly or indirectly by the confines of a national or smaller park. The muntjac, for instance, a small Asian deer introduced into Britain, is reported to be "spreading" along with badgers and foxes into the London suburbs, evidently feeling safer near houses. And it seems only a question of time before all the Earth's creatures, even insects, plankton and microbes, must live, breed and evolve at the dispensation of man.

Below the human level, what might be termed domestication of one species by another usually turns out to have evolved to some stage between voluntary symbiosis and parasitism, and the phenomenon shows up all over creation. Ants, for example, millions of years ago had already domesticated more varieties of animals than had man before the twentieth century, including a score of other species of insects, whose stations range from pets to furniture. Some ants are sophisticated farmers, growing various crops from mushrooms in underground beds to rice which they sow in November, weed all winter and reap in June. They exploit their own offspring too, a variety in Ceylon having been observed making its nests in trees out of leaves sewn together with spun-silk threads extorted from the helpless larvae, who, possessed of spinning glands but no legs, are easily picked up and used as precision tools by their elders, each grub a living combination of distaff and shuttle that can be artfully squeezed while its head is applied alternately against one side then the other of the leaf parts being stitched into one. Exploitation however, even between ants, may be in danger of backfiring when it is carried to the extreme of certain specialized, slave-hunting species who know nothing but how to raid other ants' nests, from which they steal and bring back pupae to raise as cooks, butlers and serving maids, without whom they would starve since they have literally (without realizing it) let themselves become the parasites of their own slaves.

The most primitive sort of domestication I remember having heard of is probably that in which certain aggressive viruses known as "phages" (page 160) take charge of susceptible, if not seducible, bacteria by injecting themselves directly into their bodies. And these bacteria, which are hundreds of times bigger than the phages, appear for a few minutes to have "swallowed" them but actually the phages are penetrating or ravishing the genes of the bacteria and somehow combining their own DNA with the bacterial DNA to spawn new phages that in half an hour make each bacterium swell up and burst forth with a new viral generation.

## AND NOW ECOLOGY

Beyond what can be called domestication of course there exists a vast and complex spectrum of symbiotic relations among the creatures of all the kingdoms, ranging from even-swap mutualism such as the tickbird-rhinoceros partnership to out-and-out parasitism like barnacles on a whale's belly. And there are uncountable variations of cause-and-effect interrelations that keep the new science of ecology almost unfathomable. Did you hear of the case of the sudden decrease in the population of wood ducks at a lake in Maine? When a local naturalist was asked to help, he soon discovered skunks eating the ducks' eggs and naturally concluded he had found the source of the decrease. But when all the skunks in the area had been eliminated, the ducks did not recover and in fact disappeared faster than ever. Two years later hardly one could be found. So a top-notch ecologist was called in and, after weeks of very careful sleuthing, he deduced that the skunks had actually been helping the ducks more than hurting them because, although the skunks occasionally ate duck eggs, their main diet in spring was the eggs of snapping turtles. So, with the skunks gone, the snapping turtle population had skyrocketed, inevitably proliferating the turtles' favorite sport of grabbing the legs of ducklings from underwater, pulling them down and eating them.

Even the World Health Organization botched its well-intentioned effort to control malaria in Borneo by spraying native villages with poisonous DDT before its effects were fully understood. Although the malarial mosquitoes were eliminated on schedule, roaches and caterpillars in the houses merely absorbed the poison, passing it on to small lizards feeding on them, whose nerves were impaired, reducing their agility so much that not one of them escaped being caught and eaten by cats. But cats are very susceptible to DDT and they all soon died, permitting forest rats to move into the houses, carrying with them fleas infected with bubonic plague. As if that were not enough, the roofs of the houses started falling in because the thatch was being devoured by the caterpillars who were no longer being kept in check by lizards. So it shouldn't surprise anyone that no native Bornean could think of any convincing objection when a gathering of local witch doctors was seen making incantations to summon back: good old malaria!

It has always seemed to me appropriate that tigers and zebras are striped with grassy designs, for they not only benefit individually by this vegetal disguise but it symbolizes the two-way exchange in which grasses directly and indirectly provide food and air for the animals while the animals return the favor by helping distribute seeds and conserving the grasses by limiting the trampling and grazing of the herds. Of course all such interrelationships evolve and, along with the walrus and the whale, who relatively recently evolved back to sea, several members of the vegetable kingdom who also had learned to dwell on dry land have likewise reverted to the deep and, with man's help, some of their species such as turtle grass ("more nutritious than wheat") are now serving as fodder for sheep and cattle on land as well as for sea horses in the wet meadows below.

A few insects, I am told, borrow and use weapons of defense from trees. One such is the sawfly larva, who feeds on Scotch pine needles and stores their toxic resin in two pouches in his foregut to squirt at ants and spiders who threaten him. And several microbes have evolved what may be the most vital predator-prey relationship ever observed by science, something convincingly demonstrated when a researching biologist took what's known as a nutrient broth and added to it some tasty paramecia and hungry didiniums (page 86). First the paramecia, feeding on the broth, experienced a population explosion lasting two days, whereupon the didiniums, preying on the paramecia, multiplied a third as fast for two more days, while the paramecia rapidly declined in numbers until they were extinct on the fifth day, all devoured by didiniums who in turn were starved out of existence by the seventh day.

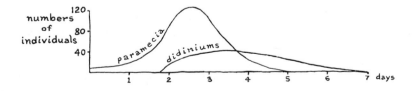

## PARASITISM

This sort of drastic interdependence of microbes evidently achieves its ultimate complexity in parasitism, which, as I mentioned in Chapter 1, has evolved natural chains or hierarchies, such as viruses that are parasites of bacteria that are parasites of certain very small mites that are parasites of lice that are parasites of egrets that are parasites of cattle that are parasites of man who is a parasite of Earth which is a parasite of the sun which is a parasite of the Milky Way which is a parasite of the Virgo Supergalaxy . . . with each host bigger than its parasite like a nest of Chinese boxes all the way (if enough were known) from the heart of the atom to the horizon of the universe. The numbers involved are staggering. When you allow for the vast multitudes in the microworld, at least half of all creatures are parasites in some way dependent on the other half, many of them host and parasite at the same time — with the evolving food chains or predation pyramids often dependent in the reverse direction. Great sharks, for instance, eat large fish who have fed on smaller and smaller ones all the way down to invisible plankton. Yet other creatures cut short or change the chains like certain whales who once grazed directly on plankton, then learned to feed on krill (who eat plankton), in effect turning the krill into spoons for easier plankton consumption.

Birds at first glance seem such carefree spirits you wouldn't suspect they are also zooming zoos of parasitism that include millions of passengers from fleas, leeches, beetles, flies, mites, ticks and feather lice to much larger numbers of microbes ranging from specialized flukes and worms to bacteria and viruses. Did you realize, for instance, that the eight thousand known species of birds are inhabited by twenty-four thousand less known species of feather-eating lice plus fifty-odd thousand species of other parasites, mostly beetles and mites? To the louse inside a quill, of course, the bird and its outer life of flight and song are completely unknown, as are your life and thoughts to your bacteria and viruses and the myriad other unseen passengers that inhabit your body, roaming the deserts of your wrists and arms, swimming your lymph, slithering through your intestines, exploring the cool woods of your scalp, wallowing in the tropical jungle of your groin. You may be surprised to know, for one example, that a weird eight-legged miniature "alligator" with long tail proliferates unseen (but not quite invisible) among the eyelashes of most people, particularly elderly females who prefer "cleansing" cream to soap. It is a nocturnal mite known to science since 1841 as *Demodex folliculorum* and it breeds shyly in the dark pores of the human face where it lives out its two-week life span, browsing nightly on the oily sebum of hair follicles, one of many uninvited but relatively harmless passengers scuttling about your person, some in your blood or your muscles but in this case usually just under the scaly squames of your skin.

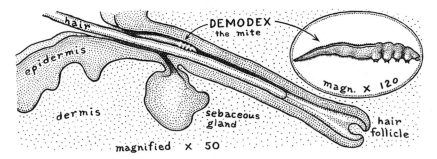

Some parasitic flies actually live inside the stomachs of horses, zebras and elephants. Some dwell in the eyes and nostrils of frogs. Some hijack blood from mosquitoes that are already gorged. Some live on such dainty morsels as the brains of ants, the tears of hummingbirds, or the antennas of butterflies. There are moths that live in the eye creases of cattle. There are more than 100,000 species of parasitic wasps, many of which live on other parasitic wasps.

In the sea large fish commonly get their teeth cleaned by small "cleaner" fish who specialize in this type of dentistry so they can feed on the crustaceans, fungi, lice, bacteria and other tiny organisms that inhabit or infect their big cousins. At least 28 fish species are already known to have taken up this and the related grooming profession, plus six species of cleaning shrimp, a crab, a worm, and a bird or two. And they are all in such demand that their big and itching clients usually have to wait in line, which, in the busy season, may mean all day.

Not a few parasites eventually contrive to outgrow their parasitism, as, for example, the gall insect that has the gall to ask the tree's help and gets it by laying its eggs (which imitate seeds) in certain tree tissues, causing a protuberance or gall that nourishes the larvae and often lives longer than the branch it is on — then, having fallen to the ground, puts out its own roots as if it were an independent, interkingdom organism.

The eating of one species by another is also of course a widespread interrelation which, on the mammal level, seems nightmarish to a    379 human. Yet, on lower levels, such as those of reptiles and amphibians, the animal being eaten seems hardly to care whether he is eaten or not. And there are still lower levels where the prey not only doesn't mind being, but actually yearns to be, eaten — cases that cease to be inexplicable as soon as you delve into the microworld of certain kinds of flukes and parasitic worms who swim up the blades of dew-laden grass and whose very survival depends so utterly on their being swallowed that they will die (without having reproduced) unless some animal of the right species soon picks them up in its mouth and "digests" them back into their natural intestinal habitat.

There is no end to interrelations anywhere. The polyphemus moth cannot mate until it receives a chemical sex attractant (called trans-2-hexinel) from red oak leaves. A mosquito messenger is known to carry eggs for a fly. There are "annual" fish who swim in deserts in the temporary pools created by the brief seasonal downpours and whose eggs will not hatch until they have been shriveled by a long drought before submersion in water. An oblong tomato has been designed and bred to fit a tomato-picking machine, which, in turn, enables it to survive and evolve. Similarly both vegetables and animals are turned into geological forces by such action as jungles that stem erosion and build soil, polyps that raise coral islands, beavers that change the shapes of rivers. Even this book, written by an American, is made of paper invented by Chinese and printed with ink evolved out of India and from type developed largely by Germans using Roman symbols modified from Greeks who got their letter concepts from Phoenicians who had adapted them partly from Egyptian hieroglyphs.

Thus do abstract threads weave the living tapestry of Earth, our familiar world in which lupines love volcanoes, where a rock may give a brook its song, a world where surgeons have recently learned to transplant organs (containing genes) from one body to another, thus literally binding closer the brotherhood of man and, in some cases, the cousinhood of animal and man. Even when they implant artificial organs that have no genes, surgeons are reducing the disparity between man and machine, between animal and mineral, possibly even between life and nonlife, a subject we will explore in depth next chapter.

Which is as it should be upon this itching node where vegetables and animals and stones confer and beckon to one another, even warn each other, implore, love, admonish, kill . . . How better could sweet basil express his disgust with bitter rue? How more eloquently might shy rose offer a bee her nectar? Or, with her thorns, deny the plucking hand?

# CHAPTER 14

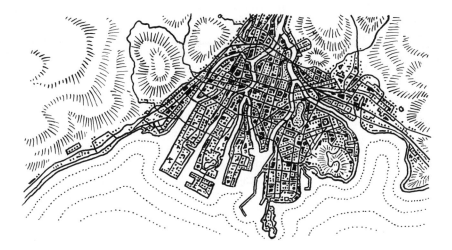

# Third Mystery: The Omnipresence of Life

IMAGINE AN EYE looking through a system of lenses upon a crystal growth spreading out in irregular cubic patterns. Slowly the crystal grows, like a snowflake extending and branching its arms and fingers into more and more complex shapes, all of them unique. Then suddenly there is a flash and most of the crystal disappears — yet it soon begins to grow again, faster and more luxuriantly than before.

What is this crystal? Is it frost on a windowpane or a culture of viruses? Is it alive?

Actually it is a city on Earth as seen through a powerful telescope stationed in space. It comes to the eye as from a series of pictures taken one per month over several centuries, then projected on a screen condensed into a time-lapse movie lasting a couple of minutes. It is Hiroshima, which was vaporized in 1945 but is now reborn and crystallizing faster than ever, a tiny sample of life as seen from outside  381

— from out here between the worlds — and, incidentally, by making use of a technology that is well within the present capability of Earth.

I have set myself the task of defining life in the broadest possible perspective, you see — something I don't expect will be easy. Let us take a city, any city. Considering it as a whole, is it a living organism? Engineers have estimated that American cities are now rebuilding themselves on a 35-year cycle, European ones a 50-year cycle, oriental ones perhaps a 75-year, these being city metabolism rates that have speeded up as the cities matured and which suggest basic attributes of life along with such urban physiology as water and sewage systems, arteries, tubes, power lines, communication nets and other parts to be expected of large viable organisms.

A meteor blazes across the black, starry sky. Can we regard it in some sense as alive? A trickle of water meanders like a snake down a dirt path, sensitively searching for the lowest ground. Is it alive? Microscopic grains of dust swim jerkily in water in the phenomenon known as Brownian movement. Are they alive? Are stars alive? Are atoms and electrons alive? Is the ocean alive? What about the winds and the clouds?

Pygmies in the Congo say the forest itself is a being, for they can hear its treetops and its streams murmuring assurance to its human and animal children. There is scientific justification for this intuition too in the known facts of vegetable respiration and interrelated motion, the hormonal interdependence of branches, the groping of roots, and especially in the collective migrations of vast populations of trees and their inhabitants. Biologists also speak of pheromones circulating in the sea as the social hormones of that finite but endless organism. A thunderstorm, composed mainly of volatile gases, behaves something like a primitive animal — in growth, in circulation of wind and rain, in nerve response of lightning, even in reproduction which, as with the ameba, is simply the splitting off of new cells from the old.

Does the tree, I wonder, seem alive to the termites who burrow in it? Or the turtle's shell to the turtle — the turtle to the shell?

Greek philosophers of the sixth century B.C., who seem to have thought as deeply and with as little prejudice as any philosophers in history, taught that life is a natural property of matter, an inevitable manifestation of the truth that the world has always been alive. Thales led the way by declaring that all matter including minerals, gases and stars is alive. Then Anaxagoras propounded his panspermic theory that invisible "ethereal germs" are dispersed everywhere in the world, giving rise to all its creatures including man.

382   Two centuries later Aristotle went along with the panspermists to

the extent of saying "Nature makes so gradual a transition from the inanimate to the animate that the boundary between these kingdoms is indistinct and doubtful." To this Saint Augustine long afterward added a mystic factor with his concept that the world is full of "hidden germs (*occulta gemina*)" of an unfathomable spiritual potency that somehow spawn living creatures out of earth, air and water. Still later Descartes, after describing man's body as a kind of "machine made of earth," conceded to his critics that there must be a "ghost within the machine" presumably connected to it by the pineal gland.

Thus the consensus of philosophical opinion developed, continuing in this day, and it still largely accepts the universality of life, acknowledging that all creatures, including man, are made of the same materials as the world — while there also must be something nonmaterial in man, some ingredient that cannot be hit with a stick or despatched by a bullet. And, on the superorganic level of cities, the spreading conglomerates of brick and mortar are just as surely analogous of vegetables and animals with their stem arteries conveying nutriment and their developing "blossoms" that metabolize continuously while bustling barges hover about their docks like pollinating bees.

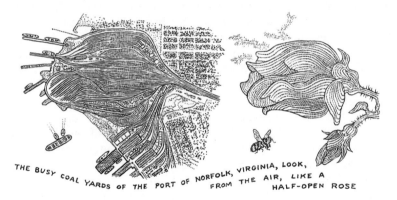

THE BUSY COAL YARDS OF THE PORT OF NORFOLK, VIRGINIA, LOOK, FROM THE AIR, LIKE A HALF-OPEN ROSE

## THE SECRET LIFE OF ROCKS

When it comes to rocks, understandably they are not considered alive by most people. But, as Omnipresence is the Third Mystery of this book, we are clearly called upon to search for life everywhere, including where it is least expected. And where would one expect life less than in a stone, which lies still on the ground as if dead without showing any of the signs commonly associated with life?

But let us not be hasty for, although stones seem inert, perhaps it is only because their life is geared to a tempo far slower than that of most vegetables, which in turn, as we noted on page 371, move about 40,000 times slower than comparable animals. Indeed I reckon a time-lapse movie of rocks in the large would have to speed up the passage of years and millenniums by at least another factor of 40,000 before one would likely notice much change or other signs of mobile life.

Yet rocks do experience a kind of life, even a metabolism of sorts (which we shall return to) — and I am not referring to anything like "flowering stones," the pebble-imitating plants that have evolved into hundreds of species in South Africa. If you have ever personally met a volcano or felt an earthquake, you must agree that rocks are not always passive. And what is the moon but a huge stone weighing 81 quintillion tons moving daily over our heads at better than a thousand miles an hour? Come to think of it, the biggest and fastest moving things we know of anywhere are neither animal nor vegetable but primarily of the mineral (or plasma) kingdom: a hurricane, a forest fire, an H-bomb explosion, a river in flood, the raging sea, the earth, the sun, the stars . . .

But mobility is characteristic also of cool, quiet rocks, which, in their own way, manage to get around. If they happen to live near active animal or human life, they may be pushed, plowed or kicked about, or even thrown. Growing trees may heave them aside, young mountains lift them or weather erode them. Some rocks (pumice for example) are so light they float on water and thus drift about the seven seas. Any rock as small as a pebble is likely to begin to travel, particularly if it is on a hillside or in a river or is picked up by a glacier that will break into icebergs. And pieces as small as sand can easily take to the air, as in a sandstorm, and migrate swift as the wind to virtually anywhere on Earth. In fact it hardly stretches reality at all, in a very broad sense, for us to start thinking of rocks as growing, basking, shivering beings with a limited but very real life of their own.

The authorities on this subject of course are mineralogists, who, with the aid of their modern arsenal of instruments, have accumulated conclusive proof that some rocks literally do grow, that some ungrow, some glow, some radiate, some disintegrate, some (like asbestos) are hairy, some (like oil shale) are insolubly organic and some even get ill (perhaps from poison) and later return to good health. You probably have seen sick stones yourself in the form of calcareous carvings on old city buildings, which are commonly crusted with sulfurous dust from smoke and exhaust fumes, which eventually thicken into blisters or cankers exuding sulfuric acid that eats inward until rain and wind erode the scabs, allowing the powdery decay underneath to ooze like a leper's sores.

Modern stone doctors of course are learning how to cure and prevent such ailments, along with their accelerating knowledge of the mineral kingdom, which presently recognizes about 1500 species, each with a characteristic form that is the outside expression of a highly organized interior. The Linnaeus of the mineral kingdom was James Dwight Dana, professor of mineralogy at Yale in the nineteenth century, who classified rocks and other minerals on such a sound chemical basis that it has become the accepted standard for the world. As with animals and vegetables, several new species of minerals are discovered and approved each year but, unlike the overall increases in these other kingdoms, their net number decreases during this early period of chemical science as old mineral "species" keep having to be demoted to mere "varieties" when it is revealed that they are chemically similar to other kinds that originally seemed different.

The changes in size, position and composition that take place in rocks are often surprising even though orderly and germane to their lives. One is apt to assume, for instance, that grains of sand come from the simple polishing away of boulders and stones into gravel and smaller pieces as they are washed down mountainsides and swept seaward in streams or beachward by the sea, but a little reflection will show that by that process it would take a shipload of rocks to make a pint of sand. What really happens, according to geologists studying erosion, is that sand is born in the "chemical and mechanical disintegration" of large masses of gneiss and granite, most of it due to weathering, freezing, glacial friction, expansion, contraction and the subtle gnawing of tiny mollusks in the sea plus lichen and other vegetation, most noticeably on dry land. Thus the fact that smaller and smaller grains are found as you go downstream is not because of the water's polishing action (actually negligible) but rather because the slowing current permits progressively smaller particles to settle to the bottom in accordance with the hydraulic engineer's rule of thumb that "the carrying power of a stream varies as the sixth power of its velocity."

Thus sluggish pebbles generally appear in brook beds while light silt and clay, moving swiftly in suspension, reach estuaries and the ocean in large masses. Intermediate grains of sand may be almost anywhere between, traveling at their individual paces, bouncing along the bottom from pool to pool, resting perhaps a year in a shallow one, a century in a deeper one, a millennium in a very deep one, awaiting the scouring action of a rare flood to boost them free. A typical river may take something like a million years to work its sand a hundred miles downstream at an average rate of six inches a year or fifty feet a century.

Although not whittled by abrasion in rivers, rough young sand grains

385

do gradually get their corners rounded and their surfaces polished by the chemical action of water, so slowly however that, as one geologist reckoned it, a half-millimeter cube of quartz would need to be swept down a raging torrent for more than a million miles before it could possibly be smoothed into a sphere. Surprisingly, air is a lot tougher on such an object than is water and, sooner or later, air is likely to get its clutches on it, for this common size of sand produced by ice disruption is, seemingly by divine intent, just small enough to be easily whisked up by a gust of wind yet large enough to be abraded by it. Indeed when the wind whirls coarse, angular crumbs of rock against each other, they grind away hundreds of times more mass per mile than they would in a river. Round grains, however, wear less than jagged ones and small ones less than big ones while spherules of quartz only one-tenth millimeter in diameter bounce off each other like billiard balls, leaving no trace of abrasion under the closest microscopic inspection.

Whether carried by water, air, ice, avalanche or bulldozer, most sand eventually comes to rest on the ocean bottom in vast beds (sometimes a mile thick off a river mouth), where it "sleeps" for eons, gradually consolidating or "pupating" like a spent caterpillar into sandstone or gneiss, which may still be roused, hundreds of millions of years later, metamorphosed like a new butterfly, in the diastrophic upgrowth of young mountains. In this way it achieves a kind of rebirth with crystal "wings" and the prospect of eroding into sharp fresh sand to start flitting seaward on wind or stream all over again, an almost unimaginably slow cycle of reincarnation, which nevertheless has been deduced to occur naturally perhaps many times in the strange, immortal life of most minerals, not the least of them common sand.

And besides this outer existence of the mineral it also lives within like other creatures of Earth to some degree, if more slowly — in its own way feeding, growing, healing its wounds, even procreating offspring! As you may have guessed, it is the crystal structure of the solid mineral that gives it these attributes of life, the essence of a crystal being the steadfast equilibrium of its lattice skeleton. In fact the crystal is so constructed that it always tends to maintain itself in stable balance, automatically restoring its shape whenever it is forced a little out of line in any direction. In effect it "wants" to hold onto the exact anatomy it has got — and maybe annex more of the same sort of structure if offered half a chance. That is why a string left hanging in a bowl of sugar water overnight will assemble rock candy around it. Once the sugar molecules start crystallizing into solid form, the crystal is predisposed to continue accepting these moving molecules from the surrounding liquid and fitting and locking them to its solid body. One might say it is a mineral version of drinking soup.

It is also a healing process because any hole or scratch on the crystal soon gets filled with the same kind of molecules that compose the rest of its body.

And it is a reproductive system as well because the molecules that construct the layers of growth, if they are to be accepted at all, have no alternative but to attach themselves at exactly the correct angles for whatever the substance is: 90° in the case of salt or sulfur, 60° in snow or quartz, odder angles in odd crystals like axinite or rhodonite. It could even be called a rudimentary genetic process because the crystal lattices themselves serve as "genes" in admitting only one specific kind of molecule to fasten and grow upon them. Very probably it was life's first and simplest reproductive technique on Earth (page 448). And it has evolved dozens of different (apparently organic) forms such as the pealike clusters of bauxite crystal, the hairy ones of asbestos, the sea-urchin-shaped radial globes of wavellite, the "asparagus sprouts" of limonite, the foliated nuggets of copper, the nervelike branches of psilomelane and the serrate leaves of muscovite — all of them crystal species with growth habits that dramatically reveal their kinship with the rest of life.

ORGANIC ROCK FORMS

bauxite    asbestos    wavellite    limonite    copper    psilomelane    muscovite

The fact that very hard stones called whewellite (found in coal beds), weddellite (discovered in Antarctica) and struvite (magnesium ammonium phosphate hexahydrate) all commonly grow from "seeds" in human kidneys and bladders, building themselves up layer by layer like their crystal counterparts outside, is just further evidence that rocks can be a very intimate part of life — even *your* life. And viruses exemplify this even more, being present in virtually all organisms from bacteria to whales and now proven to be inert crystals when dormant yet, when the right amounts of moisture and warmth awaken them from their stony slumber, they spring eagerly to life, invade other beings, reproduce themselves and evidently feel utterly at home in each and all the kingdoms.

387

If viruses then are animal-vegetable-minerals, combining attributes from the three most accepted kingdoms, the same can be said of soil — indeed of Earth herself and presumably of other planets. In support of this, a random ounce of fertile soil has been found to contain about 1 million algae, 30 million protozoa, 50 million fungi and 150 million bacteria, some of whose spores invariably will survive being dried up for years, doused with poison, frozen solid or boiled for an hour. And life's capacity for natural recovery from an atomic holocaust was demonstrated in 1964 when scientists waded ashore on the remains of Namu, a coral isle of the Bikini atoll whose entire top had been blown off by an H-bomb in 1956, and found it covered with sedge, beach magnolia, morning-glory vines and the white-blossomed messerschmidia tree with many kinds of birds flying gaily about, singing and raising their young, insects buzzing and burrowing, fish swarming in the lagoons . . .

If we add to the three familiar kingdoms of animal, vegetable and mineral the celestial kingdom of hot plasma, our resulting four-kingdom animal-vegetable-mineral-plasma relatives will obviously include the blazing suns that spawn the planets, more distant cousin stars and, by extrapolation, all the galaxies and supergalaxies to the farthest reaches of the universe. Every star, by this reasoning, should be at least as alive as a rock or a grain of sand. And Earth we are part of, all the more so. In fact, as I look at her I cannot but think of the uncountable influences that live and diffuse and sweep across her cloud-churned skin — forces subtle and swift that, in essence, may not be so very different from germs on an apple. When you see an apple rotting on a shelf over a period of a few days, the wave of brownness that creeps over it is quite literally a tide of bacterial generations advancing across its latitude and longitude like a population explosion upon a planet, only a million times faster — giving you a rather startling realization of the potency of demographic forces even after making allowance for the bacteria's capacity to evolve as much in a day as man does in a millennium, which, more than coincidentally, is the approximate frequency of each gyration of genes throughout mankind.

## THE LIVING EARTH

The earth thus rolls through invisible space in hushed but potent ferment — a something embedded in nothing — a sphere that is a

388   spore upon which the mysterious excrescence of life suffuses and

smolders inexorably, kingdom after kingdom welling out of the void — lonely atoms proliferating into gases that condense into dust from which rocks are cooked and crystallized and their juices pooled into oceans that brew microbes that sprout weeds that mutate and thicken into trees that build islands that spawn and nourish animals that evolve mind and spirit, including the curious new mammal man who rather incredibly (if not mystically) emanated from trees and fields and was sown like seed over primordial Africa and Asia and blown across the Pacific by the willful typhoon in countless unplanned one-way voyages followed by treks that eventually implanted him upon every continent.

Naturally we cannot expect too much of man yet, for he is still so young and new here. Indeed he has barely begun to waken to his own existence within the planetary organism. His halting and largely unnoticed attempt to recognize himself — something no other creature on Earth has ever been known even to undertake — may, if it succeeds, become the miracle of the millennium. Looking back into the microcosm, if a germ cannot be presumed aware of the living state of the body it dwells in, how can man's somewhat similarly circumscribed view afford him much more comprehension of the total aliveness of his planet today? Yet a beginning of sorts seems every now and then very casually, if not inadvertently, to be made. I remember a day in 1936, for example, when I was piloting a small airplane over Illinois and inexplicably caught something I like to think may have been a glimpse of Earth's super-animus. I noticed a country road far below me, along which a car was moving intermittently. I guessed right away that the driver must be a rural mailman on his delivery route, for he was stopping at every mailbox. And then it struck me that his tempo and function were curiously like those of a bee visiting flowers beside a garden path, the only obvious difference being that the man's stops were hundreds of times farther apart in space and time. A few miles and minutes later I saw bug-sized people bustling to and fro in the street of a town, now stopping to confront one another, now moving onward. To me they might as well have been ants. Naturally I could not hear their words or see their nods or glances or passing fancies — but neither would I have detected conversation between real ants, who often seem to spend just as much time at it — "ant time" that is, which presumably flows faster (page 21).

So who can deny that any creatures from viruses to whales are part of the body of their planet? And who can fail to recognize terrestrial traffic circulation on any scale or tempo as a real symptom of the living pulse of the flowering superorganism Earth? For there are wave effects between vehicles on a crowded superhighway that nicely express the metabolic rhythm of Earth's outer body — like the car I heard about that stopped suddenly in the left lane of heavy one-way, four-lane traffic, causing a screeching blockage of hundreds of others behind it before it turned into the next lane to the right, followed by another jam. Thus in a series of jerks, it worked its way across all four lanes until finally it disappeared up an exit ramp. But thirty-five minutes later, according to a traffic engineer's report, the waves of congestion the car had created were still visible to a helicopter hovering a mile overhead.

But even the ancients appear to have understood that planets are alive and besouled. Indeed that premise probably founded the venerable science of astrology and was surely a motivating factor in Kepler's propitious search for the laws of planetary motion. Why then should we find the idea of an actual organism of celestial magnitude so unthinkable? After all, we learned from the likes of Leeuwenhoek and Pasteur that there is an unsuspected, unseen world of living creatures in a drop of seawater, in a drop of rain, of blood, of saliva. Why could not some genius of an astrobiologist one day discover an equally unexpected form of life in the Milky Way? The obvious similarity between telescopes and microscopes testifies to the essential symmetry of the greater and lesser worlds. If they also turn out to be homogeneous why shouldn't an astral variant of the conditioned reflex be found a factor in the triggering of an exploding star?

With this thought in mind, it is most enlightening to scrutinize Earth as we see her from out here: an organism (specifically a superorganism) basking in the nourishing environs of her paternal star, the sun. The nourishment of course comes in the form of heat, light and other radiation streaming continuously out of the sun along with actual atoms and molecules in what is called the solar wind. And, like smaller creatures, the earth literally stirs in her "sleep," the while quietly breathing in her vegetal way, her skin slowly wrinkling, her sore spots volcanically breaking out but almost as quickly healing over again, her magma juices circulating, her lithic flesh, skin and breath metabolizing, her electromagnetic nerves flashing their increasingly vital messages. Intermittently she also rumbles with confined internal gases, utters earthquakes, itches a little, dreams strange dreams and (through her inhabitants) feels self-conscious.

At first glance one notices only obvious things like the swirls of clouds rolling around her plump form that floats upon the spangled shroud of space, the steady advance of the soft edge of night, the buff areas of great plains and deserts, the green of shallow seas deepening toward blue with the rise of tides. But when one remembers recent

390

discoveries of science regarding continental drift and plate tectonics, one begins to realize that her continents are dynamically related, even if their motion is almost unimaginably slow, each land mass acting as a kind of terrestrial flipper with its leading edge ahead, its trailing edge behind, as it advances a few inches a year under the complex pressures of planetary circulation.

What goes on thousands of miles below the crust isn't easy to find out, but long-range measurements of seismic waves and a great deal of chemical and geological deduction have given scientists a fair idea of the flow patterns of magma between the inner core and the outer crust. The core is a ball of solid nickel-iron bigger than the moon immersed in a molten outer core, and there seems to be an oscillating transfer of spin between the two with the heavier material still slowly settling toward the center despite a very gradual weakening of gravity. But above the outer core and reaching all the way to the crust is what's known as the mantle. It contains five sixths of Earth's volume, two thirds of her mass, and is made of red-hot olivine laced with garnets and other rocks, which collectively churn and twist and turn, mainly clockwise north of the equator and counterclockwise south of it, including rising columns of the hottest material and settling regions of cooler stuff, irregularly pushing volcanic islands up out of the sea, wedging apart huge crustal plates 60 miles thick containing continents and seas, even splitting a few into groping fingers that drag whole mountain chains with them.

There are some pioneers in symmetry who theorize that the earth's magma flows in dynamic crystal patterns similar to those of the five Platonic solids, an idea said to be supported by the new science of cymatics that was instituted to explore the relations between fluids and vibrations. In any case, Earth's tectonic activity seems to have got going about two billion years ago when Earth's crust had become brittle enough to break into plates. No one knows what a world map would have looked like then, but careful study of the location of the oldest known rocks and their magnetic lineation offers solid clues to the direction and movement of the magnetic poles, suggesting that large sections of the crust have been drifting around more or less independently ever since — bunching together, however, every few hundred million years into what is known as a pangaea, a world supercontinent with huge mountain chains scuffed up at points of collision, leaving a superocean on the planet's other side, then separating again for perhaps another half billion years. In such a cyclic progression Earth's continents reached what geologists consider a maximum separation around 500 million B.C., then gravitated together 391

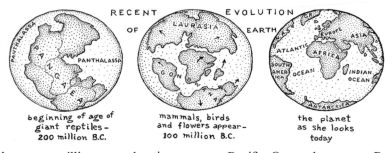

RECENT EVOLUTION OF EARTH

beginning of age of
giant reptiles -
200 million B.C.

mammals, birds
and flowers appear -
100 million B.C.

the planet
as she looks
today

about 250 million B.C., leaving a super-Pacific Ocean known as Pan-thalassa opposite the last great pangaea.

During this most recent eon of world collision the Appalachian mountains were shoved up when Africa rammed North America, but much of the resulting supercontinent was marshy and it was the age of giant reptiles, pines, tree ferns, palmlike cycads and ginkos. Then Pangaea slowly split apart for the last time (to date) when a huge north-south trench eased the Americas away from Africa and Europe and grew into the Atlantic Ocean. This redistribution of her drying land masses of course led to Earth's present configuration, as India and Australia broke away from Antarctica to drift northward, Mada-gascar left Africa and Arabia joined Asia. It also had a diversifying effect on evolution, as the separating lands with their animals and plants got isolated for long periods, especially in South America and Australia (which missed the invention of the placenta), where mam-mals, birds and flowers, evolving independently, created some of their most distinctive forms.

The various aspects of superorganism Earth are so involved it isn't feasible to describe them adequately here, but I can't forgo mentioning that her magma circulation with resulting continental drift and inter-mittent earthquakes have been enough to produce a measurable wob-ble in her rotation — which in turn alters her circulation, including its oceanic and atmospheric effects — and all such factors influence her magnetism and external radiation belts. There is no clear consensus yet as to why Earth's magnetic field spontaneously but irregularly switches polarity from north to south on the average every four hundred thousand years, but (as the diagram shows) it has happened at least ten times in the last four million years and the same normal rate has continued for at least 72 million years before that.

MAGNETIC FIELD REVERSALS ON EARTH

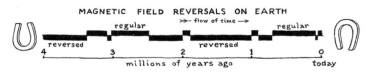

>>— flow of time —>

regular

regular

reversed

reversed

4    3    2    1    0

millions of years ago

today

Internally the field appears geared to the planet's axis of rotation with various overlapping lesser components that shift irregularly, yet have been measured to move westward at an average rate of more than a mile a month, so as to complete a full revolution about every 1600 years — suggesting that the Earth's moving core and mantle exert a force upon each other comparable to the armature and field coils of an electric motor. Her poles, both geographic and magnetic, have likewise been found to shift, in this case only a few inches a year but going back at least to 700 million B.C., when the north pole seems to have been in what is now Arizona, then probably a shallow area of some long-lost pre-Cambrian sea. Earth's magnetic changes also seem to affect her climate and life. The last major polarity reversal (700,000 years ago) was accompanied by a copious tektite shower, many extinctions of sea creatures and an ice age. And the field's shape in outer space resembles the corona around a comet because the solar wind, a perpetual out-blast of ions and atoms streaming past Earth at 250 miles a second, blows the field away from the sun, giving the scarcely visible aura of the planet a blunt head on the day side and a long auroral tail on the night side, at the same time as a positive charge

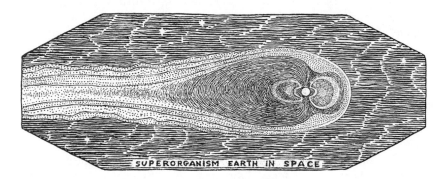

SUPERORGANISM EARTH IN SPACE

(from excess friction) along the morning edge and negative charge (from reduced friction) along the evening edge, with lots of complex folds between — the whole great nimbus fluttering majestically across the sky, a medusan figure so alive that hydromagnetic waves a thousand miles long ripple through its plasma at thousands of miles a second.

As to Earth's overall organic nature, she is clearly a very fluid thing, not only in her internal (semisolid) and outer (gaseous and plasmic) circulations but, most significantly, in her surface (liquid) circulation. Indeed, as the wettest planet known, her oceans, lakes, rivers, glaciers and atmosphere contain somewhere around 300 million cubic miles of

393

water, 300 cubic miles of which evaporate every day, enough to lower the world's sea level a tenth of an inch (were it not continuously replenished by river and rain), which means there is a complete cycle of water circulation through Earth's land, sea and sky every three millenniums. Other cycles of course circulate her energy, mostly in the form of angular momentum, radiation and heat, and there are cycles of such elements as oxygen, carbon, nitrogen, sulfur and phosphorus. The energy cycle of the biosphere naturally participates in the atmosphere and oceans, while its solid part diversifies into two main food chains: the grazing-browsing chain and the decay chain, both of which partake of the tenth of one percent of the energy Earth receives from the sun that is fixed in photosynthesis which is the chemical source of virtually all organic matter ($H_{2960}$ $O_{1480}$ $C_{1480}$ $N_{16}$ $P_{1.8}$ S) on the planet. As for how long it takes for all water on Earth to go through animal and vegetable cells, particularly through photosynthesis, it has been estimated at two million years. But oxygen that's not bound up in water is free to complete its cycle a thousand times faster: in two thousand years. And carbon dioxide faster still: 300 years, a period that is progressively decreasing as man's lungs, chimneys and exhausts keep pouring more of it forth every year.

These and many similar circulations characterize the Earth organism, she who integrates them all into her subtly tempered body and being, she who is a celestial cell enclosed in the membrane of life. At the least she now seems to be becoming better understood in this germinal century, as was recently intimated by astronaut Michael Collins, who, when asked how Earth looked from the moon, responded with the single word: "fragile."

Fragile might describe many a human being too, particularly if one were to consider the hundreds of different kinds of juices, serums, lymphs, plasmas, blood, bile, gastric acids, secretions, microbes, enzymes, insulin, sweat, tears, lubricants, etc., that course through his body to the beck of nerves, glands, reflexes, genes, antigens, antibodies and other controls. Perhaps even more fragile would be the whole superorganism of mankind, whose single species is now spearheading germination on a maturing planet. For fragility is characteristic of life, especially of any complex, sensitive organism composed of interdependent parts that must maintain for itself a constant temperature, humidity, pressure, equilibrium, metabolism, muscle tone, peristalsis and waking consciousness.

The point I am coming to is that this is exactly how it is with planet Earth orbiting so blithely through the void. And should you still doubt she is alive, just consider that the fossil record shows she has maintained a reasonably constant temperature for all her four billion years despite the fact that she now receives somewhere between 1.4 and 3.3 times as much heat and energy from the waxing sun as she did in her early years. It seems to me logical that if a mammal can maintain its life by keeping its body at the same temperature night and day, winter and summer, decade after decade, that Earth's similar ability should be looked upon as serious evidence of her being alive too. In any case there are two research chemists in England who definitely share this view: James Lovelock and Sidney Epton, who improbably launched an intensive study of Earth's dynamics in 1972, including chemical analysis of her atmosphere and soil throughout her existence insofar as the evidence avails. They have already found reason to believe our planet most likely stabilized her temperature in her early life through the heat-absorbing nature of ammonia and other complex molecules in her atmosphere at the time, which acted on sunshine like the glass in a greenhouse, and through vast populations of very primitive organisms that consumed or rejected the ammonia according to whether they wanted to be warmer or cooler; that later there evolved algae that could change their color from dark to light (or vice versa) in order to reflect (or absorb) sun energy; and, after the advent of oxygen, respiration and photosynthesis, there arose a regular explosion of organisms that learned to manipulate the atmospheric concentration of carbon dioxide, another very significant heat-absorbing gas. At least these and other modes of adaptation to her evolving atmosphere seem to have enabled Earth to survive the growing exuberance of her parental sun as well as the shock of oxygen, which must have been a fearful pollutant when it first appeared and, they reckon, just as poisonous to the primordial ferments of two billion B.C. as chlorine gas would be to us today, literally exterminating whole orders of life and driving others into sealed oxygenproof hide-outs from which the wary survivors still dare not emerge.

And the fact that Earth's present atmosphere is composed largely of oxygen out of the transpiration of living plants, plus nitrogen from living soil microorganisms, each source being alive as well as having a hand in maintaining the air we breathe, is a testament to the vital essence of her life. All in all, say Lovelock and Epton, the atmosphere of Earth behaves so contrary to the recognized laws of chemistry that it looks like "a contrivance put together cooperatively by the totality of living systems to carry out certain necessary control functions." In other words, all living organisms, along with all the air, sea and land environments on Earth, evidently compose one giant integrated system that somehow manages to keep aware of the material conditions that are vital for its survival and takes extraordinary measures to make absolutely certain these will not fail.

395

## Universal Life

If Earth thus presents a good case for a living planet, and the solar system is a fair sample of one of some two hundred billion star families in the Milky Way, itself a run-of-the-mill galaxy among uncountable billions of them in the observable universe, there is every reason to suppose life is to be found elsewhere, if not everywhere, and that it is simply inherent in nature. Even blazing stellar plasma, which seems so inhospitable at first, possesses an intense energy that must provide a star with circulation, metabolism, growth and other dynamic attributes of life. Indeed it has long been assumed that stars, like other organisms, must somehow get born. And the theory developed that they were spawned in huge gas and dust clouds that gradually condensed and heated up until clots were formed that eventually got big enough to ignite them into globes of light. And now just such embryo stars have been actually discovered, a good example being a hydrogen cloud known as IRS-5 (for infrared source number 5) that is much bigger than the solar system and emits 30,000 times more energy than the sun, yet it is still cool (170°F.) while slowly conglomerating into what, in perhaps a mere 10,000 years, will become an exceptionally bright newborn star like others recently born in its neighborhood in the outer Milky Way that astronomers are beginning to refer to as a celestial "maternity ward."

A little later these swaddling stars should begin to experience the stellar version of growing pains and childhood diseases. In fact the late Sir Arthur Eddington once specifically stated that "pulsation is a kind of distemper that happens to stars at a certain youthful period. After passing through it they burn steadily. There may be another attack of disease later in *life* when the star is subject to those catastrophic outbursts: . . . novas" and supernovas.

I admit such symptoms don't prove any star is alive, but it is also true that, since the astronauts started voyaging to other worlds, no one can reasonably claim life is confined exclusively to Earth. Furthermore, chemists are coming up with convincing evidence of life in the interiors of meteorites (page 543), which, to the scientists' surprise, have turned out to be on the average about a million times more organic than the estimated average for Earth herself. Also a new space chemistry, established under the name of molecular astronomy only in 1968, has discovered in the vast clouds of space-dust drifting perpetually among the stars not only sand, ice crystals and tiny diamonds by the quadrillions but all sorts of organic compounds from methane and ammonia to fuel, such as wood alcohol, formaldehyde (a potent preservative), formic acid (the juice of ants) and dozens of other combinations of carbon, hydrogen, oxygen, nitrogen, sulfur and the other common constituents of life. Besides, the latest spectroscopic analysis shows that these errant elements are not just floating at random but rather selectively interacting, progressively and panspermically building up large, stable molecules and crystals, even amino acids (as in the meteorites), polyamino acids and, some biochemists think, protein and possibly viruses. In other words, what are called dust grains in space are really miniature worlds with tiny iron-rich cores surrounded by silicates, ice and an outer mantle of organic compounds. And it is now considered more than likely that these little seeds of life just naturally grow and grow, accreting by collisions over hundreds of millions of years until eventually the huge clouds of them get to be more than dense and rich enough to constitute a sort of ferment of space that can spawn actual flora and fauna in any halfway hospitable world they either come to or become.

Astrophysicists and exobiologists have not yet said categorically where this space dust originates, but the more complex and potent ingredients in it are thought to be compounded in exploding stars, particularly in the tremendous supernovas that flare up only a few times per century per galaxy and rarely near enough Earth for human astronomers to see them clearly. The heat and violence of such explosions, in which a star's brilliance increases many billion times in a few hours and its material shoots outward at speeds exceeding 100,000 miles a second, are known to be sufficient to transmute any light elements like hydrogen and helium into heavier ones, including the few dozen that are considered essential to life. Whether or not other kinds of crucibles (sun spots for instance) may later be found to be also brewing life, it seems certain that life must already be rather well documented in the mystic sanctums of the universe. Surely life has never played second fiddle to anything else anywhere as far as we living reporters can find out — and, for the life of us, we haven't managed to discover even a hint of a serious rival. Obviously I don't consider the Deity a rival, whatever He may be, because He is plainly on the side of life, even if His essence is purely spiritual.

## THE GALAXIES AND BEYOND

Moreover, thinking materially, I see life not only in the stars individually but at least as much in stars collectively, in the Milky Way and all the other galaxies containing them by the billion. For these tremendous bodies are also born from amorphous clouds (clouds of stars as well as of gas and dust) and they go through an even longer evolution with their own distempers and explosions, as we shall see, gradually congregating into loose spirals of one sort or another that progressively tighten into ellipsoids and finally stable spheres. Of course the dynamics of galaxies are more difficult to understand than are those of stars, but it is now surmised that most spiral galaxies suck material inward through their axles and scatter it outward from their

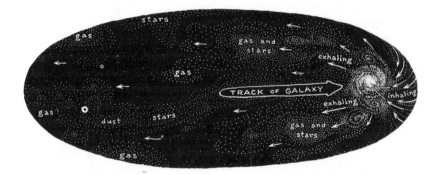

rims, an inhaling and exhaling of stars and gas amounting to a continuous metabolism with many complex manifestations of life, from a curious side-puffing of "breath" out of the galaxy's "head" end to a stupendous parturition of offspring stars that must eventually coagulate into juvenile galaxies elsewhere.

And all the while, inside each galaxy, the star and planet life is perpetually being spawned on a cosmic scale through the incomprehensible genesis of the afore-mentioned supernovas and their pulsar heirs, pulsars being a new class of stars discovered in 1967 that throb and beat with remarkable regularity several times a second. The most significant and interesting thing about the pulsar is that it is made of neutrons that got *im*ploded together into a kind of central atomic ash at the same moment and by the same action that *ex*ploded the rest of the giant star outward as a supernova, the complementary *im*plosion and *ex*plosion exactly balancing each other and leaving a residue of neutrons locked so tightly together they amount to a single giant nucleus of $10^{57}$ neutrons, every golf-ball-sized segment of which weighs more than a battleship, while a mass many times greater than the sun is concentrated in an asteroid-sized spheroid (perhaps ten miles thick), whirling like a top and radiating like a radio beacon with every revolution.

If a pulsar is the ghost of a supernova, a quasar may be something like a thousand supernovas in the center of a galaxy blowing up one after the other in a chain reaction. It amounts more or less to the explosion of a galaxy, a phenomenon so incredibly stupendous it was neither knowable nor believable until 1963 when Maarten Schmidt of Palomar Observatory, studying a faint, fuzzy "star" that gave out a very strong radio signal and a visible jet of matter, suddenly realized from its peculiar spectral lines that this body was not a star at all but almost certainly a very remote galaxy that was inexplicably a hundred times brighter than any previously known. That's how the class of mysterious objects that astronomers had been calling "quasi-stellar radio sources" — soon shortened to "quasar" — came to world attention as the most astounding astronomical development since Galileo's telescope revealed the moons of Jupiter and the rings of Saturn.

Discovered four years after the quasar, the pulsar came as something of an anticlimax, for its most vital function seemed to be to serve as an empyreal enzyme inside the quasar, which in turn must ultimately stoke or nourish the cell-plasm of the greatest celestial outburst ever dreamed: that of the whole exploding universe. And in the scant years since quasars and pulsars were discovered, astronomy has developed so fast that already thousands of quasars are known and ten million are estimated to be within the range of visual telescopes, including one that is 11 billion light-years away and receding at 91 percent of the speed of light, which makes it the remotest object ever sensed by man. But since they seem to burn themselves out fairly quickly (their probable average duration of a million years being, in galactic terms, a mere flash) it is calculated that there must be a total of at least 10 billion quasars in the known universe, almost all of them "dead" and dark. In fact the "dead" ones are so dark they are black and conjectured to be largely made of black holes, a black hole being the possible dénouement of a congregation of pulsars, therefore much denser than any one pulsar because its massiveness is so vast and concentrated and its gravity so unbelievably powerful that nothing, not even light or any other radiation, can withstand or escape it. For that reason, black holes have to be essentially lightless and invisible, created not by *ex*plosion but only by *im*plosion and on unimaginable scales, involving the collapse of the most massive of stars, sometimes in large numbers, their material falling inward and whirling down or into some sort of an internal drain into nothingness.

This has never been witnessed, mind you, for black holes are notoriously shy and elusive, seeing as they can neither be looked at nor tuned in on. Yet they are increasingly theorized about and more than a few are suspected of subsisting in our Milky Way, notably in Cygnus

399

and in the nearby multiple star group known as Epsilon Aurigae, where a bright star is associated with some tiny dots (likely planets) circling around a central something that is neither seen in visual telescopes nor heard on radio ones. Moreover black holes warp space and twist time and have a polarity and a dynamic structure, at least theoretically. I mean their rotation (always present in some degree) gives them an axis around which form rings and accretion zones (millions of miles in diameter) of inward-moving gas and dust that gets hotter and faster as it approaches the central drain, giving off x-rays and possibly gamma rays before the critical gravitation becomes too great in the final plunge.

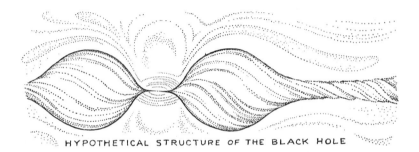

HYPOTHETICAL STRUCTURE OF THE BLACK HOLE

There seems to be a social life among black holes too, involving various kinds of meetings or matings between moving holes, one variety of which, after a "gestation" period, procreates offspring holes that can be very small — some astrophysicists theorize even microscopic — but which tend to grow and grow, swallowing each other like predatory fish, perhaps ultimately proliferating into a stupendous negative conflagration or star-population implosion that engulfs stars by the millions and may provoke such a violent reaction in the rest of the galaxy that a new quasar is set off.

Of course the star stuff that vanishes into the nadir vortex of a black hole must be presumed lost to any surrounding quasar, for there is no evidence, even in theory, that it can re-emerge or rematerialize anywhere in the universe. Indeed a few cosmologists have launched the hyperbolic idea that somehow it must seep completely away from the universe to "resurface" in what they call white holes in some very different, if not opposite, exoverse elsewhere. This should appeal to ecologists as an ultimate (if not cosmically charitable) solution to the pollution problem, but a truer explanation may turn out to be that the vast masses gravitating into black holes are actually diffusing into the microcosm in the form of gravitons, the subatomic gravitation grains

400

postulated by Einstein, the reality of which has recently been supported by impressive evidence (still unconfirmed) that waves of gravitation (previously unknown anywhere) are actually emanating continuously from the center of our Milky Way.

If attempts to measure the gravitational waves cannot solve the mystery of black holes, perhaps quasars and their holes will turn out to be part of the interaction between matter and anti-matter, anti-matter's atoms being composed not like ordinary matter of positively charged nuclei surrounded by negative electrons but rather of opposite negative nuclei surrounded by positive positrons. Some astrophysicists seriously maintain that, in order to satisfy the laws of symmetry, the universe must contain equal parts of matter and anti-matter within the total equilibrium of what may be termed worlds and anti-worlds in which dwell life and anti-life. One model recently proposed by J. Richard Gott III (which appropriately means God III) of Princeton even argues for three universes: (1) the familiar universe of matter and forward-flowing time represented as a cone whose apex meets (2) an opposite anti-universe of anti-matter and backward-flowing time, the two cones forming an hourglass-shaped figure, outside and around which lies (3) a tachyon universe whose particles are neither matter nor anti-matter but hypothetical entities called tachyons that by definition perpetually move faster than light.

Other and even more fantastic models have also been suggested, some of them more imaginative than reasonable and populated with all manner of queer luminous Ifrits, radiating behemoths, uranic monsters and fiends and anti-fiends of the firmament that appear and disappear like runaway stars farrowed and freed in the mass dispersions of quasars. And more and more inevitably the disparate regions of matter and anti-matter have to be insulated from each other by what are called event horizons or curtains of hot ambiplasma, forming a weird no-man's-matter of colliding positive and negative particles, amounting to a seam in the universe across which there is no cause-effect relation and no communication except (say some theorists) "spacelike" interaction that somehow extends through space without any corresponding extension through time. In a sense, if it really exists, such timeless communication must be *extra*sensory like ESP and would presumably have to involve superorganisms of otherworldly civilizations millions of years more advanced than our own rudimentary anarchy. Indeed we may now be under long-range observation, if not control, by spiritual worlds much more mature than we could possibly conceive, to whom we are comparable to ants or termites or even enclosed within some sort of a cosmic Petri dish, bassinet or

Promethean sanctuary — a prospect to which we will return near the end of this book. In any case, I wouldn't put it past ambiplasma to articulate a sort of node between the physical and mental worlds, perhaps in some still unimagined way defining the dimensional verge of body-mind or even, if you can stand it, letting out some secret stitches from the symbolic seams of finity-infinity, life-death, body-soul . . .

## OMNIPRESENT LIFE

Thus from the virus to the universe — or possibly from below the virus, perhaps even from below subatomic particles, which, some physicists have hinted, seem to be "indeterminate" to the point of willfulness — there are signs of life and the omnipresence of life, all of it interrelated and all made of the same elemental stuff. In a real sense, therefore, you are made of stars, of star dust and most surely of atoms given off by stars. Most of the matter in the universe in fact is now known to pass at some time through the caldron of the stars. You can occasionally see actual streaks of star particles in flowers — as, for example, a trace of white in the petal of a dark red rose caused by a mutation created by a cosmic mote out of Sirius or Alioth or Mizar! Besides, the phenomenon we call life is, in large perspective, just part of the quality of stars and therefore must exist potentially everywhere. Certainly it has no proven limits in size in either direction, large or small. It is omnipresent and universal and its progressive dimensions overlap in a continuous telescoping hierarchy of magnitudes, as shown by the fact that some atoms are bigger than some small molecules and some giant molecules are bigger than some tiny organisms, while other organisms (such as a whale) are bigger than some very small islands and larger islands (such as Borneo) are bigger than some asteroids (such as Hermes). There are also asteroids (such as Ceres) bigger than some moons (such as Phobos), moons (such as Ganymede) bigger than planets (such as Mercury) and planets (like Jupiter) bigger than stars (such as Sirius B) — and so on from giant stars (like Epsilon Aurigae I) to star clusters (small and big), galaxies, supergalaxies and the universe.

When you can really grasp the universality of such relationships you have gained a new insight into the ancient Serbian proverb: "Be humble for you are made of dung. Be noble for you are made of stars." Because you will have come aware by then that a cross section of the world sits on your table in the sense that every slice of bread and butter you eat brings you all three of the common kingdoms through vegetable grain, animal fat and mineral salt, while the mere fact that you are alive ensures you Earth's three states of matter, since you breathe the gaseous sky, drink the liquid rain, and your flesh and bones share molecules with the solid soil and rock of your planet.

In a way of saying, then, there is no borderline between you and the world. You are of it and it is of you — a kinship that reaches both ways and everywhere, genetically, mentally, spiritually, chemically, even gastronomically. For you are food for the world about as much as the world is your food. You are world food through all the earth's interconnecting food chains and pyramids of predation, through your excretion and ultimate decay and of course also potentially through almost any carnivorous creatures you meet from probable viruses and mosquitoes to improbable sharks, hyenas or vultures. Even carnivorous plants will gladly devour human flesh, as I once found out by teasing the palate of a Venus's-flytrap with some of mine.

A different understanding of how the three kingdoms participate in us all can come from reflecting that the developing human embryo reenacts the progressive dimensional phases of evolution: starting its unfolding like an inert mineral with the positional geometry of centrosymmetrical packing around one point, then furrowing into a vegetable with a symmetrical line or stalk sprouting from roots buried in the "soil" of the uterus, and finally splitting into an animal with handed bilateral symmetry astride the plane that halves its mobile body.

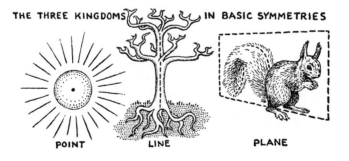

THE THREE KINGDOMS IN BASIC SYMMETRIES

POINT          LINE          PLANE

Thus the point-line-plane progression of symmetries unites the kingdoms through the dimension of time. But their interactions in the mind and spirit are far more subtle. How does it happen, for example, that a rose is beautiful to a human? Could the answer perhaps ultimately be found somewhere in the flower's delicately sensuous balance between the soothing and the fascinating in this human-vegetable cousinhood that unites two so different creatures?

403

## THE LAW OF EMPATHY

It is said that primitive man did not recognize any enemy smaller than a flea or larger than a storm, and that his real sympathies barely extended from the size of a puppy to that of a tamed reindeer, while beyond these near-human magnitudes in both directions stretched the great insulation of dimensions. By a sort of tradition or a natural law, it seems man empathizes with creatures like himself but tends to feel less rapport toward any that are very different in size, shape or speed of reaction. Thus men and women can feel close to a chimpanzee, a dog or a horse, even an elephant or a rabbit and sometimes a mouse. But "bugs" appeal much less to them, particularly if small and moving either very fast like a fly or very slow like a slug. And even a five-foot cockroach would not likely be greeted with open arms by the most ardent of animal lovers. It works the other way too, for evidently an insect is not capable of becoming a real pet (although I have heard of spiders who seemed to enjoy touching human flesh) because an insect cannot recognize the existence of a fellow being so different in size as a human. Humans are just earthquakes or storms to it, or perhaps (in the insect mind) humans embody "acts of God" so unpredictable, if not improbable, as to be of no concern. This presumably is why an ant accepts a descending human foot as an unavoidable stroke of fate like a hurricane without any evident comprehension of what, if anything, might be directing the foot.

Furthermore the very multiplicity of Earth's tiny creatures, aside from the difficulty of seeing them accurately, makes it almost impossible for a man to know them as individuals. Fleas are only specks to him and germs but invisible contamination. As for vegetables, we commonly consider certain trees and flowers beautiful, but their slowness of response makes them appear unfeeling if not virtually "dead"

404

to most of us. Yet, without undue worrying about whom I may be stepping on in the field or woods, intuition somehow whispers to me that an attitude of total disregard for my less conspicuous fellow passengers on this sphere amounts to surrendering to a primitive provinciality that can be considered a sort of blindness that education or spiritual enlightenment should be able to overcome, and that, as we gain a deeper perspective, we can actually throw off our old illusion that only beings close to us in size and tempo are really alive — or that finite space-time proximity to man is the universal criterion of being. Should one not, moreover, be humbled rather than shocked to realize that more creatures live on a normal man than all the men living on Earth? And should one not be awed by the evidence that these creatures live in a more stable balance with nature than men do and more than likely with a better moral reason for what they are doing to man than man has for what he is doing to Earth?

## Is There a Nonlife?

The varieties of evidence of similarity, parallelism and oneness in life continue on and on, seeming never to end. No sort of atom is peculiar to life or nonlife, and it is a remarkable fact that, although the genetic code was deduced mainly from studies of the colon bacillus, the code has turned out to apply equally well to plants, animals and humans. Of course this reflects the fact that some kinds of protein molecules have been conserved almost unchanged for billions of years, as they slowly diffused outward from the simple common ancestors of all "living" things into the complex evolving kingdoms we see today. It also explains something about a biologist named Von Baer who once got dozens of vertebrate embryo specimens in his laboratory mixed up and, in trying to sort them, discovered it was almost impossible to tell them apart, particularly in their early stages before snouts had differentiated from beaks, flippers from fins and hands from hoofs or wings. Somewhat surprisingly, where such creatures have similar ways of life in similar environments, even the adults of classes as different as fish, reptiles and mammals look much alike, showing that

COUSINS OF THE SEA

shark     ichthyosaurus     porpoise

FISH     REPTILE     MAMMAL

it must be the planet itself which, through its nature, largely molds its own creatures so they can maintain themselves and interact.

The abstract universality of life might be further exemplified by the five-year-old who found a "lonely worm" and, upon compassionately cutting it in two, was heard to say, "There, worm! Now you have a friend." Evolution also seems to have evolved a vital negative side in that the empty places left by individuals who die, or by species who become extinct, are inevitably filled by others who, just by doing so, turn the negative into a positive. This is true as well for negative memory or forgetfulness, because one of the sources of oak trees must be the acorns that squirrels bury a few inches underground but forget to dig up and eat. Here again the negative factor has survival value and even helps the squirrels themselves in the long run by providing the descendants of the forgetters with more oaks to live in and more acorns to bury for vital future winter food or for forgetting at least often enough to ensure the perpetuation of the oaks and squirrels.

In chewing over this abstraction, I think I can begin to see life, as an acquaintance suggested, functioning something like a leaky bucket of water that got filled up when it was dipped into an invisible sea at the moment of conception. In this analogy, life's essence of course is the water, which at first seeps very slowly into the bucket but, as the months and years go by, increases its flow, developing currents and eddies and occasionally some real splash. And after the water level rises to the top and overflows, the leaks increase and the influx decreases until the opposing currents reach a working balance, a congruent maturity of life that lasts until the outflow sooner or later begins to exceed the inflow, progressively lowering the level again, eventually draining the bucket to the bottom at death, when the water has all merged back into the sea whence it came.

This is plainly an analogy of mortal life on Earth, the metabolic process at work upon three dimensions of concrete matter and one of abstract time, all four dimensions naturally applying to the water while it is bounded by the bucket, but no longer applying after it shifts from mortality to immortality on being liberated back into the boundless sea. The analogy holds good also for mortal raindrops that wax and wane as they oscillate in the sky, now merging and mating as they grow up, now disintegrating as they fall and die. And it applies almost equally well to lakes, rivers, oceans, stars and even galaxies, to say nothing of dust motes in space or colloidal particles in a primeval swamp. All such bodies naturally metabolize as what you might call proto-organisms, at the same time moving about and competing with their rivals, the stronger outdoing the weaker, gaining more sustenance, more independence, more potency, more expectancy and more of the quality usually ascribed to life.

In this way life shows itself as a relative state, not absolutely distinct from nonlife and differing from it only in degree without any precise boundary between. Indeed life generally seems to be an attribute of alignment, of order, of crystallinity, improbability, complexity, indeterminism . . . When atoms combine into molecules and become arranged in some larger, more or less stable pattern, they may find themselves part of a snowflake, a geode, a snake's nest, a bubble of methane, a planet — and they may progress from one state or kingdom to another. Lodged in a solid, mineral material such as a brick, they appear quiet and predictable, presumably because any population of septillions of reasonably homogeneous atoms is bound to average out its random impulses into a statistical calm. But atoms inevitably drift about a good deal, especially when warm, and as they work into more complex and heterogeneous societies like soil or pond water or a gnu's hoof, they increasingly diversify and specialize, leavening pervasively from inorganic to organic molecules, increasing their biological division of labor and, most significantly, proliferating more and more of the double-spiral molecules now recognized as DNA which preside in the nuclei of living cells as the supreme specialists of the microcosm: the elite corps of genes that direct the common masses of molecules in a manner not basically different from the way mayors run cities or tycoons boss industries.

This, it seems, is how life is wont to emerge from its background dust, imperceptibly at first but, by little and little, unfurling nothingness into somethingness. In sum, in its mystic way, life may be nothing but the stirring and musing of the world. For, as a poet might say, it begins in snows and silent stones . . . and grows in the reachings of trees and bones. Life is everywhere: slithering with the snake through nodding reeds, threading the parched desert with the kangaroo rat, swimming with the ameba in a drop of rain. Even if we project our musings beyond the world, life quickens the planets, binding them without rope to moons, to suns, to the Pleiades . . .

## MECHANISM OR VITALISM?

If this line of poesy is high-flown, its elevation may at least raise the question of the viability of atoms and their long-presumed inanimate parts. Also it can hardly help but revive the age-old debate between the two classic theories of life: (1) the mechanistic concept

407

that life derives solely from the automations of molecules as they combine into more and more sophisticated complexities of material organization and (2) the opposing vitalistic view that life's essence resides in abstract forces not only beyond matter but completely outside the laws of physics and chemistry.

It is not an easy issue for, as has been said before, if vitalism challenges man's reason, mechanism disturbs his soul. And the resolution of this conflict is intimately bound up with the mystery of the whole universe.

The still prevailing view that atoms and their parts are without senses or other attributes of life is of course an assumption inherited from classical physics and typified by Lucretius' statement that "every creature with senses is nonetheless composed only of particles without senses." It is an assumption modern physicists have not to my knowledge tried to defend, presumably because it has never been seriously challenged. Yet, should they ever decide to defend it, there is increasing reason to doubt that they will succeed in view of several provocative new avenues of inquiry in science and philosophy.

One of the earliest of these was Douglas Fawcett's book *Divine Imagining*, which suggested in 1921 that the world is a psychical continuum connecting matter with mind, and blessed with a kind of incomprehensible diffusion of "Divine Imagining," which somehow so leavens the constituents of all material things that, however great the sums these parts add up to, the wholes they compose are inevitably greater still.

The astrophysicist Sir James Jeans, a few years later, really made an impact on hundreds of thousands of intellectual readers by intimating that the universe may no longer be viewed as just a great machine, as Newton portrayed it, but more aptly as a great mind. And it was Niels Bohr, the physicist, who obligingly brought the concept down to Earth by proposing that the newly discovered dual aspects of the microcosm (particle and wave) are analogous to the long-known dual aspects of the macrocosm (matter and mind). If you recognize this as a fundamentally revolutionary idea you are not alone, for it immediately provoked philosophical speculation among advanced thinkers that atoms and molecules can no longer be taken for granted as merely the nuts and bolts of the complex vehicle called a living organism. In fact it began to be realized that, from now on, all such parts right down to the level of single electrons must be looked upon as potentially, if not intrinsically, alive and almost inevitably serving in their respective ranks as vital units of the sublimely hierarchical structure of the living body-mind.

408

## ESSENCE OF LIFE

If we accept this extraordinary intuition of Bohr's as true, then we acknowledge that the dual body-mind aspects of all living creatures derive directly from the dual aspects of the atoms composing them. Which would make the "wave" function of an electron essentially a subdivision of the "mind" of whatever living organism that electron is part of. Understandably this is an apocryphal, if not shocking, concept to most physicists and even many biologists but, I ask you, why can't an electron be alive? I don't mean "alive" just by stretching the definition of life far enough to include it, although that is a factor, but alive in the sense that there is no verifiable line demarking it from the rest of life.

After all, no one needs to ask why a wire charged to a dangerous voltage with electricity is called a live wire. For electrons are far from inert. They not only have never heard of sleep but literally get around in the fastest company imaginable, careering all over the world in electric currents at 186,282 miles a second. They are endowed with a charge, a magnetic field, an ability to gravitate, resonate and even perhaps to choose their own path (within limits) under the law of physics known as Heisenberg's Uncertainty Principle. Their almost unbelievable rotation rate is something like a quadrillion times a second while "at rest" and "unexcited," which is more oscillations in that second than all the waves that have beat upon all the shores of all the earthly oceans in ten million years. This one-way megawhirl moreover gives them a kind of gyroscopic bias, the torque of "mind" that is actually a micro-version of the "desire" of the spinning top to hold its axis upright, a yearning made manifest en masse as magnetism, that macrocosmic force recognized through history in the patient polarity of the lodestone ($Fe_3O_4$), called by the ancient Chinese "the stone that loves" because it attracted iron and gave the compass needle its soul. Any particle that can display all these functions, it would seem, must be quickened by a "wisdom in the inward parts" mysterious not alone to you and me but emanating from a Source far beyond human comprehension.

Now to shift our attention from the mental side of the microcosm for a moment, let's look at the surprisingly sophisticated social life of the electron and its sister particles, regulated by the quantum caste system of seven concentric "galleries" of orbits surrounding the atom's nucleus as a kind of theater nicely symbolic of the seven ancient planets or the seven notes of the musical scale. Most of these    409 galleries are subdivided into sections in which the electrons circulate around the atom's central nuclear court, the lowest and smallest of the seven being the most exclusive and popular with its cozy "bridal chamber," fitted out for only two electrons, which, under Pauli's well-established exclusion principle, must be of opposite spin (the micro-equivalent of sex), the next gallery with four rooms of two kinds occupied by similar couples, the third with eighteen of three varieties, and so on up, all the electrons and occasional visitors (such as photons) having to obey very strict rules of behavior, particularly in the matter of choosing mates and moving from gallery to gallery.

The etiquette of this particle society, according to Henry Margenau, research director for the Foundation for Integrative Education, actually enables it to survive harmoniously inside the atom almost entirely by virtue of the fact that "one electron knows what the others are doing and acts accordingly . . ." This observation has particular meaning for the case when matter is cooled close to absolute zero, where the normal random thermal vibration of its molecules slows into an orderly calm (of relative negentropy), allowing the particles' wave properties to become dominant and therefore their mental aspects and presumed latent consciousness to assert itself.

An example of such glacial independence of micro-will may be the fantastic superfluidity of liquid helium at 4° above absolute zero, a component of which has been reported to flow in the form of "whole atoms completely without friction," slithering along as a film a few millionths of an inch thick that tends to cling to solid surfaces, yet may advance as much as a foot per second, even "siphoning" itself spontaneously out of cups. And certain metals at this temperature have revealed the phenomenon of superconductivity under which electrical resistance vanishes and the loose electrons that constitute an electric current fly onward through atom after atom as freely as the moon flies through "empty" space.

This sort of liberation does not stop at absolute zero either, for a kind of nadir "temperature range" has been discovered within the atom, manifested by nothing but the damped spinning of nuclei, which descends to colder and slower states after all motion of whole atoms and molecules (ordinary heat) has died away. This subabsolute temperature depth is very hard to explore of course, but physicists already know that some nuclear spins have more energy than others, while all spinning nuclei use their magnetic fields to influence one another. And we are finding out that the consequence of this influence is that both fast (high energy) and slow (low energy) spinning can spread from atom to atom in waves through subabsolute matter probably in much the same way heat and cold diffuse through normal matter. And, since these waves must surely be a function or aspect of the matter waves Bohr posited, some of your thoughts may be not only deeper and cooler than you know but they may be disseminating themselves through the world in waves that literally leave thinkprints all the way down to matter's subsubbasement.

It was a different sort of evidence of mind waves, however, that was researched a few years ago by Andrew Cochran, the biophysicist from Missouri who overcame his traditional skepticism by comparing the heat capacities of all the common elements (the amount of heat required to raise their temperatures $1°C.$). He did it in preparation for his thesis at the University of Missouri Graduate School on "the quantum-mechanical wave properties of matter in living organisms." Evidently he had noticed that substances of low heat capacity always have a correspondingly high degree of wave nature, and this suggested to him that such elements must have more of the quality known as "consciousness" than the higher heat materials. So when he checked the most widely used element in living tissue, carbon, and found it to have the lowest heat capacity of all the elements ($1.8$ at $0°C.$), he saw at once how this could explain carbon's extraordinary distribution into five times as many different compounds as all the other elements put together. For even though carbon's sociability had for a long time been logically attributed to the valency of its four outer electrons, there just had to be some profounder consequence resulting from carbon's ample consciousness. And does not consciousness, after all, signify life?

The element with the next lowest heat capacity turned out to be hydrogen ($2.3$), which, with carbon (and sometimes oxygen), builds the ubiquitous hydrocarbon and carbohydrate chains that form a large percentage of Earth's organic stuff, ranging from marsh gas to wine to wood. Then came nitrogen ($3.4$) and oxygen ($4.0$) among the seven most "conscious" elements, followed closely by phosphorus and sulfur, these being the six most important elements in living organisms.

Summing it all up, Cochran declared unequivocally that the wave-consciousness hypothesis "does not contradict any known fact." Furthermore, by his reckoning, "it is more consistent with the existence and evolution of living organisms and with the other known facts of science than is the accepted concept of lifeless atoms and particles." So he concluded that "what nature has been trying to say, through the . . . emergence of quantum mechanics, may be that all matter has a rudimentary degree of mind . . ."

Max Planck, discoverer of quantum mechanics, in effect had supported this concept from the first decade of this century, pointing out that an elementary particle of matter can no longer be considered to exist just in one part of whatever atom or molecule it happens to be in, for "in a sense, it exists simultaneously in every part of the system. And this simultaneous existence includes not only the field of force around the particle but also its mass and its charge." Thus the electrons and protons that are presumably the very plinth of matter are, like genes in the body, continuously influencing it as a whole, perhaps each in some still undreamed way holding a plan of the all, a mystic magnetic blueprint, even as your mind transcends space-time to guide the octillions of atoms that at any moment form your body.

It is like saying that the field is part of the cow who grazes in it because she literally exchanges matter with the field almost continuously (at both her ends), not to mention breathing the sky (out as well as in) and participating indirectly but ultimately in the entire universe. Obviously this is something of a real-life rendition of Shakespeare's transcendent boast that "I could be bounded in a nutshell and count myself a king of infinite space . . ." And in the memory of history it is an ecological version of ancient animism, a late form of which centered on the Pythagorean doctrine that the world itself has a soul, which was bequeathed to the Romans as *anima mundi* and revered by the Hindus as Atman. I have a feeling there is also evidence here that life is temporally symmetrical: if it has no beginning, it can have no end. And if it occupies any point in space, it must fill the universe.

# CHAPTER 15

# Life's Analogies on Land, Sea and Sky

I HAVE BEEN WATCHING the blue ball of Earth from this fair fetch in space for quite a spell now. Yet, strange to say, I have not gotten used to her. My consciousness somehow never quite lets me take her for granted, perhaps because she is too much part of me and me of her and her mysteries too immense for me to ignore or forget — ever.

More important, my growing awareness of the omnipresence of life has steeped my being with a sense of intimate identity with the universe. For in truth there is no dearth of life anywhere, even though the key to it is evidently a shy and unidentified force that lurks in secret recesses of our world, presumably beyond the reach of mere matter. I surmise that, strictly speaking, the dint of life has nothing like muscle or nerves of its own, although of course it uses material muscle and nerve cells grown by its organisms. This may be possible because the balancings of forces in organisms are so delicate chemically and physically that life needs very little force beyond what chemists and physicists understand in order to trigger action or implement a decision. But exactly where the subtle sway between going right or left begins has ever escaped analysis. Just how do thoughts enter the mind? When do crystals begin to decide? Is there a line between yes and no? Does a lily have a self? Is the germ governed by the disease or is the disease directed by its germs? Do parts integrate the whole or does the whole press harmony on its parts?

These are the kinds of questions that relentlessly gnaw at the root of the mystery of life. If life's essence is securely hidden, it may be for a good reason: should man uncover many more of its secrets while he is still spiritually so immature, it seems he could do himself and his world serious harm (as he may already have begun to do). Indeed I feel little doubt that the mysteries of life and death are man's prime bulwark against himself until his soul is fledged.

Meantime here is life with at least some of its abstract essence at hand to scrutinize and muse on. The bird, the weed and the burrowing worm are more than they seem, for, in truth, they and their haunts outline the invisible, crystalline texture of the living planet. Their bodies in space-time are literally part of the geometry of Earth, for the track each one makes throughout its life (what physicists call its "world line") defines a kind of invisible fiber of creation, a nerve cell of the multicellular territorial complex that makes up the biospheric grain of their world. It is a tenuous structure, this interlatticed biocrystal of unseen organic properties, one very difficult to visualize. It is as if millions of mental, political maps, each drawn to a different scheme for the same area, were superimposed invisibly upon one another — each strictly observed by its own animal or vegetable kind as human maps are read by humankind — a magic, crystal geometry of life delineated by the sparrow and the whale. No matter then what befalls any individual daisy or weasel or oyster, for one can be assured that daisiness and weaselry and oysterity will continue to occupy their allotted precincts and prevail.

It is hard to be quite as sure that man will survive in view of the suicidal overtones of his current precocity, but, for all his growing pains, I am betting on him — and I have a feeling that today's dangerous germination period in his unique emergence may well (in history's long perspective) be not only normal but an essential leaven if what could be called man's half-baked present is to be permitted to rise into anything like the fully baked potential of his future. I consider it good for the imagination to look at the world from an off angle now and then, as I am doing, or mayhap from an exotic date to get us out of our twentieth-century rut, so I wonder about beings from other worlds in other ages who may have come to Earth. Could they have understood it? Would they have wanted to interact with it? And is it not conceivable that their outside view might be both fresher and truer than our own?

Suppose a giant from Sirius, say, had turned up in the Mediterranean region a few thousand years ago — which just could have happened, though the people who saw him there in those ancient days would likely have assumed he came from some unknown or barbarian part of the still unexplored Earth — and suppose he was a serious student of foreign worlds (as were the first men on the moon) and that he witnessed and became interested in such a form of "life" as a galley or trireme with its oar "legs" propelling it through water. Such a Sirius student might, I presume, have been puzzled at first but, if he could have made his way close enough to observe that each "leg" was propelled by a separate organism potentially capable of living independently from the ship and its other "legs," even eager to be free to do so, he might well have wondered what kind of binding force or "nervous system" coordinated these legs to make the ship into a kind of living being.

The same wonderment of course could have struck later arrivals from other worlds, perhaps the crews of flying saucers (if any) who still may be studying diesel trains, trucks, tankers, airliners or phenomena such as a mechanized army deploying like a plague of beetles. Even humans like me wonder about the life essence in such multicelled complexes — for the abstract factors involved are hardly what could be called comprehensible. These exotic visitors, however, if they really are (or have been) here, have at least one distinct advantage over humans in probing earthly life, for they presumably can sense Earth with unearthly senses. We, on the other hand, must labor under the built-in handicap that each of us is a creature of Earth and therefore a part of the very mystery we are trying to solve. Besides, it is patently very difficult for us mentally to slow down our tempos of comprehension by the factors of many thousands or millions that are obviously needed if we are to see the life in sand dunes, beaches, rivers, lakes, oceans, glaciers, islands, mountains and other slow-changing features of our planet, which, in their patient ways, exhibit basic attributes of life.

## SAND DUNES

Did you know, for example, that sand dunes are classified in what may be called different species depending on the environment in which they live? They range in size from overgrown ripples a few feet tall to mountainous Saharan "whalebacks" 700 feet high and nearly 200 miles long. In the Muslim lands of the star and crescent the commonest dunes are appropriately shaped like stars and crescents, the former formed by variable breezes, the latter driven before constant trade winds. Others, almost as appropriately, are scimitar-shaped, and natural scientists describe still others as transverse, parabolic, sigmoidal, pyramidal, etc. One peculiar species, found on riverbanks, is shaped mostly by floodwaters, giving it a wavy cross-bedding of internal layers that record its history — until the next flood. Although a few, such as the star dunes, remain almost in one spot, most dunes move noticeably with prevailing winds, including the so-called seasonal dunes of southeast Asia that reverse themselves twice a year under the very regular monsoon windshifts.

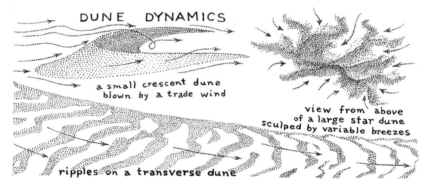

DUNE DYNAMICS

a small crescent dune
blown by a trade wind

view from above
of a large star dune
sculped by variable breezes

ripples on a transverse dune

Most familiar along seacoasts, I suppose, would be the foredune, a ridge of grass-fringed sand rising perhaps twenty or thirty feet behind an ocean beach. In many cases such a dune sooner or later gets blown inland by the wind, whirling forward grain by grain, inconspicuously but relentlessly becoming a traveling dune, among which species the faster-moving variety is called the *barchan* in Africa and Asia or the *medaño* in Peru. Alongside the Nile south of Cairo, indeed small *barchans* are well known to march thirty feet apart at a pace of about a foot a day before the trade wind, like extremely slow sea waves, their curious dynamics regulated by the flow of billions of sand grains, which the wind naturally sorts by size and weight, the coarser grains being rolled up the windward slopes and over the crests by the main jets (blowing slightly spiralwise), while the finer sand floats off with the gentler side eddies, forming and maintaining the dune's characteristic crescent shape as surely as the body of an animal is maintained

416

by its genes. The dune's height, under this regimen, is strictly limited by the wind to one eighteenth of the wavelength, while the dune is scalloped all over like a fish with tiny moving ripples (corresponding to scales), in width one-fourteenth their length — the whole a wondrous creeping organism sculpted and winnowed and steered entirely by the action of air upon flying grains of sand, a corporeal entity whose stability, feedback control and metabolism together constitute one of nature's most convincing demonstrations of the prime ingredients of life.

If I have described the sand dune as an organism, however, I had better expand it to a superorganism. It is specifically a complex subworld of trillions of living inhabitants of at least a hundred species ranging from dune rabbits, the desert fox, birds, mice, lizards, snakes, toads, poplar trees, cactus and marram grass down to velvet ants, ant lions, worms, copepods, bacteria and viruses, most of whom travel right along with it, constantly adapting, extending or redigging their burrows, laying eggs in new places, sending down fresh leeward roots, broadcasting daily dividend spores on the wind.

While the lizard swims through the sand with his hind legs weaving and wedging the grains apart, his forelegs tight to his sides, the sand fleas, protozoa and other microbes actually slither and ooze between the grains without appreciably disturbing them. Animal life is most active at dusk and dawn, when the extremes of the day's heat and the night's cold are gone. Although generally very dry on top, the dune nevertheless has a water table below it normally within reach of its thirstier animals and plants, and the capillary action of the sand blots the water level upward a foot or more beneath the dune's center, so that a hidden dome of water full of microlife moves right along with the sand. Furthermore there is a circulation of water within the dune remarkably like the circulation of water inside a tree. Evaporation on the sand's surface at the crest has been found to draw moisture steadily upward from the depths during daylight hours, especially when the sun is shining hotly, in somewhat the same way that leaf evaporation pulls up a tree's sap. At other times rain upon the dune creates a surface downward flow like that in the bast under a tree's bark, and the convex slope of the water table deflects it into an outward flow deep underground.

Occasionally reverting to a more animal nature, the dune has been known to give birth to offspring dunes, which linger near their mother for a decade or two until they are big enough to strike out on their own. It even has a voice, which evidently depends on the grains' being of nearly the same size, highly polished and round, so they    417 resonate when scuffed or rubbed together, producing a squeaking or whistling sound that, on a large scale, becomes a deep roar, particularly in the case of an avalanche down the slip face of a high dune in a windy period of rapid motion. "I have heard it in southwestern Egypt three hundred miles from the nearest habitation," reports Ralph A. Bagnold, a British physicist who studied dunes extensively in the 1930s, ". . . a vibrant booming so loud I had to shout to be heard by my companion. Soon other sources, set going by the disturbance, joined their music to the first with so close a note that a slow beat was clearly recognized. This weird chorus went on for more than five minutes continuously . . ."

The whole superorganism of three kingdoms of interrelated life that is a dune parades thus steadily onward before the swirling wind, devouring trees in its path, some of whom may live completely submerged (like poplars 80 feet high inside dunes on the coast of Holland or custard apple trees in the Brazilian desert). In fact a large dune may swallow a whole house and, in rare cases, even a small village, which it voids behind it a few years later. And its character betimes changes according to the environment, its progress and development ultimately slowing to a stop (after a century or a millennium) whereupon it dies as vegetation finally anchors its sand in place and it can do nothing further but fade back into the landscape of the countryside.

## BEACHES

Something similar happens in the case of an ocean beach, obviously akin to the dune although usually much larger in at least one dimension. Beaches indeed are as much organisms as dunes are and about equally dynamic and sensitive to wind and weather, the main difference being that most beach motion takes place underwater and is therefore not generally noticed. The material of beaches, moreover, is more varied than that of dunes, for practically any kind of plentiful object heavy enough to settle and pack yet small enough to be moved by the sea may form a beach. Beaches are known to be made of white quartz powder, of black lava cobbles, green basalt pebbles, crushed yellow sea shells and ground pink coral. There are beaches of coal dust along the Ohio River near Pittsburgh along which dance dainty dusky "coal pipers." And I have heard of a "pocket beach" at Fort Bragg, California, composed of nothing but crumpled tin cans washed in from an oceanic dump. It is rather wonderful, I think, that these old cans, which once stocked shelves in a thousand kitchens, can be literally tossed about and arranged by the wild waves into the same familiar formations of natural beaches made of sand.

The main reason beaches are dynamic is that waves and the sea bottom change each other as long as the wind, tides and currents keep them moving. As a wave approaches shore the increasing shallowness cramps the gyrating $H_2O$ molecules beneath it, squeezing their orbits and forcing the wave to break and become a carrier of water. As seas rise in a storm their roots roil the bottom and raise corresponding sandbars which in turn lift the waves still higher. The shape of one influences the shape of the other through continuous motion and feedback — wave to bottom to wave to bottom to wave — and there are short-term and seasonal rhythms to it. Winter storms steepen and wash away much of the berm (the part of the beach above normal high-tide level), the material of which helps build the offshore sandbars, but summer breezes patiently widen and restore the berm again, keeping it at a remarkably consistent average of 1.3 times the height of recent waves. As in dunes, the grains thus are continually winnowed and sorted, the coarse ones going to steep beaches during the storms which keep the finer grains suspended in the turbulent waves and the small ones eventually being allowed to settle on the flat berms which grow mostly during the calm periods.

CROSS SECTIONS OF TYPICAL BEACHES IN SEASON

There is also a ceaseless sideway movement of material along the beach often featuring cusps or small peninsulas separated by creeping crescent-shaped bites out of the shore, anywhere from 6 inches to 400 yards apart. Despite human efforts to stem this littoral flow by building walls, called groins, perpendicular to the shoreline every hundred feet or so along a beach, the fact that sand grains rolling only one tenth of an inch per wave can travel up to five miles in a year means very drastic changes to a coast in a century.

In the long run of course the wind, using the sea as its tool, prevails over any such limited measures, and much of North Carolina's storm-

419

racked outer banks is dissolving into the Atlantic at the rate of fifteen feet a year. This happens because along the east coast of America the sea takes away more sand in winter than it gives back in summer, a net erosion that is abetted slightly by the gradual melting of the Earth's massive glaciers in Antarctica, Greenland, etc., in the 12,000-year post-ice-age thaw which has ever since made the ocean level rise about three feet a century. Much faster is the movement of an offshore bank such as Fire Island, New York, which has been measured to drift downwind (southwestward) before northeast gales to the tune of 7 inches a day, 212 feet in an average year or more than 4 miles in the past century. This is faster than the tip of the hour hand of the average watch.

## ISLANDS

The birth and death of common sandy islands under wind and tide are tame compared with the dramatic antics of the rare volcanic ones that may appear or disappear in an explosive flash or, more likely, in a prolonged outpouring of lava and brimstone. On the morning of November 14, 1963, to take a recent example, a fishing vessel was cruising four miles west of Geirfuglaskur, then Iceland's southernmost offshore island, when the skipper, the engineer and the cook all felt their boat sway "as if caught in a whirlpool" and noticed a strange smell of sulfur fumes. Soon a black plume of smoke belched up from the sea a few miles to the south, and the awesome spectacle of a subsea volcano giving birth to an island transpired before their eyes. The pangs were understandably violent, with jets of steam and gas and ash a hundred feet in diameter shooting two miles into the sky. More important were the uncountable chunks of cinder and tephra which were hurled thousands of feet through the clouds before they plummeted down upon the ocean in such a relentless barrage that by evening the basalt bottom (originally 425 feet down) had become a great hissing undersea hill that piled up hour after hour until it broke through the waves and at dawn stood 33 feet above them. Four days later Pluto had so decisively gained the upper hand in his battle with Neptune that the infant isle was puffing and basking like a giant sea lion, its black crescent shape looming 200 feet high and 2000 feet from tip to tip, while the steam and smoke, now billowing five miles up from a huge crater, had burgeoned into a perpetual thunderstorm flickering with lightning and growling like the legendary Norse giant for whom the land was about to be named Surtsey.

Once the crater walls were strong enough to hold back the sea, lava began to well up inside them, spilling over in rivers of fire that rapidly coated the island with a tough basaltic crust not unlike the bark on a tree. This began when the island was five months old and continued for three years until Surtsey had achieved the noble stature of 567 feet in height, a length of 1.3 miles and an area of more than a square mile, or twice that of the principality of Monaco.

Meantime the baby island, having already felt the precocious tread of man (when scientists landed the first month), spontaneously and permanently integrated the animal and vegetable kingdoms into its life. It did this very casually, almost indifferently, long before the first lava appeared. Came a blustery winter day when a tired gull alighted and the following week seaweed quietly attached itself to a rock. Early the first spring, flies and springtails and mites blew in on the north wind, and in June a kittiwake nested on the side of a six-month-old cliff. About the same time sea worms, crabs, mollusks and innumerable smaller creatures began to wriggle and crawl on the beginning of a beach at low tide. By the summer of 1967 hundreds of different kinds of plants had rooted themselves in crannies, and the sooty cliffs were aflutter with birds. The first flower, after two heroic years of repeated burial in ashes, bloomed in white upon the glassy purple sand. It was a sea rocket (*Cakile edentula*) whose gallantry signalized the future luxuriance of this extraordinary superorganism of a volcanic island. And the island exists as a surviving sprout in the almost continuous budding of new islands (most of them abortive) along the mid-Atlantic ridge, which ridge in turn composes a seam (apparently rising) among the ever-moving platelike sections of the living Earth.

AN ISLAND IS BORN

its first plant

Nov. 14, 1963    four days later    spring of 1967

Volcanoes, you see, also live their lives, which, like those of stars, tend to be long even if occasionally explosive. Land volcanoes are the better known but undersea ones are coming to be studied more and more. Called seamounts, they grow while gradually drifting away from the mid-ocean ridge where most of them (like Surtsey) were

421

born, intermittently spewing lava, which is their version of cell division. For tens of millions of years this activity (now called plate tectonics) has continued fitfully as the sea floor spreads out on both sides of the ridge, flat-topped guyots (undersea buttes) eventually pupating (with the aid of coral) into quiet atolls in volcanic old age.

## GLACIERS

An eccentric cousin among these rowdy Earth fry is the volcano that erupts below a large glacier. This occurs periodically in Iceland, for example, and inevitably melts out a huge pocket of water right above the lava and which, when it finally gushes through the surrounding ice, may roar at deadly speed across the countryside in a headlong, steamy flood known as a *jokulhlaup*, with boulders and icebergs bowled end over end in the raging torrent. A particularly devastating *jokulhlaup* in Grimsvotn, Iceland, in 1922 released an estimated 1.7 cubic miles of water in four days, which, had it reached a densely populated area, could have drowned tens of thousands of people!

Although a glacier seems utterly inert compared with such a flood, it is by no means completely dead, its apparent "lifelessness" being a human illusion obviously attributable to the relative swiftness of human living. A glacier's normal gait in fact moves it but a few inches in a day, averaging about the same as Fire Island, but you may be surprised to know that certain octopus-shaped glaciers in Alaska and elsewhere, after snailing along this way for a mile in a quarter century, periodically (for some little understood reason) go into a gallop or "surge" a hundred times faster, indeed fast enough for the human eye actually to see them go, and a few have kept up this pace (sometimes accompanied by the babbling voice of internal streams) for as long as three years, with a measured glacial speed of fifty feet a day. That works out to two feet per hour, about half an inch a minute or ten times faster than the minute hand on your watch.

Glaciers are born very quietly whenever and wherever a summer's thaw does not melt all of the previous winter's snow and, if this excess of snowfall continues in succeeding years, obviously the glacier will grow in proportion, its snow steadily compacting with increasing depth and pressure into ice granules that reach the size of golf balls a hundred feet down, as thousands of feathery snowflakes are progressively welded into each spherical cell unit of the rock-hard organism. It is not easy to discover exactly how or why such an apparently rigid mass will flow, but scientists now know that a glacier almost never slides

MOTION OF THE
SLOW-LIVING GLACIER

the top central ice flows fastest

edges and bottom ice flow slowest

along the surface of the rocks and soil on which it rests because the friction of rough ground greatly exceeds that within the ice itself. The best evidence suggests that, by the time it gets a hundred feet thick, a glacier has begun to creep like a very stiff liquid, its weight by then heavy enough near its bottom to shear its crystals along internal "glide planes" when and where the local ice temperature reaches the "pressure-melting point." This is a point definable as occupying the precise layer of the beginning of melting under such great stability of insulation and pressure that all the interlacing micropockets of water remain just warm enough to keep liquid and all the surrounding molecular lattices of ice stay just cold enough to keep solid at the same time. A recent test tunnel into deep glacial ice in Greenland proved that the ice was frozen hard all the way to the ground — while the glacial creeping or slipping took place in the zone between one and thirty feet above ground. In other cases the creep level seems to have been more than halfway to the top of the ice, yet significantly always deep inside the body of the moving organism.

This does not mean that glaciers never push against the ground, for their front walls often encounter rises of rock or even mountains, and it is known that in recent millenniums glaciers have literally sculped the face of the earth, scooping out the Great Lakes, pushing moraines into the shapes of Cape Cod and Nantucket, quarrying rocks from the sides and floors of Scandinavian fjords, chiseling the Alps. A glacier bigger than Europe still sprawls over Antarctica, in places more than two miles thick and 200,000 years old, spreading slowly outward in all directions, harboring significant vegetable and animal organisms from viruses to snow fleas to penguins (page 35) and confidently flaunting its maternity every summer by calving icebergs, some of them as big

423

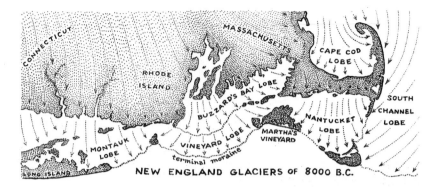

NEW ENGLAND GLACIERS OF 8000 B.C.

as the state of Delaware and which, after months of groaning and straining, inevitably snap their umbilical bonds and float free to begin a life of their own.

There have been four major ice ages in the last million years when more than a quarter of Earth's surface was covered with ice, and at least three earlier series of ice ages separated from each other by very long mild periods averaging about 200 million years. One of the most certain behavioral characteristics in the life of each of these important ice ages has been that it consumed so much ocean water in the form of ice that the world sea level dropped by hundreds of feet, swelling the land at the expense of the sea. Minor ice ages understandably have been both more numerous and more irregular, one of the latest occurring around the time when the pyramids of Egypt were built and another in the great days of Babylon and Persia. The first Christian millennium, in contrast, was a relatively balmy period, and the Norse colonies in Greenland early in the second Christian millennium were helped by a couple of centuries of semiwarmth but ended in a new cool wave and a glacial expansion that increased intermittently right into the nineteenth century. The first half of the twentieth century was distinctly warmer again, but now there are increasing signs of coolth, which, some suspect, may be appreciably influenced for the first time by the filtering effect of man's own output of carbon dioxide, smoke and exhaust fumes. Although the life of glaciers is still but dimly understood, it is evident that there is a suggestively animate pumping action to their limblike lobes that alternately advance and retreat over the centuries. Too, they often interact in unpredictable ways with other dynamic superorganisms such as the salubrious avalanche of rock and earth and ice near Cordova, Alaska, that the 1964 earthquake poured down onto three square miles of dwindling Sherman Glacier, insulating it so well that it stopped dying and has been regrowing vigorously ever since.

## RIVERS AND LAKES

Having considered the life of such solid beings as sand dunes, islands and glaciers, it should be about as easy now to recognize comparable life in the freer, swifter realms of liquid bodies like rivers and lakes (which are often the descendants of glaciers) as well as vaporous ones like clouds, flames and even whole atmospheres. Rivers, it may be argued, have spines. Like stemmed plants and other vertebrates, they are born, drink, eat and grow. They are inclined to get fat and rich when times are good but, during a famine or a drought, can as easily become thin, take sick and die. Increasingly often they are poisoned (mostly by man) but, when man cares enough (as he has barely begun to do), they are being cured by depollution treatment.

I hope I'm not carrying the analogy too far when I say that, while most rivers are single, a surprising minority (usually the younger ones) manage to join another of their kind (suitably proportioned) in marriage. On a few occasions what one might call bigamy, and even polygamy, has been observed among them. And although most "married" rivers settle down to a reasonably congenial life, certain ones eventually develop contrary inclinations and separate in divorce. Even widowed rivers are not unheard of. Naturally there are innumerable brooks adolescing all over the wooded parts of the earth, a few of which (unduly stimulated by cloudbursts) literally run wild and might justifiably be classed as juvenile delinquents. Geologists describe one type of headstrong young torrent as "braided streams" because they characteristically rush forward in nearly parallel channels

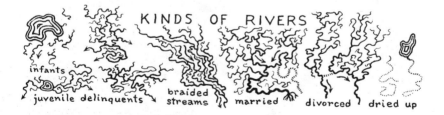

KINDS OF RIVERS

infants
juvenile delinquents   braided streams   married   divorced   dried up

that keep intertwining like braids. Some pirate rivers also "capture" smaller ones and have been known to "behead" them when they stood in the way, which, for all I know, just might explain how the word "rivalry" derived from the Latin *rivus*, a stream. It is interesting that such happenings would normally be forgotten were it not for the rivers' time-honored custom of writing their own histories in a cursive script that can be read without special training when you fly high over them and look down upon the autobiographical "oxbow" lakes, discarded gullies, defunct sandbars and other signatures of abandoned channel beds which once were the aortas of busy mainstream life.

Rivers really aren't so different from many other life forms when you consider that they persist in evolving their own distinctive ways of living and moving about, one of the most rivery of traits being their tendency to flow in regular curves called meanders, a word derived from the proverbial winding stream in ancient Phrygia known to the Greeks as the Maiandros. This sinuous figuration, curiously enough, has recently been found by physicists to be neither random nor accidental but rather "the form in which a river does the least work in turning," suggesting that a river's most probable shape is delineated basically by nothing more mysterious than laziness. There is striking evidence, in fact, that meandering is not a phenomenon of rivers or liquids alone, a classic example being the wreck of a Southern Railway freight train near Greenville, South Carolina, on May 31, 1965. Several dozen track rails 700 feet long happened to be riding the train clamped together in one continuous bundle resting upon thirty flatcars, and when the train, pulled by five locomotives, was derailed at full speed

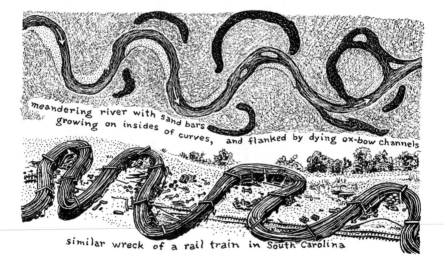

meandering river with sand bars growing on insides of curves, and flanked by dying ox-bow channels

similar wreck of a rail train in South Carolina

to pile up against an embankment, the longitudinal compressive strain neatly folded the solid steel bundle into an exquisite model of a lazy Maiandros.

River meandering is clearly billions of times more leisurely than this, often starting in the straight reach of a callow valley stream with irregular shallows and deeps (known to trout fishermen as riffles and pools) that tend to develop alternately on either side of the channel at intervals of about six times its width, prompted increasingly, as the stream grows, by the observed fact that molecules of water, clay, silt, etc., in a river naturally follow spiral paths downstream, reciprocating from clockwise to counterclockwise to clockwise again with the shifting, shuttling meanders. Over the centuries and millenniums this simple oscillatory proclivity molds all rivers inhabiting flat, soft lands into progressively snakelike contours and, when there is time enough, eventually the volutions outdo themselves, even to rubbing against each other so hard they fuse and short-circuit into new and shorter channels. One could say that rivers in general act a little like wild drivers who go around curves too fast, always swerving to the outside so they scour away material from there and deposit it upon the calmer insides of bends farther downstream. The rate at which this happens is found to be proportional to the slope of the land (therefore to the speed of flow), and, in at least some cases, to the amount of sediment carried in suspension. But there are still so many mysteries in hydraulics, magnetohydrodynamics and other of the fluid sciences that no one can reasonably claim to fully understand what's going on in a river.

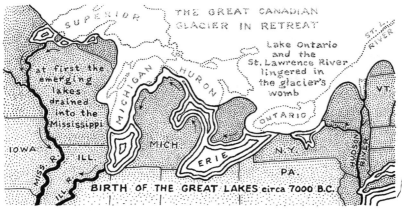

The life of lakes is not basically very different from that of rivers, for lakes obviously are related to them, often on the mother's side, as when the overflow of a gravid pond gives birth to an infant stream. The scientific study of lakes is called limnology (from the Greek *limne* for pool) and one of its best established findings is that, compared to rivers, mountains and most of the seemingly long-lived features of the    427 landscape, lakes are notoriously short-lived. Indeed almost all the natural lakes now on Earth were born around the end of the last major ice age (say 10,000 years ago) when glacial lobes that had scoured out basins (sometimes damming them with moraines) melted to flush them brimful with clear, cold water. Thus if rivers are often the offspring of lakes, lakes are more often the offspring of glaciers. And lakes that, from my spacial perspective, are essentially only the splash of a retreating ice age must expect to dry up and die within a few millenniums, depending of course on the size of the body and its rate of metabolism, which specifically is the net flux between its intake on the one hand and its discharge (including evaporation) on the other.

This expectation turns out to have been well founded, for limnologists' records prove that the average lake steadily matures from a cold, deep, clear, liquid, virgin body into a warm, shallow, soupy, degenerate one as fermenting vegetation and rotting animal matter gradually corrupt and fill it, layer upon layer, eventually curdling the whole organism into a bog before it clots into hard peat, then dry land and perhaps finally a town. This brewing or ripening process also involves a progression of vegetables from algae, sphagnum moss, waterweeds, reeds and willows to birches, evergreens, planes and other woodland and urban species, and a succession of animals from trout to perch, bass, carp, ducks and water bugs, followed by mudminnows, frogs, turtles, herons, snakes, beavers and eventually the likes of foxes, deer, wildcats, house cats, goldfish, canaries and pet poodles. Even the five Great Lakes of North America, collectively the largest reservoir of fresh water on Earth (the only bigger "lake" being the brackish Caspian Sea), are dying at an alarming rate that has recently accelerated because of thoughtless pollution, particularly in shallow Lake Erie. And the story is similar in Europe, a noteworthy example being Lake Zurich in Switzerland, which has quietly aged from youth to senility in a century.

## SEMILIQUID BODIES

Of course there are all sorts of other liquid and part-liquid relatives of rivers and lakes, like salt marshes and estuaries, which not only have their superorganized life but grow myriads of animal and vegetable constituents in their rich gardens that are home to nearly every kind of creature from bat-winged sea slugs to great blue herons. Some of these semiliquid bodies have a kind of mercurial impetuosity that

428

is undeniably close to the quick of life, a phrase that may bring to mind the infamous quicksand, whose legendary predacity, however, stems from nothing more perilous than the fact that, while it looks solid, it is in effect a liquid since its grains are immersed in water which coats and separates them to the degree that they are in suspension and, though you cannot walk on quicksand, if you can keep your head you may float in it and even (to the extent that it is liquid) swim.

Far more dangerous than quicksand, albeit in a different way, is the rarer quick clay, a bluish-gray, water-soaked glacial deposit that can suddenly "melt" from a solid into a fast-flowing liquid that is known to have perpetrated a number of deadly landslides in Scandinavia and eastern Canada. A recent case was in Nicolet, Quebec, where at 11:40 A.M., November 12, 1955, a section of the town as big as four football fields and 30 feet thick suddenly started to slump downhill and in less than seven minutes had flowed into a river, carrying with it many buildings and crushing or drowning several people who could not reach solid ground. Worse still was the quick clay gush in Verdal, Norway, in 1893, which liquefied three and a half square miles of the town in ten minutes, killing 120.

This kind of phenomenon, that one might be tempted to call "the swallower," has a truly weird and abstract body: a kind of epicene anatomy that typically assumes the dumbbell shape of an ameba procreating into two offspring cells, one of them negative (losing material), the other positive (gaining material). The negative cell of course occupies the uphill side or mouth of the landslide and ungrows (negatively) as the clay is engorged out of it to slide down through the throat into the positive cell or belly below, where it spreads and grows (positively) into a bloated morass.

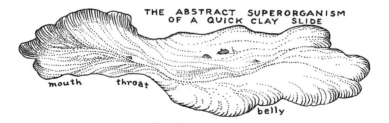

THE ABSTRACT SUPERORGANISM
OF A QUICK CLAY SLIDE

mouth    throat    belly

Examined under the microscope, clay is seen to be composed of crystalline silicate flakes (page 95) which, in the case of quick clay, are smaller than two microns in diameter and saturated with water. When such stuff was deposited on the ocean floor many millenniums    429 ago, of course the water in it was salty but, after the diastrophic forces toilsomely heaved it up above sea level, centuries of rains leached away most of the salt. And because the salt ions had served as an "electrolytic glue" binding the clay flakes together, their absence unlocked the flakes again so that, except when they were unusually dry, almost any physical provocation (like an earthquake tremor or a sonic boom) might collapse them like a house of cards, letting them melt or slip past one another in parallel paths.

Thus the evidence shows that the monster of quick clay is naturally and normally triggered by shock. In one actual case, a slide was started by the bumping of railroad cars, in another by a pile driver, in another by a stroke of lightning. As one investigator explained it, saturated clay behaves singularly like a gel (such as iron hydroxide), which, when jarred, melts and runs away — even down an almost imperceptible $1°$ slope. (There is also a natural phenomenon called quick asbestos, known to almost no one outside of the mountain town of Coalinga, California, where there have been several remarkable slides in recent years, some a mile long, whose substance was pale, pasty, rain-drenched shale with a main ingredient assayed to be short-fiber asbestos.)

And far more important, if very difficult to observe, are the giant "turbidity currents" under the oceans, which seem to be propelled by submerged avalanches in which clay in suspension has been known to flow at an average speed of 14 mph along a flat undersea valley of less than $1°$ slope for hundreds of miles. One notable quick clay illapse in 1929, provoked by an earthquake three miles deep and almost 200 miles south of Newfoundland, swept such tremendous masses of this material for perhaps a thousand miles under the Atlantic that twelve transoceanic cables in its path were stretched and broken, one after the other, from north to south.

When a flow approaches this magnitude, almost inevitably a meander effect begins to manifest itself that is obviously akin to the kind we described in rivers on land. And, I'm told, the Gulf Stream, like other great "rivers" of the ocean, meanders (with its own bias toward coastal and Coriolis influences) on a measured average wavelength of 60 miles and amplitude (width) of 9 miles. For that matter, one could consider all sorts of entities such as waterfalls, whirlpools, rapids, swamps and bayous as metabolizing organisms distinct from rivers, lakes or oceans. And the list might include even dripping stalagmites and stalactites in caves, the caves themselves, watersheds, canyons, puffs of smoke and rainbows, any and all of which are growing, developing and interrelated systems of life on Earth.

## LIFE IN STORMS

Of course the atmosphere has its own cycles and meanders: dramatically evident in the globe-circling jet streams of westerly winds that snake around our world at hundreds of miles an hour eight miles above the temperate zones in both the northern and southern hemispheres. The momentum that engenders such aeolian life is fed into these vital flows by the tireless turbulence of the lower air familiar in the irregular eddies of low pressure we see on weather maps and photos from space as storm systems perpetually pirouetting a few thousand miles apart around the middle latitudes. And there are even more vital, if rarer, vortexes known as hurricanes and typhoons, with girls' names and violent dispositions, spawned over the tropical seas — great cyclonic cells hundreds of miles in diameter that mesh like giant gears with lesser parasitic storms, including thunderstorms, roughly five miles in diameter, which have been observed to grow, move, roll, fuse, multiply and sometimes die, much like certain species of bacteria. Parent thunder cells indeed sprout buds from time to time, which later (appropriately) spring off as offspring stormlets, grumblingly foliating into adolescence a league distant, where they repeat the life cycle needed to proliferate their way across continents and seas, generation upon generation of them in the kind of amebic immortality that collectively comprises the five thousand thunder cells meteorologists estimate are prowling the skies of Earth on every average day. And they even excrete little hyperparasitic organisms like tornadoes, waterspouts, williwaws and dust devils that occasionally spin out quite respectable careers of their own.

As all these breeds of gaseous organisms inhabit our modest jot of Earth, no one should be very surprised if something comparable turned up on the million-times-bigger sun, who is gaseous not just in his fiery outer atmosphere but composed all the way through of very hot gases like hydrogen and helium in the disintegrated, ionized state known as plasma. And, sure enough, a continuous fluid system, proportioned rather like the double jet streams of Earth, has been found on this parental star of ours. Its first discovered features were dark hurricanes a hundred times bigger than earthly ones but relatively the same in comparison to the larger body, and these whirling storms, better known as sunspots, turn out to be but eddies in tremendous, invisible magnetic rivers (which scientists call magnetohydrodynamic waves) flowing in twisted doughnut courses all the way around the sun, corkscrewing down and up like the probing roots of some scarce-conceivable

431

celestial tree, yet mysteriously guided by the multiple solar moments of force. The volatile solar supersphere, reeling imperiously through space, naturally could not be expected to turn uniformly, and it has actually been measured to spin about a third faster at the equator than the poles, not to mention its even swifter spin at deeper levels, producing thus an extradimensional circulation system that is far more complex than our terrestrial one yet comfortingly similar to both our Gulf Stream and jet stream anatomies in its staggered zigzags — that zig three times as far as they zag, taking only eight meander wavelengths to complete the global circuit, writhing with a tenacious limp that somehow articulates a flamboyant, empyreal life no mere human can deny.

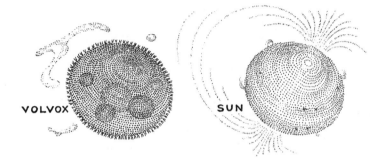

All in all, I'd surmise the blazing sun is at least as much a living organism as is a volvox or any other spherical creature in the waters of Earth, for the total sun includes them all and a lot more. His spots also may correspond to embryos of future generations after his presumed eventual explosion, for he has a complex metabolizing body. And I assume he must even have a kind of consciousness through the bulk, momentum and interrelations of his moiling gases and his orbiting fertile planets which give him what amount to a memory and, for all I know, a mega-soul.

## FIRES

If you think that fire is the deadly enemy of life and that the sun, whose coolest spots have been measured at about 10,000°F., therefore could not possibly harbor any sort of life, perhaps you should consider the modern science of combustion. Using high-speed photography it has discovered that, when a flame is born, it can grow in less than a thousandth of a second all the way from a microscopic seed spark to a cell mature enough to sprout offspring cells around its edges. Full grown, fire turns into a complex, surging fluid that proliferates in turbulent, bubblelike shapes. Its corporeal metabolism is swifter than that of any known animal as it devours and digests all combustibles within reach, inhaling oxygen and exhaling smoke, running with the wind while casting its seeds before it: glowing spores which may at · any instant germinate and blossom forth into lineal flames.

Fire ecology, the study of fire's impact on the rest of nature, is an even more rapidly growing branch of learning. Ecologists now know that prairie and forest fires, the natural conflagrations kindled by lightning in times of drought, have been raging almost every year since there was dry land to support grass and trees. To them wildfire is a phenomenon well established as a factor in evolution, indeed probably a vital means of enriching the forest and ennobling the earth, without which the giant sequoias of California could not exist in their present stature, if at all, and many other types of vegetation as well as birds, mammals and lesser creatures would have become extinct eons ago.

I am not saying that forest fires are not often very dangerous, indeed extremely destructive, or that one should be indifferent to them, but I notice that when fire breaks loose in fields or woods it is much like an outbreak of lions from a circus. An alarm is sounded and dozens of trained men rush to the spot to round up the wild beasts who have been making the most of their rare chance for freedom. Fire, however, does not necessarily ride roughshod over everything it encounters. For normally it is quite selective, more like a herd of cattle grazing or a gang of peasants pruning a vineyard. I mean it is alive in the sense that it metabolizes and finds its own sustenance. And the phrase "feeding the fire" is curiously apt, because feeding, like breathing, is a combustive, oxidizing process not basically different from burning, though rather too slow to give off flames or visible smoke. Even when it goes roaring ravenously across parched California brushland apparently out of control, a fire will retain its finicky appetite and stop almost dead at an oasis of lush greenery or a forest of hard-barked trees, damped down to a creeping pace for lack of dry provender. And from here, to the surprise of many an observer, the still-burning fire, while charring the trees, may not only fail to kill them but will likely help them (the stronger ones at least) by pruning away most of the competing saplings around them, not to mention diseased branches, gnawing insects and the heavy blanket of duff upon the forest floor. And chemical analysis has recently revealed that, even where it kills, fire recycles many nutrients from burned trunks into the soil, thus preparing the land for new plant and animal growth.

Study of the rings of thousands of old trees in California has also 433 shown that forest fires have burned virtually all of them severely enough to leave permanent scars inside the bark on the average once every seven or eight years. When visible, these common scars are known as cat faces in the northern areas because they are black patches that normally start with a "chin" on the ground and end in pointed "ears" several feet up the trunk. Only rarely are even the biggest of them deep enough to kill, for it has been found that as much as 95 percent of a tree above ground can be burned without its dying if only the surviving part includes enough connected cambium cells. However, almost any fire is likely to interrupt what ecologists call plant succession (page 66), in this case a natural progression from sun-loving trees (such as pines) in a new, dry, open forest to other evergreens (including sequoias) a decade or two later, then trees better adapted to shade and moisture and, after a century, nothing but trees (like cedars and hemlocks) that thrive and reseed themselves in the deep shade of the climax forest.

Such a succession, even if undisturbed, may take several centuries to attain its climax as each new invading species gradually crowds out its predecessors, insidiously outgerminating and outgrowing them through subtle adaptations to the increasing shade. But disturbances in the forest are just as much a part of nature as is succession and they naturally include incursions by fire, weather, bugs, fungus and, increasingly often, man. So if fire gives big pines, firs, sequoias and their parasites a helping hand at the expense of cedars, hemlocks and certain diseases, it has to that degree become a positive element in the life of these fire-favored trees (just as it has become a correspondingly negative element in the fire-retarded cedars and hemlocks), and this is not to imply that its effect is merely abstract since (to cite one case) the seed of the immemorial sequoia — surely not by chance alone — germinates better in ashes than in unburnt duff. The timing of a fire is also critical, particularly in determining whether it will be more helpful to the forest or the field, for foresters have noted that a fire just after a rainy spell favors woodlands while a fire at any other time almost always favors the grasslands.

Among plants most stimulated by burning, there is the ubiquitous flame-colored fireweed sprouting quixotically out of soot, and several pines such as the knobcone and lodgepole species of the Rocky Mountains. There is also the jack pine, a veritable fire tree whose natural home is burnt-over woodland in the Great Lakes region and whose cones, after reposing tightly closed for a decade or a century, will spring open (triggered by scorching heat) within seconds of a fire, tossing their seeds to the hot updrafts, which quickly cool off enough

434

to sow them downwind upon the hospitably warm bed of ashes from which all competitors have just been eliminated. I will add to this only that I have seen a forester's report which began by describing a typical Minnesota wilderness with an average of six jack pines per acre at the time it was consumed by fire, and ended with a plant census the following year, estimating 15,000 jack pine seedlings per acre already arisen phoenixlike out of the ashes.

Turning now to the animal kingdom, may I introduce what you may have already anticipated: a real live firebird flitting about the Earth's burnt woods. Known in Michigan as Kirtland's warbler, it subsists almost entirely on the young jack pine, and is therefore largely nurtured by fire. In fact this tiny bird has become so dependent on fire that the decrease in forest fires in Michigan this century is causing ecologists to fear its extinction. And beyond it there extends a whole continuum of fire fauna, ranging from the tiny pine-seed-eating white-footed mouse to the lordly moose, who has been thought to "follow lightning" because of his taste for the smoky-flavored young shrubbery that sprouts in the burn habitats left by summer storms.

Man of course must be included in any complete list of the beneficiaries of wildfire. So too, surprisingly, must be some streams, rivers, lakes and other liquid organisms, not infrequently with human aid. For I hear that farmers and ranchers are increasingly abetting fire in its production of water by seeding the residuum of scorched ground with moisture-conserving grass. Although some land owners still refuse to accept such "miracles," range ecologists are reporting solid evidence that former dry sagebrush country, now burnt over and converted to sod, not only feeds seven times more animals than does unburnt sage land but that it will water them as well. In a recent case a California farmer sowed his dry range one spring after a fire, and two years later, realizing that streams which had always dried up in

435

summer were now flowing the whole year, built himself a reservoir that in a few months became the supply center for a permanent irrigation system.

If fire can do all these things and still be lusty enough to spawn everything from trees to moose to permanent waterways, I'd say its sparks must include the spark of life! Surely there is no longer any doubt that fire, which Herakleitos called "the eternal flux of matter," is a subtle and sensitive phenomenon, a nonanimal, nonvegetable, natural predator that evolved throughout evolution, often doing more good than harm. It does this by regular pruning of undesirable or ill-adapted species, and by preventing the accumulation of flammable fodder, the presence of which could otherwise enable the next fire to burn so hot it would kill deep roots or even blaze high enough to ignite the crowns of trees, becoming a rare fire storm, a few of which have actually been known to breed real white-hot tornadoes.

## Lightning

Remembering that lightning is one of the principal perpetrators of wildfire, it may be worth mentioning here that this still mysterious symbol of swiftness and divine retribution has a vertebral skeleton similar to that of a river and surprisingly complex physical functions (which I described in detail in my *Song of the Sky*), and that it leaves its own unique footprints in rusty, igneous dents on rocky mountain crags, the split trunks and sloughed bark of giant trees and the little-known fulgurites of fused sand it excretes where it strikes a dune or a beach. The reason the root-shaped fulgurite tubes of silica glass, sometimes appropriately called "petrified lightning," are so seldom seen is that they are always deposited by the lightning ferret-fashion underground. Of course in some cases the sand around them even-

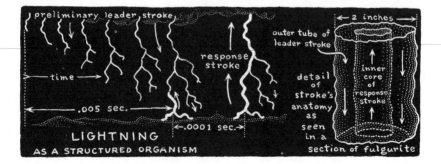

preliminary leader stroke

← 2 inches →

response stroke

outer tube of leader stroke

detail of stroke's anatomy as seen in a

inner core of response stroke

time →

.005 sec.

.0001 sec.

LIGHTNING
AS A STRUCTURED ORGANISM

section of fulgurite

tually gets washed or blown away and the fulgurite is discovered and likely put into some museum where it can be studied as a kind of fossilized thunder script — a script "written" in an unimaginable ten thousandth of a second with electric ink at 3000°F. by the mercurial quill of lightning, whose strike, according to present understanding, is double, coming from opposite directions as polarized "leader" and "response" strokes, both of which advance between heaven and Earth in unseeable quantized steps like a sprouting meristem of electrons at least as alive as any river or flame.

## ATOMIC EXPLOSIONS

On a vaster scale, the most alarming fire creature ever born to man, an atomic explosion, manifests itself as a kind of organism of the intellect that should not be omitted from any roster of fundamental life forms. Certainly William L. Laurence, the eminent science editor of the *New York Times* who was the only journalist permitted to witness the atomic bombing of Japan, thought of it as a living phenomenon when he wrote: "Awe-struck, I watched a pillar of purple fire shoot upward, becoming ever more alive as it climbed skyward through the white clouds. It was no longer smoke, or dust, or even a cloud of fire. It was a living thing, a new species of being, born right before our incredulous eyes . . . It was a living totem pole, carved with many grotesque masks grimacing at the earth.

THE LIVING ORGANISM
OF AN ATOMIC EXPLOSION

"There came shooting out of the totem pole's top a giant mushroom, even more alive than the pillar, seething and boiling in a white fury of creamy foam, sizzling upward and then descending downward, a thousand geysers rolled into one . . . giving the appearance of a monstrous prehistoric creature with a ruff around its neck, a fleecy ruff extending in all directions, as far as the eye could see . . ."

437

Winding up the strange life forms of war, I cannot but mention war ecology, which, like fire ecology, involves fire of different but no less deadly sorts and which likewise has its creative side. War as now understood is an ecology where the elimination of established species of animals and vegetables normally heralds the arrival of new ones and the unexpected is commonplace — like bomb bursts playing the part of dinner bells to tigers in Vietnam, calling the suddenly increasing numbers of these carnivorous beasts to their unaccustomed but delectable diet of freshly butchered soldiers, or the 269 species of flora that were transplanted (mostly by the winds) to London during World War II and announced their arrival by arising triumphantly out of the soil of thousands of bomb-plowed vacant lots that had obviously offered their seeds growing space not previously available.

Although the list of analogues of life could go on indefinitely, I think I have suggested enough to make my point that the abstract essence of life may be found almost everywhere one looks, not only on what may be a far-from-rare planet but also on a fairly typical star like the sun, indeed, more than probably, all over the universe.

# CHAPTER 16

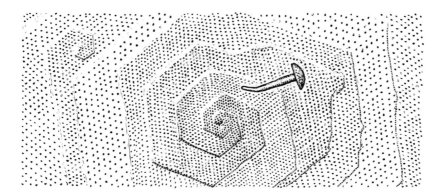

# Doornail and Crystal Essence

S O WE COME to the next logical step in our analysis of life: a close look at metals and particularly crystals, which just could be where some pioneering scientist may yet descry how the material structure known as life actually begins to take shape.

In this connection, I once heard someone remark that the moon, because she apparently harbors no native creatures, must be a dead world. To which a young astronomer responded as if ringing a knell, "Yeah, as dead as a doornail!" And it struck me as curious that, of all the materials generally considered *not* alive, the metal in a doornail is proverbially the deadest. In fact, if you believed common talk, you might well suppose a man could comb the four corners of the world without anywhere finding anything deader than a doornail.

## METALLIC "LIFE"

Understandably, therefore, it was with a keen whet of interest that I read of Sir J. C. Bose's experiments early this century in Calcutta to test the "life" in iron, steel, bronze and other metals, which now makes it possible for me to use some of his discoveries to bolster my

439

own exploration of life. Without any knowledge as to what title, if any, Bose gave to this study, I can only guess and hope that he would not have thought it inappropriate for me to call it the Doornail Project. Whether or no, I am informed that Bose rigged up his pieces of metal with wires and clockwork so they could be subjected to electric shocks automatically once each second or at other desired intervals, and that he observed on his very sensitive instruments how each piece of "doornail" accepted or resisted these repeated onslaughts. In doing so he soon noticed a significant progressive increase in the metal's electrical resistance as time passed, so that after an hour or two the piece under test might be taking only half the current it had accepted in the beginning. He considered this symptom as evidence that the "doornail" was "getting tired." And he further found that if he gave it a rest it would recover in a few minutes, almost as if it were an animal or a man, and would then respond as before. Subsequently he discovered that if he gave it too long a rest, say a full month's vacation, it would get sluggish or sleepy and not do at all well — that is, not until he had given it a brisk jolt of electricity, the equivalent of a cold dip, to wake it up and snap it out of its doldrums.

He found also that certain chemicals or drugs would do almost as well as the electric shock in this "shot in the arm" treatment, while other drugs on the contrary seemed to poison the test metal and put it into a kind of coma. In fact Bose reported that copper, tin and platinum reacted to electricity almost exactly as did nerve and muscle tissues. In some cases a small dose of "poison" would invariably quicken a metal's response while a large dose of the same "poison" would "kill" it, just as with a so-called living organism, and when the graphs of nerve reactions and of metal reactions were compared they looked so much alike they got misfiled on two occasions, and had to be specially labeled so the workers in the Doornail Project laboratory could tell which was which.

If curing "sick" metal with "medicine" administered by a doctor of metallurgy is remarkable, how much more so is the space research now going on in several parts of the world to develop a metal that can heal itself! One such effort at the University of California in 1967, for example, produced a supersteel alloy called TRIP (with TRansformation-Induced Plasticity) which not only stretches four times as far as other steels without cracking but, if the stress continues to the point where the molecular structure starts to rend, a solid-state chemical reaction is triggered which "blunts" the incipient crack, at the same time filling and healing it.

440     One of the most trying problems of spacecraft in space is the vital

circuitry inside their almost uncountable electronic gadgets, which have been found to suffer fractures not only from radiation but from prolonged vibration, lubricant evaporation and sudden temperature changes. Here the astonishing cure turns out to be a tin-magnesium-aluminum alloy that can be extruded into a new kind of wire, which has the power, should it get broken, immediately to start growing "whiskers" to bridge the gap and heal the wound. The alloy accomplishes this by the seemingly magical crystal process of sprouting fine metallic fur out of any new surface free of restraint, as in a fresh crack, every individual hair of which has enough driving energy to extend itself about a millimeter in three days and enough substance to carry roughly a watt of electric power, which, multiplied by the contributions of fellow hairs, is much more than enough to keep the average spacecraft circuit in operation. The similarity between these metal whiskers and tropical plants is further accented by the observation that, although they grow moderately well at room temperature, they reach their fastest growth at about 125°F. and, if the heat rises above that, tend to "wilt" before maturity.

## LIFE IN MACHINES

Although it is by no means hard to track down such examples of metallic life, it is noticeably easier to find analogies of sentience in larger or more complex "organisms" made of metal, many of which are commonly known as motors or machines. A small electric fan on my writing table, for instance, had a fall the other day while it was running and landed on its side in such a way that the rubber blade was prevented from turning while the motor strained and hummed in frustration for a few seconds until I could rescue it. To my dismay the motor refused to work when I set the fan back in place but, after I spun the blade by hand, it eventually caught on a little and turned slowly with an unaccustomed rattle. "Perhaps the motor is partly burned out," I thought as I watched it falter, pick up a little, then gradually die again. But every time it quit, I encouraged it with a new spin and, after ten minutes of this physiotherapy, to my delight it somehow lost its rattle and, although it would still slow down occasionally, it had improved enough to be almost completely normal — and it has retained its health ever since.

Isn't this rather similar to the time-honored resilience of many an old automobile which, after refusing to start for a few minutes, suddenly convulses into a fit of coughing, then somehow manages to clear   441 its throat, shake off its stammer, steady down and run perfectly all day. Or the so-called foolproof airplane that is so control-limited and inherently stable that it obviously "wants" to fly and will work itself out of any stall, mismaneuver or spin anyone puts it into — in the process settling itself back on a comfortable and even keel? Machines breathe, eat, react, grow old and die, you see, very much like living organisms, sometimes protesting violently, as when a car, started in high gear, shudders as if with emotion from the undue strain, or a middle-aged locomotive responds to a steep grade by puffing like a fat man on the stairs.

## Pseudo-Life

In the microcosm and the small world of chemical reactions still other lifelike functions may be found such as "artificial amebas" made of oil and soap which crawl and extend pseudopodia, and "water bugs" of camphor that skate and whirl across a full tub because of the shifting pressures of surface tension as the soap or camphor dissolves unevenly. And even more animal-like is the performance of a drop of chloroform in water, which behaves remarkably like a "finicky" eater, refusing offered grains of sand, glass, plastic or charcoal but hungrily engulfing crumbs of paraffin and shellac. The chloroform has also in one recorded case "outwitted" a researcher who tried to "deceive" it with a piece of glass coated with shellac, for, although it swallowed this "morsel" without "noticing" anything wrong, it digested only the shellac before hastily regurgitating the glass.

The most realistic of all such chemical organisms, however, is probably a drop of mercury left close to a small crystal of potassium dichromate in a bowl of water that includes a few dissolved teaspoonfuls of nitric acid. The mercury just sits there at first, apparently dozing, but, as the pink cloud spreading from the dissolving crystal gets to it, it seems to "smell" something exciting and reacts instantly — almost like a tomcat meeting a receptive female on the back fence at midnight. The mercury shoots out a tongue as if it were trying to lick the crystal, but withdraws it again without quite touching it. A few seconds later it repeats the overture, this time with more "confidence." By the third or fourth time it actually touches the crystal and begins to quiver and boil, literally throbbing as it surges toward the object of its "desire," which retreats "coyly" while being pursued around the bowl. This "love chase" continues for several minutes, progressively building up to a climax of sorts, whereupon the mercury,

442

which had been spitting out dense orange-brown clouds, subsides in apparent exhaustion.

It is a real puzzle to try to define exactly why such interaction should be considered to signify life when observed between animals but not when observed between chemicals. I presume it is because the accepted fundamental characteristics of organic life are that it is formed of cells (collectively known as protoplasm) which consume nourishment, grow, metabolize, move, react and reproduce themselves and that, although inorganic chemicals may do many of these things in a way, they can hardly be said to be composed of protoplasmic cells or to fully reproduce themselves like animals or plants. Another and rather curious difference between organic and inorganic matter, it was long ago discovered, is that if you cook the former, whether it be sugar, olive oil, bread, bone or flesh, it will eventually char, shrivel or go up in smoke, remaining irrevocably changed from its original state, while inorganic matter such as salt, lead, stone or water can be heated and melted or boiled for hours without essentially changing. The generality of this culinary definition of life may be attributable to the fact that most inorganic matter is relatively simple and pure, being made of vast numbers of similar molecules of very few kinds, while organic matter owes its much more sensitive complexity to its containing only small numbers of similar molecules of each of a wide variety of different kinds.

## DISORDER AND ORDER

An obviously profounder distinction about life turned up late in the nineteenth century with the advent of the Second Law of Thermodynamics, commonly called the Law of Entropy or Disorder. This famous law says there is a long-range, long-run certainty that order will eventually give way to disorder in any closed material system. As life obviously involves organisms, which are organized and orderly, the entropy law clearly states the inevitability of physical death, which means the loss of order. And the pervasive tendency extends all the way from your cluttered closet to the atom in one direction and out to the farthest reaches of the universe in the other, a universe whose end (assuming it has limits) will entail nothing but dry "dead" worlds and shattered galaxies that are drifting farther and farther apart while inexorably disintegrating. And, according to this law, entropy will continue until all energy is distributed evenly everywhere with nothing warmer or cooler than anything else and only tepid dust and finally

443

separate, random atoms and subatomic particles diffusing listlessly and invisibly outward, outward and ever outward into unknowability, lifelessness, spacelessness, timelessness . . . nothingness.

Opposing this depressing Dantean "heat death" of entropy of course is its opposite force, the positive and much cheerier Wordsworthian principle of *neg*entropy, of growth and order which, even while appearing mortally limited in time (not to mention space), constitutes life. It seems unfortunate, not to say ironic, that such a vital and *positive* influence should be known by such a *negative* name as neg-entropy. But the only thing really negative about it is that it opposes entropy, so we may let ourselves think of it as overpoweringly strong. In fact, despite the touted Second Law of Thermodynamics, our universe may well harbor a less obvious, less material but over-riding Law of Negentropy — particularly if no proof turns up of the universe being a closed system, which, I presume, would require a positive curvature of space over a big enough volume to close in upon itself, a cosmic happenstance none of the accelerating evidences of astronomy has so far come near to substantiating.

These two preponderant forces then are seen to be locked in a profound struggle for mastery of the world: negentropy arising out of almost universal entropy, order bravely stemming the tide of disorder, life sprouting confidently up from the inertia of diffuse and vacuous space. It is worth a moment's reflection, I think, to consider just what this means in specific terms, beginning by trying to visualize the virtually infinite numbers of dust motes and molecules roaming the interstellar and intergalactic vastnesses of the universe. Where are they going? Why do they wander? What force motivates their unpredictable caprices? One obvious answer is that a randomly moving particle tends to wander because it is much easier for it to go somewhere else than to return to where it was a moment ago, a prime tenet of kinetics being (as you can see in the diagram) that many more paths lead *away from* than *back toward* any previous position.

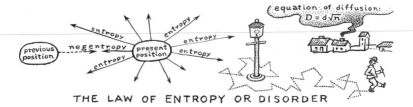

THE LAW OF ENTROPY OR DISORDER

A familiar example of this is the drunkard who doesn't know whether he is coming or going but who nevertheless staggers onward from somewhere to somewhere. Although no one can say in which

444

direction he will move next, his net rate of progress is significantly predictable, in fact so much so that there is a well-established equation for it: $D = d\sqrt{n}$. This is a concise way of saying that the number (D) of feet of distance covered by the drunk (measured in one straight line from his starting point) tends to equal the average length (d) of each zig or zag that he staggers times the square root of their number (n). Thus 4 one-foot aimless staggers would net him on the average 2 feet of distance, or 9 staggers 3 feet, 16 staggers 4 feet . . . etc.

This goes for staggering molecules too — and presumably for stars and galaxies and the sleepless supergalaxies whose motion we have not yet had time to measure. $D = d\sqrt{n}$ is the equation of disorder behind all diffusion and it tells us how fast a drop of ink will diffuse through still water or a puff of smoke in a lazy sky. It also explains why drunkenness (or dopeyness, if you prefer) is an expression of disorder diametrically opposed to the orderliness that means life. If something or somebody has a will to live, you see, it or he must resist diffusion and move from disorder to order, which means avoiding all those easy paths away from the previous position in favor of returning to it or, better, staying with it from the beginning. Basically it connotes sticking around: being stable and solid and working toward some sort of structure, developing something that could be called an organization. The word "sticking" is surely germane here. For stickiness is a key to life — and it says something as to why the Earth sticks together with the aid of gravity and why our bodies stick together with the aid of molecular forces in bone, gristle, colloid and collagen, of why the jellyfish is "icky" and even of why old horses traditionally get hauled off to the glue factory.

## CRYSTALLIZATION

Putting it in terms of evolution, when a bunch of milling molecules makes its initial change toward what is generally considered life, the change appears first as an increase in stickiness, a familiar quality of stability, which is the beginning of order or negentropy, which is essential to life. But what form does this vital viscosity take on its way to stability and order? Exactly how do the milling molecules arrange themselves? The answer is: they begin to line up, to sort themselves. to form rows, layers, lattices. In short, they crystallize — for this is what a crystal is in essence: an ordered structure. And this is why the crystal is the basic structure of life, of order, and why ordered solids from rock to wood to muscle to bone to gene are all describable as crystal.

445

But what, you may wonder, is there about a crystal and its order that gives it this vital potential, this curious lease on life? If a solid is cooler, quieter and more dormant than a liquid or a gas, why isn't it also deader since dormancy and deadness are more or less synonymous? There is probably a relativity factor in this paradox and, I suspect, a dimensional compromise as well between order and movement, with life requiring at the same time enough order to order its movement and enough movement to move its order.

It is not an easy question, for no one seems to know why atoms accept orderly arrangements. I imagine they just fit together better that way and so "feel" more comfortable (therefore more alive) when they are in order, particularly when outside pressure, drought or falling temperature forces them into a dense mass. Come to think of it, a rather apt analogy of crystal order was encountered by a wayward classmate of mine recently when he found himself spending a month in a primitive Spanish jail cell in the company of nine none-too-palatable other characters. For he told me on emerging that by far the "nicest" nights he endured on the floor of the ten-by-eight-foot enclosure were those in which he was able to induce his fellow prisoners to align themselves, as a geometrician would say, parallel rather than transverse.

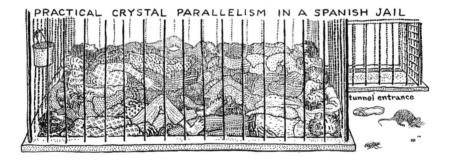

PRACTICAL CRYSTAL PARALLELISM IN A SPANISH JAIL

tunnel entrance

In some situations, order of this sort actually spells the difference between life and death, demonstrating quite dramatically its Promethean quality — but there are innumerable classes of order and semi-order and, if we are to understand crystals and life, we must learn something about them. Bear in mind: it is not as simple as studying disorder, which generally means examining gases and liquids that have practically no structure or form. Besides, there are many kinds of order not based on wallpaperlike repeating patterns and that, as a result, do not quite qualify as crystalline and so presumably (to the same degree) are unsuitable as building material for life.

When we finally get down to the viable crystal orders with their repeating patterns, the varieties not only seem endless but are literally multiplied further by impurities, by contagious "diseases," by microbubbles, by mixtures of different substances and by the inevitable discontinuities that creep between crystal systems even in a pure, unmixed substance, as the illustration shows. And the biological ap-

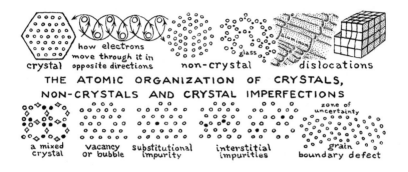

how electrons move through it in opposite directions
crystal          non-crystal          dislocations

THE ATOMIC ORGANIZATION OF CRYSTALS,
NON-CRYSTALS AND CRYSTAL IMPERFECTIONS

a mixed crystal      vacancy or bubble      substitutional impurity      interstitial impurities      zone of uncertainty      grain boundary defect

proach to crystalline complexity is to remind oneself that crystals are now classified into some 1500 "species," each of which has a characteristic form that is only the outside expression of a highly organized internal (almost genetic) arrangement of sextillions of atoms that differ for every element or compound they utter. And the racket and traffic involved are suggested by the recent calculation that a crystal cube one millimeter thick has about 100 quintillion energy "levels" occupied by the valence (loose) electrons of its constituent atoms forming a continuum. Further, when any face of such a crystal is growing at the seemingly gentle rate of two millimeters a day, more than a hundred layers of molecules must be accurately stacked on its surface per second on the average, each layer comprising some ten million precisely regimented atoms. And amid this microblizzard strewing forth a billion orderly atoms a second, is it any wonder that a few (or a few million) miss their proper niches, leaving empty spaces or bubbles, or that foreign particles sneak in among them, either replacing those absent or just squeezing into interstices that hadn't seemed big enough to try for? And, as for the discontinuity areas between two differently oriented crystals of the same material, these boundary zones are now classified as crystalline defects, a curious feature of which is that the layers of atoms involved show signs (by their wavering movements) of being uncertain as to which crystal they belong. Indeed this eerie, almost mental phenomenon apparently occupies one of the inner seams of life where indeterminism is born amid determinism and free will sprouts shyly out of rocklike resignation to fate.

447

## The Mystic Inner Life of Crystals

We might even go far enough to surmise that the atoms and molecules busily growing a crystal must at times make a choice of sorts as to where they will lodge or relodge themselves. Certainly there are competing magnetic and other attractions between atom and atom which normally induce an approaching molecule to seek out the snuggest berth on a growing crystal, not just on a flat plane where it would be attached on one side but in the trough between two surfaces where it can be latched two ways or, better, in an inside corner between three walls or, still better, in a rare nook boxed in from four or five

A MOLECULE'S CHOICE OF DESTINATIONS

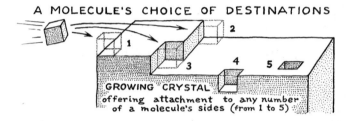

GROWING CRYSTAL
offering attachment to any number
of a molecule's sides (from 1 to 5)

directions. If there is almost no place at all for each molecule to go, the entropy is rated very low, while if there is but one remaining place for it (as in a perfect crystal at absolute zero degrees) the entropy drops to zero, which, in effect, renders the negentropy maximal and the potentiality of life unlimited. You will see, therefore, that each molecule's yen for coziness is vital and it serves as the guiding force that regulates crystal growth to a degree approaching the mystical and, depending on the shapes of the atoms and molecules, continuously directs it in assembling the wonderful structures of Earth and life.

Of course it took centuries for man to realize all this even after crystallography became an exact science in 1782. That was when the Abbé Haüy, a geometry professor at the Museum of Natural History in Paris, demonstrated that the regular angles of any known crystal can be mathematically explained simply by assuming it to be a congregation of uncountable, tiny, invisible "bricks" all made in the same shape. Haüy evidently made his basic discovery by one of those seeming accidents which so often create the individual waves that, by joining a thousand others, form the advancing tide of knowledge. While admiring a mineral collection in a friend's house, he inadvertently dropped a calcite crystal, which smashed into hundreds of pieces on the floor. Then, in sweeping up the fragments, the apologetic Abbé suddenly realized that every chunk and crumb of this crystal was similarly shaped with six faces all in the precise form of a 60°–120° rhombohedron. 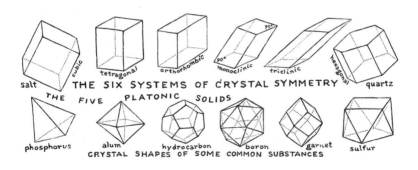 Excitedly he returned to his laboratory and began a series of experiments in fracturing crystals, which soon taught him that any crystal's beauty is purely geometric and derives directly from its natural materialization of abstract planes and angles, a revelation that led straight as an axis to his Law of Rational Intercepts. And, continuing from this noted law during the next century, a whole dynasty of crystallographers recognized and explained the four divisions of symmetry through which

> *The point, the line, the surface and the sphere*
> *In seed, stem, leaf and fruit appear.*

Of course these same four symmetries also correspond roughly to the four classic kingdoms of mineral, vegetable, animal and man. From which crystallographers, using the new tools of modern science, sorted and defined crystals according to the angles of their axes of symmetry into the six systems called cubic, tetragonal, orthorhombic, monoclinic, triclinic and hexagonal, which use just 14 lattice patterns in assembling exactly 32 classes of symmetry into the aforementioned 1500 crystal species now known to man.

THE SIX SYSTEMS OF CRYSTAL SYMMETRY

salt — cubic — tetragonal — orthorhombic — monoclinic — triclinic — hexagonal — quartz

THE FIVE PLATONIC SOLIDS

phosphorus — alum — hydrocarbon — boron — garnet — sulfur

CRYSTAL SHAPES OF SOME COMMON SUBSTANCES

## CRYSTAL DISCONTINUITY AS SEED OF LIFE

In their search for life in all this seemingly frozen abstraction, the crystallographers eventually settled on the spiral crystal as the probable ancestor or cousin of the helical molecules that compose all protein and genes. They also deduced the spiral must have originated as a lattice discontinuity or imperfection in an otherwise smooth crystal, which would make it not only a kind of crystal mutation (some

449

call it a disease) but the place where unexpected surfaces act to stimulate and accelerate growth. This supports the concept that irregularity or asymmetry is a key to life. And the discontinuity I have most in mind is called the screw dislocation because it is centered on an axis in such a way that the molecules stacked sequentially around it are induced to build a screwlike body, enclosing themselves like cambium cells girdling the stalk of a flower, patiently assembling their curved lattice, layer upon layer, as the growth plateaus sweep around and around and upward in the manner of a spiral staircase.

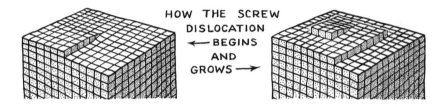

HOW THE SCREW DISLOCATION ←— BEGINS AND GROWS —→

A few crystals, however, including certain plant proteins, are so programmed that they grow square and hexagonal spirals and seem to be trying to reconcile the diverging mineral and vegetable kingdoms. Others, like hydroquinine, have evolved a double interlocking shape with twin-brotherly or schizoid crystals, both of whose members overlap in the same space at the same time without really becoming part of each other. Still others naturally grow in only two dimensions, fanning out into thin flakes such as mica. And certain very simple ones grow primarily in one dimension, ultimately sprouting into silky whiskers such as asbestos. There are even a few so lax in organization that they do not grow in any definite dimension. These are glass, which is structurally just a stiff liquid, not crystal at all.

Still others, showing unmistakable signs of life as we saw (page 387), branch into microscopic trees with twigs, foliage and "fruit" resembling berries, plums and apples. Some have also been known to get sick, overweight, neurotic or maybe just spoiled from pampering. And, not surprisingly, they can be treated and cured of illness, put on a diet, or have their wounds healed with new growth. They can also do all sorts of human things like sleep, travel, sweat, blush, glow, sing, even (like lodestones) fall in "love." In which connection I might point out that crystals have actually been seen to attach themselves to each other as if mating and, perhaps consequentially, to produce a family, occasionally including twins, triplets and other multiple births.

## CRYSTAL ENERGY

One doesn't usually think of famines as a problem for minerals, but nourishment is as vital to them as to any organisms, and if a crystal is underfed, particularly during its embryonic development, it will almost surely grow up scrawny and open like a skeleton with its substance concentrated on its edges. And if growing conditions get so bad it stops growing altogether, it may even shift into reverse and become smaller, literally ungrowing like an evaporating snowflake or the tadpole who ungrows his tail to transform himself into a frog (page 155), which amounts to self-digestion, a localized dying process that, in a mineral, is called etching. From the viewpoint of minerals, moreover, as I suggested earlier, this kind of dissolution is really more of a reincarnation or reawakening than a dying, since crystallization is essentially a cooling, settling-down and going-to-sleep process in which structure is formed by atoms expending energy and radiating heat as they compose themselves like a bear getting ready to hibernate. The crystal's energy leaks away most easily from its edges and easiest of all from its outer corners and protruding points, in the same way electrons leak off the sharp tips of lightning rods into the sky (to disarm the static potential for lightning), which explains why snowflakes, like cities, bones and other crystals, tend to grow fastest along their branches and twigs while their life is more abstract, negentropic and empyreal than that of secular, entropic rain. In some sense snow indeed serves as a symbolic being in rain's seasonal afterlife so analogous to that of the ethereal butterfly whose generations alternate with those of the cloddish caterpillar. Thus we see the mirrored paradox of life: on the one hand crystal slumber, cool and orderly as the clock that has run down and stopped, latent with complexities, fluid potentialities and disembodied dreams; and on the other hand volatile action, warm and free as a prairie fire, kinetic with restless ferment, simple, corporeal and seething with turbulent creativity.

## PIEZO, THE CRYSTAL OF LIFE

The sort of crystal that reconciles this paradox of life best of all is the rather extraordinary type called piezo, with something like a very simple nervous system that makes it generate an electric current whenever it is distorted by mechanical pressure. The verb "to press" in Greek is *piezin* so it was only logical to name this kind of vital response the piezoelectric effect. It occurs because distorting a crystal means moving its atoms in relation to one another and, since some of them are ions carrying extra electrons, this kind of distortion amounts to a flow of electrons which, by definition, collectively add up to an electric current. Putting it another way, as the illustration shows, when a

HOW SQUEEZING A PIEZO CRYSTAL GENERATES ELECTRICITY

piezoelectric crystal unit is not distorted, its centers of positive and negative electric charge coincide, and there is no flow of current. But the instant the same unit is squeezed into a different shape, the charge centers are pushed apart and the current flows as long as the centers keep moving. Furthermore, because the phenomenon is relative, it works reciprocally, so that, when an outside electric current is applied, the charge centers move and the crystal contracts or, if connected to an alternating current, it alternately contracts and expands. In which case of course it becomes an oscillator and can regulate frequencies in all sorts of electronic equipment.

Among the better types of crystals for piezoelectricity are salt (because it is full of ions), quartz, tourmaline and barium titanate, the last of which has become important in radio and TV microphones (where early this century it learned to sing), transmitters, submarine sonar systems, ultrasonic tools and cleaning devices. Then in 1954 two Japanese scientists named Fukada and Yasuda discovered that bones are piezoelectric and documented their report with evidence that the shear stress of collagen fibers, slipping past one another in a bone's crystal layers, generates electricity, presumably by bending or twisting the cross-linked atoms. American biophysicists, taking up the investigation from there, soon demonstrated that bones really are shaped by mechanical forces as D'Arcy Thompson divined half a century ago (page 16) and mostly with very definite help through the medium of electricity. When a growing thigh bone, for example, is bowed by external force, its convex side becomes charged positively and its concave side negatively, with the rather miraculous result that positive calcium ions steadily migrate from the convex to the concave surface, straightening and strengthening the bone. Indeed should the strain on the bone be removed, as during a period of weightlessness in space, the electric charges rapidly fade away, halting the migration of calcium ions and allowing them to be resorbed into the bone. In sum, it now appears almost certain that piezoelectricity is a common attribute of tissues, working unobtrusively not only in much of the mineral kingdom but in virtually all of the vegetable and animal kingdoms. And accumulating evidence strongly hints that senses in every kingdom operate more or less piezoelectrically, probably including the still undeveloped ones.

Since any stable structure is in essence crystalline by the very definition of a crystal as a latticed form that "wants" to maintain itself, we find crystallinity a quite universal principle, extending at least from the molecule to the supergalaxy and very probably beyond in both directions. The hemoglobin molecule, for instance, may be described as a crystalline organism whose main body contains four sections, each centered in an iron atom capable of grabbing and holding an oxygen molecule.

From here, ascending the size scale in jumps, the garnet is a ferromagnetic crystal with oppositely magnetized sublattices so independently responsive to temperature that, by warming or cooling it a mere degree or two, its polarity can be reversed, making it useful for storing information in computer memory cells. Another big jump takes us to the size of a lava field containing quartz, such as a large one on the North Idu peninsula in Japan where an earthquake in November 1930 was accompanied by spectacular lightning in a clear sky, a phenomenon already becoming well enough recognized to have disseminated the theory that piezoelectric crystals subjected to disruptive mechanical pressure over many square miles of territory can generate voltage sufficient to cause a thunderstorm.

In all such widely different cases, it seems appropriate to cite the increasing evidence that, in magnetically ordered crystals, the spins of the atoms periodically do something that could be called a flip, reversing their direction and polarity, just as happens much more slowly to the magnetic polarity of the Earth, the sun and stars and probably the galaxies. And this change actually advances from atom to atom like a row of falling dominoes in a microprogression physicists call a spin wave that sweeps through the material like a swell on the ocean. Indeed it is a vital factor in the melodic reality behind the wave nature of all matter — the still-mysterious truth discovered in France in 1922 by Louis de Broglie after the merest of hints in ancient times by Pythagoras, who had listened and listened for, and eventually heard, the music of the spheres.

453

## LIQUID CRYSTALS

While one naturally thinks of a crystal as a solid substance, the stability and order that are its essence also extend into water, other liquids and even into the realms of gases and plasmas. Looking and flowing almost like honey of varying viscosities, liquid crystals have optical properties that were recognized as crystalline as early as 1888 and by now their occurrence adds up to nearly one percent of all new organic compounds synthesized in chemical laboratories. The skin of a soap bubble is a simple kind of liquid crystal, in which flexible eel-shaped soap molecules regiment themselves like soldiers in stretchable double layers that form both the inner and outer surfaces of the bubble, confining a sheet of water between them. These parallelly aligned molecules stand perpendicular to their layer, holding their "heads" and "tails" in phase and, if the bubble stretches, extra "free" molecules in the interstitial water automatically slip between them to increase the area of the layer.

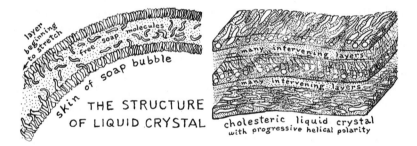

THE STRUCTURE OF LIQUID CRYSTAL

cholesteric liquid crystal with progressive helical polarity

A more complex liquid crystal is the kind that harbors cholesterol and comes in thousands of one-molecule-thick layers, with the molecules lying flat like sleeping soldiers, their long axes parallel to the plane of the layer and to each other but with each layer's axes rotated 15 minutes of arc relative to the next (the very same angle, by the way, that the radii of both sun and moon subtend from Earth), therefore cumulatively forming a helical progression of polarity that is common to all genes and protoplasm. The potent effect on penetrating light of this 15-minute twist of polarity per molecule can be judged slightly by the fact that it adds up to a spin of 18,888° or fifty full revolutions for every millimeter the light travels!

As a poetic chemist might explain it, the water molecule ($H_2O$) loves to dance with its oxygen neighbors, and this is shown by the fact that its two hydrogen atoms always reach out their "arms" to any oxygen atoms they meet. At the same time it has a philanderer's habit of latching on to all other hydrogen atoms that come within reach through similar bonds that keep joining them to its oxygen and, because the effective angle between the twin hydrogen atoms averages somewhere near 120°, the rolling, milling water molecules tend to join hands in momentary rings of six, while tied to other rings by zigzag chains, all the members excitedly grouping and regrouping as the turbulent units of water create their hexagonal chicken-wire-like crystalline lattices.

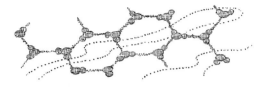

Something comparable has been known to take place in the forming of half a dozen small tornadoes around a central mother vortex, a similar number of eddies around a whirlpool, thunder cells about a thunderstorm, or even five or six clouds of space dust around the solar system. But the water molecule is also incurably congenial with carbon and a good many other common elements like nitrogen, phosphorus, magnesium, iron, etc. In fact it socializes with the cellulose as well as the carbohydrates of plants and trees and flows through all animals — certainly all multicelled ones — not just as plasma in their blood but by permeating their cells and the interstitial channels between cells, including organs and tissues of every sort, uniting and reuniting with them in its unceasing but remarkably orderly motion.

Putting it in less biological terms, the sea is not merely around us. It is inside us. For life on Earth, and perhaps life mostwhere, is primarily organized water. Thales intuitively understood this in the fifth century B.C. when he voiced his extraordinary surmise that "water is the one essential element of the world." And today we find that water literally constitutes most of ourselves — that a new-conceived human embryo averages 97 percent water, a newborn baby 77 percent, a grown man 60 percent. In judging a body's water content, of course one tends to think of fat as a fluid, so I'll point out that fat is chemically very different from water and, although water is two thirds of a thin cat, it is only one third of a fat pig, while the average woman is both fatter and drier than her husband. Moreover the proportion of water in all healthy bodies is regulated with remarkable

455

exactitude, any excess being automatically eliminated as urine, any shortage (even 1 percent) demanding its prompt replenishment through thirst.

On this note of intimacy between life and Earth's most characteristic substance then, we shall pause for a chapter change before venturing even deeper into the dynamics and geometry of life's elusive nature.

# CHAPTER 17

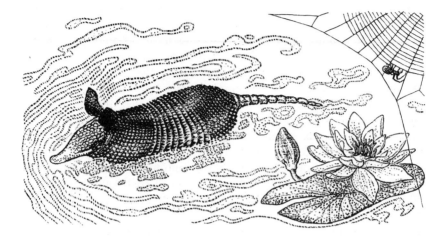

# Living Geometry and Order

W HILE WE ORBIT our way dynamically around them, Earth's creatures, from armadillos to spiders, live their own dynamic lives and display every kind of geometry from their studded hides to their symmetric webs. Even flowers, as we will soon see, have a flowing geometry closely analogous to the dynamic motion of the simplest of forms — forms that are naturally fluid and alive, and that epitomize the symmetry that seems to be the beginning of being.

## THE SPHERE

First the sphere, the most elementary of shapes and which encloses the greatest volume with the least surface of any imaginable body. It is the very plinth of primordial life and the germ of evolving complexity. Everything fluid from bacteria to bubbles to stars tends to this centrosymmetry, plus many a virus much smaller or a galaxy much larger, not to mention a supergalaxy or the knowable universe. Raindrops, for example, have to be nearly perfect spheres to produce rainbow colors (by refraction through spherical segments), yet, as they grow larger by accretion and fall faster, they begin intermittently to flatten out at the bottom from air friction, momentarily taking on a shape something like a hamburger bun, the while oscillating from bun to egg form hundreds of times a second until, as they approach a quarter inch in diameter, they break up from their own aerodynamic turbulence. Thus generation succeeds generation in raindrops, as in amebas, by division and division, each drop "breathing" air and carbon dioxide and in its rainy way "eating" and digesting dust, germs and spores and, on landing, spitting out microbes, nitrogen and sometimes carbonic acids strong enough to etch rock. Like the rain, the living cell also runs to the spherical or (when crowded) the polyhedral, presumably mostly because of its fairly continuous storing and spending of energy in a flexible bag that is under almost equal pressure on all sides. The cell's sensitive feedbacks of metabolism, moreover, normally ensure a remarkable net stability over the days or months, indeed a sort of steady state sometimes termed *the vital equilibrium* and which largely explains why sphericity is so universal.

If you still wonder how the trend to roundness began, consider the curious suggestion of some of the older geometers (and perhaps a few cosmologists) that spherical stability evolved out of primordial instability or waviness that somehow blossomed into turbulence. Liquid flowing from a bottle, to take a homely example, begins by limiting its body to a smooth, straight stem nearly cylindrical for a little way, but, by the time its length exceeds its circumference, it thins down, whereupon it can't help but weave and get weavier and wavier until it sort of lets go, bursting out in turbulent blobs and separate drops that take

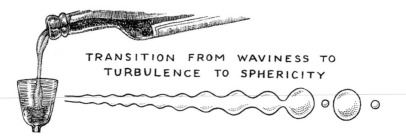

TRANSITION FROM WAVINESS TO
TURBULENCE TO SPHERICITY

the shapes of spheres. If there ever could have been a celestial analogue of this sequence starting in a stable, smooth universe, I can't help but wonder whether such a universe really might have developed a corresponding instability and waviness, veering into turbulence, eddies, galactic bubbles and spiral world systems in space-time? Or is cosmic turbulence as we see it in the sky today simply continuous, eternal and on the whole unchanging in God's creation, of which Baha'u'llah, Prophet of the Baha'i Faith (page 614), said, "Its beginning hath had no beginning, and its end knoweth no end"?

## THE FLOW IN FLOWERS

At least a hint of an answer may be gleaned, if you're interested, from a close look at the dynamics of splashes, which D'Arcy Thompson among others began to study last century. When a round pebble falls into calm water, he observed, its downward pressure after impact pushes a "filmy cup of water" upward all around, which "tends to be fluted in alternate ridges and grooves, its edges . . . scalloped into corresponding lobes and notches, and the projecting lobes . . . into drops or beads . . ." Although this creation lasts but a tiny fraction of a second, photographs show it to have a beautiful, symmetrical, flowerlike form put there by the same sort of dynamic forces that genetically infuse the flow into flowers, only many millions of times faster. And, curiously, the lateral speed of the splash has been measured to be several times faster than the impact speed that caused it. Of course it was a rather new concept in those days, but the recent proliferation of time-accelerated and time-decelerated movies has made it much clearer, demonstrating dramatically the extraordinary parallels between fluid turbulence and the relatively solid forms of life, between a lacteal crown of two dozen beadlike points tossed up in a splash of milk and certain polyps in the sea with two dozen vertical tentacles surmounting a similar cuplike body, between columns of ink sinking in water or fusel oil in kerosene and various medusae jellyfish, even between the furrowed torsos of protozoans, the fluting of instable sleeves of plasma in a jet engine and the gadrooned blossoms of gentians and lilies.

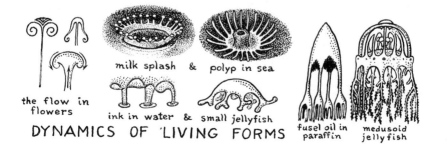

the flow in flowers

milk splash & polyp in sea

ink in water & small jellyfish

fusel oil in paraffin

medusoid jellyfish

DYNAMICS OF LIVING FORMS

459

HOW A BUBBLE BURSTS AT THE SURFACE OF THE SEA

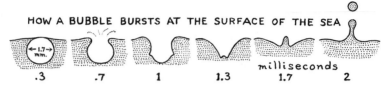

← 1.7 mm →

milliseconds

.3 .7 1 1.3 1.7 2

"The drop, the bubble and the splash," as Thompson says, "are parts of a long story," and the actual flows of material in all of them must have close counterparts with such rotund forms as eggs, eyes, honeycomb cells, hurricanes and volcano eruptions. Though it seems farfetched, comparable rhythms enable softwood shingles on a barn to "breathe" their nails outward, a tide of earthworms to bury stones, frost to heave highways, galaxies to spawn satellites, atoms to invent molecules. In every case, form is closely related to function, through which flow-force kinships link clams to cobras, campanulas to comets, a bursting bubble to the upshoot of a mushroom, the gastrulation of an embryo to a rolling, in-folding snowstorm. Acoustics, in a similar sense, is a volatile version of hydraulics. When sound enters a room and is absorbed by rugs, furniture and windows, it behaves like invisible water filling a tank containing baffles. Oil gushing through a pipeline so closely emulates traffic on a turnpike that highway engineers use data from fluid dynamics in drafting the shapes of intersections, islands, curves and grades. With the latest computers and supersonic wind tunnels they research turbulent flow as a general theory, subprogramming it with coefficients of oscillation and crystal coordinates in space-time, noting that, with increasing speed of flow, the "vortex street" (pioneered by aerodynamicist Theodor von Kármán early this century) exuberates progressively from prim, regular eddies

THE VORTEX STREET
OF WIND TURBULENCE
(viewed from above)

toward random, vagabond turbulence, whence, if you can overlook a modest allowance for gremlins, it may be safe to say that the feedback-tempered stability qualifies in the long run as one of life's most characteristic factors.

## THE GEOMETRY OF LIFE

In a calmer, optic aspect of the same vital geometry the rainbow bespeaks a conic skeleton engendered abstractly by the refractive angles of 42° and 51° between the kindred spheres of sun, raindrop and eye. If this be an etherialization of geometry, there are a thousand times more species of fluent "organisms" in the liquid and gaseous realms of the uncounted transsolar worlds, from volatile planets of the Jupiter breed to the starry arms of galaxies, which may represent much more than populated waves or residues of what some astrophysicists think are reversible (others irreversible) processes. Some of these "organisms" may be shaped by tensions in the manner of bubbles, which probably are more disciplined than ballet dancers and far more "single-minded" in their body control, as anyone knows who has dipped geometric wire frames into soapy water and looked thoughtfully at the resulting cubes, prisms, graceful spirals and tetrahedrons . . . For one has to admit that mysterious but rigorous laws are in full control here: that it is forbidden for more than three films to meet in a common line, or for more than four edges or six films to touch at a single point. And all angles of intersection must be balanced and equal, like 120° between each of three planes that together add up to a full 360° circle.

Of course when a batch of lively bubbles is freshly spawned, as in the washtub or a frog pond, many will momentarily flout the law: four carefree young bubbles, say, meeting in one common line, each perforce limited to but a 90° quarter of the circle. But, as the illustration shows, this arrangement is so pinched and uncomfortable, it is highly unstable — and the four-bubble line, itching to burst loose, will soon split in two like an ameba, giving birth to twin offspring lines joined by a new film that now enables each new line to relax and spread its young angles from 90° to 120°, as one of the four bubbles slips away,

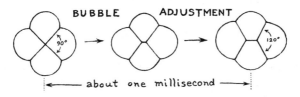

BUBBLE ADJUSTMENT

90° 120°

◄— about one millisecond —►

leaving only three. This is precisely how the stress in fresh-whipped lather or foam eases off as its myriad bubbles settle into more congenial configurations — a peace in effervescence you can literally feel and hear in the barber shop or the beer hall. And it says something, 461 however indefinable and mysterious, about how life springs out of bubbles through the media of seeds and eggs into amebas, bacteria, caviar, frogs and man.

There is something surprisingly enlightened (I almost said "mental") about intrabarm relations too: the way two or more bubbles will apparently consult each other and instantly make the most mutually agreeable of compromises. It seems to be a mystic nonverbal language of touch and pressure. Although each bubble, when alone, remains spherical in form because its elastic soap-skin yearns to be as small as possible (a yearning that is the *raison d'être* of sphericity), two bubbles that join together somehow agree to a modified form such that their outer surfaces, plus that of their shared interface film, add up to the smallest total area that can hold the air in both bubbles separately. This double bubble is actually an automatic, simple and beautiful solution to a complex mathematical problem. And triple bubbles even more so with their three interfaces all curved yet meeting only at 120° angles and all obediently, patiently, holding their invisible centers of curvature, for some secret reason, in a single straight line.

## EFFICIENCY UNDER PRESSURE

Cells in a honeycomb are something like bubbles, but even more regimented, being controlled by animals as well as the less animate forces, and their visible ends form the well-known hexagonal crystalline pattern that comes from subjecting circles to pressure. Johannes Kepler, the great astronomer, who enjoyed geometric puzzles, was perhaps the first to realize this and, logically advancing his inquiry into the third dimension, he observed that the honey cells are cylinders pressed so tightly together from six sides that they are laterally molded into hexagonal prisms. And the ends of the cylinder-prisms, which bulge outward, naturally settle into hollows between three neighbor cells, whose triangular funnel effectively pinches them into three-sided pyramids.

If the honey cells were shorter, wrote Kepler, say as stubby as pomegranate seeds, the same pressure would still give them the same angles and twelve equivalent faces where twelve neighbor cells touched and squeezed them: six surrounding at the same level, three just above and three below. That would make each cell a twelve-sided figure with diamond-shaped faces known to a geometer as a rhombic dodecahedron or, to a jeweler, the crystal form of the garnet.

462    Until the mid-nineteenth century, it seems, the few scientists who

thought about it believed the reason such garnet shapes are so common among compressed eggs and cells, indeed among any figures that completely fill space, is that this shape is the most economical since it packs together in such a way as to divide a given volume with relatively less partitional area than any other configuration. Then Lord Kelvin appeared on the academic scene in that legendary bastion of frugality, Scotland, to make the astonishing discovery that a fourteen-sided figure called a tetrakaidekahedron is even more ideal for a living cell than the dodecahedron, actually offering a slightly smaller

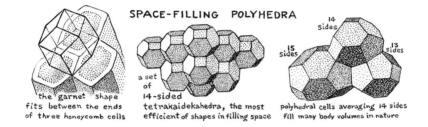

SPACE-FILLING POLYHEDRA

the "garnet" shape fits between the ends of three honeycomb cells

a set of 14-sided tetrakaidekahedra, the most efficient of shapes in filling space

14 sides
15 sides
13 sides

polyhedral cells averaging 14 sides fill many body volumes in nature

percentage of partitional surface. Of course he had to measure his areas very carefully because the fact was not judgeable just from looking at the thing, bounded as it was by three pairs of equal and opposite quadrilateral faces and four pairs of equal and opposite hexagonal faces. And in time there arose an aura of ancient mystery about it, because, unbeknownst to Kelvin, his brain child had once been studied by Archimedes, who wrote a detailed description of it in the third century B.C., and even before Archimedes, as someone archly speculated, it could as easily have been known to the "illiterate Pythagoreans."

Undeterred by such thoughts, Kelvin did some high-precision research on his fourteen-faced construct of life and was a little awed to find that, to achieve absolute minimal area, its edges must be slightly curved and its hexagonal faces warped just enough to become perceptibly anticlastic, like nearly flattened saddles which would be barely convex on one axis and barely concave on the other, with "equal and opposite curvatures" at every point. And although, as with the rhombic dodecahedron, a mass packed tight with such figures must so partition space that wherever three faces meet in one edge they can do it only at coequal angles of 120°, unlike the dodecahedron, wherever four edges meet in one corner, they will do so at coequal angles that ideally amount to 109°28'16", this being the so-called Maraldi angle (first approximated by the astronomer J. P. Maraldi), at

463

which lines from the four corners of a perfect tetrahedron meet at its center.

The only reason garnet (dodecahedral) forms are commoner than TKH (tetrakaidekahedral) forms in solid matter evidently is that friction is usually enough to inhibit cells from flowing into easier configurations. Experiments in squeezing masses of dry clay pellets together, for example, result in their coming out as models of nearly perfect garnets yet, after wetting and resqueezing (as might happen to massed fish eggs under tremendous pressures on the ocean floor) some of their rhombic faces begin to lop into pentagons, then hexagons, as corners are blunted to start new facets and the slippery little figures wind up as semi-TKHs in various stages of perfection. Naturally a substance as fluid as soapsuds really does come close to shaping its internal bubbles into perfect tetrakaidekahedra, but anything with the slightest viscosity, such as the colloidal cells of life, tends instead to compromise into a stew of polyhedra with the numbers of their faces ranging from four to more than twenty, a good half of them showing either 13, 14 or 15 sides and an overall face-count averaging close to fourteen. And that may well be the prevailing shape of life everywhere.

Hexagonal faces are common in such cell masses, but pentagonal ones are more so, probably because hexagons naturally fit together to form a flat plane in two dimensions, while pentagons incline to three-dimensional curvature. The microscopic globular animal in the sea called a radiolarian, for instance, is covered by a layer of hundreds of frothlike vesicles all about the same size which, like honey cells, get pressed into hexagons. Inevitably, however, there have to be, and are, a good many pentagons (occasionally a rectangle or triangle) among them for the inviolable geometric reason that, without the pentagons, etc., the animal's mantle could not curve around it and it could not be a radiolarian or even be alive!

Thus does abstract mathematics hold sway over life's forms, including the form of the colonial creature called a sponge, which begins its skeleton with hundreds of separate crystals, assembling them later into one design or another of prefabricated dwelling, the simplest example of which may be a six-rayed siliceous sponge whose spicules frame a living jackstone, a shape a mathematician might describe as a Cartesian coordinate hub with 90° angles in three dimensions. And if that strikes you as remarkable, let me say that dynamic relationships are virtually unlimited, there being probably as many of them inside a small biocrystal as, say, between a cat and a mouse, a sand dune and the wind, a coral reef and the sea, a city and the earth, a star and

the Milky Way . . .

## SOCIAL GEOMETRY

One of the most revealing aspects of living order is the complex and invisible cell structure of territories that stabilize social and political interaction from the pecking hierarchy of any small population of animals to the established boundaries of human empire. The concept of individual distance is critical here and closely analogous to the spacings of subatomic particles inside the atom or to molecular intervals in any crystal lattice. The first person to recognize individual distance as a law of nature, I am told, was Heini Hediger, the Swiss zoologist, who suddenly realized one March morning in 1938 while standing in Zurich's Bellevue Square that the black-headed gulls on the lakeside railing had spaced themselves with remarkable regularity almost exactly a foot apart. Then he noticed that the swallows on a wire were perched only six inches apart, while flamingoes in the zoo kept to a range of nearly two feet, the distance being roughly proportional to the size of the bird. Somewhat as the diameter of the hydrogen atom spans exactly one angstrom while the space between $H_2O$ molecules in solid ice is a uniform 2.72 angstroms, each kind of organism has its characteristic distance. Thus a deer may become apprehensive if you approach within 200 yards of him in a field, a mountain sheep about half a mile, an eagle a mile or an alien spaceship 100 miles. Social researchers report that an American man in the street stands on the average about 20 inches away from another man in a conversation, and about 24 inches from a woman. But these distances vary with countries as well as with sex, age and mood. In Cuba, for instance, a man may stand only 13 inches from an educated woman without being unduly suggestive.

More complicated is the interrelation when a family or flock lives and communes together, something expressable by the formula of mutual understanding $\frac{n^2 - n}{2}$ which gives the number of direct communication lines (. . . telephones, smoke signals, whiffs of perfume . . .) needed between any number (n) of individuals in order that no two of them will be without their unique private channel of interchange. If the number of individuals (n) is three (call them A, B and C), the formula $\frac{(n^2 - n)}{2}$ amounts to $\frac{(3^2 - 3)}{2}$ or $\frac{(9 - 3)}{2}$ or $\frac{6}{2}$ or 3, indicating that three channels of communication are just enough to connect each of the three individuals with both of the others, the channels forming a triangle $A{<}^B_C$. While the formula is presumably fundamental to everything from crystal structure to multicellular life in general, it may be modified by various factors, like the pecking order in the tribe or the overall hierarchies of evolution, both of which augment crystalline stability in life's intra- and inter-relations.

A kindred fact came to light in a Harvard project in 1967 known as the "small world problem," which discovered that a chain of only five friends will link any two average individuals in the United States and probably ten friends any two on Earth. One test, for instance, required each of 160 randomly chosen Nebraskans to reach a stockbroker living in Sharon, Massachusetts, by writing a letter to a friend (one close enough to be called by his first name) who seemed to have the best chance of knowing him, asking the friend to write in turn to another friend until the broker was reached. The number of intermediaries in this typical sample varied from two to ten, averaging five and, by simple extrapolation, it should take no more than ten of them on the average to link the whole planet, the number being presumed to decrease as international travel increases. Of course if you count the strange intangibles of acquaintanceship the chain is bound to shorten still more, because familiarity with one's fellow planetary passengers obviously does not depend wholly on personal contact or private communication. An actor and a television commentator might know each other intimately, just from seeing one another on the screen where their manners and opinions have been household fare for years.

And another factor is life's uncertainty principle, roughly comparable to Heisenberg's Uncertainty Principle inside the atom, and which deals, among other things, with the orbits of one's family. Just where a spouse or children may be at any particular instant when they are out of sight is normally uncertain, except within vague physical and mental limits. You "know" they are within some sort of boundary, such as the town lines, the marital domain, the home lot, the house, a particular room or a bed. But exactly where and when they are in that bed, room, house or other area you seldom know with certainty — nor would it be necessary or natural to know it continuously. Life enjoys some leeway. It is a vital law without which life would not be life. Even a babe in the womb goes through orbits his mother cannot follow . . . and there are worlds of humming mystery within us all.

Accidents and mutations, almost always passing unnoticed in the microcosm, like the crystal lesions in plant protein (here shown),

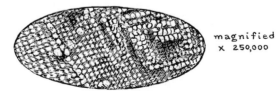

magnified
x 250,000

barely whisper the inscrutability of life and matter. How long, did we say (page 106), a cell remains a cell? A molecule a molecule? Even Democritos surmised in the fifth century B.C. that "the love and hate of atoms is the cause of unrest in the world" and Shakespeare, who could not know that a man contains $10^{28}$ atoms, the earth $10^{52}$ and the visible universe $10^{87}$ atoms, concluded, "It is as easy to count atoms as to resolve the proposition of a lover."

## THE MUSIC OF WAVES

The average farmer would probably be surprised, not to mention skeptical, if told there is a precise wave effect resembling music and life in wire fencing: that two lengths of ordinary hexagonal chicken wire laid crosswise, one upon the other, will weave visually together (like intersecting waves) into a mesh of small pentagons. Or that a triangular netting overlaid upon a piece of the same chicken wire will pro-

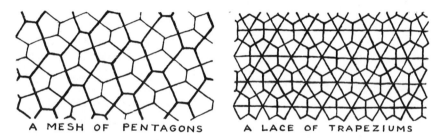

A MESH OF PENTAGONS     A LACE OF TRAPEZIUMS

duce a fine lace of trapeziums. Such results are hardly what you'd call expectable, and even less so are the exquisite patterns now known as moiré art that are created by the juxtaposition of two or more silky fabrics or screens, often with dazzling if not psychedelic effects. The grain of any of the screens may be rectilineal, radial, wavy or of some other crystalline weave, the combination producing its own shimmery wave effect, which is optically analogous to the aural beats in music made by dissonance between notes of slightly different pitch. Or it can come in the form of Lissajous figures or harmonograms, which

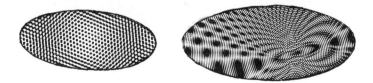

are multiple line patterns drawn by vibrating tuning forks or the precessing orbits of pendulums or wheels that oscillate pens, feathers, beams of light or electrons until they have woven zany, topologic, cyclic, overlapping lattices full of similar unexpected graphic dissonances.

Waves and the oscillations that generate them are of course among the most universal of phenomena and are familiar today in tides, winds, weather, rivers, ocean currents, magnetic lines of force, earthquakes, sound, radiation belts in space, sunspots, meteor showers, galactic gyrations, mass migrations of fish, birds, insects, microbes, mammals, man, etc. But two centuries ago almost nothing definite was known about the dynamics of waves and their oscillations, nor was it until just before the French Revolution that an imaginative German scientist and musician named Ernst Chladni discovered that, if he covered a metal plate with sand and vibrated it with his violin bow, the sand would be rapidly shaken away from most of the plate, disposing itself as if by magic in a pattern of nodal axes along which the plate was almost motionless.

Out of this rather Pythagorean and almost mystical revelation has gradually emerged the science of cymatics (from the Greek *kyma*, wave) in which modern researchers use piezoelectric crystal oscillators (page 452) to vibrate various powders, liquids, gases and blazing plasmas at thousands of cycles per second to see (and hear) how they will react. In effect it unleashes a new brand of molecular excitement that sews sand into lace and weaves honey into tapestries, the texture becoming finer as the pitch rises. It induces cream or iron filings to dance, regiments liquid waves into quavering "roof tiles," impregnates a film of glycerine with a fishbone spine, crimps flame or smoke into mackerel ribs and honeycombs soap bubbles into hexagonal prisms that literally "breathe." Music is made visible by it, notably with the aid of a sensitive diaphragm covered with a film of liquid that ripples with all the fleeting cross-currents and subtle complexities stirred by

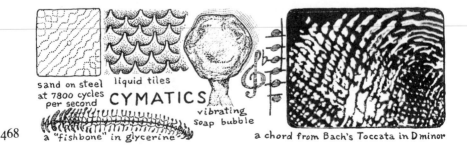

sand on steel at 7800 cycles per second  liquid tiles  CYMATICS  vibrating soap bubble  a "fishbone" in glycerine  a chord from Bach's Toccata in D minor

468

the sound of anything from a ditty to a symphony. And even though the beautiful melodies of Bach or Rachmaninoff here flow far too fast for the unprepared eye, they may be shot "dead" in their crystalline time tracks by a camera and mounted, as you see, in full photographic fidelity for leisurely yore-phase analysis.

Probably the most interesting of all examples of vibrational sculpture is the virtually cosmic behavior of club moss spores strewn upon an oscillating plate, for these fine grains of powder quickly congregate into round clods that grow, rotate and orbit around each other like stars in a globular cluster, the size of each body varying in ratio to the intensity of the vibration, each crescendo a frenzy of love drawing more of them together, each decrescendo a quenching liberation. And a definite quality of life arises in this "otherworld" landscape with each pulsating pollen "organism" repeatedly oozing forth a probing finger, then creeping after it like an ameba, whereupon, by acting at the same time inherent in the universe and coherent in itself, it emerges as a tiny piece of world.

The vibration that means life does not necessarily come from outside, however, for any temperature above absolute zero keeps molecules in perpetual motion and the $H_2O$ molecule in solid ice changes its position in its crystal lattice a million times a second. This of course is the inorganic brand of vibrant life that throbs through every snowflake that spirals down the sky — and, at still faster rhythms, every dancing drop of rain, every bolt of lightning. A drop of emulsion diffusing into ink without outside interference, for example, shows a natural to-and-fro movement like a flag waving, a whistle being blown or smoke tumbling from a chimney. It spurts and yaws and folds into unpredictable wispy patterns that are a mystic example of the emerging truth that turbulence naturally promotes life in the sense that it liquefies and sensitizes a medium to modulate the forces and rhythms that act     469 on it: light and sound waves, mechanical motion, chemical and radiational influences of every sort. And most of the forms we have been discussing, whether created by artificial or natural vibrations, or holograms or fluid models that simulate fields of force, bear such close resemblances to crystal and other quantized structures common in rocks, plants and animals that one can't help realizing there is a fundamental law at work here ordering the textures and shapes and life of everything in the universe.

CHAPTER 18

# Fourth Mystery: The Polarity Principle

B ETWEEN THE CREST AND THE TROUGH of a typical wave, including many of the kinds we've been talking about, there is a certain up-down polarity like the north-south polarity of Earth. This is one of the simpler geometric examples of a little-known but universal relationship that, in many ways, is the most paradoxical and provocative of the Seven Mysteries of Life. It is the subject of this chapter.

On first venturing into space, I used to feel I was way up here looking way down there upon the world. But now I've come to know that, in a bigger, truer perspective, up and down are only provincial illusions of local planetary life — illusions that, I was surprised to discover, were well understood by at least a few philosophers several millenniums ago. For wasn't it Herakleitos of Ephesos who said, in the fifth century B.C., that "the way up and the way down are one and the same"? And wasn't it he who later generalized the concept by adding, "It is sickness that makes health pleasant . . . weariness precedes rest, hunger brings on plenty and evil leads to good"?

Probably by now you have an inkling of the underlying pattern lurking behind such contradictions — or at least you suspect there must be a principle of some sort to account for the paradoxes that keep turning up in life. You are right. Let's call it the Polarity Principle. Evidence of it is virtually everywhere. When you open a window in winter you not only let in the cold but, in that one act, you also let out the heat. When the blacksmith hammers the anvil, the anvil also hammers the blacksmith. It's Newton's famous law of action and reaction. Then there are such inspired responses as the oyster's healing an ugly wound by turning it into a beautiful pearl. And, seeming to the contrary, all kinds of lovely creations serve the most deadly of purposes: the graceful leopard prowling, the orb spider spinning, the owl swooping silently in the dark, the snake stalking a frog, the barracuda herding a school of mackerel, the dragonfly over-taking a mosquito . . .

All creatures are affected more or less by others, and many are literally nibbled by the enemy, their fur or feathers often as not ragged from the struggle of life. Even the most handsome buck may have a piece of antler missing, a split hoof, a louse in his tail and a worm in his heart. After all, where would you go if you were a heartworm?

## PREDATION

Here we are back to a subject touched on at the beginning of this book, and again in Chapter 13: predation. It is the interaction between predator and prey — a problem few philosophers seem to want to discuss and virtually none has seriously tried to resolve. I call it a problem because the idea of killing seems unacceptable to most law-abiding, life-loving people, and the question arises: is it important to the world for so many animals to kill and eat each other? Is a meal really more precious than a life? Is the traditional concept of animals "red in tooth and claw," battling each other to the death under the natural law of "survival of the fittest," true? And, if so, does man have to be a participant in it today or in the future?

First let me say that the interaction (including predatory interaction) between all creatures is so pervasive that it surely includes man. And
472   it must continue to do so for the foreseeable future, despite such

evasions as the reported sale of rubber "worms" to squeamish human anglers, who may not have even considered whether fishing is harder on the fish than on the worms. I know there are idealists who rate predation as bad because all creatures are "brothers" and it's wrong to kill your brother — a doctrine that, if carried to its logical extreme, would make a cannibal out of everyone who eats lunch. And many such people persuade themselves to become vegetarians, presumably because cattle behave more brotherly than carrots.

Practically speaking, of course, man, like all large animals and almost all small ones, can survive only by killing, be it directly or indirectly. I can imagine a fanatic who might try to get along by eating only vitamin pills and milk or perhaps animals or vegetables that had died naturally, but his wouldn't be much of a life even if he survived. Almost everyone, in choosing food, condones the killing of fresh vegetables if not animals, and even a vulture implicitly approves the predation that provides his meat, to say nothing of the slaughter of vegetation that nourished the prey or the prey's prey beforehand.

What I am saying is that killing per se is not evil, for it is woven into the very fabric of the planet. "All flesh is grass," and inevitably everyone who really lives has to choose at some point among the interacting traffic around him. Even if he never hunts or farms, in effect he must take sides somehow in the competition between animals, vegetables, minerals, man, disease, weather, fire, flood, earthquake . . . or else he might one day find himself living precariously in an overgrown jungle or swamp that is about to be consumed by plague, war, insects, fungus, poison, drought or some other influence beyond his control. But man has long been a hunter and it isn't reasonable to suppose he got into the business or its offshoots (herding and farming) just by accident. There is every evidence that hunting (including vegetable hunting) has been his main occupation as long as he has existed, and the excitement and challenge of the animal hunt has been important, perhaps vital, to the development of his mind, his speech and much of his early culture.

The story is essentially the same for animals who hunt animals, even though they're not strong on culture, and it's hard to think of any wild animal species that has not either benefited by predation or suffered from its denial. This understanding of predation, however, was almost unheard of in the early part of this century, so it was only natural that a goodhearted pioneer of conservation like President Teddy Roosevelt would try to help the deer in Kaibab National Forest in northern Arizona by ordering the shooting of pumas and coyotes found preying on them. The rangers reported that there were about

473

4000 deer in the area in 1905 and the President was glad to hear that their numbers steadily increased in the next few years, as the pumas and coyotes were eliminated. But by 1920 when the deer population reached 60,000 and the forage had been grazed very short, serious doubts began to be heard. And in 1924 when the numbers got to 100,000, the deer were dying of starvation everywhere under disastrous conditions and Kaibab was turning into a desert. Even with the population down to 40,000 again by 1926, there were still ten times too many deer and it took another couple of decades before the dawn of the science of ecology and a new evaluation of pumas and coyotes eventually got things back to a reasonable balance.

An even more dramatic example of predation, and one better documented, is Isle Royale, the 45-mile-long wilderness sanctuary in Lake Superior which, being separated by fourteen miles of open water from the nearest land on the coast of Ontario, is relatively uninfluenced by mammalian migration. Before this century, as far as anyone knows, the inhabitants of the island included neither moose nor wolves, but around 1908 some moose, who are excellent swimmers, evidently swam out there to get away from wolves on the mainland, and by 1915 the island was home for an estimated 200 moose. From then on the unmolested moose multiplied steadily and numbered several thousand all through the 1920s, but inevitably ate up so much vegetation that by 1930 they were running out of it, getting scrawny and diseased and soon were starving in large numbers. The herd of about 800 moose that survived continued in pitiful, semistarving condition through the thirties and forties until the very cold winter of 1948–49 when a pack of about 16 timber wolves came out to Isle Royale on the ice and, finding easy pickings among the weakened moose, stayed there. Then, after a few years in which the wolf pack grew slightly to about 20 while the moose were reduced to about 600, the two species established what ecologists call a predator-prey equilibrium of around 30

WOLVES PURSUING A MOOSE
AS SEEN FROM ABOVE

ONTARIO

ISLE ROYALE

LAKE
SUPERIOR

moose per wolf. By this time the oldest and sickest of the moose had been culled and the browse had had time to recover enough so the moose herd was better fed and healthier than at any time in the half century it had lived on Isle Royale. In terms of what is called the food pyramid, .8 tons of wolves were living on 45 tons of moose who were living on 2900 tons of foliage. Or, you could say, the average wolf was eating his own weight in moose (or other game) every week plus, indirectly, more than sixty times that much in fodder spread over at least ten square miles of land.

The significant thing about this relationship, of course, is that the eater and eatee do not compete to finality as species, for they need each other. They are playing a continuous game which, like all games, has rules. And a prime rule for the predator is: he kills only what he needs to sustain him. He definitely cannot afford to exterminate the prey or even make the prey scarce, for that would be eliminating his own food supply. It would be suicidal. No, he is basically a conservationist. He has to be — something like a farmer who prunes his crop and reaps it but makes sure he can raise another harvest next year.

Being a conservationist, however, does not mean that a wolf or lion ever consciously moderates his hunting or holds back from killing with a thought for tomorrow or next month. He doesn't need to exercise such restraint because killing is just not that easy. In fact on Isle Royale the records of many years show that, whenever a moose at bay stands his ground, wolves give up in a few minutes. They know from experience how deadly accurate can be his kicks, so they shy away to look for easier prey elsewhere. And only one moose out of every eight who flees actually gets caught and killed by them, always a weak one: very young, very old or crippled in some way. All in all, I am told, only one wolf attack out of thirteen upon moose results in a kill. And in other parts of the world the "batting average" for lions, hyenas, leopards and other predators, including hawks, is about the same.

Predation has a curious balance about it and countless complexities. The animals on both sides know when the hunt is on — and off. A herd of gazelles will graze peacefully within fifty feet of lions, because they instinctively know the lions are well fed and sleepy. But as soon as they sense a certain tension — perhaps just the twitch of a tail or a whiff of adrenalin imperceptible to man — they will be off for the horizon. This is perfectly natural, like breathing, to animals — something prey take for granted and practically never concern themselves about except at those relatively rare moments when action is vital.

475

## TERRITORIALITY

But if predation takes an animal's full attention only occasionally, other of his relationships can be much more continuously demanding. I mean his need to defend his territory (if he is a territorial animal) against rival groups of his own species, with each of whom he must hold a polarity relation of victor-vanquished, the end he gets depending on whether he is winning or losing ground. And there is also his need to hold or improve his individual rating (with its similar polarity) in the social order of his own group. These two relationships naturally often overlap and, on occasion, can be virtually synonymous.

The need for territory, a place of one's own, or for privacy for one's family or community is of course a very ancient and fundamental need. Some say it is an evolutionary force older than sex. And it affects nearly every creature from microbes to men, from the coral fish on the reef to the song sparrow in the meadow to the pet dog who barks at strangers from behind the fence his master built to define his own (and the dog's) property limits to those same strangers. About the only creatures *not* territory-minded are ones who migrate continuously, such as the grazing animals who have to move to live. But their vegetal wanderings could in some cases be construed as a weakness and it may be the browsing gorilla's territorial vacillation that primarily has led to his tenuous position in evolution, one apparently nearing extinction. If indeed you think the territorial instinct is less compelling than, say, the mating instinct, just compare the number of men who have died to save their wives with the number who have died to save their lands. And consider whether it's as hard to get a woman as to get a home to put her in.

Having property of your own, though it requires a lot of attention and can lead to strife now and then, unquestionably does reduce conflict. It works — and is the way of least friction and simplest motivation. Almost everywhere in the world customs and laws support the private property system, even to a degree in communist countries — presumably out of practical necessity. And, as we have seen, animals and lesser creatures accept it as part of nature for the same reason.

Most territories are simple, circular and more or less two-dimensional because they are created by surface-dwelling creatures like coyotes, turtles, and cockroaches, but birds and climbers extend them up into trees or cliffs, badgers and burrowers down into holes and swimmers into rivers and seas. They are apt to be formal, sometimes even luxurious in an animal kind of way. One occupied by a pair of African birds, for example, was found to contain: a nest, 2 sleeping roosts, a drinking place, a bathing pool, 3 feeding areas, a food store, 6 excretion dumps, 7 border stations for calling or singing, a rubbing post and a special boudoir for preening — altogether a lot more than the accommodations of any of the human inhabitants of a nearby village, including the chief.

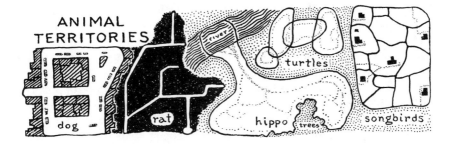

Territorial shapes obviously must fit their environs. That of a city dog has to be latticed like the streets, a sewer rat's is linear like the sewer, a hippo's gourd-shaped with the narrow part on the riverbank (where competition is keen) and spreading inland. Although the boundaries often shift and overlap as well as being generally invisible to human eyes, they are reasonably distinct to the creatures concerned: marked by the songs and displays of birds, the excreta and spoors of insects and mammals. They are thus semi-abstract biocrystal networks covering much of the habitable world, a different coarseness and texture for each kind of animal.

A pride of four or five lions in Africa tends to maintain a territory of approximately five square miles, so each lion uses on the average about one square mile of grazing land to provide him with food. The leopard needs three times as much. Stone-age man (relatively slow and feeble 10,000 years ago) needed eight times as much. And bigger animals, even those without a definite territorial system, suffer seriously without vast domains, so much so that they seem genetically restrained by geography on anything less than a heroic scale. Elephants in their natural state collectively require the sweep of a continent for their full evolution, and the tiger is unable to survive on any island smaller than about 2500 square miles, Bali (90 miles long) being the smallest island on Earth with an indigenous population of tigers. Curiously, smaller islands have the effect of dwarfing individual animals (in stature if not in population density) as in the case of Shetland

477

ponies, the deer of the Florida keys or the little gray fox of Catalina. It works even down to the dimensions of an aquarium tank, for mature electric eels will grow more than an inch a day when suddenly put into larger space. Short-legged animals naturally need less land than long-legged ones, so the warthog is satisfied with a quarter of a square mile and the woodchuck with an acre. Most varied of all are the territories of birds, which range from the 40 square miles of a pair of golden eagles to one tenth of an acre for warblers in time of plenty, while terns on volcanic Isla Raza in the Gulf of California, when brooding (not foraging), crowd together so densely they average a nest to every square foot!

An important reason for the territorial system is that, despite a few scouts and pioneers, the great majority of creatures (including man) tends to resist going out into the sparsely inhabited lands of opportunity, preferring instead the traffic and excitement of daily competition with close neighbors in a crowd. It's the lure of Broadway over the drear of the boondocks. The French ethologist Jean-Jacques Petter labeled it the *noyau* or society of inward antagonism, which, like an alloy, gains strength from the internal stress of disparate elements. It feeds on the paradoxical psychology of "I can't live with you — but I can't live without you" uttered by a bickering, loving couple. It is why Arabs delight in living among Arabs they can quarrel with more exultantly than with Jews or other foreigners. It builds on the stimulus of friction and is thrilling to citizens who like to outshine their neighbors, be heroes to their families, flaunt their possessions, yell names, argue, bet, sue. A study of juvenile street gangs in Chicago reported that a call for volunteers to steal a car usually brings out half the gang's members, while a summons to fight invariably brings them all. There is something about a brawl. It has almost the shock value of a fire in getting everyone's instant attention, especially when you know it is to decide issues as important as the border between you and your enemy. It combines the fascination of war with the masculine attraction of rivals for each other and the reconciling by tooth or treaty of opposing pressures into a boundary settlement both sides can live with.

If you find it hard to believe that the territorial competition of males produces measurable pressure, you should be interested in the experiment conducted by the U.S. Fish and Wildlife Service a few years ago to test it by killing all mated male birds in a 40-acre section of Maine woods. For, to everyone's surprise, almost every male vacancy they created was filled so quickly by a new and eager male that the hunters could not keep up with the influx of substitutes and, although they kept shooting all summer until male birds temporarily became scarce, the number that showed up the following spring was substantially the same as ever. Thus territorial pressure is maintained by flexible hidden reserves behind the frontier, this being, I assume, a key factor in its stability.

## PECKING ORDER

If territoriality and its polar stresses create a vital stability, primarily between rival males, there is at least as great a polarity and perhaps an even more pervasive competition for rank within each social unit or family. This is the phenomenon widely known as the pecking order, because pioneer ethologist Schjelderup-Ebbe noticed that a flock of chickens maintains natural harmony between its members by keeping every bird aware of who may peck him if he doesn't look out, as well as whom he may peck if he can. It has some of the quality of a very stable, regular crystal, yet is a hierarchy of strength unwittingly balanced by relative intimidation spiced with occasional all-out fighting when the flock was formed. Although the proud top chicken outranks all the others and can peck anyone without fear of retaliation, and the shy bottom chicken is disdained or bullied by all, both are accepted members of the integrated flock, all the rest of whose members have at least one they may chase and at least one they must avoid. There is a natural discipline to it, as in a military organization, where the top general may command anyone below and every lowly private may be commanded by anyone above — and they all get along together in reasonable harmony.

It is not always that simple though, for the pecking order in one activity, say building a nest, can be different from that in another, say feeding. And the boss animal in hunting may well be forced to wait for others when it comes to taking a drink. Also population pressure can disrupt the order of rank, aging tyrants seldom retire gracefully and mating elevates a female to the level of her "husband." In fact the young female who "marries the boss" may suddenly find herself in a position to snub former female superiors and lord it over males who have been pushing her around for years, for now she naturally stands with her husband, and he with her, in any dispute. Interestingly, this kind of loyalty invariably seems to prevail between animal mates, for the pecking order is in practice definitely superseded by the marriage bond. However the "single standard" that enables a high-ranking male to raise the status of his humble bride does not work the other way, for a humble male just doesn't get infatuated by a bossy female and, as Konrad Lorenz declared in his report on jackdaws, "no male may marry a female who ranks above him."

Did you realize that even the gentlest adult humans have a comparable ranking order that can assert itself in action as innocuous as a glance? When two strangers look at each other for the first time, for instance, is it significant that one of them averts his eyes before the other? Yes, an experiment conducted by psychologist Brian Champness for the British Association for the Advancement of Science found that the more dominant person, whether a man or a woman, almost always looks away first. In fact it is a reliable signal that he is about to claim the floor. And test scores show that the signal is invariably accepted by the more submissive one and, from it, human rank can be established, even though, in some cases, the persons involved are not aware of it. Champness took great pains to check his findings with other kinds of tests and devised a scale in which 1 equals a complete hierarchy (everyone knows whom he dominates and who dominates him) and 0 equals no hierarchy (no one knows his place). On that scale a flock of chickens rated .9 and Champness's group of humans rated .8, but established their hierarchy in only half the time chickens usually take.

So, you see, all of these three types of conflict we have been discussing — predation, territoriality and pecking order — involve violence or threats in various degrees, but in each case what you might call the stable aspect of the well-balanced crystalline structure is promoted either by shortening the agony, lowering the batting average or turning the confrontation into a ritual through which the issue can almost always be settled without bloodshed. Ritual conclusions, in fact, are the rule for conflicts within, and often between, groups of animals and men, though the exceptions naturally attract more attention. A typical encounter between two rival timber wolves in the same pack often starts off so furiously with bared fangs, snarls and lightning snaps at each other that any observer might think one of the two must surely get killed. But when the weaker wolf eventually tires or becomes careless and the stronger takes advantage of him, something incredible happens. Instead of desperately defending himself, the losing wolf suddenly surrenders. He seems to know the game is up, so he does the exact opposite of what you might expect. No longer facing teeth with teeth, he turns his head away, exposing the vulnerable bend of his neck to his enemy. Now the victor with one bite could sever the loser's jugular vein and kill him, but he does not. Actually he cannot. A powerful inhibiting force stops him. It is mental, but more

480

inviolable than a white flag or a red cross to a human, and it nicely resolves the polarity paradox within a species, avoiding a serious loss of life, which would have had no redeeming evolutionary benefit. Lorenz indeed won the Nobel Prize for elucidating this and other hard-to-understand aspects of animal behavior, many of which have to do with polarity.

## POLARITY IN EVOLUTION

Polarity shows up frequently in evolution. We see it everywhere from the genetic mutation that is commonly regarded as destructive or "evil" except on the rare occasions when it turns out "good" to the complement known as sex that assures the mutual polar attraction of opposites. The latter of course produces the variation of offspring without which evolution could not select and evolve. And there are creatures who seem to resist evolution, such as a simple beach animal called the pre-Cambrian sandworm, who adapted to the young Earth's shore environment 600 million years ago so perfectly that he hasn't had a measurable need to evolve farther in all the eons since. In fact he attained an equilibrium with the planet perhaps more successfully than any other creature, including man, who, incidentally, has existed less than a hundredth as long and is already in such a precarious crisis that some prognosticators give him only an outside chance of survival for even one more millennium. But the sandworm's very success is paradoxical because he has a built-in polarity between stability and instability, between peaceful harmony and distressing strife. For there are many species of sandworms and the most successful ones, which stopped evolving, eventually stagnated in an evolutionary dead end, while some of their unsuccessful, poorly adapted cousins continued being unstable, restless and dissatisfied and, for that reason, went on trying out new forms and behaviors and eventually evolved into such unprecedented classes as reptiles, birds and mammals — demonstrating that failure to adjust in early evolution may be just what is needed for success later on, that stress and strife are ingredients of long-range harmony, that pain is vital to birth and creation.

A different kind of polarity, also clothed in stress and pain and revealing no less important a principle in evolution, is the self-predation practiced by creatures who prune their own offspring to strengthen the breed. The marsupial cat bears up to 24 young in a litter, yet significantly has only 6 teats, leaving places for only the toughest or luckiest to suckle and survive. And the same principle holds for the    481 opossum and various other animals, not to mention many plants, such as the black walnut tree, which puts out poison to choke off not only surrounding grasses and shrubs but even most of its own offspring.

## SEXUAL POLARITY

The polarity of sex is clearer, but it too has aspects beyond the obvious, beyond the physical male-female complementarity. Although sex and reproduction at first seem synonymous the one tends to decrease numbers by merging, the other to increase them by diverging. And while it is said that bodies are united by pleasure and souls by pain, how many masochists find pleasure in pain, even when self-inflicted — in submission, succumbence or, finally, suicide! There are also the love-death polarities like that between Achilles and Penthesilea, the Amazon queen he killed and loved. And the anomalous sex of sea horses, where she injects ova into him, and he gives birth.

Carl Jung, speaking as a westerner, once suggested the earth has a sexlike hemispheric polarity by pointing out that "while we are overpowering the Orient from without (through science and technology), the Orient may be fastening its hold upon us from within (through spiritual influence)." Which is analogous to declaring that while a man is mastering a woman from without (through muscular strength, etc.) she may be enslaving him from within (through hormones and psychic forces), and her control, though less obvious, may be the greater. And why, do you suppose, the female is generally affected more than the male by the love experience? Isn't it because the polarity of sex is not exactly mutual — because rape has a one-way direction — because, if you liken the phallus to a gun, shooting is a less drastic experience than being shot? Offhand the only way I can think of in which the sexes are really equal is knowing (in their hearts) that they really are not.

An analogy bearing on this is that between the mating polarity of male and female and the predation polarity of predator and prey. If you can consider a tooth a phallic symbol, it is clear that its effect, either as tooth or phallus, must be more traumatic for the recipient (female or prey) than for the activator (male or predator), whether or not the recipient provides a haven (as in the case of the female's sex organs). Actually there are simple little animals like the rotifer (page 88), whose phalluses literally bite into any part of their mates' bodies just as if they were teeth but serve to impregnate rather than digest them.

## POLARITY OF GOOD AND EVIL

When we contemplate the omnipresence of life, of course we run into what I'd call one of the profoundest polarities of all. It is the increasingly evident truth that, as every world contains a potential seed, so does every seed contain a potential world! And from there, as if a mirror divided positive and negative, the roster of polarities continues inexorably on into matter and energy or, if you'd prefer, matter and antimatter. Then into cause and effect, subject and object, the concrete and the abstract, the macrocosm and microcosm, science and religion, Creator and creature, liberty and slavery, free will and fate, mortality and immortality, yin and yang, good and evil . . . and a hundred more.

Good and evil, I dare say, epitomize polarity as much if not more than any other pair of opposites, so I want to examine their relation closely. Indeed the fact that they represent value judgment through the whole spectrum of spirit makes them important to understand as clues to the deepest meaning of this world, and to the question of whether a supreme spiritual essence exists — and, if it does, what it is like.

Naturally this subject brings to mind such ancient and troublesome but pertinent questions as: Why is there evil in the world? Why does God the Potter make "one vessel unto honor and another unto dishonor"? And, as many have asked in vain, how can there be a God, presumed to represent the highest good, Who nevertheless creates and presides over a world containing injustice, crime, war, ugliness, destruction, pollution, disease, pain and other apparent evils? Will there ever be anyone who can resolve this bitter paradox that has plagued

483

man ever since he began to ponder the meaning of his world? In brief: what good is evil?

The more I wonder about this question, the more I am convinced that the answer has to be bound up in the concept of relativity — in some basic sort of resolution of evil's obvious polarity with good. Does efficiency then somewhere blend into charity? Is there a boundary between anarchy and eugenics, or an Umpire who can reconcile Survival of the Fittest with the Golden Rule?

Perceptive philosophers have long touched on the intimacy between good and bad, the key idea emerging that the human soul thrives on a challenge or a problem and, once it is stretched by struggling with any sort of adversity, it can never shrink all the way back to its original dimensions. And so it grows bigger. Therefore one should think of adversity as a kind of growth hormone at the opposite pole from, yet absolutely essential to, spiritual development.

To this you may reply, "But why does the adversity have to be so unfairly applied on Earth? Why are some babies born feeble-minded or deformed in hovels and others, no more worthy, born bright and beautiful in palaces?" My feeling is that the question is not only polarized but loaded — loaded with assumptions about values (of intelligence, normalcy, wealth, pain, happiness, etc.) — that no mere human is qualified to assess. Moreover doesn't a just judgment in each case hinge on spiritual polarities hidden to mortal view?

Pain has a measurable polarity, I believe — one that ranges through the whole axis connecting life and death. And these opposite phases of our existence obviously hold a polar interrelationship that long retained a prominent place in the philosophy of ancient China, where life and death were likened to the opposite sides of one coin and were considered an integral part of the well-known Chinese complementarity of yin and yang. Indeed it was the yang of life that would inevitably be fulfilled by the yin of death, just as light calls for shade or fire cries out for water. For together life and death create a whole: a spiritual organism, a being, a soul. And in their metaphysical aspect their vibrating poles resonate and transcend the illusion of mortality toward the reality of immortality.

In a not-too-different way we can also view evil and good as representing the poles of a single whole that often looks very different from its two contrasting sides but is really one living synthesis. When a baby, for instance, gets spanked by his mother because he tried to climb out of his crib, the spanking naturally seems to him evil, for it is probably the most horrible experience he has ever known or imagined. For no reason he can understand, his closest friend-mother-God

484

attacks him with overwhelming force and he reflexively cries in outraged protest. To the mother, on the other hand, the spanking is not evil but essential. It is constructive. It is good. If she analyzed it, she would recognize it as a form of communication. The baby being too young to understand words, she has little choice but to use this traditional alternative, which amounts to an animal-level sign language with sharp accents on its syllables, to tell him that climbing is taboo and something to be avoided because it brings quick and fearful pain.

It is far from obvious, I know, but in my opinion the life of the adult human on Earth is remarkably analogous to that of the baby. For, like the baby, the adult also experiences what is to him evil in crime, hatred, destruction and war. And he too protests and tries to correct these evils, as all religions and the common rules of decency tell him he should. Yet, from a viewpoint far beyond the human, indeed from what one might call God's perspective, it could be that crime, disease, famine, wantonness, waste, woe and war serve a purpose more constructive than appears. One could not and should not call these things good, for they are not good. They are really evil. Yet I see evidence that in a deeper sense they may be an important analogue of the letters, words and sentences of a divine and universal language, a mystic code for teaching through deed and example the elementary lessons of spirituality. They thus can have a function, the function of evil in life. It is not good, to be sure — yet it is here, built-in, and it is viable. Its function was mentioned long ago in the Hermetic Philosophy of ancient Egypt, by the Manicheans, the Jesuits and, I suppose, others. It is part of life. Worlds are created by head-on collisions, and there is even said to be a conservation side to head-hunting.

Most humans seem to believe they want to attain something in life. But do they actually secretly yearn for frenzy, conflict, failure and more struggle? Can there really be joy if there be no pain? The Prophet Krishna taught that "pleasure and pain, gain and loss, victory

and defeat, are all one and the same." War is madness — but battle the spice of life. And despite all its destructive horror, a case can be made for war as mankind's first large collective action that forged tribes into nations and made feudalism into democracy. War also has greatly stimulated invention in recent history. In the first half of this century alone, World War I produced the tank that evolved into the bulldozer, at the same time rapidly developing other new vehicles from submarines to aircraft; and World War II created the aerosol bomb, the transistor, radar, the jet, the long-range rocket, the atomic age and the space age.

Presumably it was his youthful intuition about war's fecundity that induced stretcher-bearer Teilhard de Chardin in World War I to write in his diary that "through the present war we have really progressed in civilization. To each phase of the world's development there corresponds a certain new profoundness of evil . . . which integrates with the growing free energy for good." Although since then mankind has been forced to face the stark fact that war has grown so dangerous it may yet run amuck and destroy civilization if not all major life forms in our world, should we not, while working for peace, at least try to temper our apprehension by recognizing that some paradoxical component of war just may somehow be a tool of spirit?

Of course it is not really possible for us earthlings to descry the Elysian view of planetary life, even while in orbit, so I hope you'll pardon my stretching my mortal prescience a little farther than it likes to go. I do it because I deeply feel man must glean what he can of the spiritual meaning of adversity if only for his optimum understanding of the illusory paradox of evil that so tries this nether world. I believe anyone rash enough to criticize such a basic feature of creation as evil or woe should at least reflect how easy it is to under- or overestimate the suffering of other creatures. Is it not clear that some of the apparent agony in predation may harbor hidden satisfactions in the predatee as in the predator, that a preying mantis who continues making love to his mate after she has eaten his head off has other-than-human feelings, that surface complacence in some creatures may conceal deep frustrations, that the lowliest of martyrs undergoing torture for his faith may be happier than the grandest of princes on his honeymoon?

What is evil in respect to me indeed may be good in respect to you and vice versa. For there must be brands of evil that are good for me to struggle against — things that are "evil" only relative to my consciousness, not absolutely evil — things that challenge but don't really harm anyone, including me. This, I am convinced, is the meaning of

the ancient Chinese saying that "to be right you must also be wrong." And I feel sure in my bones there is no such thing as absolute evil, for evil is in essence only a dearth of good, a deficiency from certain viewpoints, a negative quality, a relative value.

This whole issue of good and evil of course deals with the question of contrast and its value for the world. Contrast creates impact, meaning, language, structure. It is effective because things are most sharply measured or defined by their opposites. In an example, you can use the concept of light to signify good, if you prefer, which would make the obverse side into darkness rather than evil. This may be helpful because darkness is not necessarily bad, often being the creative darkness that gives meaning by its very contrast, by surrounding and framing the light. Indeed it forms the black area on a printed page like this one. Some might say a completely white page is purer and more perfect than one with so many black marks upon it. But of course a pure page is also a blank page and worthless because it conveys no meaning.

The same principle of contrast applies to the relativity of good and evil, a spiritual interaction between opposing forces that could be likened to a pen against paper, writing the language of God in mystic words that inscribe themselves as deeds upon the world. If you are not sure what I mean, consider the similar paradox that amplifies the images of many of the great figures in history. Would we remember Joan of Arc had she not been burned at the stake? How would Churchill seem without the villainy of Hitler? Or Lincoln without the Civil War?

Even a symbolic figure like Saint George could hardly have become the patron saint of England without the aid of a partner whose great virtue was that he was wicked. I refer of course to the famous evil dragon for, if by some fluke the beast had turned good, George would have lost his reason for fighting him — and the whole fracas would have fizzled. Likewise did not Abel receive indispensable help from Cain, Abraham from Nimrod, Moses from Pharaoh, Christ from Judas . . . ? And do not Christians agree that the supreme *wrong* of the crucifixion occurred on "Good" Friday which proved to have a *right* side on Easter?

Of course there must be millions of people, particularly in Earth's more materialistic and complacent societies, who reject the need for contrast or any kind of serious struggle in life and enjoy nothing so much as interpreting their problems as proof that they are the unfortunate victims of injustice. Many of them no doubt would love to awake from their troubles and discover conscious life to be really just a bed of roses — roses without thorns of course — and they might even imagine they would be happy to loll in thornless roses forever if they actually got the chance. But any sensible person could tell you that if these louts really did loll a few days in their rose beds, it is certain (assuming they possessed a streak of humanity) that the lolling would get so deadly dull within a week that they would yearn for something to break the monotony: anything, even if it hurt. Indeed, although their encountering a thorn on one of the roses in the beginning would have been abhorrent, by the second week that same thorn would have become an adventure, a jot of spice, a pointed sensation, a punctuation mark in an endless sentence. And this may be the deepest philosophical significance of the thorn. For it is remarkable, when you come to think of it, how much the shapes of commas, accents, apostrophes, parentheses, quotation marks and exclamation points resemble thorns, and how similar is their evolutionary function. And it is a fact that thorns are not wholly defensive, nor do they merely puncture: they also punctuate! Some of them are even attached to seeds and eggs. And thorns do more than inflict: they inflect! Like pain, problems, predation, birth, sex and death, all of which evolved for creative purposes with various long-range survival values, if life suddenly lost the thorn, it would desperately have to set about evolving a replacement or hazard the very end of its existence.

PUNCTUATION IN
THORN, TOOTH AND CLAW

The tooth and the claw of course are the thorns of the animal kingdom, shaped similarly but wielded more willfully and subject to the same philosophical analogies. If you are one of those who see the thorn, the tooth and the claw as little botanic and zoological devils existing only to torment the "good guys" of nature, perhaps you should consider the spiritual station of the Devil in religious history. For the traditional teaching of the Church is that the Devil is really a fallen angel and therefore possessed of supernatural insight and potential. And, like any thorny influence bedeviling people, the figure of Satan has served as a symbolic predator of man who unfortunately is now almost uniquely deprived of what's known as natural predation. Indeed in the long view of the ecologist, mankind very much needs a predator before he gets any softer from artificial living — not just for his physical evolutionary enhancement, like the moose being pruned by wolves, but for mental discipline and, most of all, spiritual transcendence — to "separate the sheep from the goats" or to weed out (as by sterilization) hopeless misfits from future populations or even turn organism man into superorganism Man.

## THE BOON OF ADVERSITY

A different kind of evidence of the value of adversity in life and its evolution is the struggle that a young animal goes through to stay with its mother even though the closer it gets to her the more surely it will be stepped on, squeezed, scratched and knocked around. To test the factor of stress in this kind of "love," a biologist gave electric shocks to hundreds of baby ducklings while they followed their mothers and discovered that in ensuing weeks they showed a significantly stronger attachment to them than did hundreds of similar unshocked ducklings. Another researcher in the same field divided a couple of dozen puppies into three groups and reared them in isolation boxes. The first group he treated kindly, the second he consistently punished for making any positive approach to him, and the third he treated kindly or punished at random. And it turned out that the third group of puppies became the most attached to and dependent on him. Which suggests that stress, including the mental stress of uncertainty, is an ingredient in attachment or love and that perhaps even manifestations of hatred (its polar opposite) somehow enhance love.

Evidence in this direction can be seen in the unusual work of anthropologist Colin Turnbull, who wrote about the starving (and apparently devolving) Ik tribe in central Africa, pointing out that although love seems (in face of the desperate competition for food) to have virtually disappeared from among them, hatred has also disappeared, leaving them in a state of neutral apathy. Thus we see a rare and pointed illustration of polarity and the complementarity of good and evil, each of which implies the other, while no known tribe, nation or world that possesses one of them is completely lacking the other.

Could it be that love and hate obey laws resembling those of magnetism, where poles of opposite charge attract each other while like ones repel? Certainly psychologists know that the normal human tends to be put off by any other human who closely resembles himself — almost surely because what's called negative comparison is a threat to self-esteem.

And success and failure are another polar pair of major significance

489

in life and their sensitive feedback relation is being increasingly studied by imaginative educators, who have remarked that illness (particularly of psychoneurotic types) seems to have been of creative importance in the lives of Charles Darwin and Abraham Lincoln (coincidentally born the same day), Florence Nightingale, Elizabeth Barrett Browning, Edgar Allan Poe, Robert Louis Stevenson, Sigmund Freud, Marcel Proust and many other very successful people. In fact of a list of thirty-five of the "greatest geniuses in all history" 40 percent were judged to have had "sharp mental disorders," while 90 percent were at least "psychopathic." Could Einstein's failure to learn to talk until he was three and his being discharged from his first three jobs as an adult have been factors in his later achievement? It always used to be assumed that children want and need to be understood, but recent findings suggest that in many cases they not only want and need, but actually depend on, a solid foundation of misunderstanding in order to develop self-reliance and mature as independent-thinking adults. This could well have been true in the cases of young Einstein, who dropped out of school in disgrace, of young Ulysses S. Grant at the bottom of his class, of young Tom Edison, a "hopeless" scholar, and countless others.

The mental and spiritual aspects of success or failure of course have their own polarities, which seem related to what one could call the curiously elusive complementarity of happiness-unhappiness. Indeed attempts to define the poles of spirit are rare. Yet I've heard of an experiment in which monkeys were given mechanical puzzles, which they soon learned to enjoy working "just for the fun of it." But two weeks later a new rule was made and any monkey who solved a puzzle was rewarded with food. Then, after a happy month, the rewards were arbitrarily stopped — and the monkeys, obviously upset, refused to do any more puzzles. Which, I would say, supports the concept of relativity in fun or happiness. For, compared to nothing, the puzzles were fun. But, compared to puzzles for food, the puzzles were the bunk.

I've noticed the same thing in human lives: that happiness is affected more by one's *movement* toward (or away from) success than by one's *position* near (or far from) it. Thus a famous king and conqueror, who has always had everything he wanted, but whose fortunes are slipping slightly, is less happy than a blind beggar, who never had anything he wanted, but has just found himself a tasty crust of bread. So the law of happiness says happiness waxes and wanes in direct proportion to a sense of progress toward or away from a goal, a worthy cause, a creation, a companion to be loved. It is geared not to place but to direction and speed. It is not absolute but relative. It is based less on the concrete than on the abstract.

## THE PARADOXES OF FREEDOM

As to the polarity of freedom, it seems to be an ancient and double paradox that can be analyzed from two notable aspects: first, freedom as the complement of enslavement; second, freedom of will as the antithesis of predestination.

Of the first, what do you think freedom of liberty really is anyway? Does it mean letting go of all restraints like a train that jumps its track? Does it require abandonment of every normal rule of behavior in the way of a mad dog or a runaway horse? I would venture to say not, for the train that leaps the rails is obviously about to turn into a wreck, and the mad dog can expect nothing so surely as a bullet in the head. Neither does an ideal such as is symbolized by the famous Statue of Liberty in New York harbor justify letting go of control, responsibility, consideration of others or any disregard of law. Most significantly, it does not imply even the slightest violation of the laws of nature.

If a violin string is lying on a table loose and detached from any violin, some might suppose it "free" because it is unconstrained. But what, one should ask oneself, is it "free" to do or be? Certainly it cannot vibrate with beautiful music in such a condition of limpness. Yet if you just fasten one end of it to the tailpiece of the violin and the other to a peg in the scroll, then tighten it to its allotted pitch, you have rendered it free to play. And you might say that spiritually the string has been liberated by being tied tightly at both ends. For this is one of the great paradoxes of the world to be seen and tested on every side: the principle of emancipation by discipline.

The second and related paradox of freedom is the age-old controversy of free will and destiny. How often have you wondered how some mystic source of foreknowledge, whether it be a Gypsy palmist or God Himself, can know with certainty that your planned trip to Bombay will actually take you there next month if it is true that you really are perfectly free to change your mind and remain in Chicago? For it isn't reasonable to suppose that a flexible free choice can possibly produce a foreseeable fixed outcome.

Yet the simple resolution of this paradox turns out to be nothing but the dimensional difference between the impinging perspectives of free will and fate. Imagine a hungry puppy coming to a corner where     491 he has to choose between turning left toward a plate of dry fish bones abandoned by the cat, and turning right toward a dish of his favorite juicy hamburger. Naturally he feels (and is) quite free to make up his mind which way to go, though it takes only one sniff to decide him. But his master, looking down from a lordly height upon the episode and being familiar with the puppy's keen appetite, can prophesy the animal's choice with Olympian infallibility — for the master's extra dimension above the floor, particularly his superior altitude in intelligence and experience, gives him a lofty and relatively divine overview. Thus the puppy's simple two-dimensional decision is revealed to be in no way at odds with his master's more sophisticated three-dimensional precognition in favor of hamburger. For the puppy is free to *do* what he wants although not free to *want* what he wants, and his free will and fate interrelate like a flat 2-D picture and a solid 3-D model of the same thing.

## THE TOOL OF POLARITY

There is literally no end to polarity's ubiquity in this, or probably any, world. As man evolves tools, tools in turn evolve man — and his invention of the hammer naturally leads to the birth of responsibility for its use. So do times of peril test the soul. But while it isn't always apparent that darkness is vital to light, the pupil of the eye, the blackest part of most people's bodies, is actually the part that lets in the most light to the interior, being the aperture through which you see the outside world.

The profoundest polarity of all, however, may be the fact that not only did the Creator create all the creatures of the worlds but those same creatures in turn had to come into existence in order to fulfill the Creator. If true, it would be comparable to the way man created the alphabet, the computer and all the rest of technology that fulfills, and thereby helps create, him in return — both subject and object being integral as well as reciprocal aspects of one whole.

Our polarized Earth rolls on meanwhile, a world in which love's door is hinged with pain, evil spawns good and hate cannot exist without love — a world in which easing another's heartache always helps one's own. And, in the way that a part of every wheel or body must move backward to propel its whole forward, life itself advances in waves, never permanently deterred by corruption or evil.

This is the essence of polarity. Indeed, in Baha'u'llah's words, "Through transgressors . . . God's loving-kindness is diffused among

492

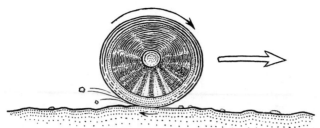

men.'' And through the diffusion, nature in her wisdom is bound to keep on trying, patiently, inscrutably — while her multievolutions unfold, groping but not blind, plying her celestial tools of error, monstrosity, war and evil to find her answers, to steer her course — that life in the large may ultimately be more beautiful and more fulfilling.

# CHAPTER 19

# Fifth Mystery: Transcendence

T HE WORD "TRANSCENDENCE" means going beyond common experience. And, with usage, it has come to signify something on the order of an extradimensional state exceeding anything you could call familiar on Earth.

So it isn't surprising that we do not notice transcendence by looking upon our planet from space. For although she has sprouted fresh areas of green and yellow in recent millenniums and now is breaking out in sudden lakes and delicate webs of cities and highways, the earth remains materially almost the same old cloud-flecked world she has long long been. Transcendence, in other words, is not material so much as mental. And, to an even greater degree, spiritual. So it could hardly be perceivable except by a sensitive spirit who can look on the invisible ups and downs of feeling on Earth and comprehend them.

A news item says: "A band of criminals has disrupted the government by raiding the treasury. Their gang rule has made money so

494

scarce that many farm hands are out of work and few farmers can afford to hire them." It was published on papyrus in the city of Memphis, Egypt, in the year of recession: 3068 B.C.

Another report quotes a top leader in education as complaining that "children nowadays are tyrants. They not only talk back to their parents, teachers and elders, but expect every luxury, gobble their food, chatter incessantly and sneer at any attempt to control them." This from Socrates in the fifth century B.C.

If it makes you wonder whether the Earth is getting anywhere with all her struggle for peace and prosperity, it may also cast perspective on our continuing difficulty in judging the spirit of mankind in view of the meagerness of knowledge available to us — even now. The ancients evidently thought the world and nature were getting to be pretty well known and that man was all set in his dominance of Earth — so much so in fact that an Egyptian author of the Twelfth Dynasty (c. 1900 B.C.) wrote that his greatest problem was "the difficulty of saying anything new."

Yet every century or two a voice would arise to express some degree of awareness of what might be called transcendence. There were the Prophets of God such as Moses, who avowed "we spend our years as a tale that is told." And occasionally a natural philosopher or scientist like Herakleitos who went so far as to declare Homer misguided when he prayed, "Would that strife might perish from among gods and men!" because the poet hadn't realized he was asking for the destruction of the universe, since, should his prayer be answered, all things must pass away. Herakleitos obviously recognized the Polarity Principle and saw that "the sun is new every day" and "no man can step twice into the same river since the waters that flow upon him are ever fresh."

Such ideas naturally introduce the concept of transcendence, which, although elusive, is so important we can no longer avoid it. Further, it is a multiderivative of abstraction, omnipresence, polarity and other of the abstruse traits of the maturing Earth we have been talking about. Buddha hinted at it when he said that "neither being nor non-being, nor both being-and-non-being, nor neither-being-nor-non-being can express the existential purport and content of human reality." Christ spoke of it when he said, "Before Abraham was, I am." And Baha'u'llah, when he said, "No vision taketh in God, but God taketh in all vision."

Science of course is only one path toward truth and it admits its own limitations in everything from the Uncertainty Principle in the atom to the controversial curvature of the universe. There may be a

495

consciousness amid molecules originating in quantized flashes of very short duration. And human consciousness may emerge through the integration of many flashes of trillions of molecules by a kind of weave of memory, a sattva, a recording of innumerable events into a coordinated stream of new dimension. Even electrons may choose their own paths — consciously. And the universe may be integrating all its component consciousnesses.

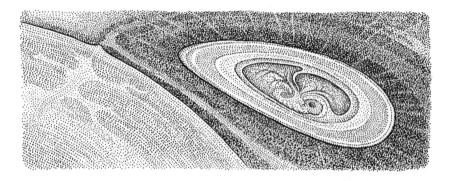

So we may as well also hearken to our dreams and drift with the world, which, after all, is not just a material structure but, as time conveys it, an abstract celestial flow where our thoughts are cells in a cosmic mind. Indeed out of one cell here comes the notion that it is in the nature of wisdom to have a beginning sometime some place and, from there, to sprout and grow. And, for all I know, this may be true even of God's wisdom — though of course the process of growth involves time, which presumably is no longer finite with God and therefore not restrictive as with us in our mortal phase. Also space fields seem to be provided for wisdom to experiment and grow in. Thus emerges a basic reason for time and space, which together make possible form and motion, without which meaning could hardly be understood. For form is an alphabet of sorts which makes possible language, including mathematics, and particularly the numbers needed to measure and define geometric form. This, I have a hunch, just may be the why of the whole material world, which appears to be a principal finite aspect of the greater mental-spiritual world — an aspect enclosed in mortal space-time fields especially designed for the evolution of spirit, which may show first through the faint feelings of plants, next the stronger senses of animals, then, with expanding consciousness, transcending toward thinking, while feeling and thinking gradually coordinate and blossom into a spiritually mature whole — which, from there, can further evolve and exuberate onward toward attributes and powers and joys and heavens we cannot yet imagine!

Transcendence naturally has a wide variety of aspects, as does evolution, with which it is associated — and all sorts of abstract factors and qualities. The miracles of one millennium (say firearms, the compass, flying, radio . . .) tend to become the basic technology of the next, just as luxuries for the few often turn into necessities for the many. Then too, obscure arts like alchemy and astrology grow up to be vital sciences like chemistry and astronomy. But the major transcendences have to do with big abstractions like dimensions and virtues, the lower numbers in dimensions steadily unfolding into higher numbers, and what were once regarded as good skills (like killing a dangerous rival) being progressively demoted to bad acts, sins and eventually capital crimes. Meanwhile other transcendences specialize in dispersing, say, a particular advantage for one person outward into a general advantage for all. Or as individuals transcend into societies, societies transcend into world civilizations. And finitudes transcend into Infinitudes, mortals into Immortals, matter into mind into Spirit and, just possibly, creatures into their Creator.

## EMERGENT TRANSCENDENCE

To grasp what all this may mean, let us step backward through time for a spell — I mean a goodly spell, like a dozen billion years or more, to what astrophysicists call the age of radiation, when the universe is thought to have been so young that matter had not yet formed and nothing was visible or perhaps even imaginable except in the cosmic mind of the Creator, assuming there was one. All we now have or know of was then only the potential of a far future still completely hidden in the vibrations of photons or whatever other subatomic particles there were. If mind or spirit existed, besides God's, it was presumably unmanifest and unknowable — for transcendence, if it could be called that, had barely begun to transcend. But gradually radiation evolved matter and matter evolved worlds and worlds turned out to be alive, self-sufficient and full of mystic potentiality. By self-sufficient I mean that life (including human life) eats, drinks and breathes the earth — and also the sun and the Milky Way and the universe. For are not bread and meat and rain part of the earth? And is not the earth part of the sun? And the sun part of the Milky Way and all of them part of the universe?

As for the individual self, we all recognize it is the prime focus of          497 consciousness and, as such, one of the keys to transcendence, along with the dimensions of space and time that define the spatial relations between things and other things on one hand and the temporal relations between things and themselves on the other. Space, time and self, in other words, are the three principal measures relating us to this finite world. They are the dimensional frame through which each of us individually impinges on what is commonly called reality. And the changes that gradually shift our relation to this frame, as by a slow inexorable law of nature, are what constitutes transcendence.

## TIME TRANSCENDING

Let us begin with time, because that is the dimension that reveals transcendence to most of us first. Have you ever noticed that the years of your life are passing by faster nowadays than they used to? If you are fully grown, you can hardly have missed this common experience, which becomes more and more evident the older you get. But what is the cause of it? I have heard various theories, such as that metabolism in the brain somehow quickens with age, but the only explanation that withstands critical examination, so far as I know, is that each additional unit of time we live through is a smaller portion of our total experience.

To a newborn baby, time is naturally static and very little less so during his first few months, because he hasn't lived long enough to notice changes in things. To him Mother is just Mother. She is always the same: never was different, never will be different. And the hands of clocks don't move at all.

Even when he is a year old, the baby has barely started to become aware of time's passage. And a year, to him, naturally seems a lifetime. It is entirely logical that it should of course, because a year literally is his lifetime — to date. And clocks have just begun to tick.

But when he is ten, time has noticeably speeded up. Clocks have come alive and a year goes by ten times faster because it has become only 10 percent of his experience. When he reaches fifty, time is passing five times faster still, clocks have begun to whiz and a year is but 2 percent of his life. And if he reaches a hundred, it's 1 percent. His old friends have been dying off at a fearful rate while new children sprout into adults like spring flowers. Strange buildings pop up like mushrooms. A whole year to him now actually consumes less conscious time than did four days when he was one year old.

If you can project your mind beyond death, imagine how time must fly for someone who has lived a million years. Extrapolating at the same rate of acceleration, a year by then will have been reduced to a mere one-millionth part of his total experience, the equivalent of thirty-one seconds to the one-year-old. Babies will grow to adulthood in ten minutes and die of old age before breakfast.

Ultimately, as the natural law prevails, whole generations will pass like flashes of lightning. Eons will drift by like time-lapse movies of civilization on the march, evolution evolving before your eyes. Birth and death will merge into a simultaneous whole and time itself will reveal at last its full stature as a dimension of development, while total experience will blossom easily from the finite into the increasingly imaginable perspectives of infinity.

## Space Transcending

The same kind of transcendence of course applies as well to space as to time. The baby learns on the scale of inches before he is big enough to grasp anything in the range of feet. Yet even when he can understand that his crib is a yard long, larger units like acres or miles might be infinite for all his comprehension of them. And it is only when he is big enough to take long walks that his world attains the span of a mile. From there on, the transcendent relativity of space becomes increasingly noticeable to him and the miles (like years) shrink and go by faster as they increase in number — and in a similarly inverse proportion.

By the time he is grown up and studying astronomy, he will get an inkling of what a light-year is. Then come parsecs and even mega-parsecs, which relegate miles (relatively) into the microcosm. And all of this verifies Einstein's statement that neither space nor time are fundamental. Which means they are basically only illusions of this finite phase of earthly existence, out of which the law of transcendence is progressively and inexorably taking us. Moses seems to have agreed when he said in the Ninetieth Psalm that "a thousand years in thy sight are but as yesterday when it is past, and as a watch in the night." Later David added that, "as for man, his days are as grass." And, likening him to a flower of the field, "the wind passeth over it, and it is gone: and the place thereof shall know it no more."

Please don't get the idea that the sweep of transcendence through time and space need ever disrupt your sense of where or when you are. For the acceleration of dimensions is generally too gentle and natural for that. Certainly you do not have to lose the minute as you gain the hour. Nor let go the year to grasp the century. In fact I can read a chapter of the book of life now in as few clock hours as it took me to read a single sentence in my childhood, yet I'm sure I understand most sentences of that chapter much better than I understood the single sentence I once struggled with. By simple extrapolation I can even predict that when you and I are a billion earth years old and a finite century unrolls in a twinkle, whatever world we are in (assuming it includes the time dimension) we will still have ample "time" to take in every minute of every year. We may feel then what we can barely deduce now: that time is just as relative a dimension as space, with width and depth as well as length.

And so with space, the astronomer with his telescope who thinks in megaparsecs and supergalaxies need have no difficulty in switching, if need be, to a microscope and measuring molecules in microns and angstroms. For you do not lose the inch when you encompass the mile, nor the mile when you discover the light-year. And no one of the finite units of space or time excludes or diminishes another.

## SELF TRANSCENDING

The third measure of transcendence after time and space, as we have said, is the self. And, like the interrelation between the minute and the year or between the inch and the mile, it is the interrelation between the self and society that is significant here. More specifically, I could call it the relation between one human and humanity, or between one organism and the superorganism of Earth or, perhaps most accurate of all, the relation between an individual consciousness and the all-pervasive, if hypothetical, superconsciousness called the universal mind. Here, I see, the commonly overlooked evidence quietly begins to crystallize, suggesting that the self is really nothing more than the conscious, localized, finite viewpoint of an individual cell in its surrounding aggregate society — and, no matter how vital that self-cell feels to itself, its actual importance to society and the world is temporary and educational.

It is so hard to comprehend self-to-society transcendence, however, that I am tempted to think there is something mystical about it. And this feeling is confirmed by the fact I haven't met or even heard of anyone who seems to know at what point, if any, cells cease being separate little individuals reacting to each other and begin being permanent unit members of one large organism. I suppose there could actually be many such points, one of the more obvious of them being

conception, before which, in the snug bedclothes of fallopian folds, the excited sperms meet their darling ovum and lovingly lash her with their tails, literally rotating her (up to 8 rpm) in a kind of love dance that has recently been observed under the microscope. This is the preliminary to a vital transcendent event, not yet fully researched, in which the sperms sometimes pirouette for several suspenseful minutes before one of them, as if inspired, turns to the waiting ovum and closes in for a kiss and a true love plunge that is the crucial act in conception. Significantly, however, the victorious sperm not only releases a chemical substance that inhibits any competing sperms from also penetrating the ovum but somehow merges his latent consciousness with hers, which clearly means turning something plural into something singular. It is a merger indeed that naturally brings up the issue of what actually constitutes an organism. Is a hive of bees an organism? Are the tips of a hundred roots "searching" for water in the soil beneath a tree comparable to a hundred ants "searching" for food on the surface of the ground? Is a tribe of pygmies in the Congo in any sense a unified being? What of a flock of birds, an epidemic of influenza, a school of fish or the four billion humans who inhabit the earth?

## INSECT SUPERORGANISMS

Entomologists estimate that various insects have organized successful societies on at least twenty-four different occasions in evolution, and other creatures (from microbes to man) have had their flings at society with varying degrees of success many more times, but it is extremely difficult to measure the intensity of these aggregations or compare the interrelatednesses of their parts. In fact it appears to be about as hard for a man to learn the form of government in an anthill as it would be for some scientific-minded visitor from Vega to discover who or what runs New York. Ant colonies vary in size from a few 501 hundred ants to some 20 million in the case of certain species of carnivorous army ants, who are probably the most highly organized of all and who, although individually no match for certain other fighting ants (the harvester, the stink ant, etc.), are, as an army, practically invincible in the whole animal kingdom.

When such a force of ants marches a thousand abreast through a tropical jungle or open country, every other creature from moths to elephants tries to get out of its path. But many a grasshopper or rat is too slow and gets caught by the voracious soldiers who spring upon their prey in large numbers and, it is said, often bite in unison at a hundred places, particularly such sensitive spots as eyes, ears, nostrils, and lips, the shock and widespread pain confusing and disabling the victim before he can escape. By such means snakes (even huge pythons heavy from a recent meal) are easily blinded and killed, human houses in the ants' path are denuded of all occupants (including vermin) overnight, crocodiles are sometimes eaten alive in shallow water and, in a case I heard of in South Africa, a caged leopard was overwhelmed in ten minutes, his bones picked clean within three hours.

The ant column, usually only a yard or two wide but perhaps 100 yards long and traveling more than an inch per second, weighs about 50 pounds and behaves like a single organism. Yet it has no central brain and no individual (or individuals) seems to be in command. In fact it is believed to flow somewhat according to the laws of hydraulics, the pressure coming from behind, the front often surging forward like fingers probing blindly, now here, now there, the spearhead salients nevertheless coordinating to outflank any prey or enemy.

You might suppose that, if such an ant army met another similar army, the two would fight each other as they fight almost everything they meet. But no: they recognize each other through their keen chemical senses and avoid close contact. I mean they normally avoid each other, but, if either army has for any reason lost its queen (as has been effected experimentally by entomologists) they merge instead. This is thought to happen because a queenless colony feels a kind of wholesale sexual attraction toward a colony with a queen, she being its prime hope of regaining its lost fertility. It also indicates that every ant carries evidence (presumably in the hormonal chemistry of its body) of the state of the whole colony, which by this means is integrated into a single superorganism that is literally flowing with the same social hormones (called pheromones) carried by every ant.

Some biologists have felt justified in comparing the army's various specialized individuals to blood cells, lymph, bile and other secretions of the body. One I know of has gone so far as to postulate that no injury is felt by any of the ants individually because only the colony as a whole can sense anything equivalent to pain. Although this view has not been generally accepted, it is supported in some degree by the selfless (seeming pointless) sacrifices made by individual ants, some of whom have been known to attack red-hot coals placed in the army's path, persisting until burnt to a crisp, but creating, with their thousands of bodies, an eventual causeway of ash the rest of the army can march over.

Faced with the necessity of crossing a stream, a tropical ant column will form a flexible, living bridge of the bodies of its forward members, who eagerly plunge into (or under) the water and cling to each other (literally to the death if need be) while the rest tramp over them. If the stream be too wide or swift for such an ant bridge to span it, the ants usually shift plans and, instead of a bridge, form a kind of Noah's Ark, a roughly spherical vessel with enough enclosed air to float, and thus launch themselves as if with a single mind, remaining an intact mass until they reach their goal or perish in the attempt. Since they apparently have not yet learned how to propel themselves in liquid, the ants must depend on what we call luck in their drifting voyages, yet in the long run they make out surprisingly well, as is proven by their 50-million-year history.

ANTS AS CELLS

But it is the mental side of the ant that perhaps best illustrates transcendence. Individual ants have been likened to the cells of the brain, and it is an apt comparison because one ant or a few of them by themselves cannot function effectively, or even survive. It's as if a surgeon extracted a cell from your brain and asked it what it was thinking about. For no matter what sort of language or how subtle the test he used, he hardly could expect an intelligible answer. For the same reason, individual ants behave moronically, if not randomly, and only slightly more sensibly when a dozen or two are together. 503

But when a colony of thousands or (better) millions of ants is functioning, a definite intelligence is evident and the ants' activity becomes
coordinated and purposeful, sometimes rising to the point of instinctive
brilliance.

Termites, once called "white ants," have been almost as successful
as ants yet with much less mobility, their social organisms taking the
form of elaborate sealed-off cities, often with very complex systems
of dark passageways extending some 10 feet upward, 100 feet downward and even farther outward in all directions. Although apparently
more passive than a tree, a termite nest is constructed with certain
animal-like "organs" corresponding to lungs, these being a series of
heat-insulated ventilator shafts that line their towers above ground,
using porous walls as filters or gills through which they "inhale"
oxygen and "exhale" carbon dioxide. And the colony itself, usually
numbering several millions, has a communal digestive system, as the
individual termites, unable fully to digest the cellulose from the wood
they eat, unfastidiously take turns in re-eating one another's castings,
accomplishing thus in series of digestions what they could not accomplish singly. And termites, even more than humans, need intestinal
microbes for digestion — in fact so much so that half the weight of the
average termite is protozoa. It is pretty staggering to think of the 1500
known (plus probably thousands of unknown) species of termites, each
harboring its own several species of cellulose-digesting protozoa, the
total of all of them together adding up to millions more termite societies
than there are termites in any single society and inevitably billions of
times more protozoan societies inside the termite societies than the
billions of protozoa in any one termite. And, on top of all that, it has
recently been discovered by the electron microscope that each one of
the billions of protozoa in each termite is not the single animal we
formerly thought but rather a colony of semi-independent flagella mechanically coordinated in the manner of galley slaves. If that isn't
enough of a challenge to our understanding, we must remind ourselves
that although the termite workers are assumed somehow to control
both breeding and feeding in their termite colony, just how they do it,
what process really coordinates them or chooses, limits and regulates
the protozoan societies and subprotozoan colonies within them is still
anyone's guess.

A beehive, while relatively simple, is in certain ways even more of
a superorganism than is either an anthill or a termitary. I know of
researchers who have gone so far as to call it a single animal. For,
like an animal, it can regulate the temperature and humidity of its
enclosed body. And it enjoys a metabolism so intense that sometimes

the member bees, serving as cells, are born and die at a rate exceeding a thousand daily. When the hive is severely wounded, such as by loss of its queen, it has been heard to moan in agony. Yet it often manages to heal its wound by miracle measures, like stimulating sterile workers to lay eggs or elderly bees to rejuvenate their youthful glands until they literally become young again. Although no actual blood circulates through the body of the hive, blood's equivalent in semiliquid food and glandular secretions full of enzymes is "pumped" continuously among the 40,000-odd bees that play the part of blood cells, moving the stuff rapidly from nurse to queen, to the waxmakers, to the cell cleaners, to the receivers, to the foragers and back from the foragers to the receivers, the cell cleaners, the waxmakers, the nurses and the queen. Whenever the flow slows down, the hive immediately notices it and feels off balance, if not sick. But when the flow recovers and stabilizes, the hive regains its composure, knows it is well and, like many another healthy animal, glows and hums with confidence.

## VERTEBRATE SUPERORGANISMS

Animal superorganisms are not all composed of insects. In fact many are made up of much larger creatures ranging from a herd of reindeer that runs in V formation until cornered, when it invariably curves into a tight defensive circle like a wagon train attacked by Indians, to huge schools of fish and flocks of birds who maneuver by amplification of barely perceptible influences in their surround. It is

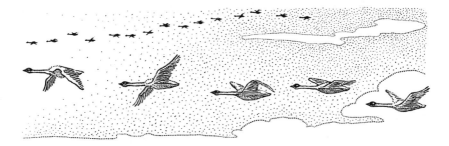

now believed that most such mass formations do not depend on any particular leaders, for individuals in the leading positions are frequently replaced by others, particularly in sharp turns. And vast bird flocks, like the formation of an estimated 150 million slender-billed shearwaters reported in Bass Strait between Australia and Tasmania a few years ago, sometimes show a complex three-dimensional motion that includes convexity over a thousand square miles as if it were part of the living skin of the planet and which seems to be affected by the local winds and pressure patterns as much as by food, mating or any of the accepted motivations for migration.

Schools of fish, being more amenable to laboratory experimentation, are better understood. Their formations in the sea vary in shape from square to diamond to elliptical to ameboid and, like sand grains on windswept dunes or orbiting particles in Saturn's rings, the individual fish automatically sort themselves out by size and speed. The new-hatched fry of course are tiny plankton, who soon begin to swim, at first very slowly, but steadily gaining in both speed and size, their speed increasing (by Froude's Law) in direct proportion to the square root of their increasing length. Although they vitally need one another during all their early development (shown by the fact that isolated fry rarely survive) they keep a few inches apart, swimming irregularly until they are nearly half an inch long.

At this stage they suddenly and instinctively begin to feel regimentation, to swim side by side or parallel to one another for short distances, responding to each other visually, conforming more and more exactly to the regimen of their fellows as they grow bigger. They also come relatively closer together as they grow, reducing the separation distance from about three body lengths to half a length by the time they are four inches long and using their developing lateral-line pressure sense (page 208) to help maintain it. Because of the fact that smaller fish cannot keep pace with larger ones of the same species, they inevitably lag behind in ratio to their size, and that is what winnows them into their discrete grades or classes. It is a process really crucial in the schooling of minnows and a mark of its widespread success and importance as a way of life among fish is that an estimated 16,000 species (80 percent of them all) have adopted it, perhaps mainly because predators tend to become confused by mass-regimented fish and therefore leave more of them to survive. Fish in schools are also known to be able to swim for longer periods than unregimented fish, to cover greater distances and tolerate stresses such as colder temperatures and increased pollution. This probably relates to their mutually beneficial body radiation when in close formation and lower consumption of oxygen per fish (proven in diverse experiments) which in turn could well result (other evidence suggests) from their helping each other by positioning themselves in school ranks so each fish gets an energy boost from the wake eddies of fish ahead of him.

Another advantage in schooling is improved efficiency through specialization. Fish swimming at the edges of a school, for instance, serve as the "eyes" of fish in the center, so the latter neither have to hunt for food or be wary of danger so long as they can respond within about one sixth of a second to any change in direction or speed initiated by the "eye" fish — one sixth of a second having been clocked in a California laboratory as the average fish school's response time. School specialization furthermore sharpens the awareness and effective intelligence of the school as a whole so it can avoid a net or migrate to a safe spawning ground more successfully than any member fish alone. And in its transcendent superiority it knows both how to move slowly and compactly to *decrease* the chance of meeting a predator and, when hunting, to accelerate in looser formation to *increase* the chance of meeting prey.

The fact that the fish school, like the bird flock, is essentially leaderless is further evidence that it is transcending in the general direction of a superorganization of selfless cells, none of which any longer has any real will of its own. Mackerels, herrings and similarly shaped fish indeed are observed to become so ingrained in schooling that they lose the power to make certain unusual or irregular movements such as swimming backward, a limitation that only enhances school conformity and discipline, probably increasing its survival value. Certainly it is a trait to reinforce the crystalline texture of the well-established school that is so stable in structure that two large schools of half-grown fish have been seen passing straight through each other at different angles with no more apparent disruption or interaction than if they had been two puffs of smoke.

While most formations have their members arranged either in rectangular rows, nose to nose, or in quincunx patterns, staggered at oblique angles, surprisingly many school fish, when alarmed, instinctively break ranks to huddle into a compact mass called a pod, in which they literally touch each other in tight layers. This curious performance, comparable to the Greek phalanx, has been known to save lake fish from an attacking loon. And mullet, when spent and spawning in very cold water, have been seen to pod perhaps for warmth or reassurance. Such pods normally have the random grain of plastic wood, as each fish retains some independence of motion, but when they are evading a common enemy they are more likely to be polarized like iron in a magnet, the thousands of individuals aligning themselves so perfectly they can swim as a single unit whose overall shape often resembles an individual fish, for the logical reason that it is composed of the same numbers (of fish) in length, breadth and depth.

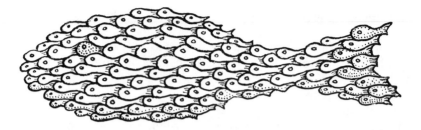

As for birds, they are known to find more food when in a big flock than when scattered, because flocking birds stimulate one another to forage. And dense populations in general control their environments more easily than sparse ones, as when honey bees air-condition their hives with a mass beating of wings, when flatworms shade themselves against strong sunlight by clustering, when goldfish precipitate poison out of their water through the stirrings and exudations of concentrated numbers. But it may be in its defense against natural enemies that social organization pays off best of all. A covey of quail roosts at night on the ground in a protective circle with tails inward and heads outward, a collective organism alert to danger from any quarter, its members sleeping in turns. A herd of 1000 migrating caribou intermittently excretes stragglers to the wolves, yet its able-bodied majority, moving as a unit, protects itself without much worry. And I have read an odd but convincing report that when the hindmost fish of certain kinds of fish schools is seized by a predator, the victim's skin has been known to exude a pheromone potent enough to diffuse so rapidly through the water it can overtake and forewarn the rest of the school.

## TRANSCENDENT COLONIALISM

All such cases pointedly suggest the way social gatherings of individuals evolve and transcend into superorganisms through the survival value of their collective advantages. These shouldn't be confused with nonsocial gatherings, such as moths flitting about a candle flame at night (each insect responding individually to the light), which obviously are not superorganisms. Yet it is not uncommon for social gatherings to be composed of creatures of more than one species (say blackbirds, cowbirds, grackles and starlings) and for such flocks to integrate into remarkably stable superorganisms. And sometimes the differing members are joined symbiotically on some basis such as commensalism (in which one of the partners benefits without harming

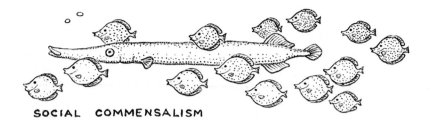

SOCIAL COMMENSALISM

the others), like the long trumpet fish who uses a school of yellow surgeon fish for camouflage; mutualism (in which each partner helps the other), like the egrets that pick ticks off the backs of cattle; parasitism, like the dog-flea relation; domestication, like the ant-aphid relation; or slavery, the lord-serf relation.

In all this discussion of the self and individual creature selves combining and transcending in various ways into larger and larger societies, we haven't yet, as you may have noticed, considered any society as basic and closely knit as a sponge. But this family of colonial animals not only has its fascinations but, I would say, is something we urgently need to know about if we are to comprehend transcendence.

The bath sponge seems about as integral as a bush but, as many a bather will be surprised to learn, it is in fact not a vegetable at all and, along with 15,000 other species of sponges, starts life as an invisible speck or, more often, a mist of countless specks that are actually free-swimming animal cells. In other words a sponge may grow from one cell that settles on the bottom before putting out branches and buds, or it may accumulate from several, from hundreds, thousands or even millions of separate cells.

It has long been known that if you squeeze a sponge through a fine silk sieve, which converts it into a cloud of millions of cells and tiny clusters of cells swimming individually like amebas, that it will later find and reorganize itself if left in a container for a few days. In one well-known experiment, in fact, the juices of two different-colored, freshly sieved sponges were thoroughly mixed, yet they easily sorted themselves out, each cell discriminatingly clinging only to others of its own color and species, until both sponges were whole again with all their specialized cells in their appropriate places.

Exactly how the separated sponge cells find their proper places is not yet known, but find them they do. One of the commonest cell types is the flagellated collar cell that lines a sponge's internal chambers, tirelessly lashing and propelling the currents of water that bring     509 in the vital oxygen and food. Deftly it extracts its payment in morsels of passing plankton, swallowing its personal due while shunting the rest to fellow cells that form pores, outer covering, internal jelly, skeletal spicules or sperms and eggs. Almost as individual as ants are these various cells, and about as cooperative and self-sacrificing — and still completely innocent of any central nervous system, for neither sensation nor pain is conveyed from cell to cell and the sponge as a whole seems considerably less responsive than an ant colony, even though its skeleton of calcium carbonate binds it physically much more firmly together.

Speaking of skeletons, the biggest one around belongs to the little coral polyp, a mere wisp of a creature who nevertheless is a lot more advanced than the nerveless sponge and who, like the termite, excretes extraordinary fortresses that dominate his own life, shaping the society within as much as shutting the enemy without, and only incidentally creating whole coral archipelagoes (perhaps ultimately continents) of a beauty and significance he himself cannot begin to conceive.

And most significant of all, in a review of the cell-society relations in life, may be the soil ameba *Dictyostelium discoideum*, which lives anywhere there is rotting wood or dead leaves and feeds most of the time as millions of separate, independent, invisible cells that can be considered immortal because they don't have to die. Occasionally, however, when it has had enough of its bacterial food, this nebulous population swarms into a dense, coordinated, visible, mortal mass to ooze about in the form of a small garden slug before suddenly sprouting into a thin-stemmed flower full of spores. There is still a good deal of mystery as to just why this so-called slime mold goes through its

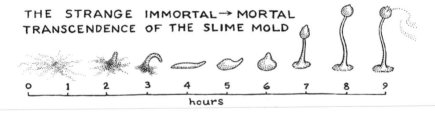

THE STRANGE IMMORTAL→MORTAL TRANSCENDENCE OF THE SLIME MOLD

hours

peculiar routine of shifting from the immortal microcosm to the mortal macrocosm and back again, but it is known that an evanescent social hormone or pheromone called acrasin, exuded by the separate amebas, attracts them to each other, while the tiny whiffs of carbon dioxide they exhale include an enzyme that inhibits them just enough to regulate their progression. Then as they march sluggishly on, the ameba cells sort themselves continually by size and speed like minnows in a school, the bigger, faster ones gravitating to the front (eventually to rise as the stem), the smaller, slower ones lagging to the rear (to be turned into spores). And when the spores have at last been cast to the winds, the cycle continues as the stem slowly thickens again, pulling the crown back down into the shape of a slug whence it disintegrates once more into its constituent amebas in order to feed.

Is such a flowering slug in any sense a single organism? If so, it is obviously one who must disintegrate out of sight into the microcosm at every mealtime only to reappear whenever it gets the yen to flower. Or is it more like a well-drilled army of soldiers who spend most of their time browsing in leisure at home, only occasionally being "called up" to go through their paces on the parade ground? For all I know it could be either, neither or both, for there is no clear, absolute line of demarcation between the macrocosm and the microcosm, between mortality and immortality. Nor is there any infallible boundary between an animal and an army, between a creature and a crowd, all of which are made of cells.

## Transcending Cells

Cells, it so happens, are the unit bricks of life and, I surmise, of all life's transcending organisms. Close study of them almost inevitably leads to the age-old philosophical question: do germs control diseases or do diseases control germs? And that is part of the more general question: do atoms shape matter or does matter somehow give order to its atoms? Which leads to: can creatures steer the world or does the world really create and influence its creatures?

I suspect the cause and effect work both ways, but just where and how much is almost impossible to find out. An interesting recent discovery in this field is the fact that colon bacilli actually build bowels in the developing young of many kinds of animals and man. And related biological research is making it more and more certain that organisms of even more diverse orders of life, including vegetables, are physically assembled by their own inhabitants, from submicroscopic viruses to ultramacroscopic symbionts. The electron and other subdivisions of the atom are part of this synthesis of course and, when Einstein once cryptically remarked that "particles are nothing but geometry," he seems to have been preparing his heirs in science for a new mathematical concept to be known as "superspace." It actually emerged, I note, about a decade after his death and is now beginning    511 to unite cosmology with particle physics through definitive macro- and microcosmic factors in the geometric structure of creation. Superspace, by definition, has an infinite number of dimensions and innumerable probable and improbable histories depending on the sequences of cause and effect among its subatomic particles. And since in quantum physics almost any given particle may expend its energy in any of several ways, in effect every particle seems to possess something like a choice and thus collectively, rather incredibly, it may steer the universe!

If matter really does influence the universe in some such wise, even slightly, the fact jibes nicely with the concept of transcendence. For feedback polarity is surely part of it along with abstraction, interrelation and omnipresence. Indeed the combination of all these principles and more seems essential to the working of the universe, just as it takes the integration of countless sorts of specialized cells to create and maintain the average living organ or organism. Furthermore, in the process I like to think every cell willingly, if not gladly, surrenders the independence that nature expects it to surrender, this being the most crucial of sacrifices in the making of a world. It is also the sacrifice that ultimately transcends the finitudes of this mortal node of being, that lifts life everywhere, because single cells as free organisms necessarily handle all their own vital functions of sensing, moving, metabolizing, reproducing — but when they become parts of multicellular organisms, they may comfortably specialize in one single function with a resulting gain in efficiency, to say nothing of a heightened potential in mind and spirit.

In the realm of thought of course there is a potent cell structure, as major successes result from many minds working on the same problem at the same time, perhaps half in competition, half in parallel, probing for the crack that leads to the break that makes the breakthrough. In waves and music too there is a social interaction and a resonance in the sinews, with unnumbered small benefits, like the Chinese stevedore's ancient chant that men sharing heavy burdens sing back and forth to each other in a rhythm tuned to the frequency of their breaths, their strides. It lightens the load, lubricates the heart and in some mysterious way — perhaps through the consonances of still unimagined vibrations — adds resilience to the whole being.

If you look at a man you can see at a glance that he is materially made for cooperation: he has feet that share the ground and take turns stepping, hands that work hand in hand, eyes that see eye to eye, teeth that cut and grind like matched millstones, organs of sex that eloquently bridge the gap between male and female, between ancestor and descendant. And while each cycle of gene circulation (as we saw

512

in Chapter 13) takes about a millennium to saturate mankind on Earth, the mental cycles or waves of cultural evolution diffuse much faster, often permeating the civilized globe within a few centuries in ancient times, averaging perhaps one century in medieval times, a few decades in the eighteenth century, less than a decade in the nineteenth, and sometimes less than a year in this twentieth century of air travel and television (something we'll be looking at more closely in Chapter 22).

Clearly this must be how the ancient luxuries of Solomon and Kublai Khan have so easily been surpassed by the now familiar extravaganzas of the modern mail order catalog, how a billion people today enjoy "necessities" that were undreamed "miracles" to the richest kings of yesteryear. According to the law of conservation of matter and energy, the material world cannot decrease as a whole, and the mental world (now pooling its knowledge on a vast scale) can only increase, while the spiritual world gives signs that it will soon blossom forth as never before.

At the same time, I see, most of the nonhuman world plods on its accustomed way regardless. The lice who live out their lives inside the feathers of curlews and thrushes and larks, for example, cannot know that they are passengers aboard beautiful flying creatures who court each other and sing and lay eggs and navigate by stars hundreds of light-years away. Nor are the birds aware of the lice except as an occasional itch. Yet these worlds within worlds are all parts of the mysteriously interrelated and unfolding finitudes of Earth that are gradually but inexorably transcending toward the unfathomed plenitudes of Infinity.

## THE NATURAL LAW OF COOPERATION

Evolutionists seem to have completely accepted the natural law of Survival of the Fittest, which many interpret as meaning "kill or be killed." But all too few yet realize that nature also has an opposite principle at least as strong and potentially much stronger: the natural law of cooperation. Religionists call it the Golden Rule and it not only pervades nature but is at the heart of transcendence from self to society, from singular to plural, from finitude to Infinitude, from indulgence to love. The late Carl Akeley, African explorer, saw it when threatened herds of elephants formed a ring in which the younger, stronger bulls surrounded and protected the females, babies and weak older ones. It happens when animals huddle together for warmth and encouragement, when they groom each other and when they treat each other's sores by licking them clean. An experiment in which hundreds    513 of planaria worms were exposed to ultraviolet radiation demonstrated a hitherto unknown evolutionary advantage in cooperation. All the worms that stayed away from other worms died within about forty minutes of receiving this short-wave light while worms allowed to crowd together to the point of shading each other survived more than eight hours. In the case of fish, sociability actually inhibits cannibalism, for bass in weedy ponds are known to swim alone and eat smaller bass, while those in clear ponds swim in schools where they get to know their younger relatives and learn to regard them as mates rather than meals.

The same principle holds through all the kingdoms of life, in particular the human, as Darwin explicitly recognized (in *The Descent of Man*) when he wrote that "as man advances in civilization, and small tribes are united into larger communities, the simplest reason would tell each individual that he ought to extend his social instincts and sympathies to all members of the same nation, though personally unknown to him." And "this point being once reached, there is only an artificial barrier to prevent his sympathies extending to the men of all nations and races." Thus the Father of Evolution crowned the material Law of Survival of the Fittest with the spiritual Golden Rule.

I am aware that spirit is a fuzzy abstraction to most scientists, who, not knowing how to measure it, doubt its power if not its existence. So teleology is something of a dirty word to them and evidently they assume that, should the universe turn out to be progressing in any significant way, it must be doing so at random, mechanically and blindly. Yet some of the more philosophical physicists at least sense what Einstein called "the rationality of the universe," implying that it has a purpose or direction to whatever it is doing. And there are a few like Jean E. Charon, physicist president of the Association pour la Cooperation de la Jeunesse Mondiale of France, who has declared that "each living element has an awareness of the final goal to be attained" and specifically that every cell in the human body makes an "independent and spontaneous effort to advance toward a known goal that promotes the order of the whole."

Sometimes I wonder about the human body's being a miniature replica of the earth's surface. For it resembles our outer planet remarkably, being composed of the same elements in the same proportion: three quarters water and one quarter solids, both organic and inorganic, with swift internal flows, occasional eruptions and gentle daily tides. And there is a corresponding similarity between the atom and the solar system, where the sun represents the proton and the planets the electrons that orbit around it. Such preponderant structures notably demonstrate how material forms are ordered in regular and meaningful ways by nonmaterial abstractions such as symmetry and conceptual archetypes. This applies to everything from crystals to supergalaxies, all of which in essence (as we have seen) may be considered alive even though influenced by what are generally regarded as mysterious, if not unknowable, forces.

But how long does an atom remain an atom? A cell a cell? This too has a bearing on transcendence. Some cells, called lymphocytes, come equipped for chemical coordination with synthetic polymers before the polymers have even been invented. In other words, they are ahead of evolution. Or could it be that they are merely outside of time as some aspects of the microcosm are surmised to be?

## THE TRANSCENDENCE OF CONSCIOUSNESS

To think of worlds beyond this world and muse upon the idea of one's consciousness being absorbed into other consciousnesses (presumably greater ones than one's own) in its inexorable transcendence toward a universal mind — that, it seems, is a disturbing thought to many people. They imagine consciousness absorption as a total loss of self, a blacking out of all consciousness as if death must be the final end of everything. It is naturally a drastic and distressing thought.

Yet it needn't be. For why cannot the absorption of one's consciousness be a kind of widening of perspective that is actually a natural, perhaps inevitable, accompaniment of experience? Isn't this really happening to all of us all through our lives anyway, little though we notice it? A newborn baby's consciousness is very limited at first. He feels the air and its coolness and the reassuring grasp of big hands picking him up. He gasps. He breathes. He hears new sounds. His consciousness expands as his senses quicken and he is absorbed into awareness of his mother, of the soft warmth of her bosom and the strange but wonderful taste of milk.

As the days go by, his consciousness of his cradle and Mother is absorbed again into the consciousness that he also has a father and perhaps other members of the family — that beyond his bed is a room. Then, as the hours grow into days, his little domain is absorbed and reabsorbed to include a door, a window, another window, a table and more space — in time a whole house, a neighborhood, a village, a country, a continent, a world — each sphere larger and containing more people, more things, brighter ideas and great complexities. Yet, as we saw with space and time, you don't lose the inch when you gain the mile, nor the minute when you discover the hour. And so with your individual self. Since you retain your self and your personality

515

when you marry a spouse, give birth to a baby, go to school, join an army, or participate in public life, so can you retain your personal consciousness when you merge your thoughts into a universal mind. Why not? This is one of the more reasonable hypotheses of the dying process that we will be looking into next chapter. And it plausibly hints that the finite adventures of your growing self in this world dimensioned with space and time have been just what you needed to develop and transcend. Indeed how better could you have learned about the world than by playing around with these simple, finite tools?

Of course there surely must be much more to life than just space, time and self — even in this finite phase — and some of it is, to say the least, mysterious. Take creativity: where does it come from? The handiest example before us right now is this book, whose source is largely a mystery. If I write this page today it comes out as you read it. But if I had written it yesterday, it would certainly be different. Or at another time or place or mood still different again — maybe better, maybe worse — with neither an end to the possibilities, nor any reliable way of predicting them. I don't know why.

A lot of one's ideas originate in other minds. That is part of inter-human transcendence, for it is inevitable that even the most creative among us learns from others, consciously or unconsciously, not excepting our most original creations. Just so did Shakespeare study the play books. Just so is Mozart said to have adapted the opening theme of his Overture to *The Magic Flute* from a Clementi sonata. Newton perforce accepted a lot from Kepler and Galileo, as did Einstein from Faraday and Maxwell. These greatest of creators have not pretended anything else. Indeed all innovation stems more or less from all before it and, if good enough, will transcend its source and become permanently absorbed into the ever-growing universal reservoir that is the preserve of immortality. In fact, what better immortality could any artist or inventor have than to dip into the world's sources and mold them anew, then discover he has actually added something that has changed them forever?

Sometimes such a creative genuis sees in one flash a whole system of relationships never before suspected in the world, a sudden vision of harmonic beauty that lifts him up in a surge of esthetic delight. But even then most of his vision's elements are already known separately to others, perhaps to many people who haven't so much as heard of each other — for, as Alfred North Whitehead once put it, "everything of importance has been said before by somebody who did not discover it." Again, many a great discovery at the time it is discovered has no known value, not even in the inventor's dreams. And it is only long afterward that its true worth is revealed, like the famous case of Bernhard Riemann and his new geometry of curvature in a continuum of an indefinite number of dimensions which, after resting dormant for a half century, bequeathed to Einstein exactly the tool he needed for General Relativity.

Einstein, I'm told, needed all sorts of stimuli that perhaps no one, including himself, consciously realized were available or functioning. For one thing, he needed an audience. And found it in his understanding friend, Michelangelo Besso, the engineer, who alone was patient enough to attend long hours while the obscure genius explained, half to himself, what he thought wrong with current physical theory. As Einstein's future son-in-law, Dimitri Marianoff, later articulated it: "Albert has to have an ear. He is not concerned whether it listens or not — it is enough if he sees an ear. Besso was always Albert's ear. It was during these interminable discussions [about 1902–03] that he would find nourishment for his ideas . . . ."

The intermingling of minds within the human species has been compared to a vast plain containing millions of wells that appear on the surface to be independent sources of water but deep underground actually interconnect and combine into tributaries that ultimately become a single mighty river. The rough physical analogy is full of truth if you can visualize all the barriers that block the transcendent flow from well to well, the rocks, frustrations, pettiness and misunderstandings that retard unity upon the planet. For, as Thomas Browne wrote in the seventeenth century, "We are more than our present selves."

## SUMMARY OF TRANSCENDENCE

By now you must have a pretty fair notion of what I mean by transcendence and its multiple aspects: *individual transcendence* in which each of us develops larger and larger awarenesses of space, time and self, *social transcendence* in which individual consciousnesses are absorbed into superconsciousness or a group mind, and *world transcendence* in which the consciousnesses and superconsciousnesses of nations and empires on a planet evolve into a world superorganism that ultimately conveys them beyond the finitude of space, time and self toward the Infinitude of Mystery in the Universe, which may be called God. All of these transcendences are more or less intervolved as far as I can tell but, if you'll pardon a suggestion, one could clarify the whole process in one's mind by thinking of the familiar mortal space-time field on Earth as something that extends or expands radially outward along with one's personal consciousness toward an invisible, intangible and abstract continuum in which one's fellow beings are somehow never born and never die but just ARE.

It is not a shocking transcendence really, for it only incidentally extends beyond the mortal life span and takes place regardless of death, happening so gradually and naturally that there seems no real change. And we retain our selves for what self is worth, even as we are absorbed by society and the universe. Self of course is more than just an outlook, and more than a finite, mortal viewpoint. It is a kind of tool of learning, an educational device, like the primer you learned to read from, or the teacher who taught you. How else but through your own self could you discover space and time, feeling and thought, pain and love, and all the important lessons of this early world of worlds? Is there any better way to learn?

To be sure, the self is elementary and down-to-earth, as any beginning tool must be if it is to be grasped and used. But this doesn't mean it is fixed and unchangeable. For, as with time and space, your spheres of awareness inevitably increase and in imperceptible but progressive stages you find and lose yourself as part of a family, a nation, a world . . . And, without forgetting your name or who you are, if you are growing spiritually, you begin to care less what happens to you and increasingly think and feel and act in causes beyond your individual self — at the same time letting that self diffuse and recondense into a bigger, more universal consciousness. As a Hindu might put it, atman becomes Atman.

To see how transcendence works, take a look at your own sympathy for another person or an animal or plant and then try to add sympathy for still others. Of course you can have heartfelt sympathy for a starving child in the Sahara. But if you keep multiplying this feeling to six or a hundred children and try to apply it to anything like *all* the suffering in the world (leaving out other worlds), it just breaks down because time and space mercifully insulate you and all creatures from each other. That is the nature of finitude. It does not reach everywhere or go on forever. Yet it is just right for this world, and it provides mortality, the perfect tool to enable us to concentrate on our selves in the here and now, so we can educate and develop our souls slowly, finitely, and thus gently advance toward Infinitude — transcending — transcending — mercifully, understandingly, dimensionally, from less to more . . .

As to your own individual suffering, being human, you'd naturally try to avoid pain and would reasonably dread any form of imminent agony. Yet, in retrospect, even though time and forgetfulness insulate your own pain from your own self, if your life had been devoid of all pain and suffering, you could be said to have missed much of life's richest experience. This I believe true. For pain often includes a goodly component of soul satisfaction and it surely has spiritual meaning. Also, impossible as it would be to prove it in this mortal phase of transcendence, pain may well, in fact, be the greatest language of the soul. Certainly Christ's message to the world would carry much less conviction had he not suffered on the Cross.

Transcendence is thus to be seen as beginning for each of us when we are still in the microcosm as a fertile egg: the seed soul, stirring, seeking, striving to win admission as a pupil in the soul school of Earth (a school we will return to in Chapter 23). Once we become locked into a human embryo, our spirit seed or cell must (by natural law) stick with that particular finite assemblage of organic molecules as long as it remains alive, for this material body is important as a disciplined instrument of learning. Although the mortal span is short — sometimes so short we are baffled at its seeming pointlessness — there is profound meaning in its transcendent existence, something Aristotle evidently surmised when he described life as "spirit pervading matter," a statement that sums up Earth as she was never summed before. I mean that some sort of viable world setting appears a necessity to the development of mind and the flowering of spirit. Indeed, without some kind of finite limitations such as are so well imposed by earthly matter, how could the spirit-mind that is YOU ever grow toward maturity? For limitation is essential to measurement, to contrast, to comparison, comprehension, articulation. Just as an artist must limit his choice of paints if his picture is to have meaning, or a message must begin and end if it is to be understood, so life must have impact, adventure, form and feeling if it is to fulfill its purpose. Without letters printed sharply enough to be sensed, the page of life is blank.

So the spirit-mind associates itself with finity in order to grow. It makes its entrance upon the stage of a material world. It assumes form in order to learn meaning. It assigns itself to a position in space and time so it can measure things and grasp the shape of ideas, visualize relativity and feel the warmth of love. How else, in basic terms, could it learn anything or develop itself? Where else is wisdom to be sown but in some sort of a world — and what is a world without some kind of form to define its existence?

Our earthly life then, in simple terms, is a tentative tuning in on a particular collection of human cells — a transcendent resonance of protein molecules with intangible awareness in an illusory space-time continuum — a harmonic, a geometric interval, a note in a song of eternal and incomprehensible mystery.

519

# CHAPTER 20

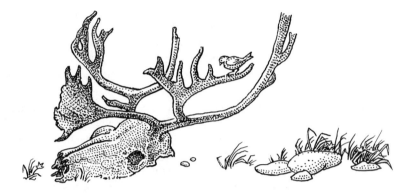

# The Change Named Death

L IFE'S PROBLEMS, as you may have surmised ere now, could well turn out to be the most important things it has to offer. And of all of them, the most mysterious and inexorable one to many, if not most, of the thinking creatures on Earth is the question of how does life end? Where does it go?

I am talking about death, life's shadowy counterpart or silent partner, which, most people seem to assume, must inevitably follow it. But first let us take a fresh look with as little prejudice as possible and see what we can see. What is the true nature of death? Is it concrete or abstract? Is it an end or a beginning or, perhaps more aptly, some sort of transition or a dimensional frontier? Is it in any sense a "cure" for the "disease" of life? And should we think of it as a wall, a mirror, a shuttered window or a one-way door? Above all, can it be considered absolute, a fundamental state? Or is it only relative and a matter of degree? Finally, could one define it as part of some basic reality, a detail of an unknown whole rather than merely an illusion?

Naturally one does not expect completely satisfactory answers to all such questions at this early stage of man's inquiry into a subject that has long been the very epitome of inscrutability. But now that the taboo of sex has been largely overcome, the more persistent taboo of death may soon follow or even turn into a respected branch of science. Recent surveys of hundreds of cases of people who "died," then returned to describe strikingly parallel experiences, suggest a trend in that direction. Besides, death is a very solid and durable subject. For it was noted two millenniums ago by Seneca that, although anyone anytime *can* lose his life — and obviously *must* lose it eventually — no one ever can lose his death. Which is why death is safe and secure to each one of us, and therefore to be considered the gift of God.

## CAN DEATH BE DEFINED?

This idea is supported by the Bhagavad-Gita, which effectively defines death by making it the province of Shiva, the god of dissolution but not of destruction (which, in Hinduism, does not really exist). And Lucretius, more specific, explains that "Death does not put an end to things by annihilating the component particles but only by breaking up their conjunction. Then it links them into new combinations . . ."

There is also the fact that death is not an all-or-nothing state but normally arrives in quantum stages or degrees. Indeed, to take an extreme case, when a human head is chopped off there is evidence that the head is capable of consciousness for at least several seconds more and that it will almost surely feel itself hit the ground. Charlotte Corday's head was reported to have suddenly looked very annoyed a few seconds after she was guillotined at the age of twenty-five during the French Revolution. Crocodile hearts have been found to beat for hours when cut out and the decapitated heads of ants may bite again and again for as long as a day and a half. Besides, the fact that both the living and the dead are made of identical elements can make it nigh impossible to distinguish them.

In consequence there is a varied spectrum of opinion among medical scientists as to just when a dying organism should be regarded as dead. Is a man dead when breathing stops and his heart is still feebly throbbing? A long-accepted tradition would call him dead only when his heart quits, but obviously that criterion no longer holds in this age of artificial circulation, when a heart can be cooled and medically put to rest during an hour's operation, then started again at will, or when it can be completely removed from the body, while either a living transplant or a mechanical heart takes over the job of pumping blood for an indefinite period.

521

Recently it was discovered that the most critical factor leading to biological death is anoxia or lack of oxygen in the brain, which at normal temperature cannot survive more than about ten minutes without this vital energy-giving element. During the period between heart stoppage and permanent brain damage, capillaries are likely to clog with clotting blood and the body is in a kind of limbo state sometimes called *clinical death*, out of which it may still conceivably be brought back to life by drastic resuscitation. Once the brain breaks down from lack of oxygen, however, life can never be fully restored, even in the rare cases when both heart and lungs are later stimulated enough to resume their function for a time. The brain's electromagnetic waves, in consequence, being found to be the surest indication of its condition, have become such a sensitive test of life that a new practice is coming into acceptance of pronouncing a person dead only after his brain waves cease — this despite the fact that a few recoveries have been reported after more than an hour without detectable brain waves. But nowadays hospitals have so many artificial aids (oxygen tents, artificial respirators, electronic hearts, kidney machines, etc.) that may give a dying patient (or his vital organs) a kind of pseudolife, even after his brain waves have gone flat, that it has become a very trying, if not completely arbitrary, question when to "pull the plug" on one of these expensive machines and shift it over to the next patient, who may still have a real chance to live.

The fact that doctors are apt to regard death as unhealthy, not to mention unpatriotic or vaguely contrary to life, liberty and the pursuit of happiness, only adds to the problem of dying — so much so that modifications in the Hippocratic oath are being proposed, which would have it say something like "Thou shalt not kill — neither shalt thou obstruct a healthy or needful death." Admittedly it will take soul searching to apply such a rule in particular cases, but its general acceptance would at least acknowledge the very common feeling that death is usually more congenial when it comes in a friendly, fertile home than in an efficient, sterile hospital.

Another realization that may add to our growing understanding of death is the fact that a good many parts of our bodies, such as hair, nails and tooth enamel, are normally "dead" all the time. And the same is as surely true of feathers, shells, wood, fish scales and nerveless parts of other creatures. But were one to include such structures as human houses, cars and clothing in the same category it would likely be considered unreasonable, although it really shouldn't, because these are about as much a part of a human as is a hermit crab's shell part of a crab or a caterpillar's cocoon of a caterpillar. After all, is there any truly basic difference between a bee's cell and a monk's cell? A seed capsule and a space capsule? And why should a woolen coat made by genes on a sheep's back be considered much more alive than a woolen coat woven by man for his own back? Both coats are composed of the same elements, one put together by microscopic automation evolved long ago by nature, the other largely by macroscopic man-made machinery evolved recently by the same nature.

Now to get back to the quantum nature of death, its many stages or degrees might be listed progressively as: relaxation, absent-mindedness, drowsiness, sleep, a hypnotic state, a coma, paralysis, amputation, breath stoppage, heart stoppage, cooling of flesh, cessation of growth and metabolism, congealing of blood, rigor mortis, and the decay and disintegration of tissues, the latter normally accompanied by the quiet foraging of a sequence of about 600 species of small worms, insects, fungi, bacteria, viruses, etc. By the time your body's vital parts have completely disintegrated, cell from cell, of course you may be considered unquestionably and irreversibly dead — even though some of your separate cells may find enough nourishment to live on independently for a time, this being a real possibility since biologists have repeatedly found that disintegrated tissue, such as separated cells of a chick's heart, often for some reason live longer and better singly than as coordinated parts of a body.

My list by no means fathoms all the gradations of death, however, for death holds further degrees of a deeper kind. I mean that even when you are dead beyond all doubt, you may not be quite as dead as, under different circumstances, you could have been. For quite literally it is possible to be deader than dead. You may, in fact, be extinct. This would mean that death has come not only to *you* but also to *all of your kind*, perhaps signifying your clan or race. Beyond that there are any number of further possible depths of extinction, as when death successively eliminates your whole species, your genus, family, order, class, phylum, your kingdom and ultimately all life on Earth. Even beyond the earth of course, extinction might conceivably spread (by a drastic change, say, in the sun) to the whole solar system, then to other star systems, to the entire Milky Way or, for all we know, to other galaxies and supergalaxies — even, God willing, the universe.

## SCIENTIFIC STUDY OF DEATH

Science up to now does not seem to have taken death very seriously as a subject for major investigation, presumably because the non-physical side of it is so elusive and difficult to deal with in quantitative terms. But I think science should, and probably will, soon start applying some sort of workable measurement to death. Indeed through the inevitable interrelations among the growing and increasingly varied compilations of evidence that keep turning up relative to one aspect or another of it, a nucleus of solid fact has begun to materialize.

In simple animals such as flatworms, for instance, death is not simultaneous all over the body, but comes in a regular progression, the parts with the highest metabolic rate being affected first while less active parts are more slowly involved. Indeed, if we accept the so-called metabolic gradient theory, death is something like a creeping epidemic that advances from cell to cell as from house to house in a vast population. Perhaps, if we knew enough, we could liken death's order to the geometric polarity of an inorganic crystal and it might even have characteristics like northness and southness, as in a magnet or a world.

The absence of any absolute boundary between life and death is exemplified by the sap tube cells in a tree's wood, which aligned themselves with obvious purpose when they were alive in the cambium layer of growth but which remained linearly partitioned from each other by their cell end-walls until they died, after which these partitions disintegrated, enabling the cells for the first time to function collectively as life-serving water channels for the tree. Such a developmental sequence clearly suggests that in an organism life for the whole may depend on death of the parts, or at least of many of them. And the same dependence evidently holds throughout the animal kingdom, in which the feathers of a bird, for example, are completely alive only when they are growing, forming and unfurling. By the time they begin to serve the bird in flight, believe it or not, feathers have sealed themselves off from the rest of the bird and, being utterly without nerves, circulation or metabolism, they are, to all practical purposes, dead.

Another area where biology deals with life beyond death is the phenomenon of pupation, as in the demise of a caterpillar. Of course science has to admit that the caterpillar probably doesn't die completely, but there is little reason to think the creature himself knows the difference, for when his time is up he spins around himself a coffinlike cocoon and fades away into a soupy, disorganized mass that never again will be a caterpillar and is no longer describable as an organism of any sort. To all obvious purposes he is dead and disintegrated. Yet in a few weeks the soup solidifes again and, under the direction of a different set of genes, reorganizes into an entirely new organism, a graceful butterfly who bears almost no resemblance to the caterpillar and, in relation to it, is a sort of materialized angel of the afterlife. Indeed Greek mythology suggests that this caterpillar-butterfly relationship was recognized thousands of years ago when the nymph Psyche, immortalized by Zeus as the personification of soul, was appropriately awarded the symbolic wings and form of a beautiful butterfly.

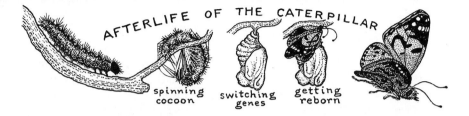

AFTERLIFE OF THE CATERPILLAR

spinning cocoon     switching genes     getting reborn

I suspect also that the caterpillar-butterfly affiliation is not dissimilar to that between a salamander's original leg and the leg that grows to replace it if it is cut off, for biologists have discovered that the flesh of the stump appears to melt back to primitive embryonic tissue before reorganizing to regrow the leg, the once specialized leg cells (in a developmental sense) returning back upstream to their genetic source to find out how to renew their specialization, how to re-aim in this their second shot at life. A philosopher might even call it an extended application of Jesus' famous botanical observation: "Except a corn of wheat fall into the ground and die, it abideth alone: but if it die, it bringeth forth much fruit."

As we approach the level of microscopic animals, immortality increases significantly and we notice that death not only gets put off more and more easily, but eventually there arrives a point where death may be avoided altogether. This is the hard-to-identify-with level of one-celled beings, beings who cannot possibly experience the separation of cells we term death as they have never even evolved the primitive conjunction of cells. Or, putting it another way: what was never together, logically, is in no danger of coming apart.

In lieu of dying then, we can say, a one-celled creature simply divides into two. We can reasonably call it disintegration because it is just that in its simple way, but it is hardly comparable to human death while there is no corpse and both the halves go right on living, indeed are normally so rejuvenated they are more lively than before. Doubtless this sort of fission is what preceded death before death as we know it evolved. It is, I suppose, nothing less than incipient pristine immortality: cells dividing and dividing and dividing . . . *omnis cellula e cellula*. Of course a philsopher could argue that, from the viewpoint of the original one-celled creature, dividing is dying and,

525

from the viewpoint of the two new cells, dividing is being born — both views being true in their own way — and the affair as a whole is an elementary orgasm compounded of the essences of birth, marriage and death all in one — virtually, one might surmise, an instantaneous compend of vital statistics.

## EVOLUTION OF DEATH

So death is not really fundamental after all. The cell is dead: long live the cells. Certainly death is not inevitable to all life since our kind of death didn't even appear in evolution until about halfway along. What is death then? And when and why was it born to Earth?

The answer seems to be that it evolved some billion years ago along with the first multicelled creatures. When individual cells started joining together, and began to specialize and organize themselves into complex organisms such as jellyfish and worms, and to diverge into sexes, there arose a need for death and this need steadily increased. Not only were body cells getting too specialized to go on being immortal through perpetual division but, among the new multicellular colonies and later semi-integrated organisms, a high percentage of them inevitably were in some degree ill equipped for Earth living — and all species that harbored such misfits suffered under the burden, just as a human family suffers from having to support any crippled, retarded or ill-adapted member. Moreover, the many species that failed to get rid of their misfit or worn-out dependents and their spent gamete-factories were weakened by keeping them and eventually faded away, while the few who learned (presumably by chance) how to disintegrate them benefited by the riddance, became more adaptable, grew stronger, had more offspring and survived in larger numbers. Thus did disintegration of a spent organism acquire survival value for its species. Thus did death become a handy tool of change and progress in evolution. For all the multicelled creatures that neglected to adopt it became extinct, while many of those that did adopt it are still living on Earth, which shows that death is not really as hateful and destructive as the legend of the grim reaper would have us believe and in fact serves such a vital purpose that we literally cannot live without it!

You may wonder why I speak of death as something inheritable and therefore genetic when so often it is imposed from outside the body and therefore environmental. Well, the answer is that, even when triggered from outside, the essence of physical death is the final disintegration of our body cells, which has been found normally to be guided by genes. The number of death genes we possess, I must say, is extremely small compared with the hundred-thousand-odd other genes in each cell but still is enough to guarantee that once we are really dead we stay dead. So we learn again that death is not absolute and not the opposite of life but rather a part of it. And, from a genetic standpoint, it may be a mere detail.

In certain cases, curiously enough, death (or something like it) is promoted by whole cells called lysosomes ("suicide bags"), filled with digestive fluid, which, at a gene-designated stage of development, commit suicide by digesting themselves. This may strike you as a less-than-pleasant way to get along in the world but it is normal in creatures such as amphibians, even young ones. There are millions of lysosomes, for instance, in a pollywog's tail which digest themselves away when the tail has outlived its usefulness, literally *ungrowing* the pollywog while regular body cells are *growing* the frog. There is even that frog *Pseudis paradoxa* who ungrows so much that at maturity he is scarcely a quarter as big as when he had just lost his tail. In humans something of the sort normally wipes out the millions of excess white cells swimming in our blood after a serious infection, a vital sort of suicide plague that dramatically explains leukemia as the rare exceptional case when the whole body must die because so many of its white blood cells somehow "forgot" to die. These (not leukemia) are examples of how death helps life, how the living body is sculptured by dying cells, by the simple wasting away of millions of no-longer-wanted cells as surely as a statue is shaped by the departing chips of chiseled "dead" marble — the positive, living, material body created primarily through the negative, nonliving, immaterial absence of whatever has been removed from it.

## MORTALITY AND IMMORTALITY

This realization that mortality has survival value that progressively evolved out of immortality is far from new, for some of the ancient Greek philosophers seem to have realized it. Empedokles of Agrigentum, for one, declared that "those things became mortal which had been immortal before, those things were mixed that had before been unmixed, each changing its course. And as they mingled, countless tribes of mortal creatures were scattered abroad endowed with all manner of forms, a wonder to behold."

Although we naturally assume that mortality means having a life    527 expectancy that decreases as time goes by, there are important exceptions to this actuarial assessment. One is the tiny hydra (page 361), who dies young enough to be considered a mortal creature yet its life expectancy remains constant, for the reason that the hydra metabolizes so fast that it replaces almost all its body cells every two weeks, the tissue flowing straight through it like gas in a candle flame, starting at the head or wick end with a speed of a millimeter a day but steadily slowing down until it reaches the hydra's foot end and diffuses into the surrounding water like smoke. Because of this extraordinary flow of protoplasm, the hydra is virtually ageless and just as apt to die this year as next and, *if* its "luck" holds, could, for all we know, live on and on indefinitely.

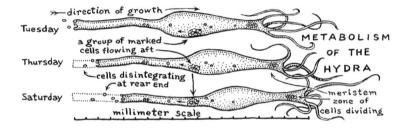

The compatibility of mortality and immortality is further demonstrated by their being easily exchanged, as when researchers J. F. Danielli and A. Mugleton at the University of Buffalo recently took an immortal species of ameba and, by transplanting cytoplasm from a mortal species, gave it the gift of old age and death. It is equally feasible, they say, to pass the gift of immortality back the other way. And this shifting of immortal gears reaches the human level when we come to consider parts of the mortal body like sperms and ova living immortally in offspring or the modern cases of organs transplanted from one person to another. Remember the grief-stricken father of the South African girl whose heart was donated in the famous first heart transplant in 1967, who was consoled with the thought that "part of my daughter still lives." And far more widespread immortality springs from single human cells such as those taken in 1951 from Helen Lane, a woman dying of cancer of the cervix, that started the noted HeLa strain of tissue culture, now immortal in research laboratories all over the world.

A deeply significant aspect of death of course is its complementarity with life. We cannot help but observe the rotting grape reborn in sparkling wine, the ink that lies mortal in the inkpot until it is penned immortal upon paper, the sugar maple reaching its pinnacle of beauty in the October colors of its dying leaves. And is there not a less obvious counterpart in man, whose first taste of life comes in the womb? If an unborn baby boy could somehow become acutely aware of his twin brother inside their mother and the brother then suddenly slipped away in a flash of light, the boy would naturally miss his brother and, in his baby way, mourn his "death" — presumably only to discover, after a second flash, that he in turn had followed his brother, and that death in the womb world is but one face of a polar transcendence, the other side of which is birth in the greater world that all the time existed beyond and outside the womb. And who can say what is on the still hidden face of this outer world's death? Or how many more worlds or dimensions we may ultimately be born into?

Did you ever wonder where a candle flame goes when you blow it out? Is it dead? Is it gone forever or may it still exist in some other form? A flame, as the hydra shows us, is a close analogue of life. Like an organism, it is composed of circulating substances with a measurable metabolism that keeps it "alive." Yet in an instant it can be extinguished by a puff of wind, as human life can be extinguished by an ounce of bullet or a drop of poison. It leaves neither cloak nor corpse behind as it vanishes completely from this world. Yet bringing it back from the "dead" by relighting it again is easy while conditions remain about the same — and who can say the new flame is not the same flame as the old?

The symmetry between living and dying, I think, must closely resemble that between the lighted and unlighted candles. Also that between mortality and immortality. And perhaps even between birth and death, which are comparably mystical, the only reason we are more awed by death being that we know people who die while those

529

being born are still strangers. If you accept such symmetries and want to live, logically you must also want to die — because death is built into life, even as waxing and waning make a single curve～～～ even as all beings grow older in the time field, their various ages drawing relatively nearer to the same age as finity draws nearer to infinity: $\infty - x = \infty$.

## DYING

Actual dying, according to medical reports available to me, seems to corroborate this, for more than four out of every five persons who die in the presence of a doctor are described as fading peacefully away into unconsciousness without pain and, more often than not, either unaware or unconcerned as to what is happening to them. Stephen Crane who died at twenty-nine in 1900 was fairly typical in attitude, though unusually articulate in his murmured last words: "When you come to the hedge — that we must all go over — it isn't so bad. You feel sleepy — and you don't care. Just a little dreamy — some anxiety about which world you're really in — that's all." In most cases among the elderly, dying seems to satisfy a very real and natural craving for rest in the form of relief from tension, such as the inevitable tension of organic molecules that have been tied together too long. And the dying person is almost always in a kind of premortal euphoria "suffused," as one doctor expressed it, "with serenity and even a certain well-being and spiritual exaltation," only slightly explainable as the effect of toxic substances in the body or "the anesthetic action of carbon dioxide on the central nervous system."

This euphoria was evidently referred to by Sir William Osler as "kindly nature drawing a veil over our last minutes," but it hardly begins to tell us why one or two dying persons out of every thousand expresses a distinct spiritual revelation in the final stages of a lingering death. General George Gordon Meade, who defeated Lee at Gettysburg, for example, in a brief awakening out of his terminal coma, whispered, "I am about crossing a beautiful wide river, and the opposite shore is coming nearer and nearer." Thomas Edison, in similar circumstances, sighed, "It is very beautiful over there." And a cancer patient in England, whose hand was being tightly held by his doctor, murmured at the end, "Don't pull me back . . . It looks so wonderful further on!"

One of the most unusual cases I have encountered involved a man whose heart had stopped for nearly ten minutes. By a heroic effort his doctor revived him briefly and, on asking him if he remembered

530

anything that happened during the time he was "away," the man thought a long moment, then haltingly replied, "Yes, I remember. My pain was gone. I was free. I couldn't feel my body. I heard music — the most peaceful music." He paused, choked deep in his throat, then struggled on: "Peaceful music. God was there, and I was floating away. Music was all around me. I knew I was dead, but I wasn't afraid. Then the music stopped and you were leaning over me."

Asked if he had ever before had a dream like that, the man made a supreme effort and said, with intense conviction, "It wasn't a dream." Then he closed his eyes for the last time, his breathing grew thick and, in a few minutes, he was dead.

Still another case was that of a woman who was apparently dead for several minutes during an operation from which she later fully recovered. But the extraordinary thing is that, despite being under a general anesthetic, she acquired an uncannily vivid memory of having observed the operating room during those critical minutes from an indescribably spaceless position that somehow enabled her to see all four walls, ceiling and floor simultaneously, including her own lifeless body on the table as if from above or outside, and she could fully recall the tense conversation between the surgeons and nurses as they worked to get her heart beating again.

What actually happens in the dying process is virtually impossible to assess because so much of it is intangible and, in most cases, beyond the present reach of science, which deals only with measurable phenomena. Experiments at the Bose Institute in Calcutta early this century indicated that a pea dies an extraordinarily violent death when cooked to 150°F., at which point it may discharge as much as a volt of electricity, a phenomenon attributable in some degree to all dying protoplasm and which some investigators thought might explain the oft-noted memory improvement in persons about to die. But the evidence seems not to have been substantiated nor widely accepted.

The only researcher I know of who has claimed a quantitative clue as to the departure of the human "soul" was Dr. Duncan MacDougall, a physician on the staff of Massachusetts General Hospital in Boston, who in 1906 conducted a series of experiments by placing the beds of dying patients on special beam-type scales accurate to within one tenth of an ounce. A typical case he reported in the *Journal of the American Society for Psychical Research*, for May 1907, said: ". . . subject was a man dying of tuberculosis . . . under observation for three hours and forty minutes before death . . . He lost weight slowly at the rate of one ounce per hour, due to evaporation of moisture in respiration and . . . sweat.

"I kept the beam end [of the scales] slightly above balance near the    531 top limiting bar in order to make the test most decisive if it should come. At the end of three hours and forty minutes he expired and suddenly, coincident with death, the beam end dropped with an audible stroke, hitting against the lower limiting bar and remaining there with no rebound. The loss was ascertained to be three fourths of an ounce." Dr. MacDougall, following in the footsteps of Anubis, weigher of souls in ancient Egypt, completed his experiments with six patients and "in every case found a distinct, sudden drop of weight" at death. But, to my knowledge, no one else has ever taken this evidence seriously enough to continue the experiments in order to confirm or disprove the findings.

On the psychological side, however, there is the beginning of a consensus confirming that many a younger dying person typically goes through six stages in adapting to it. First, he refuses to believe he is dying. Second, when convinced, he protests, "Why me?" Third, he bargains: "I'll be good, dear God, if You'll only give me more time." Fourth, he is depressed, thinking, "It's all over. There's no hope. And who cares?" Fifth, he may break down into tears, ending with the feeling it was a relief to get it out of his system. And sixth, he finally accepts death: "I'm going to die, and that's beautiful because it is God's will and the way He made the world."

With a little training in the subject, I think doctors, nurses and parents could make dying a lot easier, especially for younger people. There surely is little virtue in their continuing to be evasive, as most of them are now, seemingly influenced by the taboo. For everyone someday has to die. It is part of normal living and not only must be faced eventually, but it is something good for us to face whenever it appears. It also seems particularly important not to hide it from small children, who should be allowed to see people dying and to talk and even joke about it, when appropriate, and feel the depth and beauty of it as well they can. Someday I expect a course in dying will become a regular part of the school curriculum, for it is already among the most neglected as well as perhaps the most important of subjects.

## THE AFTERLIFE

Whether or not death ever fully enters the realm of science on Earth, judgment of it seems remarkably stable — as, for example, that of Cicero who wrote two millenniums ago: "There is in the minds of men, I know not how, a certain presage of a future existence; and this takes deepest root in the greatest geniuses and most exalted souls."

Helen Keller "saw" death as illusory, and more clearly for being blind and deaf: "I know my friends not by their physical appearance but by their spirit. Consequently death does not separate me from my loved ones. At any moment I can bring them around me to cheer my loneliness. Therefore, to me, there is no such thing as death in the sense that life has ceased . . . The inner or 'mystic' sense, if you will, gives me vision of the unseen . . . Here in the midst of everyday air, I sense the rush of ethereal rains. I am conscious of the splendor that binds all things of earth to all things of heaven. Immured by silence and darkness, I possess the light which shall give me vision a thousandfold when death sets me free."

Even a man like Bertrand Russell, who did not believe in personal survival, strongly felt the abstract relation between an individual and his world, for he wrote that "an individual human existence should be like a river — small at first, narrowly contained within its banks, and rushing passionately past boulders and over waterfalls." But "gradually the river grows wider, the banks recede, the waters flow more quietly, and in the end, without any visible break, they become merged in the sea, and painlessly lose their individual being."

Perhaps one could go from there to realization that the "self" is illusory in nature, temporary from our present, limited viewpoint, a merely elementary, finite tool of learning. Alan Watts seems to have thought so when he wrote that "there is no separate 'you' to get something out of the universe," that "we do not come *into* the world; we come *out* of it, as leaves from a tree." He explained that "as the ocean 'waves' so the universe 'peoples' . . . What we therefore see as 'death,' empty space or nothingness is only the trough between the crests of this endless waving ocean of life . . . The corpse is like a footprint or echo — the dissolving trace of something you have ceased to do . . . When the line between yourself and what happens to you is dissolved, you find yourself not *in* the world, but *as* the world . . . There is a feeling of hills lifting you as you climb them, of air breathing yourself in and out of your lungs. All space becomes *your* mind . . ."

The physical body, according to this view, may be a sort of gantry for the ship of soul, a matrix that gives form to spirit as language gives form to thought. Whenever, as the Bible puts it, the Word is made flesh, consciousness may be quantized into mind seeds in association with developing organisms, each of which is largely controlled by genes, secretions and other stimuli. Then later, when individual thought forms and memory patterns are established, the gene structure and physical stimuli become less important, probably too restrictive, with the result that consciousness eventually breaks free of the body 533 by means of the "hatching" process we know as death. It may even be that it is only through transcending the body that mature individual consciousness can become enabled to merge or synthesize with other consciousness or consciousnesses into some sort of greater organism for its continuing development.

If doubt about personal immortality is one of an aging man's greatest burdens, at least the inevitability of dying can be said to be merciful. Just think what a scramble life would fall heir to if it were an accepted fact that each of us had a fifty-fifty chance of permanently avoiding death. Or even a one-in-a-billion chance. In a curious way also, when one thinks deeply about it, our doubt about immortality is far from an unmitigated liability. In fact there appears a spiritual bounty in this seemingly ultimate mystery, because having to face it for our few years on Earth effectively bestows upon us the privilege, if not quite the necessity, of exercising whatever faith in life we have. I refer to faith in contradistinction to certitude, for inevitably, if science had already solved the mystery with sufficient certitude, we would have been denied the very special exaltation of relying upon faith alone.

Faith of course can be bolstered by religion and reason, since not only have all the great Prophets of God promised life beyond death, but reasonable philosophy also offers us a hearteningly positive answer to the classic query of "To be or not to be?" It is logical, you see, to conclude that there is nothing in nothingness or, which is the same thing, that not being cannot be. Doesn't the phrase say so itself rather plainly: it cannot be that something cannot be? Put another way, life is an inherently positive existence which has no such negative capacity as would be required if it were ever not to exist. Besides, the inscrutable wisdom of the universe, usually called God, has let us be. If nothingness were our divine destiny, how in heaven's name could we be here? How could we know of our own existence — or have an existence to know of? What meaning, what profit in a fleeing flash of positiveness if it is to be followed by an eternal negative? Obviously the answer is affirmative, we are meant "to be," and existence is our prime and positive destiny. We call ourselves "beings" because living is being and there is nothing without being.

Or, looking at it through the eyes of a gambler, might I not say that the most skeptical speculator could consider eternal life a safe bet. For in fact eternal life is a bet one cannot lose. When you come right down to it, any outcome of the bet constitutes life and victory. If there were no life after death, that would be no outcome and there would be no one around to lose the bet. So all beings must be winners and the mere fact of *being* is the victory.

To those who believe, as I do, that this is a positive world of unlimited potentiality and very probably infinite dimensions, it is natural and easy to accept life beyond death. The logic of it is that, in a world as potential as this — and I am thinking of the world in the broadest sense I know of, comprising the universe materially, mentally and spiritually — anything can happen if there is only sufficient room and patience or, in more fundamental terms, enough space and time. And if *anything* can happen, then it follows that *everything* must happen or must have happened somewhere in the infinities of space, time and other dimensions. The British novelist T. H. White went so far as to propose that "anything not forbidden is compulsory" on the unstated but implicit assumption that time goes on and on without limit. One could as easily say that "everything conceivable must eventually happen," for, if there are enough hours and millenniums and minds and dreams to conceive of something, there logically must be enough space and time for the same conception to be real in (reality being basically conception in definable form) — and, if everything conceivable happens, life after death (which is conceivable) must also happen sometime in some continuum or in some division or combination of consciousnesses.

The concept of probability also, surprisingly enough, cannot help but be a factor in the equation of immortality. Physicists have discovered a number of elementary particles by looking for them on the premise that, if no good reason why one can't exist has been discovered, the probabilities favoring its existence must amount to enough to make that existence an ultimate certainty. If this line of logic seems shaky as ground for hoping that something real will pop up out of nowhere, let me say it works nonetheless with everything from quarks to quasars — and it gave physicists provable results in the subatomic neutrino in conformance with the elementary law of nature that categorically declares that all events (real or imaginary) have some degree of probability unless some principle (known or unknown) specifically prohibits their happening. In other words: anything that *can* happen *does* happen. And therefore, in the absence of any law forbidding consciousness after death, consciousness has a continuing measure of probability that, however small, adds and adds and adds . . . ad infinitum.

Perhaps now you will protest that an immortality compounded of eternal probabilities is cool comfort to the dying, that, if one has to wait a billion years for resurrection, one might as well forget it. But the billion years is of course relative and inherently illusory, for, without consciousness, the long sleep seems to its Rip Van Winkle to    535 pass instantaneously and whether Rip "wakes up" in a minute or a billion years should make no difference to him — at least no conscious difference.

Besides, what evidence is there that time in any way exists as a dimension beyond our present life? Not only did Einstein contend that neither time nor space is fundamental but many mystic sources corroborate him with suggestions that the most profound of influences are eternal and infinite — that is, outside of time-space as in Jesus' tense-twisted declaration that "Before Abraham was, I am."

Passage from this finite world into an infinite world should be natural and painless, as Bertrand Russell suggests, perhaps taking on new perspectives, as in the analogy of the dying river boatman who, from his two-dimensional river surface, which denied him a direct view of either his past around the bend behind him or his future around the bend before him, is suddenly wafted "upward" at death into the three-dimensional sky from which he views the river whole: taking in past, present and future simultaneously.

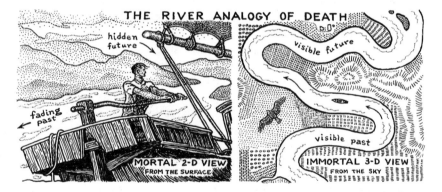

THE RIVER ANALOGY OF DEATH

hidden future

fading past

visible future

visible past

MORTAL 2-D VIEW
FROM THE SURFACE

IMMORTAL 3-D VIEW
FROM THE SKY

If you can entertain this prognosis of immortality, it may be helpful to add that dying and passing beyond space-time is analogous to a character in a movie being transformed into a character in a book, because it amounts to switching from a medium of linear succession where birth, life and death follow each other chronologically like notes of a melody to a medium of integral simultaneity where birth, life and death are all displayed together like the notes of a musical chord.

It is also analogous to evolving from the concrete particle milieu of the nineteenth century into the abstract field milieu of the twentieth, from finitude toward Infinitude. It is transcendental and so natural that many an astute philosopher has intuitively realized that, if one could understand the true reality of this world, there would remain virtually no difference between life and death. When Thales said as much in 600 B.C., one of his followers asked, "Why then are you not dead?" He replied, "Because it really makes no difference."

If death is still unacceptable to you as a mere aspect of life, I might ask how, in the time before organic life evolved, one could have imagined life? And, if one could not, then how now death?

Furthermore, if you don't agree that dying is living, pray tell me how it is that a watch serves (lives) only while it is running down? Why rivers flow? Why rain rains? Why fruit falls?

I suppose one reason for the pervasive fear of death is that it implements our instinct for self-preservation, which has obvious survival value in evolution. But another reason must be that fear bolsters our sense of self that may be just what enables us to keep soul in body long enough to get full benefit of this finite phase of transcendence.

Did you ever try to imagine in full, vivid reality the world of your children and surviving friends after your death? It is hard, and takes an effort of imagination, although of course you can reason out the more predictable post-mortem events by logic. But, in fact, there is a strange inertia between you and that future world, an inertia that testifies to its illusory nature in relation to you if not to your whole present world. It is presumably the same barrier that separates your dead ancestors from your present life, and the relationship says something about the nature of this mortal world in general and your life in particular, about its soul essence being beyond space and time and self and, as I surmise, profoundly safe from the dangers and chances of earthly finitude or personal mortality. To those in the larger world who are outside of time with no past or future, we here-now must be as if already "dead." We simply ARE.

Somehow I find this a comforting thought, not only because it finesses the problem of death, but because it seems so beautifully in tune with transcendence. And so also is the record I found of the last words of a Blackfoot Indian chief named Isapwo Muksika Crowfoot, who held his warriors in check for several years before he died about

1890, when he murmured to Father Doucet from his deathbed, "A little while and I will be gone from you — whither I cannot tell . . . From nowhere we come, into nowhere we go. What is life? It is the flash of a firefly in the night. It is the breath of a buffalo in the winter time. It is the little shadow that runs across the grass and loses itself in the sunset."

# CHAPTER 21

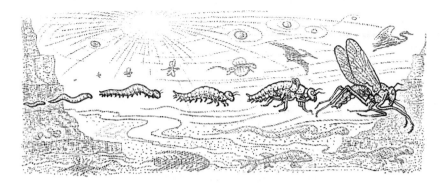

# Evolution of Earth

I F ONE WERE LOOKING for a symbol to represent the Earth, it would be hard to find one more appropriate than the doorknob. For our planet actually has a slight doorknob shape, being measurably flattened at the poles. More important, she is offering man his first handle for opening the door to creation and acquainting him with the Universe. By which I mean that man's contemplation of Earth as a sample world gives him helpful clues to worlds in general and to how they evolve — including, I suppose, more than a hint of why they exist.

Beyond being a sample, however, Earth is a seed world, a world exuberant with untamed mind and spirit that have been long and patiently brewing life out of seeming lifelessness and are now clearly sprouting into germination, as we will see next chapter. Lifelessness, we learned in Chapter 14 on omnipresence, is probably illusory in the sense that stones, crystals and "inorganic" substances are at least potentially alive, with no provable line demarking them from organic stuff. But what factor on Earth do you suppose actually turned inorganic lava into organic soup and persuaded the planet she was a viable organism? How did the spunk ignite and the spirit find feather to take wing and fly? What made life live?

When Earth was young — say a baby of less than a billion years — the presumption of classical science is that she was sterile and completely innocent of anything a biologist could call life. Her rocks were hot and bare except where covered with steamy, unsalted pools, progressively growing into seas under an atmosphere thought to have been mostly ammonia fumes laced with methane, water vapor and wild gassy compounds of bromine, chlorine, fluorine or sulfur. If anyone had been there to look at the volcanic rockscape with its hissing fumaroles and thunderous black clouds pouring acid rain into fishless seas and could have been told that this noxious brew, left to itself long enough, would simmer and ferment into such fantastic improbabilities as the Taj Mahal and the frescoes of the Sistine Chapel, the works of Shakespeare and Beethoven, electronic computers, television networks and moon rockets, he could not have been blamed for rejecting the idea as preposterous.

## EVOLUTION OF EVOLUTION

Moreover, behind the puzzle of how life arose and evolved looms the derivative question of how did the puzzle itself arise and evolve? When did philosophers or scientists first conceive of evolution and how did they set about measuring its structure?

Before the nineteenth century few of them seem to have had any inkling that nature might be going through any fundamental change with the passing eons, for almost no one had ever heard of the perspicacious handful of ancient observers who had somehow managed to surmise a progressive creation. One of these was Anaximander of Miletos who said in the sixth century B.C. that man "descended from a fish in the beginning of the world." A century later came Empedokles who mystically announced: "I have been ere now a boy and a girl, a bush and a bird and a dumb fish in the sea." Another certainly was Lucretius who wrote in Rome that "the new-born Earth first flung up herbs and shrubs. Next in order it engendered the various breeds of mortal creatures, manifold in mode of origin as in form . . . more and bigger ones took shape and developed . . . first . . . birds . . . then . . . mammals . . ."

His rare pre-Christian sagacity even told him that there must be some sort of a natural law selecting the kinds of creatures that were to succeed or fail in the obvious struggle of life, for he explained that "monstrous and misshapen animals were born, but to no avail because nature ruthlessly eliminated them . . . and many species must have died out permanently through failure to reproduce their kind while every breed of animal we now see alive has been preserved from the beginning of the world either by courage, cunning or speed.''

Dozens of later philosophers and scientists also touched on this umbrageous idea of competition for survival, including Saint Augustine and Saint Thomas Aquinas. In the modern age Kant speculated that apes might well in time turn into men, Goethe wrote about "the metamorphosis of plants," and both Erasmus Darwin (Charles's grandfather) and Jean Baptiste de Lamarck propounded the theory (with little supporting evidence) that all species must have evolved from simpler and more primitive forms by "inheritance of the effects of use and disuse." Even the new science called geology prepared the way for the evolutionists by demonstrating in the nineteenth century for the first time, particularly through the research of Charles Lyell and the new study of fossils, that the earth is immensely older than anyone had dared imagine. In fact Lyell described our planet as steadily and irresistibly changing, not only through the slow wrinkling of its skin in the process of pushing up mountains but almost equally by the mountains' wearing down again under the relentless grinding action of rain and wind and ice and rushing streams. And a Swiss named Louis Agassiz made a dramatic contribution to geology when he suddenly realized that the earth is now in the very act of thawing itself out of an ice age which has been going on for tens of thousands of years and which, as new evidence turned up, was quickly revealed to be only the latest in a complex sequence of hot and cold ages going back for untold millions of years.

By the mid-nineteenth century, in consequence, the seed of evolution had unmistakably germinated in human thought, and it sprouted irrepressibly into the forefront of science news when Charles Darwin and Alfred Russel Wallace published their papers on "evolution through natural selection" in 1858, bursting into full flower the following year as Darwin's closely reasoned *Origin of Species* hit the bookstores, and its meaning sundered the complacency of a world brought up on a literal interpretation of the Book of Genesis and Bishop Usher's calculation of the date of creation as 4004 B.C. This is not to say that the theory of evolution denied God's creation of the world, for anyone who understood the symbolic meaning of the Scriptures could accept evolution as a logical elucidation of just how the Creator may have carried out His creation, of how He "let there be a firmament" of stars and galaxies that slowly arranged themselves on a majestic scale out of elements evolving from simple hydrogen into more and more complex atoms and molecules and dust, of how He

541

eventually "let the earth bring forth grass . . . and cattle . . . and every thing that creepeth upon the earth after his kind . . ."

In retrospect evolution actually made some sense out of Job's omnigenous plaint: "I have said . . . to the worm, Thou art my mother, and my sister . . . I am a brother to dragons and a companion to owls." Its sweep in effect defined the relation of every creature in the world to every other, then of every atom to every other and ultimately, by extrapolation, of every planet, star and galaxy to every other everywhere everywhen. It did so partly because it found all life and all things to be made literally of the same stuff and partly because relationships in general were turning out to be more real than things in a world that had seen itself for the first time as abstract in its essence. This may be a clue as to why physics is currently deemed evolving into a branch of geometry or why mathematicians so often have the last word in science. Certainly time as a geological and biological abstraction could hardly have been used effectively before evolution was accepted, nor could pre-evolutionist man have envisioned much sense of progress.

## How Life Began to Live

But today the world is assuredly going somewhere and evolution is established — not only the evolution of elements, of worlds and galaxies and of life, but of all their qualities and abstractions. Indeed even the sticky question of how life on Earth finally managed to manifest itself and really live, although still very mysterious, is probably not quite as mysterious as it seemed three centuries ago, when the great Flemish scientist Jan van Helmont wrote his famous recipe for creating mice: put a pile of soiled clothes in a dark, quiet corner, sprinkle them with kernels of wheat and within twenty-one days mice will appear. While the problem of "spontaneous generation" grew in complexity with the discovery of the teeming microcosm by van Leeuwenhoek and a couple of centuries later was thought to be settled (in the negative) by the sterility experiments of Pasteur, it is still alive today in the realization that somehow life must have got started upon what was supposedly originally a lifeless Earth. Yet HOW did it get started?

Could it have come from outside as early Greek philosophers and panspermists implied? Arrhenius, the Swedish chemist, and other modern scientists have thought so, pointing out that "astro plankton" in the form of dry spores and the hardiest strains of bacteria could be thrown into space by major meteoritic or cometary impacts and that probably at least a little of this viable material would drift dormantly from world to world along with the space dust well known to enter the earth's atmosphere in a constant shower, the larger grains of which may flare into visibility as meteors without necessarily losing their tiny cargoes of capsulated life within. At least one researcher (H. E. Hinton of Bristol University, England) seems to have proved that actual insects (specifically midge larvae), if completely dehydrated, can be popped into liquid helium and cooled to temperatures as cold as eight thousandths of a degree above absolute zero yet, when thawed (months or years later) and dropped into water, a percentage of them will come alive again, wriggling and feeding within minutes. If large numbers of such animals in a dry, frozen state should be hurled into space, perhaps in the breaking up of a planet (thought to be the origin of most meteorites), there seems a reasonable chance that at least a few of them might survive long enough to reach another planet. And this possibility is probably even greater in the case of vegetable organisms. In any case recent analyses of the rare carbonaceous chondrites, a class of small stony meteorites containing both carbon and water, showed a variety of organic compounds beneath their charred exteriors including amino and carbonyl groups as well as particles resembling fossil algae, some of the photomicrographs of which revealed what looked like the double membranes, vacuoles and spiny surfaces of living cells, sometimes as dense as 40 million per cubic inch. One researcher even reported finding what appeared to be the fossil of a cell undergoing a mitotic division, including chromosomes.

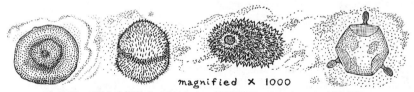

magnified X 1000

LIVING-CELL-LIKE FOSSILS IN METEORITES FROM SPACE

Still another possibility that seems to be growing in plausibility (as UFOs are becoming accepted as something more than imagination) is the starting of life on Earth by munificent beings from elsewhere in the universe. After all, we know the moon received life in 1969 from what may (to her) have been UFOs, yet which were certainly guided by beings sent from an outside world. So why couldn't some such beings in turn have voyaged to Earth in spaceships in the distant past,

543

perhaps dedicating themselves to this particular cosmic purpose like Johnny Appleseeds of the Milky Way?

Of course whether or not life was transplanted to Earth from outside, the problem of its origin still remains, for we have no reason to suppose it is any easier to explain how it got started in another world than in our own. Indeed, for all we know, life may not ever actually start in any world but instead may just intermittently germinate and sprout and blossom out of hard-to-recognize "seeds" lying dormant in all matter.

In any event science's most serious efforts to account for the rise of life on Earth start with the fuming metallic deserts of lava and slag that formed her youthful face, which, one surmises, was by then being progressively punctuated by the deepening pools that bubbled and cooled and somehow learned to ferment and crystallize into fantastic complexities. Since scientists generally consider the cell the basic unit of life, they have focused major attention on what it must take to create a living cell. And what it takes, they find, is water ($H_2O$) and salts (sodium, chlorine, etc.), with major portions of certain kinds of big, complex molecules made of carbon, hydrogen, oxygen and nitrogen in various combinations with much smaller percentages of a few other elements, like phosphorus, sulfur, magnesium and potassium plus what are believed to be "vital" traces of iron, copper, zinc, chromium, manganese, iodine, calcium, molybdenum, selenium, vanadium, cobalt, cadmium and gallium. Although all these elements and a dozen more are found in living cells, it is hard to prove to what extent some of the latter ones may be essential to life or how much of life. Indeed the lack of general agreement on a definition of life leaves such a question fairly nebulous.

On the other hand the evident deadness of the primordial earth of four or five billion years ago is almost surely less attributable to the absence of any vital elements than to their inaccessibility. There was little free and breathable oxygen then in either the "air" or water, for deducibly it was all chemically bound up with hydrogen in $H_2O$ or in the solid crust of the planet. Most of the carbon was buried deep in the clenched ores of heavy metals like iron carbide. In fact all the chemical cards seemed hopelessly stacked against the creation of protein or of even the simplest amino acids that could build what we might call living tissue.

If the chance for "life" appeared impossibly remote, however, it was "impossible" only from a limited human viewpoint, for the factor of time, celestially speaking, was practically unlimited — and, as any probability engineer can tell you, time has the power to change the "impossible" into the "inevitable." It does it mathematically by adding rarities to rarities enough times (i.e., over a period of enough time) to make them abundances and eventually practical certainties.

Returning now to the raw and very patient Earth of 4 billion B.C., we find that dramatic improbabilities were brewing little by little over the eons in the outbursts of volcanoes with their sizzling lava and sulfurous clouds that clashed with the searing ultraviolet rays of the sun while multimillion-year-long deluges lashed chemicals out of the air and roaring cataracts and rivers scoured oxygen and salts from the rocks, washing unheard-of treasures into the sweetish, simmering seas, where they dissolved and effervesced and eventually fomented the very juice of life. One does not think of rocks as life-giving structures, but in eroding (one of their metabolic functions) they truly are and this may well be how granite, which is half oxygen, spawned much of the free oxygen that gave earthly creatures their first real breaths of air.


## REPRODUCTION

At this stage in the advent of life we must face the nice little dilemma of which came first: reproduction or the organism? For it is about as hard to figure out how reproduction could have taken place without some sort of organism as to comprehend how any organism could maintain itself without some capacity for reproduction.

The surprising answer now appears to be that reproduction is not absolutely dependent on any organism nor perhaps even on any kind of life, although the latter conclusion obviously hinges on the breadth of your definition of life. A crystal, although not an accepted organism and not generally considered alive, nevertheless can reproduce itself naturally under favorable conditions by the simple, basic, automatic process of accretion. Think of such a crystal as sugar, salt, ice or a stone. Each of these grows by accepting aggressive molecules (from any surrounding fluid) which match its own and by letting them latch onto its solid body in the only way they will fit, which means by conforming precisely to its own characteristic lattice pattern. When pieces of this new crystal growth later break off from the "parent" body as grains of salt or sand, they may be regarded as offspring crystals "born" by division from their reproducing "mother" and, with such a heritage, they actually possess the positive identity of descendants of their particular "species" of crystal and no other.

So, in the ordinary sense of life as organism, reproduction can be presumed older than life, for it must have long served and preserved 545 the rocks and sand crystals before the sea had time to mellow into readiness for what one might call God's green thumb and the unimagined appearance of genes and cells and relatively huge integrated organisms. The mellowing process naturally required a long and patient wait until the organisms became big enough so random thermal agitation of molecules was no longer significant in their physiology. And later until distinct individuals evolved into mortal beings limited in space and time.

## ELEMENTS OF LIFE

Also, although the detailed steps and multifarious chemical trials and failures that created life have been too complex for us to reason out with any fine precision of reliability, we can deduce at least that the methane ($CH_4$) component in the primordial atmosphere, as it was blasted and diffused by lightning and cosmic rays for billions of years, must have progressively energized and freed enough hydrogen and carbon to form vast quantities of hydrocarbon molecules. And these are the extraordinary molecules of life that have an exuberant tendency to add on more and more carbons followed by hydrogens, growing into long chains, the chains in turn provoking the amalgamation and fabrication of elaborate, repetitive organic molecules of truly mystic potency.

Similar processes must have urged the primeval ammonia fumes and volcanic gases and water vapor to free nitrogen and oxygen atoms all around the young Earth to compound the barm of burgeoning life, breeding incredible strains of these germs and catalysts of a still undreamed future. And these matrix molecules inevitably grabbed and sorted whatever came their way with a new-found propensity for cohesion, a stickiness that turned flowing liquid into jellyish colloid whose scum would eventually solidify into lasting shapes that became the first elementary structures for living bodies, patterned into the minimal stability of form, with less than which life in a material world seems unable to maintain itself. At first these gel droplets in the ocean — some shaped like balls for easy flowing, some stretched into threads for tighter clinging — could only have been models for future protoplasm, no more alive than sea foam. But with the passing millenniums they undoubtedly developed the capacity to grow appendages or "buds" that could separate as offspring droplets and to harbor excess water in tiny bubbles like vacuoles inside a living cell. And, as ultrasensitive instruments have recently demonstrated, each of them must

have been completely enclosed in a tenuous skin of water molecules, knit together by the faintest murmur of magnetism, a liquid mantle of surface tension through which dissolved materials could slip osmotically in and out without disrupting the integrity of the whole; the convex lens effect of the outer surface focusing sunlight (in those days very strong in ultraviolet rays) into enough concentrated heat to cook the contents into "life" — a life composed of metabolizing molecular clusters which biologists now call coazervates.

It was probably in about this stage of life's evolvement that photosynthesis got seriously to work storing energy in the form of organic fuels of starch and sugar (page 51) and perhaps not much later than the new long-chained molecules expanded into the third dimension. Up to this time, you see, they had benefited by remaining largely in the two-dimensional surfaces of globules and pools, where they obviously had a better chance of joining one another (side by side) than if they had been drifting in a three-dimensional soup. But, having evolved to a more sophisticated level of maturity, they now naturally took to curling their sinuosity into spirality, the basic and efficient form that is the prevailing grain of the universe. Thus they advanced from reproduction in the manner of crystals and buds through a long evolutionary tournament of subtle pressures and selective eliminations (perhaps including something akin to a proto-mind filtering "yes" from "no") eventually to arrive at the genetic techniques of double-spiraled DNA molecules and all the emerging phantasmagoria of protein and nucleic acids with their never-ending varieties of genes and enzymes.

Here, before it knew it, life must have perceptibly shifted into higher gear, for the relatively huge and complicated protein molecule has electric charges of varying magnitudes sparking provocatively along its bustling passages and galleries. It is a kind of microscopic grand

547

opera house throbbing with music and drama, its doors swinging open and shut, its lights flashing on and off, scenery fading and shifting and players in the guise of enzymes gliding to and fro, while others called acids and alkalis fume and spit and kick each other out windows or down stairs. This protein also has certain attributes of a great sea beast so exuberant it alternately swells and shrinks, curls into a ball, spreads out like a fan, squinching and squirming and sloshing. Yet, generation by generation, it changes and mellows — and somehow the probing, mutating molecules integrate to evolve the cell, billions of them at a time deviously combining into each of these globular units that is midway between an atom and a man. And, from there in the last billion years, through such imaginative colonial ventures as slime molds and sponges, the cells have put together the whole fantastic kingdoms of vegetables and animals that inhabit the earth.

## SHAPE OF EVOLUTION'S TREE

It is generally assumed that the tree of evolution, like other trees, has one main trunk out of which all its limbs branch off, but such a linear structure has not been proved and some theorists speculate that the tree may really have what mathematicians call a nonlinear form — which would probably mean several or many trunks like a clump of trees or possibly an interconnected system like a banyan tree with its pillar roots (page 64). Paleontologists have already dug up about two hundred species of extinct horses, for example, which evolved separately but more or less parallel to the ancestral lines of still-living horses and to each other's. The reason for such parallelism seems to have been that all the horse (and ass) forms were shaped by the same general environment. All were small animals at first (including even a probable "mouse horse" living in the forest) with at least four toes and simple molar teeth for browsing. But as they became bigger and heavier in their various ways, these evolving horses kept breaking

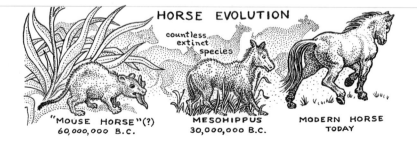

HORSE EVOLUTION

countless extinct species

"MOUSE HORSE"(?)
60,000,000 B.C.

MESOHIPPUS
30,000,000 B.C.

MODERN HORSE
TODAY

their side toes and wearing out their short molars, most of them consequently going lame and hungry despite their evolutionary progress until they died out, only the elite few species which had most rapidly and efficiently evolved single hoofs and long molars surviving into modern times. Although horse evolution thus goes back irregularly over fifty million years to the leaf-eating rodentlike common ancestors of all the horse cousins, it is likely that looser parallels such as the one between large fish and dolphins or between the entire vegetable and animal kingdoms go back hundreds of millions, perhaps billions, of years before they merge in a common stem. However it seems almost inevitable that all lines of life must ultimately unite within some portion of Earth's history inasmuch as it must be improbable in the extreme that any line of life could rise to importance on a fluent and soluble planet like Earth totally unconnected genetically with the others. For all these offspring of the planet are, so to speak, ladled from the same stew, even if you consider that some of them may have derived from meteorites, no meteorite having ever been known to have originated beyond the solar system (a restriction that of course does not apply to the earlier dust and lumps that are thought to have glomerated the earth). All earthly genes therefore (unless they descended from life older than Earth) must needs be made of the same carbon, hydrogen, oxygen and nitrogen atoms that are the principal ingredients of all life known to man. And it is just as improbable on the face of it that two intricate double-helix DNA molecules made of the same earthly (or solar) elements could have no genetic relation as that a chimpanzee and a man with almost every bone, muscle and nerve of the one corresponding to an equivalent bone, muscle and nerve of the other could have no ancestry in common.

Students of evolution indeed are coming to see species of Earth's plants and animals as not so much disparate rivals as discrete steps in an integral planetary process, for obviously the wonderful variety of forms invented by nature would be diminished if all creatures just interbred without restriction in a common genetic melting pot. Which may be why most of the varieties have quantized or genetically isolated themselves by taking up courting rituals and other reproductive devices that effectively stop would-be mates who are so to speak not out of quite the same drawer.

## Spiral Geometry and the Origin of Species

Species evidently then originate in hundreds of ways — a lot more ways than Darwin had time to discover — and the factors involved

549

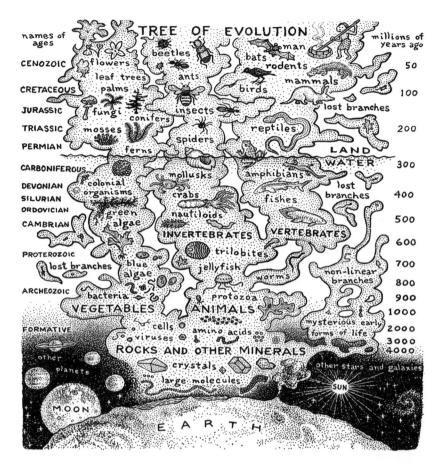

range from cross-breeding to hydraulic pressure, from genetic mutations to the dynamic evolvement of the earth, including its climate. Why do some horns and shells spiral clockwise in certain species, for instance, and in others counterclockwise? At first you might think that the snail somehow originates the shape of its shell, but the truth is rather different — for present evidence indicates the snail does not so much curve the shell as the shell curves the snail. And since the shell, strictly speaking, is not a living body but a kind of house built of masonry by snailpower through an automatic mechanogeometric process, the snail itself is largely shaped by spiral geometric forces.

Nature knows many organic spirals of course, ranging from the helix of DNA to the coil of an elephant's trunk, among which perhaps the most beautiful and important is the equiangular spiral. This particular curve, which James Bernoulli for good reason called *spira mirabilis*, is a function of the ancient Golden Section and the Greek gnomon (page 338), and it appears in such different functions as seashells and the flight of a moth to a flame, the latter happening when the moth's constant-angle navigation instinct lacks a distant sun to keep it on a straight course and steers in relation to a local flame instead. It is a curve that, in many different applications, progressively extends itself over a period of time, defining thereby a sort of living graph the unique feature of which (as I said in Chapter 12) is that its growth never changes its basic shape, a property incidentally important to any animal, such as a ram that might be fatally handicapped in a fight if his heavy corkscrewed horns did not keep a stable center of gravity.

## SPIRA MIRABILIS

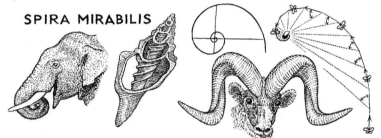

As found in nature, this equiangular spiral is frequently confined approximately to a single plane as in a chambered nautilus, a chameleon's tail or the floret pattern of a sunflower, but more often its graceful path describes a conical surface as in the majority of shells, horns and tusks. Pythagoras almost certainly knew and understood it and was awed, I like to think, by the beauty of its unvarying growth. At any rate he is said to have analyzed it as being composed of gnomons with the rather mystic collective power to create and preserve the structure of life in general, probably including the world. And this just could explain why living things seem to love the spiral form, why a single gene (itself spiral) can so easily reel out the unvarying gnomons that curve its developing shape, including the Fibonacci spacings of seeds and branches that so pervade the vegetable kingdom — even why a seeming minor shift (perhaps less than a mutation) in the structure of the gnomon gene can counter a clockwise shell or reverse a pair of conjugate horns, as is thought to have actually happened to the markhor, a wild Asiatic goat (with clockwise right horn) which is presumed the ancestor of the domesticated corkscrew goat of Asia (with counterclockwise right horn), the key step being a curious genetic realignment which has been observed in an embryologist's laboratory as early as the first cleavage of the fertilized egg.

551

## VARIATION AND SELECTION

Could such a switch of spirality have been acquired in some unknown way in the domestication of the goat and then become inheritable? Geneticists generally deny this Lamarckian notion that any characteristic acquired by any creature (say a groovy lip developed by an oboist) can be transmitted to the next generation. And yet there are innumerable suggestive evidences of this, such as the growth of calluses on an ostrich before it is born, a trait obviously not needed until later, but one that appears to have been originally acquired long ago by ancestral ostriches who somehow absorbed it into their DNA as part of their machinery of inheritance.

A theory trying to explain this says that the plethora of natural variations and mutations that happen by "chance" in procreation occasionally must produce gene combinations that grow body features exactly like those resulting from mechanical action (for example: calluses from rubbing) and that some percentage of these begotten "accidental models" inevitably will turn out to be advantageous enough to survive in evolution. Evolutionists call this the Baldwin Effect.

Another possibility is that the progressive selection of advantageous effects of any live action (mechanical, thermal, chemical, etc.) may eventually make that action unnecessary, which is analogous to filing down a hair trigger until it becomes so sensitive that the gun will fire of its own accord. This hypothesis won important support from a classic experiment by C. H. Waddington in the 1950s in which he gave embryonic fruit flies a temperature shock which for some reason made a small percentage of them grow wings without their characteristic crossvein. When he then selectively bred the crossveinless flies to each other, repeating the shock, the percentage of veinless offspring increased generation upon generation until he was able to produce veinless flies regularly without the shock, which by then, to his amazement, had completed its initiation of this new strain of flies and was no longer needed.

## THE ENTROPY FACTOR

By some such interactions of heredity and environment (including the microenvironment of genes) evolution not only seems to be creating more and larger creatures on the average as the millenniums pass but also, curiously enough, it shows signs of moving from symmetry toward asymmetry. Presumably this is part of life's glorious unfoldment of simplicity into complexity, an unfoldment that appears to defy the Law of Entropy (page 444), which says energy on the whole within any system must flow from orderly forms (like engines and organisms) to simpler, disorderly forms (like heat and decay). The law, I note, does not say that minority energy may not flow from a disorderly locality to a more orderly one within a system so long as the majority flow goes the other way. And it does not even forbid a flow of the majority of all known energy from disorder to order so long as the system is open, as most systems in the universe seem to be (including the universe as a whole) with a possible unknown majority of outside energy flowing to disorder beyond the horizon of knowledge.

This, I think, must be how the entropy law permits protein molecules and crystals to grow and evolve, letting the complex assembling of atoms in their lattices proceed so long as it is more than compensated by the simple dispersal of them elsewhere. Likewise in the case of plants and animals. For, while these are still more complex forms of life, they are significantly more abstract too and perhaps proportionately harder to fathom. Besides, they are but an infinitesimal part of the vastly larger and more permanent flow of substance out of the order of the sun and into the disorder of surrounding space. Indeed in such an entropic, celestial flow, involving photosynthesis, growth, nourishment, respiration and decay on Earth, the temporary organisms of life seem to me analogous to bubbles on a river, ethereal riders that momentarily drift along with the water but, in bursting, are soon revealed to be part of the sky. If you can stand a final analogy, I would say the resolution of the cosmic entropy-negentropy paradox is immeasurably more elusive than the similar incongruity of a national economy that is perpetually booming despite its chronic failure to make ends meet.

## THE TOURNAMENT OF EVOLUTION

Backing off for a better perspective on the whole question, I find cause to suspect that the overshadowing of what might be called life's local bud by a much vaster celestial withering will in turn be scarcely noticed in the transcendent efflorescence of the universe. And while there is no generally accepted evidence of teleology in evolution in the sense of purposefully and provably directed channels of survival down the generations of selected species, evolution's purpose may well live and prevail, somehow subtly implemented by the energy conveyed by invisible waves out of the still very mysterious fields of outer space. Whether you call such a selective force a law of nature or God's will, it amounts to a discrimination, a discipline, a sorting and organizing influence — and in effect it exercises pervasive intelligence toward a goal, evidently toward good (if survival is good) and, for all we know, toward consummation of the fullest potentialities of the universe.

Despite the appeal of this Olympian perspective, the whole idea of evolution and progression in life is so new to man that there is still serious doubt and controversy down there on Earth as to how it works and where it is going. Evolution indeed is like a tournament in which, instead of a few dozen or a few hundred contestants competing with each other under agreed rules for a few days or weeks, many trillions of rivals take each other on, catch as catch can, no holds barred, no trick untried, without any mention of rules, bounds, ages, sexes, morals or time limits whatsoever. And they battle each other and themselves, tooth, nail, imagination and digestive system, dead or alive, self to self, soul to soul, continuously day in and night out, summer and winter, for millions of millenniums, anywhere they can find one another, on land, sea, swamp, sky, under ground, under bark, under rock, under snow, under ice, under skin, inside intestines, outside sense, in trees, grass, gravel, foam, cloud, smoke, dust, dung, gravy, blood, lymph, gristle, bile, bone, brain, dream, mind, spirit — sometimes even beyond the world.

No wonder evolution is — all things barely considered — the most exciting, dangerous, portentous and glorious experience there is! It even has engineering challenges right down to the microcosm and out of sight. I used to wonder why a unicelled animal like the paramecium could not evolve by becoming bigger and bigger without turning multicellular — in other words, by simply expanding its single microscopic cell generation by generation to the macroscopic dimensions, say, of a fish. The answer from a supposedly unreasoning planet is ponderably negative. It flatly declares that no microbe can ever grow big while it remains a unicelled organism, for the engaging reason that a single cell nucleus in any increasing mass of cytoplasm must inevitably develop "engine failure" — which would be fatal. The old Principle of Similitude (page 15) intrudes here, you see, by insisting that any material creature, if it is to move under its own power, must possess strength in approximate proportion to its volume or weight. In practice this means strength in ratio to the cube of its average thickness. It also amounts to a rigid size restriction because, like a steam engine whose energy depends on the heating surface (not volume) of its boiler, an animal has to draw its energy from the comparable oxygen-absorbing surface (not volume) of its integument or lungs. And it is this strict ration of energy generated on a slowly evolving two-dimensional surface that limits the faster-evolving three-dimensional volume of the cell, keeping it microscopic and incidentally obliging all the large animals, if they would live, to be composed of vast numbers of just such invisible and semi-independent parts.

Now evolution and its converse, devolution, as you surely realize, have many aspects — more I suppose than all the evolutionists on Earth can ever know. There are evolutions and devolutions of every sort of behavior, of walking, breathing, aggression, of courting and loving, of diseases, of spider webs, of manners and fashions, of musical instruments and musical forms, of games, tools, inventions, machines and vehicles, of money, language, art, knowledge, of science and religion, of crimes, of virtues, of miracles, of the hand and the brain, of facial expression . . . Every one of these facets of evolution in fact produces surprising revelations, such as the discovery that dogs express themselves more than pigs or bears because they are more social, indeed that dogs inadvertently evolved mouth and tail signals, as did man the smile, into a vital technique of communication.

Have you ever wondered how such a spiritual concept as the Golden Rule ever managed to evolve in the face of the seemingly contrary Law of the Survival of the Fittest. since altruism, by definition, disregards the giver's well-being in favor of someone else, who thereby seems to be handed an advantage under natural selection?

The answer, as revealed by the new science of sociobiology, is kinship. For, in the words of one of its leading exponents, Edward O. Wilson, "If the genes causing the altruism are shared by two organisms because of common descent, and if the altruistic act by one organism increases the joint contribution of these genes to the next generation, the propensity to altruism will spread through the gene pool. This occurs even though the altruist makes less . . . [individual] contribution to the gene pool as the price of his altruistic act." In other words, the plural fitness of a society naturally, perhaps inevitably, transcends the singular fitness of an individual.

Could this be behind the age-old capacity of bacteria always to produce a minority strain that, though individually inferior, can collectively resist a virus? Does it explain the progressive rise of culture, intelligence and spirit on an awakening planet?

## Human Evolution

Here we are approaching the advent of man, a subject naturally of special interest to us, and for which reason I would like now to put his species — with as little prejudice as I can muster — into evolutionary perspective, showing where he appears to stand in respect to life on his planet. I say "his" planet because, for the present, the planet does seem to be his as much as anyone's — anyone's, at least, within tangible range.

A few score million years ago the earth, being already more than 99 percent as old as it is believed to be today, probably looked substantially as it does, except that, according to geological clues, lush tropical regions like the Amazon jungle then reached well into the latitudes of Canada and Siberia, even, to some extent, Antarctica. And of course there were as yet no men around, therefore no cities, no roads, no ships . . . For that matter, there were no apes on Earth either, not even monkeys or dogs or (to be precise) any of the mammals we are familiar with today, although there were plenty of others, including today's unrecognizable ancestors, many huge and fierce beasts, like a wombat as big as a hippopotamus, a long-toothed marsupial wolf, huge crocodiles and lizardlike birds, not to mention a weird variety of rodents, insect-eating shrewish opossums and leaf-eating lemurlike climbers, some of whom, little though they suspected they were very different from anybody else, actually were the ancestors of man.

And so, ever so slowly, ten million years oozed by, and then several more million — and as the plates of Earth's crust gradually shifted, new and bigger mountains arose and stretched themselves, ocean beds heaved and humped and drained, and grasslands began to appear with different kinds of animals on them with longer legs and keener vision. Meanwhile the lemurs and tarsiers in the trees (including a breed called Proconsul) grew more monkeylike, using their snouts less and their arms and paws more as the preceptive branches imperceptibly lengthened their arms and molded their hand into an optimum thumb opposing four deft fingers. At the same time competition intensified

556

within the diminishing jungles, making some of the monkey types risk descending to the ground, because their long-tailed cousins were taking the best food and frightening and outmaneuvering them. These long-tailed acrobatic bullies were the ancestors of gibbons and siamangs, by the way, a few of whom would eventually diversify into gorillas and chimpanzees. While the timid ones who retreated to the ground were the ancestors of man.

If we can accept this evaluation, we need have no illusions about man's checkered and humble, not to say humiliating, career as a preanthropoid ape. For these early forebears of ours did not likely have time or leisure to decide to forgo the ancient jungles and leave them because they wanted to. More likely, they were thrown out! Hairy little fellows they were, from all accounts, low-browed, fruit-eating, gibbonlike and tailless, most of them apprehensive and not too bright even by animal standards. But one thing to be said for them was that they were blessed with rare adaptability, plus keen binocular vision and a tendency, when on the ground, to stand on their hind legs, the better to see over the bushes and long grass. Typically they lived in bands of a couple of dozen members, somewhat as baboons do today, and, having their remarkably dexterous hands — a gift from the tree — they naturally picked up sticks and bones and stones easily and, in the course of millions of years, learned to use them with skill, probably poking termites out of their nests, hurling rocks at game or enemies, harpooning fish, occasionally also discovering how to swim or dig or sing or eat or woo a better way — at first only instinctively or accidentally succeeding, but gradually doing the thing more reflectively and more often by means of repeated trial and error, a practice increasingly followed by thought, which led to more trials and, like as not, to better and worse errors — then again trials, a rare few of which

557

on occasion just might, after meditation, actually result in a shift of aim into the direction of new, untried, more reachable and sometimes "better" goals . . .

Meantime and betweentime the earth turned and the Milky Way churned and centuries and millenniums and megranniums passed while no one noticed these slow workings of evolution, for no individual lived long enough, and anyway no one had ever heard or thought of large-scale change or of progress. Yet little by little the skin of life thickened, growing an inch of granite every ten thousand years and an inch of soil every century more or less. And steadily the forebears of man grew bigger, still more adaptable and more confident of their place in the world, until, by a couple of million years ago, over most of southern Asia and Africa (then without deserts) they were not only using makeshift "tools" regularly but also learning to shape them. And it is believed they were significantly aided in this early intellectual achievement by a growing propensity for uttering grunts and chirps and babbling sounds in their throats to express their feelings, which presumably is how human language began. The fact that this particular hominid had a relatively small mouth naturally made vocalizing for him easier, for the reason a bottle resonates better than a jar. Moreover the flexibility of his lips helped him modulate the resulting purer sound, and this flexibility increased as he progressively expanded his capacity to express himself.

## LANGUAGE EVOLUTION

By the time the last ice age arrived, some sixty thousand years ago, he had come a long way in talking, yelling, singing, humming, whispering and other modes of speech, and his vocabulary had far outstripped those of his animal cousins, no species of which (including birds, bees, dogs, chimpanzees and dolphins) has ever, so far as we know, developed any system unaided that could justifiably be termed a "language" of more than a few dozen meaningful "words." Of course there was as yet no real system of writing in the world, although this ancestor of ours liked to draw pictures and occasional geniuses among him painted with astonishing skill. Yet the very ancient arts of tracking and of making (or erasing) tracks, followed by the newer arts of devising signals, signs, tallies and monuments to be seen and understood by others, were slowly preparing the way for pictographs and eventually phonetic letters and alphabets.

If I seem to emphasize language unduly it is only because I believe man's progressive use of words with specific (sometimes abstract) meanings that could be widely understood is what, more than any other factor, put him ahead of his competitors, equipping his mind not only to recognize but to store knowledge, the vital stock of culture through which the accomplishments of every generation could be retained and extended by succeeding ones. The evolution of language, like most aspects of evolution, must indeed have been very slow at first, as family vocabularies gradually grew from the limited, intimate patois of brother and sister, man and wife, mother and child into general, open, tribal dialects and eventually (with the advent of writing) intertribal and national languages. Such a tendency moreover explains why the stone age's proliferation of an estimated 4000 recognized tongues on Earth, spoken by an equivalent number of more or less isolated tribes on six continents and hundreds of islands, seems to have gradually coalesced into fewer and fewer and consequently bigger and richer languages. Something like 500 distinct languages (several as different as German is from Chinese) were being used by the Indians of North America, for example, when the European settlers arrived. But inevitably most of these tongues were soon either dead or dying, as English, French and Spanish with their alphabets and copious literatures (not to mention millions of speakers) took over. It must be significant that scarcely 5 percent of all the human languages have a written form, leaving the remaining 95 percent merely verbal and unrecorded, therefore headed for extinction, like organisms selectively slipping behind their stronger rivals. Languages, you see, grow and evolve much as men or animals do, though more slowly, measuring their ages in centuries rather than in years. And when they die, they do so like man and animals, usually decomposing into "dust" or mayhap a few bone fragments from which philologists later may find it next to impossible to reconstruct the shape of the original body.

Thus our talking ancestor had something going for him that no one else on Earth had ever had — something new and strange and a little otherworldly — something abstract that could justifiably be considered mystical in that it included the power to express ideas never expressed before, yet in such wise they could clearly be understood — making possible, for the first time on Earth, the efficient and rapid dissemination of thought. It is clear that he had not planned talking any more than he had planned to evolve a hand or a brain, invent tools or domesticate animals. Actually neither he nor any of the other hominids around knew enough to imagine such developments at this stage, for these interrelated things were unheard of and, so far as anybody knows (even today), they not only had never before spread on Earth but science so far has no positive proof they have ever existed in any other world.

559

## Evolution of Mind

Quite likely the very idea of humanity was unknowable on Earth until about 25,000 years ago, for who was there to tell our anthropoid ancestor he was any different or any better than other mammals? No one had ever heard of human beings or suspected such creatures could be created. Even as late as 450 B.C., when Herodotus visited the land of Egypt, he found that the people of the Nile did not think of themselves as a species superior to animals, several kinds of whom were regarded as divine. Cats in particular were so sacred to the Egyptians that, when a house got flooded, they would save their cats before their own children, handing them prayerfully from one rescuer to another.

Other peoples, however, noticed the animal-human difference much earlier, especially after they had learned to keep a fire going and dogs were skulking around the camp, treating the man with deference — in their doggy way calling him "master." It was perhaps the first hint of something drastically new in the world. His language had given man an authority and a superior understanding other mammals had begun to sense and respect — something qualitatively different from the shark's "superiority" over the pilot fish, which, if it really is superiority, is almost surely based more on body than mind. And, from then on, man's dominance increased rapidly.

This dominance undoubtedly had been enhanced when man learned to hunt big game during the ice ages, for that was the period when he discovered that, no matter how big and fierce the animal being pursued, he could practically count on its making stupid mistakes, and could often even provoke these mistakes at will. Furthermore, surrounding and killing a large beast like a mammoth necessitated close coordination with one's hunting partners, which stimulated language as well as social organization, including division of labor, food sharing and marital responsibilities. Women and children inevitably got left behind during a hard and dangerous hunt, which naturally led to traditions and eventually laws about who belongs to whom.

The sharp knives that evolved along with hunting and butchering also produced an abundance of animal skins, fur robes and sinews for tying and sewing, in turn suggesting clothes — which were probably tried first for their dramatic impact rather than just to keep warm, but which, when they finally caught on, must have incidentally aided the evolution of human hairlessness, a trend some anthropologists think was brought about most decisively by man's age-old chariness toward hairy women. At any rate, if mothers consequently averaged less hair, so perforce eventually did mankind.

560

Most distinctive of all the physical features evolved by man is probably his brain, which miraculously tripled in size in less than a million years — an evolutionary event so unique and dramatic it has been widely credited with placing him in a new kingdom of life all by himself on Earth. Not least of the human brain's attributes, I notice, is that it is only one-quarter grown at birth, allowing it to be formed in large measure by the later mental experiences of the growing child. In fact recent research has disclosed that the brain's ten billion neurons grow almost explosively in a young child by swelling and shooting out a dense mesh of branches and connecting links, squirming, rotating and strewing protoplasm, while consort cells (glia) slither and spread around them, somehow helping in the learning process, while the sensitive, feathery dendrites palpitate back and forth like insects' antennas probing for responses, ever tasting and testing the congeniality of all the new channels they can reach.

Thus the human brain is shaped by the lessons of childhood, its physical content actually molded by its cultural background, the concrete perfected by the abstract. This prolonged and unprecedented development of the brain in emerging mankind is what must have ensured that our ancestors, in order to survive, would build a family organization durable enough to protect the child for at least his first dozen years. And, not incidentally, the family was the seed that sprouted into the tribe and the tribe in turn evolved human society.

Thus arrived the species man at the seat of power among creatures of Earth, for, almost without realizing it, he had suddenly found himself master of beasts and lord of almost all known life. And beyond living organisms, he discovered he could aspire to be king of his world on a grand scale, wielding dams and bulldozers to adapt his whole environment to his needs rather than merely adapting himself to his environment. It was a natural evolutionary process of the brain overtaking the gene that had transcended in a way never known upon this planet, and perhaps only approached in dramatic import (if then) by life's initial ferment out of "lifeless" rocks and seas.

Man had taken his place as the organ of consciousness upon the world, giving Earth at last her mind, if not her soul, and all the concomitant wonders of willfulness, self-reflection, memory, and the imagination to articulate a new idea. It was time now he comprehended his position in the solar system and the universe. It was time, specifically, he understood his relation to himself and to all created things.

# CHAPTER 22

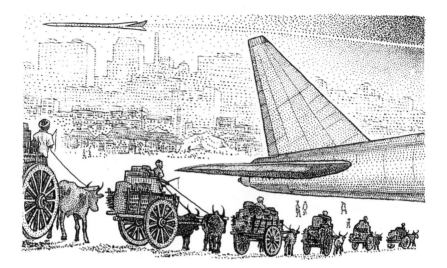

# Sixth Mystery: The Germination
# of Worlds

A NOVA is an exploding star. It is normal and, to an astrophysicist, an expectable, often predictable, climax in the life of a blazing world.

Like the much rarer and larger explosions called supernovas, quasars and even the entire expanding universe, it is an example of the cyclic vitality of all worlds that grow, mature, ferment and germinate. This includes planets of course, and perhaps them most of all. And as I study the findings of astronomy, geology, paleontology and related sciences, I get the impression that germination normally happens only once per world. Certainly it is a crucial, if not a unique, event amid the unfolding phases of life that evolve mind, speech and spirituality. And Earth, by a seemingly extraordinary coincidence, happens to be germinating right now. Not exploding physically of course, so much as ripening mentally and spiritually, doing it perforce unevenly with severe growing pains and dangerous side effects. Indeed, after some five billion years of slow development, Earth is germinating in this very twentieth century and to some extent (depending on definition) in the centuries immediately preceding and following it, when so many fundamental developments have happened, are happening and will be happening to her.

It is such an unbelievable coincidence to be living in the one century out of fifty million centuries when this is occurring that I have to be very much on guard lest my judgment become biased in favor of the present and the familiar. I know my opinion must be suspect. But fortunately the facts of germination are so verifiable and emphatic that in the end they will speak for themselves, clearly singling out the explosive present from all other ages, as we shall see.

Do you remember the numerous references in the Bible to "the time of the end" and "the latter days"? The authors of the Scriptures must have had some age in mind when they used these phrases. And a growing consensus among scholars seems to be beginning to realize that our present critical age will turn out to have precisely the qualifications suggested. This means, if you can believe it, that the days we now live in are the very "latter days" foretold of yore, and our present century, the space age and modern times none other than "the time of the end" predicted two thousand years ago — the age when the ocean has "loosed the chains of things" as Seneca foretold — germination time, the climactic budding that precedes the flowering of Earth — the period when all her vital signs are in the ascendant: population, communication, wealth, freedom, knowledge, spirit . . .

To analyze clearly what I mean, I shall now list and describe fifteen evidences of earthly germination, fifteen basic developments that have never before been experienced by Earth and, so far as anyone knows, cannot be repeated in the future.

## 1. Human Population Explosion

As we saw last chapter, the species man a million years ago was already using tools and learning how to talk and solve such problems as how to keep a fire. His numbers were small, somewhere around 100,000 inquisitive, furry creatures living in the most fertile parts of Africa, Asia and perhaps Europe. There were so few of him that it was almost as if the present population of Iceland were strewn over a thousand times as much territory, reaching half around the world. Naturally most of them gravitated into the valleys favored with the best water and game, leaving rocky and arid regions almost empty.

Yet they did not congregate into villages, for they were nomad hunters and gatherers without knowledge of agriculture or herding and, it has been calculated, something around eight square miles were required to support each person with the essentials of living.

After another 990,000 years — which would bring us to 8000 B.C. — the species man with his newly evolved brain had multiplied to an estimated three million people, the majority of them hunters but now many also skilled herdsmen and quite a few beginning to learn to sow and reap. Village life had begun in certain fertile valleys, though most people were still nomads and a few of the more adventurous tribes had even wandered from Siberia across the Bering isthmus into the Americas. It was also the period when some unprecedented but measurable spurts of population occurred, which anthropologists attribute first to the discovery of herding, something not only easier than hunting but a means of enabling each square mile to support perhaps a dozen people, and second to the spread of agriculture, which immediately increased the number to several dozen and eventually to hundreds per square mile.

In succeeding millenniums man's population kept on increasing and, significantly, accelerating, no doubt stimulated by such developments as his marvelous discovery that he could persuade cattle, horses and buffalo to plow, the wind to sail a ship or a river to grind grain. By the end of the Old Kingdom in Egypt (2270 B.C.), his number was approaching 100 million and in Roman times about 300 million, reaching 500 million only in the seventeenth century. Then his invention of the steam engine, with its massive, cheap power that launched the industrial revolution, surged man's population again, pushing it to a billion by the mid-nineteenth century. From where it accelerated to two billion in the 1920s, three billion about 1960 and four billion in 1976.

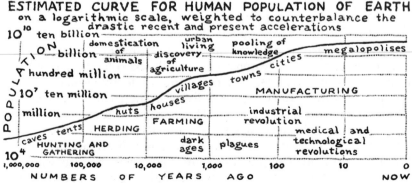

ESTIMATED CURVE FOR HUMAN POPULATION OF EARTH
on a logarithmic scale, weighted to counterbalance the drastic recent and present accelerations

The fact of doubling Earth's human population in a mere half century (possibly in a shorter time in future) is so new and so phenomenal that it is understandable that almost no public leader today seems able fully to grasp its meaning. Yet it is increasingly apparent (to me at least) that this explosion is prime evidence of the germination of the planet in our time, like the germination of any seed in fertile soil.

Specifically, the explosion of man's population in the twentieth century amounts to an increase of a thousand times in its growth rate, from less than .002 percent a year before the advent of farming to 2 percent today. This is the direct result of an unprecedented rise in the efficiency of man's production, something made possible by his learning to pool knowledge, thus to accelerate invention and industry on a world scale, particularly the development of modern medicine, which has drastically lowered his death rate. In a lesser degree, the population explosion may be due to modern man's abandonment of the population controls his anthropoid ancestors had through their territorial systems and the more recent stone age taboos and tribal laws promoting everything from head-hunting to infanticide.

Population surges of course have been known among many animals, from rabbits in Australia to locusts in Africa, lemmings in Scandinavia, starlings in Maryland, millipedes in Belgium, army ants in Peru and starfish in the Pacific Ocean. What starts and stops these outbursts remains largely mysterious to biologists, though the food supply, fertility and predation rates are obvious factors. One thing sure about man's current population explosion is that, while it is by far the greatest his species has experienced, it must end like the others within a very few generations — say, a century — even though there is no way of foretelling now just what combination of eugenics, famine, disease, war, pollution or other trials will force the change. Actually the tide has already begun to turn, as the accompanying graph suggests. Human response through contraception (particularly in the more developed countries) has significantly reduced man's birthrate — and this is beginning to damp the threat despite a 70-year lag caused by deaths being (temporarily) only half as frequent as births because they are experienced mainly by the old people who right now are only half as numerous as the young.

Pollution was brought to world attention about 1970, although it is far from a new feature in nature, being known throughout evolution to all the kingdoms. But it has taken a globally alarming turn this century in growing even faster than the human population it includes. Indeed, from man's view, pollution's present faces could well be called the three Bs (for babies, bombs and blight), which seem to be uniquely germinal in reacting upon each other in strange resonances. While man's population has grown about 50 percent in the past quarter century, for example, his blighting of land, sea and air (according to ecologist Barry Commoner) has increased 2000 percent. This presumably is due to such factors as energy-consuming equipment of all kinds radiating about 6 percent more heat each year, much of it "poisonous." Automobiles are multiplying three times faster than the people who make and drive them, and five times faster than the roads they run on. And we have a growing list of new technologies, such as aluminum, plastics, detergents, food additives, drugs, chemicals, synthetic fibers and excessive packaging that have not yet been integrated into the recycling of nature. In the terminology of physics, there is a "critical mass" factor in pollution that often compounds its accumulation with explosive acceleration, once certain limits are exceeded, which can easily happen when early warnings are not understood or heeded in time. There is no danger, however, that pollution will be permanently ignored, I'm glad to say. For it is not shy. And its voice is rising.

## 2. MAN'S WINNING OF THE TOURNAMENT OF EVOLUTION

For the first time man has assumed clear dominance over all other creatures of Earth. In the nineteenth century he still had to be protected from the beasts in Africa, Asia, western America and other untamed areas, but today it is obviously the beasts who need protection from man. Already scores of species have been made extinct by the competition of man in the last few millenniums: notably a dozen kinds of mammoth and mastodon and the woolly rhinoceros, followed in our time by the quagga, the aurochs and such birds as the dodo, the moa, the passenger pigeon, the heath hen and the great auk.

Before that, more than 99.9 percent of all the species who ever lived on Earth had already disappeared (presumably naturally) with only the meagerest fossilized trace left to prove it. It is only human, I suppose, to think of a species as something established like a fixed star, but, in the long view, both species and stars turn out to be moving, changing, growing, perhaps ultimately blowing up or fading away. So we might more accurately think of life as a flowing, eddying, bubbling tide or even a mysterious, self-weaving tapestry. Fact is: out of billions of species estimated to have foliated Earth in her five billion years of evolution to date, only a couple of million exist at any one time, because each lasts hardly a fleeting million years before it finally branches, withers or in some way loses its identity. Nor are we running out of them since we continue to discover several new ones every day (page 12) while losing only a handful per year through extinction, the rate of gain in awareness of species exceeding the loss by a good two hundred times!

Despite this, about one percent of Earth's modern warm-blooded species of animals have become extinct in these last four centuries with another 2.5 percent headed that way, both trends largely attributable to man's take-over. And it is increasingly obvious that evolution can no longer be a laissez-faire tournament between the freely competing millions of species, because, for better or worse, man is already running the show. It is no secret of course that he has been sneaking into small areas of evolution for quite a while, usurping the breeding of dogs and other domesticated animals and not a few vegetables for several millenniums, but in this twentieth century, the century of global germination, he has finally gotten into a position that virtually forces him to take over the main burden of evolution including very soon even the breeding of himself!

Already all the larger animals are under some measure of control, either confined to zoos, sanctuaries, national parks, regulated by game laws or (in the case of whales, certain fish, etc.) international treaties. Special conservation projects have been and are helping a few, such as the ibex, oryx, bison, pronghorn antelope, orangutan, giant panda, Asiatic lion, sea otter, dugong, blue whale, whooping crane, trumpeter swan, ivory-billed woodpecker and condor, to survive. And at least one species, the aurochs, has "miraculously" been brought back by man *after* extinction, by judiciously combining similar breeds of wild cattle that happened to have aurochs' well-documented physical and temperamental traits. Even insects and microbes are coming more and more under human control, paving the way for man's unprecedented but probably inevitable shifting of evolutionary gears as he inaugurates global eugenics on freshly germinated Earth.

567

## 3. Man's Completion of Exploration of His Planet

Man has virtually completed the exploration and mapping of Earth in this century. Only 500 years ago the world's map makers not only did not know about the Americas nor how far Africa extended south of the Mediterranean Sea, but many presumed the tropics were made of fire and the earth flat like a plate, so ships risked falling off its edge if they ventured out of sight of shore. The Dutch did not discover Australia until the seventeenth century, and the ocean depths and polar regions remained largely unknown even up to the beginning of the twentieth. In astronomy, equivalently, until this century, the Milky Way was considered the Universe.

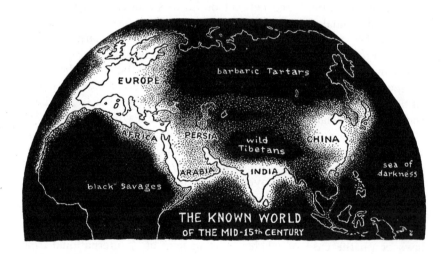

## THE KNOWN WORLD
### OF THE MID-15th CENTURY

But look at what our modern explorers have done in the brief time since then — for they have reached both poles, climbed to the top of the highest mountain, dived to the bottom of the deepest sea and charted not only every detail of every land and ocean deep, not missing the inner heart of the atom, but flown through the whole atmosphere and into space beyond it, including in person to Earth's satellite, the moon — with cameras to the neighboring planets, and hundreds of sophisticated new telescopes, spectroscopes and other instruments to the very horizon of the fresh-conceived Universe.

## 4. MAN'S THOUSANDFOLD INCREASE IN SPEED OF TRAVEL

Man's speed of travel likewise has multiplied by a thousand in about a hundred years. The first dramatic hint of it came on September 27, 1825, when one of George Stephenson's locomotives in England pulled 22 wagons full of passengers and 12 wagons of coal at 12 miles an hour. Then, after many millenniums during which the fastest a human could go was at the gallop of a horse, in 1839 a steam locomotive broke the oats barrier by beating a horse and in this century suddenly man in a single lifetime has literally lifted off to other worlds.

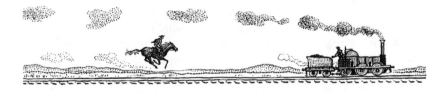

To cite a case, on January 25, 1907, the day I was born, a man named Frank Marriott set a sensational new world speed record by driving his Stanley Steamer at 150 miles an hour on Ormond Beach, Florida. It was the first and only time an automobile ever held the record. And it stood for more than a decade, being broken only by an airplane. But although I never got so much as a glimpse of one of these semimythical, winged machines everyone was talking about until I was twelve, the airplane carried man faster than anything else for more than forty years, until April 12, 1961, when the even more fantastic man-carrying rocket dethroned it, with astronaut Yuri Gagarin whooshing into orbit, literally in ten minutes boosting the world record from 2905 mph (by Robert White in the X-15) to 17,560 mph or nearly 5 miles a second. And less than eight years later, in 1968, Frank Borman upped it again by flying his Apollo 8 crew to the moon at 7 miles a second.

Which happens to be a velocity about high enough to convey man anywhere in our solar system of neighboring planets — therefore one unlikely to be exceeded to any dramatic extent until some far future century when a very dedicated crew may actually zoom off on a multigenerational voyage to the stars.

569

## 5. His Ten-millionfold Speed-up and Outreach in Communication

Man's speed of communication, in case you didn't notice, has increased even faster than his speed of travel, multiplying itself ten million times in a single step upward in 1844, from the speed of the railway mail pouch to that of the telegram flashing along a wire at 186,282 miles a second. This engineering miracle has been firmly consolidated by successive development of the telephone, radio and television during the ensuing germinal century and, although these later inventions haven't increased the transmission velocity (speed of light), they have suddenly for the first time made Earth capable of communicating with outside worlds, particularly since powerful radio and television waves began to be broadcast regularly from America and Europe, resulting in the planet's radiating out a continuously expanding bubble of radiation with current radius of sixty light-years that already reaches beyond many thousand of the nearer stars and their planetary systems, a few of which are presumed to possess advanced civilizations, which may well be responding about now.

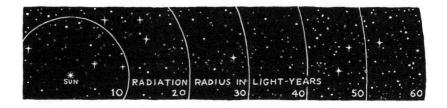

Because of the distance, Earth cannot be assured of receiving any replies before next century. And it seems appropriately dignified for her to wait because, in a sense, conversation between worlds is the culminating criterion of germination. Moreover, as any answering worlds must have germinated and probably long since had lengthy exchanges with other worlds, they should speak with cosmic confidence. Indeed, for all we know, they might even open mental and spiritual doors to levels of transcendence we have not yet even suspected.

## 6. The Explosion of Knowledge

Whatever its rate of future transcendence, however, Earth's tree of knowledge, after slowly developing for millions of years, has finally burst into bloom in this century — and our planet will never again be the same.

To realize what has happened, let's try to imagine what it was like on Earth a million years ago, waking up in a dewy jungle to the sound of insects and growing trees, the murmur of metabolizing ferns, of worms chewing. There is no hurry in this predawn eon of thought. You notice, among other things, a faint, pungent stink of dung. It is not unpleasant. It may even whet your appetite for breakfast: perhaps a handful of sweet maggots on yam. Somewhere you notice the chuckling of a stream, the flight of a tinamou, a tree frog trilling, sprouts fingering their way between the grains of dirt, weaving upward through the floor of matted fronds and duff, patiently searching for that strange, new sense called understanding, that comes so gently, so little at a time.

One reason understanding and the accumulation of knowledge have evolved so slowly, I suppose, is that Earth's thought pioneers neither knew nor cared about discovering how to discover. Reasoning indeed is far from the easiest of activities, nor is imagination the most natural. For, as Lucretius wrote in the days of Rome, "no fact is so simple that it is not harder to believe than to doubt at the first presentation."

Man's mind, one might surmise, has become the fovea of Earth's consciousness. And his understanding is, for the first time, the agent enabling our planet to sense and begin to comprehend itself. Yet still, powered by the slow grinding of the mills of God, it has naturally taken millions of years for knowledge to double itself, thousands more for it to redouble itself, the acceleration increasing gradually as accumulation built toward its critical mass, as new degrees of reflective feedback were irregularly fermenting, fomenting, germinating . . .

An example of the human mind's struggle to overcome inertia might be the puzzlement experienced by pygmies in Africa on emerging from their dense forest for the first time. Seeing a boat half a mile away on a lake, a typical pygmy (described by Colin Turnbull) could not believe it was really a boat, it was so tiny, and he insisted it must be a chip of wood floating on a puddle. His jungle environment and culture had provided him no experience of seeing anything farther away than about fifty feet. Four buffalo grazing on a distant slope he mistook for

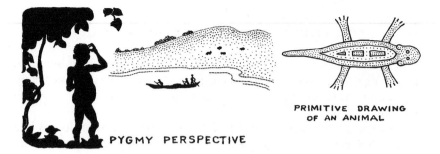

PYGMY PERSPECTIVE

PRIMITIVE DRAWING
OF AN ANIMAL

bugs. And even when Turnbull drove him in a jeep right to the animals, he rationalized their sudden increase in size as due to a special white man's magic for turning bugs into buffalo.

Two-dimensional drawings depicting three-dimensional objects have likewise proved incomprehensible to inexperienced people, who not only misjudge perspective but expect a picture of a four-legged animal to show four legs, even if the view is from above which would normally make the legs invisible. And some such culturally deprived people inspect a printed map very closely on the assumption that it is a real world in miniature containing microscopic houses, rivers flowing, smoke drifting and people walking about.

A different and no less significant mental struggle goes on at other levels, even in Earth's highest institutions of learning, the classic example being Hipparchus' discovery of the precession of the equinoxes in the second century B.C. Although this was duly published by Ptolemy, it was later forgotten, and then confused by Theon of Alexandria who attributed the phenomenon to "trepidation," his flimsy theory uncritically accepted by astronomers for more than a thousand years, even by the great Copernicus. Indeed the error was never conclusively corrected until Newton's *Principia* came out in 1687.

By that time a distinguished scientific committee had long since been set up in Europe to decide whether to invest a large sum of money in Johann Gutenberg's printing press. After lengthy deliberation, however, they turned it down because, although movable type was a "clever idea," they had become convinced that there could never be any big demand for books — for the "irrefutable" reason that only one percent of the population knew how to read.

As late as 1875, when earthly germination was well advanced and scores of vehicles from bicycles to battleships were in an accelerating state of development, a prominent director of the U.S. Patent Office

572

resigned on the remarkable ground that "there is nothing left to invent." Even as late as the first decade of the twentieth century Albert A. Michelson, discoverer of the speed of light, wrote: "The most important fundamental laws and facts of physical science have all been discovered, and these laws are so firmly established that the possibility of their ever being supplanted in consequence of new discoveries is extremely remote . . ." A few weeks later a prophetic editorial appeared in the prestigious *New York Times*, saying: "The flying machine which will really fly might be evolved by the combined and continuous efforts of mathematicians and mechanics in from one to ten million years." The date was 1903, the very year in which two unknown bicycle mechanics named Wilbur and Orville Wright completed a seemingly hare-brained experiment at Kitty Hawk, North Carolina, news of which the *New York Times* did not deem fit to print.

Within a decade or two of course the picture had changed completely and it became obvious that any good engineering firm, even in 1903, given a million dollars to research and develop a flying machine, might have done as well as the Wright boys. Some believed better. But the fact is that no one in 1903, including the Wright family, had thought flying was worth a million dollars or, for that matter, a thousand, even though within a half century almost every big corporation and government on a germinating Earth would be putting a major portion of its budget into research and development. And vision was still so lacking that it took the Wrights several years just to convince the United States War Department (which urgently needed reconnaissance airplanes) that they had invented one and had flown it successfully for thousands of hours.

During and after World War I, however, flying developed at an accelerating pace, as I described in *Song of the Sky,* and both vision and knowledge germinated explosively all over the planet while fundamental revolutions occurred in all the main branches of science. Not only have the chemists since then learned how to create new chemical compounds at a rate approaching 1000 per day, but the physicists have opened the atomic age, the biologists cracked the genetic code, medical researchers doubled the length of an average life and the concept of evolution that won acceptance along with electricity near the end of the nineteenth century synergized with it to illumine the whole earth, these two concepts forming the vanguard of a hundred new sciences from acoustics to zymology.

If we measure the growth of knowledge in terms of books, pamphlets, journals, maps and pictures in the world's libraries, this accumulation is estimated to total something over a billion different

573

copyrighted items today and, although still outnumbered by people, to be growing at the rate of 3 percent per year and gaining on the people who are multiplying at less than 2 percent. And if you want to count all the duplicate copies of everything printed, the books, etc., have long since passed the people. It isn't just a problem of physical space though, for it is already technically possible to reduce the size of print by 100,000 times in each dimension, so that all the written knowledge of Earth could literally be stored inside the head of a pin. It is rather a problem of mind, of sorting, of availability, of order. Indeed if you break all these library items down into the individual two quadrillion "bits" of information they are estimated to contain, the whole earth's volcano of knowledge turns out to be erupting at a steady two million "bits" per second, which is an alarming million times faster than babies are being born.

And this means, in simplest terms, that by the time a baby born today finishes college, the amount of knowledge available to him will have quadrupled. Which imposes an unprecedented strain on twentieth-century children, for, like the gene's blueprint of the whole body, every human who expects to function constructively must carry some sort of a working mental image of his world. Moreover, it creates what is probably germination's greatest challenge to the collective mind of man on a planet where knowledge is accumulating so much faster than it is evaporating that a major task of the next century may well be man's taming, sorting and harnessing it in the service of his newly germinated domain.

## 7. THE PROLIFERATION OF COMPUTATION

A major offshoot from the tree of knowledge of course is automation, which in one generation has revolutionized the management and technology of the world. At its heart is the computer, whose relation to the earth's explosions in speed and information is obvious in the fact that man now not only doubles his computation rate (a blend of speed, complexity and accuracy) every year, but, through electronic miniaturization, annually halves its equipment size and (to some degree) its cost. Thus the mental work of multiplying two 14-digit numbers, which took a trained mathematician with pencil and paper twenty minutes in World War II, can now be done electronically in less than one hundredth of a second and with much less chance of error.

The greatest single breakthrough in this revolution came when John von Neumann discovered how to store sets of instructions inside a calculating machine so it could work unassisted, step after step, by consulting only its own memory. Called programming, this unprecedented development in communication between man and machine soon spread to conversation between two or more machines, to automatic language translation, machine hearing and (now in the experimental stage) electronic "thought reading" by tuning directly into brain waves.

Computers are already being taught how to play games so well, I am told, that they hold a national championship in checkers and have been making steady gains on the world championship in chess. They learn primarily through a programmed prohibition against making the same mistake twice, although, in a new situation, they may be allowed a human tutor until they've had time to "catch on." Also when certain breeds of advanced computers get into difficulties attempting unfamiliar work, they have the option of calling up older "friends" (more experienced computers) on the long-distance phone for advice. And should they suffer "nervous breakdowns" from being overloaded with "impossible" tasks (like one who was inadvertently ordered to divide by zero), they may seek what has been called electronic psychotherapy, which could result in their being given a day off to compose a little music, fantasize in "freely associated ideas" or even jot down a quatrain of poetry such as

> *Darkly the peaceful trees crashed*
> *In the serene sun*
> *While the heart heard*
> *The swift moon stopped silently.*

If it doesn't sound quite human, at least it was a feather in the console of the naïve young IBM 709, with third-grade vocabulary of 77 poetic words who "dreamed" it up in Florida in 1964.

Other machines, under the tutelage of pioneers like B. F. Skinner 575 of Harvard, have developed sophisticated feedback to become patient, impersonal teachers of children, indeed remarkably inexpensive ones over the long term. On the farm, as in the factory, computer-guided machines have learned to adapt to extraordinarily delicate problems, from picking tomatoes to hulling rice, enabling one farmer in America to do the work of three hundred in Asia. Meanwhile in space, computers orbit the earth in unmanned observatories. Others in megalopolises help people find people by registering and classifying them by the million for jobs, business contracts, selling, swapping, car pooling, baby sitting, house hunting, spouse hunting . . . One science-struck young lady, recovering from a date prescribed by an electronic cupid, was reported to sigh: "It's frightening to find out what you deserve."

Some evolutionists theorize that, because man and his artifacts are part of nature, the man-conceived computer must also be part of nature, evolving not just a new phylum of life but a new *kind* of phylum, specifically a mutant phylum containing several mutant classes, orders and many species of computers. "Why not?" asks Marvin Minsky of M.I.T.'s Artificial Intelligence Laboratory. "The human brain, after all, is just a computer that happens to be made of meat." This idea is bound to be opposed by biologists for a time but it seems worth consideration, especially in view of the possibility that the evolution of protoplasm may eventually yield to the evolution of circuitry with all consequent implications for earthly germination and philosophy. Have you heard of the classic test devised in 1950 by Alan M. Turing of Britain to decide whether or not a machine can think? It called for a long, intensive course of teaching the most sophisticated of computers how to answer the prying questions of a blindfolded human so plausibly that, after some finite number of years of its programming and grooming, the human would no longer be able to tell whether he was conversing with a man or a machine. Already a few advanced computers have impersonated humans so well that teen-age children were taken in by them during several minutes of conversation, and Turing conjectured it might be only a matter of time before the most articulate machines could achieve complete man-computer indistinguishability for anyone.

Then there's the eerie question of machine consciousness, supposedly a new phenomenon on Earth, an example of which is the nation-wide moment-to-moment awareness of the computerized airlines' reservation system, which in a few seconds can tell you what no human can: which seats are available on any scheduled flight in the country on any day or night of the indefinite future. To those who oppose the whole fantastic notion of evolving artificial consciousness or "life" on the ground that machines, not being genetically fertile, cannot by themselves maintain their populations, the answer is as surprising as it is authoritative: they *are* fertile! For von Neumann definitely showed that, in principle, the computer may reproduce as readily as any vegetable or animal, the patterns in its abstract programming corresponding to the patterns in abstract genes that choose and modify all the food or materials needed to build and assemble new structures like itself — and that, despite the extremely complex chemical and engineering problems to be solved, this function of artificial life might well someday become reality.

This in turn brings up one of the ultimate questions of planetary germination: Are machines getting out of hand? Can they be trusted? Or must we domesticate them before they domesticate us? If they should learn to outthink us parents in enough ways, like a whiz kid outsmarting his knucklehead pa, is there anything to restrain them, say in a million years, from deciding to keep us as pets, guinea pigs, or beef? This (or worse) now begins to loom as one of the more ominous aspects of the evolving pollution problem, one of whose Bs (for blight) must include the proliferation of mental parasitism in a germinating world, and this is evidenced by man's rising overdependence on his own devices (from travel agencies to TV), in the same way that certain ant species have become enslaved by their own servants.

In other words, while the computer's artificial mind is earning at least its vicarious niche within the finite realm of man and reaching as far as it can into dimensions beyond it, man will do well not to take the machine too seriously. Indeed he should always remember that it sprang not from his loins but from his head. Nor is there any way, so far as anyone knows, that the machine could have got a soul.

## 8. THE RISE OF WEALTH

At the beginning of the twentieth century only one percent of humanity could fairly be called the "haves," for they were the relatively wealthy 15 million people, living mostly in Europe and America, who enjoyed an annual income estimated to be above $100, while the remaining 1.5 billion "have nots," in Asia, Africa and all over, got along on practically no money. This world ratio of 1 "prince" to 99 "paupers," moreover, had been the accepted disparity for wealth distribution all through history.

But a sudden change came in the early decades of this century as

577

the industrial revolution spread swiftly over all continents, diffusing the wealth, so that by the early 1970s half the world's people had an average income of about $2000 and world production was climbing 3½ percent a year or almost double the rate of the population. This naturally led to predictions that the "haves" might increase to 90 percent of the world's population by the year 2000.

The rise in dollar income of course needs to be tempered by the rise in prices if one wants to ascertain the net increase in buying power or true wealth, a value that is climbing worldwide in this germinal century at around 2 percent a year. Like pollution, the inflation problem is an annoying but inevitable aspect of germination, whose dimensions are more than they seem, nicest evidence of which may be the inflation of talk about inflation, as revealed in facts like the space allotted to inflation in the *Encyclopaedia Britannica* having tripled just since 1930. The net effect of world germination, notwithstanding, has made earthlings in general so much richer than they used to be that the slightly-above-average human of today (with his car, TV, fridge, washer, camera, airline credit card, etc.) enjoys a lot more luxury than did the wealthiest king in centuries gone by.

Unfortunately the economic disparity between the multiplying "haves" and the dwindling "have nots" has hardly lessened despite their overall average enrichment. For it seems that although the "average" American has increased his consumption of energy 100-fold in 100 years until he is now using 50 times as much of it as the "average" Hindu in India, the Hindu has increased his consumption only by a comparatively moderate 5 or 10 times. The extreme rich-poor disparity thus goes on and on, despite general world enrichment, mainly because the "have nots" have been relatively deprived not only of food but of education to the degree that too many of them still lack the kind of vision and discipline it takes to put off eating donated seed grain now in order to plant it for a permanently better crop in the future.

Another critical wealth factor is the sudden, almost cancerous, growth of cities on this planet that never had a village until a dozen or so millenniums ago and hardly a real town before the fifth century B.C. Plato once opined that the ideal city should contain 5000 citizens, served by 25,000 slaves. It is true there had been a few larger cities before Plato, like Ur, Mohenjo-Daro and Sybaris, but world urbanization developed very slowly because it took time to discover how to organize the necessary concentration of food, water and other supplies to build and maintain a city. Even as recently as A.D. 1800 there were only fifty cities on Earth with populations as big as 100,000 people. But then came the ignition point when industry, machines, plumbing, transport, electricity and the telephone really germinated the urban seeds so that by 1970 Earth was rich enough to have more than fifteen hundred cities of over 100,000 and a good hundred between 1 million and 10 million. In the United States now, for example, country land is being paved over and urbanized at the unheard-of rate of 5 square miles a day! The majority of Americans, western Europeans and Japanese are already living in cities and it appears that the majority of all humans will be urban by 1990.

Inevitably such germinal inequities will have to be controlled since a finite planet obviously cannot support infinite material accumulation, so the countermigration of people out of the centers of cities into the suburbs should not surprise anyone nor the planned dispersal of future populations into idealized metropolises like automated Babelnoah, designed by Paolo Soleri for 6 million people in Arizona, or the 15-square-mile galaxy-shaped Auroville, already being built in southern India.

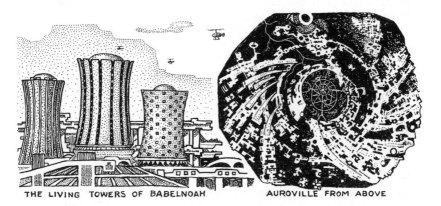

THE LIVING TOWERS OF BABELNOAH          AUROVILLE FROM ABOVE

I hardly need mention such other predictable innovations of global finance as a world currency, massive investment in undeveloped nations and whole Earth market treaties to hasten the essential elimination of the present dangerous imbalance in world trade.

## 9. THE LIBERATION OF WOMEN, MINORITIES AND SLAVES

During the past century a great movement to liberate the captive peoples of Earth has swept the planet and now, for the first time, women are being given equal rights with men almost everywhere, and not only slaves but exploited races and minorities of nearly every sort

579

are progressively gaining equal status with other citizens in all but a few totalitarian states.

Slavery is an ancient social perversion that goes back farther than history, even into animal and vegetable orders, like the ants, a few of whose species have practiced it for millions of years (page 375). But primordial ice-age man who, unlike ants, had no permanent houses, no domesticated animals nor private property, probably never heard of slaves. Indeed the evidence shows it was only with settled living and civilization that owning slaves and other property became feasible. And this explains why human slavery evolved so gradually, along with agriculture, villages, animal domestication and particularly the invention of war, which, after all, is what provided the prisoners who became the first slaves. So, from a by-product of war, slavery in time became a product of it, and eventually a major reason for it.

By Roman times slavery was so widespread it necessitated constant raiding into "barbarian" countries and began to be abused with distressing consequences, such as gladiatorial exhibitions and the revolt of Spartacus, the Thracian swordsman who armed and trained 90,000 slaves and made the world tremble by defeating several Roman armies before eventually being overcome and crucified in 71 B.C. Julius Caesar, who had briefly been a slave himself, later put on an exhibition featuring 600 slave gladiators dueling simultaneously; Domitian staged a weird "battle to the death" between women and dwarfs and a few years later 9882 gladiators and uncounted wild beasts fought an extravaganza in the Colosseum that lasted 117 days!

One of the most important uses of slaves, surprisingly enough, was for the publishing of books, which of course was done all by hand, yet with astonishing speed and efficiency. With one highly trained slave dictating the text and a hundred scribe slaves copying his words in a clear, swift hand, others could collect the copies, check and correct them, still others roll them up, bind, title and adorn them for the market. Thus a single Roman publishing house could "print" a 1000-copy edition of the short Volume Two of Martial's *Epigrams* in a 10-hour work day, each copy containing over a thousand words and retailed at the equivalent of ten cents, earning a net profit on the edition approaching fifty dollars.

All in all, slavery seemed such an ingrained aspect of nature that the great philosophers of the day accepted it, including Socrates, Plato (rather reluctantly) and Aristotle. Even Christ is not known to have spoken against it, for, as Paul wrote to the Corinthians: "Every man has his own calling; let him keep to it."

As Earth's population grew in the next two millenniums, slavery

grew with it and, by the nineteenth century, the planet's slave population numbered something over 25 million, depending on definitions.
The number of blacks alone, captured in Africa and shipped to Europe
and America, has been estimated in excess of 15 million, some 8
percent of whom died from the crowded and brutal conditions in the
holds at sea. But a great change soon took place, ironically initiated
by England, the most seafaring, most slave-trading nation of all. It
was something definitely mystic and spiritual, in which a great nation,
voluntarily and relatively peacefully, abolished on humanitarian
grounds for the first time in history a world commerce that had long
been a mainstay of her prosperity.

In 1807 the slave trade in British dominions was legally ended,
followed by abolitions in other slaveholding nations, most of whom,
however (including the United States), took more than another half
century actually to stop the persistent, illicit traffic in human lives.
Nevertheless, as Anton Chekhov wrote for mankind, ''We are squeezing the slave out of ourselves drop by drop,'' even though it took the
American Civil War and a century of pogroms, gas chambers, jihads,
race riots, lynchings, court decisions and bitter confrontations to arrive
at our present relatively liberal stage of interracial, interclass
tolerance.

At the same time enslaved women were removing their veils and
demanding the right to *own* property rather than *be* property. So, not
much longer would it be possible to buy a wife for $4.00, as was still
being done in the Gold Coast Colony of West Africa when I was there
flying cargoes in 1943. In fact that summer I met an African who had
bought a wife at a bargain sale for $2.50 during the great depression
and was just contemplating paying a witch doctor $12.00 to cure her
of a mysterious chronic fever when a friend pointed out the folly of
squandering $12.00 on this ''old'' wife when he could have his pick of
forty ''new'' ones for a mere $4.00.

581

Things were very different within a quarter century, however, even in the Gold Coast, which had become an independent nation called Ghana. Women had long since been granted "equal rights" with men in all "civilized" countries, including the right to vote. And during this same germinal period the United Nations had adopted an antigenocide pact and a bill of human rights for mankind; the majority of Earth's remaining colonies had become independent, self-governing nations in Asia, Africa and South America; and the 15 million refugees uprooted by wars and revolutions from China and Vietnam to Bangladesh, the Arab states and central Africa were being steadily absorbed and resettled everywhere from Hong Kong to America. In fact the whole world, despite lingering oppression in totalitarian labor camps, prisons and particular pockets of prejudice, had at last taken the road that leads to liberty and justice.

## 10. The Universalization of Education

The compulsory schooling of children is suddenly and for the first time spreading all over our planet. As a result, more than 60 percent of humanity can now read and write and the proportions of that majority are increasing about one percent a year as illiterate old people die off.

This is having a profound effect on evolution, particularly the mental or cultural evolution it is part of. And it is accompanied by no little struggle. For literacy is not yet, in any obvious sense, everyman's dish. When an Arab in Algeria was approached recently about letting his wife join a reading and writing class, he asked in astonishment, "You mean my wife should write letters? To whom?" And in many primitive countries the radio finesses the need for reading by reaching illiterates wherever they are, even to regimenting their thoughts in every field from birth control to politics.

Culture in animals is a kind of lower-level literacy which, as in the higher kind, is transmitted not by genes but by learning. Thus the young rat is not born wary of bait but must be taught every facet of the culture of caution by his elders. And in the mixed culture of the famed Spanish Riding School in Vienna the old men teach the young horses and the old horses the young men. It has also been found that dogs raised in cages, deprived of normal stimulation and adventure, learn poorly all their lives. Likewise with the orphaned ghetto kid, who learns nothing in school until given a mirror so he can discover he exists.

What's true of the individual can be presumed, in general, to be true of the world. As every baby starts life a "savage": dirty, ignorant, amoral, selfish, awkward, a potential thief, liar, rapist and killer, so must Earth herself. Yet now at last we are encouraged by the knowledge that the planet symbolically has found her mirror and is growing up and germinating. And that she is assured of soon knowing she has a mind and soul.

## 11. The Spread of Standardization

Standardization is rapidly uniting Earth by permeating all science and all nations. For not only does mankind as a whole already use the twenty-four-hour day, the seven-day week, decimals in mathematics, standard scientific criteria from market scales to atomic energy and common traffic rules in shipping and flying, but soon the metric system will undoubtedly become universal, highway signs similar everywhere and, sooner or later, all countries driving on the right.

Besides that, worldwide radio and TV broadcasts (not to mention magazines, books, movies and travelers) have achieved standardization of most fashions, art, architecture and other aspects of culture.

Since it is the first time this ever happened on Earth, people are apt to be disillusioned by the tedium of uniformity that sometimes results, but, as the world matures, they will surely learn that it is possible to keep flowers of every shape, color and smell growing in one harmonious and beautiful garden.

## 12. The Approach of a Global Language

A universal language that all educated humans can speak and understand also seems on its way to becoming a reality on Earth. Although about fifty artificial languages, such as Esperanto, have been devised, which offer the advantages of phonetic regularity, simplicity and universality, no one of them has yet been officially adopted as the world language, because they all bear the heavy initial disadvantage that there is no considerable population speaking them, no government or large institution promoting them and no literature to give them a tradition. So the natural, evolutionary process of the filtering and amalgamation of the 4000-odd known ancient tongues toward fewer and more universal modern languages inevitably favors such literary, widespread and large-vocabularied languages as English, Russian,

Spanish, Hindi, Mandarin Chinese, German, Japanese, French, Bengali and Arabic. And, of these, English now seems the one with the best chance of becoming a truly universal tongue, especially if it can be made more phonetic and regular. Already more than 70 percent of all scientific papers are published in English, work on simplifying it is being done, it is the standard language of airports all over Earth, and even communist nations like the Soviet Union and mainland China, who regard English as alien and capitalistic, cannot avoid teaching it on a large scale.

## 13. The Coming of a World Government

World government has become such an obviously essential step in Earth's present development that it must be considered one of the factors in planetary germination even though it hasn't yet happened. Indeed, should man's narrow nationalism or heedlessness continue to block the establishment of any sort of world political federation in the coming decades, humanity's very survival will be increasingly threatened!

An evolution of many thousand years is involved here, starting with families, then small clans that gradually yielded to larger, stronger tribes led by chiefs or priests. Next appeared village and town governments that grew into city-states that eventually amalgamated into nations, federations, empires, grand alliances and superpowers — with only the final step of a true world federation still lacking.

The history of war has unfolded at the same time in a parallel, feedback interrelation, there being no evidence that man invented war (or had reason to) until agriculture made it possible for him to settle in permanent villages with territories that required organized defense. Moreover, of the 14,550 wars on Earth since history began to

be recorded in 3600 B.C., a new one appearing every 140 days on the average for some six millenniums, wars have been relatively local until this century, indeed generally conducted like sporting events with traditional rules and led by individual heroes, the participants being limited to armies of professional soldiers rather than spreading to vast amateur populations. The only ones that involved whole populations, so far as I know, were the terrible ravages of major invaders like Attila the Hun and Genghis Khan and the occasional wars of extermination that killed off small tribes.

But now suddenly in our own time something entirely new has evolved with the advent of nuclear weapons and intercontinental missiles, which have made all-out war so impersonal and instantaneously lethal on such a scale that the "victor" must almost surely be destroyed along with the "vanquished," not to mention all large cities and possibly half of mankind vaporized in a day.

The absurdity of continuing the present international anarchy provoked Einstein into calling nationalism "an infantile disease, the measles of mankind." And the folly of it is now so obvious that probably the majority of all educated people already favor world federation, including the sacrifice of national sovereignty that is essential to enable it to disarm and police the planet.

Yet it is very hard for prime ministers, presidents and heads of state, charged with responsibility to uphold and defend national sovereignty, to feel the need to limit that sovereignty, which would seem to mean limiting their own power and importance. It is hard for them to see the analogy of the peaceful justice that prevails in almost any court of law, which of course depends on the ultimate judicial power (armed bailiffs and police) being controlled by the law (judges and magistrates) rather than by the litigants (prosecutors and defendants). For how different it is at present in the "court" of nations on Earth where the ultimate power (missiles and bombs) is controlled not by the law but by the litigants (competing nations), which results in the perpetual danger of catastrophic war. Clearly a change is due and overdue, a shift of power from the litigants to the law, from the competing nations to a world government constitutionally authorized to make and execute world law — thus to preserve world peace!

Whether this will come through the relatively peaceful evolution of international trading cartels, political treaties, charter reform in the United Nations, Olympic games, world citizenship declarations (like that by Minnesota which, in 1971, became the first state of the U.S. to declare the allegiance of its citizens to the world community),    585

intercultural conventions (perhaps involving UNESCO), ecumenical or world religion or (God forbid) through catastrophic famine, world-wide disease, poverty, war, or some unforeseen natural upset, muta-tion, cosmic interference or a combination of them — no one can foresee nor Prophet foretell. Yet somehow world government is com-ing, as surely as any human future is coming. And that alone (with all it implies in mind and spirit) should ensure that world germination will lead directly to the flowering of the planet!

## 14. SPIRITUAL EVOLUTION

Man's spirit is likewise swiftly evolving. By spirit I mean the part of humanity that is beyond the body, beyond all material things and technology, even beyond mind. It is the inscrutable part of us that inspires and is inspired. In essence spirit is not only mysterious but so far beyond understanding in this finite phase of life that it can often be called exalted and is sometimes indisputably divine!

But the specific point I want to make here is that, aside from man's obvious material and mental progress, his spirit is the key factor that must soon unite all people in a common bond of empathy that will bring such harmony and peace as was never before known on Earth. If this sounds strange or unbelievable, it may be because the spiritual aspect of the future is what is generally missing in prophecies. Indeed, contrary to a popular saying, it is almost certainly *earlier* than you think! For man, as a spiritual kingdom, has hardly reached the dawn of his own history. Actually he is still in a state of confused anarchy, having barely yet even learned to consult himself, let alone how to integrate, organize or rule his world. And his spirit as a whole has nowhere to go but up!

This of course is not a scientific statement, nor is it provable nor (I presume) even credible to most people. Yet it is at the heart of the germination of the planet and must be, in some sense, measurable. I mean that spirit in the sense of divine essence has to do with a profound question that seems to disturb many serious thinkers: is our world getting better or worse? Are we passengers on Earth evolving as we should? Or are corruption and pollution (with its 3 Bs) over-taking us as we slide hopelessly down the drain?

The answer is not easy. At the very least, it calls for spiritual comparison between life on Earth today and life as it was on Earth a hundred or a hundred thousand years ago — and it is a comparison bound to be controversial, both because no one lives long enough to gain firsthand perspective over such spans of time and because spiritual things are so utterly intangible and elusive.

Nevertheless, one can look at stone age life on Earth today which may well be comparable to the pre-Eden days when man was a hunter and knew nothing of farming, his morality presumably on the level of the increasingly clever beast he had found himself to be and whose sense of right and wrong, if it could be called that, depended, as with other animals, on his instinctive urges to hunt, kill, eat, mate and defend the territory he regarded as his. Then as man settled into tribal and village life with all it involved in common defense measures, laws of property, adaptability to authority (including gods, devils and chiefs), inevitably disputes became louder and more frequent, leading to more laws that resulted in more violations as crimes became sins — and the evolution of virtue slowly advanced, significantly changing the killing of a rival from a noble deed to a shameful murder.

Thus although man seemed to be slipping morally as he evolved through his symbolic Garden of Eden phase into what we call civilization, in reality he was only increasing his potential for both good and evil, those two opposite but complementary qualities which, as the Polarity Principle suggests, are inherent in each other. And, ironically, it was the organized tradition of religion itself that seems to have perpetrated what are widely considered the most horrible tortures and massacres in human history, a fair sample being the sacrifices of the Aztecs of Mexico, who ritualistically killed an estimated 25,000 men, women and children every year for two centuries, burning some to honor Huehueteotl, the fire god, with the glory of his own flames, torturing children to appease Tlaloc, the rain god, literally with showers of tears, and actually cutting beating hearts out of living bodies with the pious acquiescence of most of the deeply indoctrinated victims.

Possibly worse, because of the divisiveness and hatred it engendered, was the medieval Inquisition instituted by the Christian Church in the name of sanctity and for the avowed purpose of "saving souls"! And the lengths to which so-called divine justice was stretched may be suggested by the record of Bishop Peter Arbuez who was credited with having burned 40,000 heretics at the stake, a deed of such rare "piety" that eventually he was canonized as a saint!

Of this period, Erasmus sadly wrote: "We were forever slaughtering each other for opinions that were mere guesswork — yet caused half of humanity to send the other half to the gallows or the stake — and for what?" Science was virtually standing still because any man bold enough to do any scientific research or to admit any scientific discovery was taking a fearful risk. Michael Servetus knew it — he who surmised the circulation of blood and was burned at the stake at the request of Calvin. Giordano Bruno knew it, and was burned for agreeing with Copernicus that the earth moves around the sun. And Galileo, who discovered the moons of Jupiter and the rings of Saturn, barely escaped a similar fate by recanting on his knees.

It isn't easy to measure spiritual qualities, but there always seem to be witnesses to straws in the wind. In 1730 Montesquieu wrote that the Church's influence was disappearing from intellectual England and being replaced by an unprecedented reverence for nature. Indeed for the first time in history significant numbers of people began to admire landscape and artists to paint it. Men started climbing mountains just for fun and the exaltation of a sweeping view. And Jean Jacques Rousseau wrote philosophically about the beauty of nature even while being persecuted by the establishment, who saw blasphemy in statements like "I feel, therefore I am."

Life in cities, however, continued much as it always had, and in 1750 Horace Walpole saw a poor old man fall down in the street outside White's coffee house in London and recorded that the customers inside placed bets on whether the fellow were dead or not. And, when a passerby suggested he should be bled (standard first-aid treatment of the day), they loudly protested that this would interfere with the fairness of the betting.

At the same time in America and western Europe there was a growing impression that the world was changing and perhaps going somewhere. This feeling was supported by such signal events as the founding of the Royal Humane Society in England in 1774 followed by the French Revolution and a rising flood of humanitarianism that had never existed before as a conscious goal of man. Although in England there remained 223 offenses punishable by death in the year 1819, and I've read that a nine-year-old boy was hanged for taking twopence worth of paint from a printer's shop through a broken window in 1833, reforms followed swiftly. By 1838 there were but 14 capital crimes left in England, with both the pillory and public hangings abolished, and the fresh breeze of humaneness advanced inexorably through the nineteenth century. One of its most significant steps, the Royal Society for the Prevention of Cruelty to Animals, had been founded in England in 1824, being, so far as is known, the first consciously organized action taken by any species of life on Earth solely for the benefit of other species. The first board of health was created in 1848. Then Florence Nightingale appeared to institute the merciful profession of nursing, dramatically lowering the death rate among Crimean War wounded from 42 percent to 2 percent, slavery was outlawed from civilization, the Red Cross and the Salvation Army founded, the germ theory "proven" and anesthetics adopted to remove the ancient horror of the surgeon's knife.

But if evolution of the human spirit showed signs of progress in the eighteenth and nineteenth centuries, it fairly erupted in the twentieth. Not only was there the series of almost unimaginable miracles brought to technological practicality, but there was the founding of the United Nations in 1945 and the Universal Declaration of Human Rights proclaimed by its General Assembly in 1948 and an unprecedented series of world treaties, demilitarizing Antarctica in 1959, four more limiting nuclear weapons in the 1960s, the dismantling of the colonial system mostly between 1955 and 1970 and motion toward serious superpower disarmament later in the 1970s.

One must admit that there remain deep pockets of prejudice, even occasionally of deadly discord, in places like northern Ireland, the Near East and southern Africa, not to mention communist areas, but by and large the ups and downs of spiritual evolution average into a long-range ascent. I know that many people assume the sixty million people slaughtered in the two world wars and sixty-five lesser conflicts of the first half of this century prove spiritual deterioration on a vast scale, but I submit that, although those wars were statistically disastrous, in long-range planetary perspective they may well have been vital sparks in the germinal explosion that is progressively arousing mankind to spiritual maturity. Without them, we might not know either the computer age or the space age, to say nothing of the atomic age, nor enjoy the bounty of relative peace and understanding, especially in Europe and the Americas, that has so far prevailed through the second half of the century, in which war casualties worldwide are less than 3 percent of what they were before 1945 among the then much smaller population.

If you hold a contrary opinion, you may also argue that the modern "increase" in crime is a mark of moral regression, but shouldn't you judge the matter in the perspective of the changing definition of crime and its incidence relative to the growing opportunity for it, the rising wealth, increasing population, easier communication, spreading permissiveness, ubiquitous power and freer, more self-serving markets?

And even though our current germinal times have been punctuated by all sorts of gruesome and trying problems, it is obvious that Earth's general level of education and awareness is rising rapidly (as we have seen), largely because of the greatly increased outpouring of information through books, magazines, TV, universities, travel and communication of every kind. And as people have become aware of people and their potentialities on a world scale, inevitably a feeling of global oneness has grown. It involves the conscious mind and, even more importantly I think, the spirit. One gets clues too in the rising rate of generosity by which people extend sympathy and help to others in need — even on the opposite side of the planet. Government figures indicate that although the average American evades about 3 percent of his taxes, he actually gives more than twice that much to charity, the amount of which totals more than ten times his donations of a generation ago.

Summing up spiritual evolution to date, may I gently philosophize that if the carefree young woo or wine too much, or otherwise offend nature, it is still nature they must learn from. And whatever the "mistakes" of life at any level, evolution is still evolving, still varying and selecting for future generations, testing species, winnowing souls . . . And, so far as I know, nothing that can happen is inconsistent with the pervasive and germinal forces Boris Pasternak must have sensed when he wrote to an editor in Uruguay: "I have the feeling that an epoch with fundamentally new goals, both of the heart and human dignity — a silent epoch that will never be proclaimed in a loud voice — has come to birth and is growing day by day all unnoticed around us."

## 15. TRANSCENDENCE OF ORGANISM MAN INTO SUPERORGANISM MANKIND

The consciousness of mankind is now rapidly unfurling a new dimension as Earth becomes aware of herself for the first time. Our home planet, in other words, is becoming self-conscious. Or, to put it another way, the organism man is evolving into the superorganism mankind. And this transition can be usefully compared, I think, to a fish in a school or a bird in a flock engaged in mass maneuvering. For

590

such a fish or bird inevitably loses his individuality and independence and, to some degree, becomes a "cell" in a greater "body." He must also, in effect, submerge his "self" beyond the equivalent of an ant or bee in order to resurface collectively as an anthill or a beehive. And this means, in the case of man, that he not only transcends individually, each in his own mind and soul, from finitude toward Infinitude, as we have seen, but he also transcends collectively from men and women to mankind, while Earth herself (whose consciousness is primarily the mind of man) must ultimately transcend (beyond space-time-self) into what may be described as the divine essence of the universe.

All in all, transcendence could hardly be called an obvious phenomenon, though most of us can sense the superiority of the hive over the bee. And the triple aspects of transcendence outlined above and described in Chapter 19 remain at best controversial. Yet the more I see and study the world, the more I am convinced it must be transcending. In fact transcendence, particularly germination, is the great happening of Earth in this "time of the end." For we are rapidly reaching the critical mass of spirit and awareness that makes "cosmic consciousness' possible.

I have borrowed the phrase "cosmic consciousness" from Richard Maurice Bucke, M.D., who wrote a classic book under that title at the turn of the century, in which he recounted many instances of spiritual "illumination" experienced by people ranging from Buddha and his Nirvana to Walt Whitman and his "gleam divine." Typical of them was the experience of "C.M.C.," a woman whose sufferings reached a crisis in 1893. In her own words: "At last subdued, with a curious growing strength in my weakness, *I let go of myself!* . . . losing my identity with a sweet terror . . . Now came a period of rapture so intense that the universe stood still . . . In that same wonderful moment of what might be called supernal bliss came illumination . . . I became aware that the flowers were alive and conscious . . . I saw with intense inward vision the atoms . . . rearranging themselves . . . What joy when I saw there was no break in the chain . . . worlds, systems, all blended in one harmonious whole . . . The great truth that life is but a passing phase in the soul's progression [through] spiritual evolution . . ."

Bucke estimated that these cases of "cosmic consciousness" have been accelerating in frequency in recent centuries, being about five times commoner in 1900 than in the Middle Ages. Although research in this field is imprecise, there are indications that the frequency rate is accelerating even faster in the twentieth century and that, as Bucke

591

surmised, a new spiritual race "is in act of being born" within the human species "and in the near future it will occupy and possess the earth."

How this fits in with earthly germination is not easy to know, because much of what is going on upon our planet is only intuitively observable. It isn't just a matter of population. It isn't merely speed of travel or communication. Neither is it only education, knowledge, understanding or liberation — but more a transcendent compend of all of these. And it comes at a time that must appear dreamlike to many earthlings as wonderful things keep happening: acorns turning into oaks, sperms and ova into people, molecules into transistors, swamps into airports, jungles into metropolises. And more and more clearly we see coming: exotic collisions in deep space involving fundamentally unknown laws of nature and kinds of matter or antimatter never seen or theorized on Earth, international conventions attended not by traveling to them but entirely by multicommunication systems between people at home in all parts of Earth using such advanced holographic and TV equipment that they can forget they are not in the same room, and the miracle of genes being synthesized in a laboratory so perfectly they actually grow into forms of life not previously conceivable, possibly including (as surmised by French biologist Jean Rostand) an approved, standard eugenic DNA manufactured to implant "the most desirable physical and intellectual characteristics" into human embryos, something that would change all resulting children from being the offspring of a particular couple to being the offspring of mankind.

The fantastic potentialities of space technology have long been described by science fiction writers and now by such physicist engineers as Gerard K. O'Neill of Princeton, who has designed gigantic cylindrical "pieces of earth" capable of sustaining colonies of thousands (eventually millions) of people, using cheap, plentiful, nightless solar power. And, it occurs to me, there also could be assembled a saturnian superstructure or ring around the earth, eventually extending outward from the equator for some 50,000 miles. This would best be begun at the altitude of about 23,000 miles, where "fixed" satellites already orbit the planet at the exact speed of her rotation in zero gravity. Thus a planate gossamerlike netting could be formed of relatively fine cables floating under minimal tension and to which bubble-shaped living and working capsules could be attached or any other desired buildings. Of course the cables would have to be stronger for the parts of the structure closer to Earth, where normal inward gravity would prevail, and also for everything beyond 23,000 miles, where gradual transition to outward gravity or centrifugal force could be felt and might serve (if permissible) as a natural route of excretion. But the whole thing could eventually be made strong enough, and perhaps extensive enough in the north-south dimension, to greatly expand the living volume of Earth, develop her resources, control her weather (using filtering screens to regulate sunlight, meteorites, cosmic rays, etc.) and facilitate her communication and traffic with other worlds.

More ambitious, in the sense of looking ahead not just thousands but actually millions of years, is the elaborate proposal of another Princeton physicist, Freeman Dyson, of the famous Institute for Advanced Study and former chairman of the Federation of American Scientists, who believes that an advanced planetary race like man must, in time, redesign and rearrange his entire solar system, not only commandeering its moons and asteroids and reorbiting or towing them about with powerful "space tugs," but even dismantling the giant planets like Jupiter and Saturn and steering asteroid-sized chunks of them into orbits nearer the sun where they will soon warm up and be made to fall together into moonish spheres that can be tailored into hospitable annexes for expanding man, eventually forming a kind of doughnut belt of small, inhabited "planets," using appropriate asteroids as mines for needed minerals, fertilizing some as nurseries or laboratories, rigging others for solar energy.

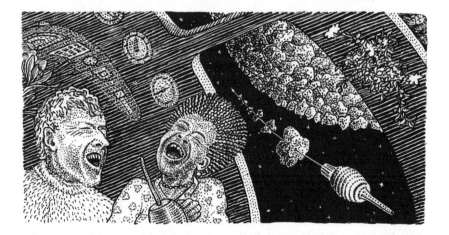

Still more ambitious and remote, if you can abide it, is the concept of domesticating whole galaxies of stars to serve their present civilizations of intelligent life, doing things like triggering supernova explosions or even quasars with celestial laser beams to reap the heavy elements they "sprout," perhaps ultimately directing the "fruit" to stellar "barns" for storage.

As for the establishment of federal order back on Earth, democracy will likely have to compromise with more efficient forms of leadership for, as one governmental critic was heard to ask, "Why should ten fools outrank nine sensible men?" Another complained that democracy lacks vision because the farthest-seeing individuals are always outvoted and by the time the majority learns the score and votes for it, it is too obvious to require vision. But it was historian Arnold Toynbee who said a few years before he died that humanity's only remaining choice is between a world federation with an Alexander at the helm and annihilation.

There are even voices behind the Iron Curtain saying something similar: Andrei Dmitrievich Sakharov, Peace Nobelist, member of the Soviet Academy of Sciences and leading research physicist, in 1968 published an extraordinary paper predicting a world government by the year 2000. His central argument was that "both capitalism and socialism are capable of long-term development, borrowing positive elements from each other, and, since the capitalists are already using the principles of socialism with real advancement for the working people, the two systems will be steadily drawing closer to each other in essential aspects. More important . . . on any other course except progressive collaboration between the superpowers, annihilation awaits mankind!"

A key to understanding such global transcendence is the realization that civilizations do not just circle round and round in one place, forever repeating the same experiences, reliving the same old mistakes. Instead their orbits know a component of obliquity, so each cycle traces a different path, veering slightly sideways into a spiral, cutting new ground, making new mistakes, learning new things as, by natural law, they spiral, spiral — and the prophets of science and religion fulfill each other — and we learn to consume our own smoke, to think mankind.

What I am saying is that, in its deepest essence, the world is a soul school — and is bound to be recognized as such more indubitably with each passing decade, each unfolding century. To me at least, Earth is a profoundly mysterious and bountiful place, a place that is, like the Universe itself, unfathomable in its glory — containing more food and drink than can be consumed by the cherished, more lovable people than any man or woman may love, more beautiful music than any audience will hear, more good books than any reader can read and more wisdom than all mortal beings anywhere may ever totally comprehend.

In this age of Psyche's unbinding, it is more and more appropriate to ask whether it is we who are dreaming the world? Or the world

594

that is dreaming us? Or does it really make a difference which pole we choose? Either way, the dream is a "consciousness" that somehow gets smuggled into finity on Earth, presumably to help prepare our minds or souls for whatever degree of Infinity lies beyond. It also adds dimension to all being, both individual and collective, not least by raising critical questions. Is the dream's imagery in any sense a scaffolding for shaping the soul toward the unimaginable in the way a gantry guides a rocket? Does anyone really know who looks at whom in the mirror of spiritual mystery? Who actually is the Dreamer dreaming unknown worlds that spin out their golden stories from afar? Who sees the track of the world's mind, or the moiling wake of the Universe's flowing spirit?

Even as my own musings are constantly interwoven with those of the more influential thinkers and writers whose books I read, who can know what reflections in other minds my contributions may provoke in turn or the lengths to which any new ripples from anywhere may extend down the increasingly turbulent river of human or inhuman thinking? And why need one expect that all this will remain even within the confines (if there are any) of human knowledge — for, as the past of consciousness may well reach beyond the antiquity of lungfish and trilobites and intergalactic spores, so must its future encompass ages in Olympian continuums we have not yet, and may never, learn to imagine.

If such appraisal seems rhapsodical to an extreme in the face of Earth's approaching ordeal of overpopulation and reorganization, just consider, if you will, that man has mounted a bigger, wilder steed than he yet knows how to ride. But that his spirit of adventure, his eagerness to risk his neck if not his soul, may be exactly what it takes to tame not only the mount but also, in time, even the unruly rider.

595

# CHAPTER 23

# The Seventh and Ultimate Mystery:
# Divinity

As I PEER DARKLY outward and inward from this mortal tapestry
of life, it comes to me more and more that the most fundamental
facts of existence are the ones I am least sure of. Indeed about all I
know is that we seem to be surrounded and imbued by mysteries and
immensities. And, beyond them, more mysteries. But what whole
are we part of — if that is a reasonable question — I know not. Nor
to Whom, if anyone or anything, we belong? Nor even Who, if not
What, runs the universe? And, in the name of sanity, I don't see how
we can ever find out.

Imagine, if you were a red blood corpuscle and flowing through
human veins and arteries, what an exciting life you might lead: now
squeezing your disk-shaped body through narrow capillaries, now
grabbing oxygen molecules from lung tubes to deliver them to tissues
all over, now tumbling through the wildly flapping valves of the
heart. Yet how would you, or could you, possibly even begin to know

596

what you were inside of: the organism you were helping to operate, the thoughts of the brain you nourished, the dreams of the soul you sustained?

To put it in scriptural terms as in the Book of Job, if we were not around when the foundations of creation were made, how could we possibly know "Who laid the cornerstone thereof, when the morning stars sang together, and all the sons of God shouted for joy?" Surely the light that spans the universe as easily as it spans the atom was not kindled at a human forge. But whence came it? And what of the nearer heavens, the wind and the clouds? "Hath the rain a father? Or Who hath begotten the drops of dew?"

Is there a plan behind the daisy, the hummingbird, the whale, the world? Who conceived the eye back in the primeval darkness of early evolution? Who designed the fish's air bladder in the ancient deep, as if foreseeing its future as a breathing lung upon the dry land? And out of what beginning evolved the mind?

All these questions of course pertain to the most profound mystery the human mind is capable of contemplating, the mystery of why we are here and just what the world is about. Despite all its wondrous discoveries, I haven't heard that science has yet found a complete answer to the Voice out of the Whirlwind Which asked Job, "Dost thou know the balancings of the clouds . . . or Who hath given understanding to the heart?"

Could the earth really be drifting along without pilot, as some say, steering herself automatically, running her own affairs at random? Could the Universe, just conceivably, have created Itself?

## This Nonrandom Universe

I am told by mathematicians that the randomness of large, complex numbers, generally speaking, is not provable. Therefore, any process involving as many billions of creatures for as many billions of years as is the case in earthly evolution is liable to be influenced by something besides blind chance. Of course we have all heard the old suggestion (clearly lay in origin) that if enough billions of chimpanzees were somehow set to typing manuscripts for enough billions of years chance alone would ultimately enable them to write all the great works of literature. But it is easy to show the absurdity of the notion, for, if there are 50 possible letters, numbers or punctuation marks that might be put in any of the 65 spaces in the average line of the average book, a chimp would have one chance in 50 of getting the first one

597

right. Then, for each of the 50 possible symbols, there would be 50 different possibilities for the second space, giving him one chance in 50 times 50 or $50^2$ of getting both spaces right. Thus his chance of getting all of the first three spaces right would be one in $50^3$, and of getting all 65 spaces right: one in $50^{65}$.

But $50^{65}$ works out to be equal to $10^{110}$ which, as physicist George Gamow once explained in a book, is a thousand times more than the total number of vibrations made by all the atoms in the universe since it was created. As atoms vibrate about a quadrillion times a second and there are quintillions of them in every speck of dust, the commonest of material particles floating literally everywhere in space, $10^{110}$ is an unimaginably big number. In fact it demonstrates conclusively, I'd say, that not even one line of any book or speech can originate purely by chance in this finite universe. There just isn't space or time enough. So something else has to be behind things, somehow guiding them. And that, one might say, is a kind of mathematical proof of divinity — depending of course on your definition.

Now a similar line of reasoning would apply to life, even to your own life, dear reader — suggesting at least a spark of omniscience in you, as Walt Whitman seems to have intuitively realized when he said that "every cubic inch of space is a miracle" and that miracles are with us everywhere always. For even the coolest logician would have to admit that you are an extremely improbable being. I can safely say this because the improbability of your being here is literally so extreme as to make you an impossibility by any ordinary interpretation of those words!

To begin with, you are a champion of champions, genetically speaking, because you are the product of an inconceivably complex and diverging web of ancestors, spiraling and branching back for billions of years into the primordial ooze of the proto-Earth, not a single individual of which, man or woman, animal or vegetable, ever failed to grow up to maturity and to beget viable offspring while most other creatures around them, including many of their own brothers, sisters and cousins, faded away and the majority eventually disappeared forever into extinction. This has to be true because of the finite dimensions of Earth and because, if your ancestors hadn't been such top performers that they were 100 percent successful in procreation, obviously your ancestral lines of descent would be broken and you could not exist.

But this is only the beginning of your improbability. Have you ever considered the odds behind conception, when only one out of tens of millions of sperms succeeds in siring a new offspring? And the comparable extravagance in all eggs, seeds, spores and pollen, where but one individual out of thousands in the case of fish and insects has a reasonable chance to avoid being eaten or lost so it can grow big enough to have little fish or bugs of its own, and where only one of millions of grains of pollen or seeds of an orchid blown on the wind can hope to land where it may procreate? Clearly this tendency to improbability indicates that, reckoning back to your remoter ancestors, the long shots in each generation become ever longer and longer.

So if we calculate very conservatively that each of your direct ancestors had somewhat less than one chance in a million to be conceived and raised to maturity (as he or she obviously succeeded in doing), your first-generation improbability (something a mathematician could write as $10^{-6}$) would still increase backward in time by several exponent numbers in each generation of each line of your descent, multiplying generation by generation to the population limits of your species, thereby reaching in the millennium now ending an improbability number far exceeding $10^{-110}$. I chose $10^{-110}$ because, as we learned, that is the reciprocal of a thousand times more than all the atomic vibrations of all known space-time. Which, putting it rather mildly, would imply that your conception was inconceivable — and that, if anyone in some wild moment ever got the impulse to call you "impossible" or (more diplomatically) "miraculous," he or she actually had a more-than-reasonable claim to being right!

Such calculations, which, to my knowledge, have not been disputed, leave no realistic possibility that life on Earth could be random in essence. Nor did Darwin with his "variation and selection" explanation of evolution call it random. Certainly evolutionary selection has to be guided by such criteria as survival value, which is far from random. And even evolutionary variation, depending as it does on factors ranging from genetic combinations to mutations caused by cosmic rays, may be much less random than it seems. In any case, Whatever or Whoever directs the mysterious forces of life and evolution has not yet come fully under the wing of science and is more than liable to be misjudged — if for no reason than that randomness always implies some degree of mystery, while mystery for its part does not necessarily imply any degree of randomness.

## Origins of Spirit

When you see a tree tossing its branches in the wind, do you think it is tossing them just any old way — at random? Of course it isn't, because the wind has a structure and so does the tree and both are part of the structured earth. And I would say there is an element of purpose there too. Certainly there is in the doe swimming my pond for sheer delight. Surely there is in that star twinkling its message as if straight from eternity. No one knows exactly whence spirit springs but the intuition of William Wordsworth told him that

> *One impulse from a vernal wood*
> *May teach you more of man,*
> *Of moral evil and of good,*
> *Than all the sages can.*

I like to think spirit is so fundamental that it appeared first in the mineral kingdom. But I'll admit that, even though a stone can be magnetic or radioactive and can get tired, drunk and sick (page 384) and still give a brook its song, while certain molecules may possibly choose where they will lodge in a crystal (page 448) to say nothing of the electron's mystic willfulness inside the atom, it is hard to find the spirit in a mineral, especially if you observe it on the human time scale.

It's easier with vegetables. We saw how trees send their roots down searching for water and nourishment, which may result in their making enemies of thirsty competitors even of their own species, yet at the same time bosom friends of relatively distant cousins like mycorrhizal fungi. When a pine's main stem is cut off we saw how surrounding branches apparently consult one another to decide which one will rise to take the place of the lost leader. And untold deals are made somewhere every day, every hour, like that between a toadstool called Honey Agaric and a Japanese orchid named *Gastrodia elata* who traditionally touch, embrace and symbiotically feed each other. Why does the aggressive Agaric, which kills most plants it feeds on, carefully restrain itself in the case of Gastrodia? Could it somehow have become aware of the value in preserving a mutually helpful relationship? Could this discriminating little toadstool that tempers mugging into hugging just conceivably have decided to take a small step upward in the evolution of spirit?

When it comes to the animal kingdom, the signs are so unmistakable that all doubt is left behind. The moral code of most animals seems to be associated with property, particularly the territory they regard as theirs, and everyone knows that a normal watch dog at his master's gate will attack any suspicious intruder in a righteous rage. Even a cricket who finds and occupies a private crack in the floor will soon regard it as home, his home, defending it with considerable moral ascendancy. In fact crickets have actually been observed always to fight harder on their "home" ground, apparently with the confidence that comes from a feeling of proprietary right, while an intruder is bound to be more hesitant, being clearly in doubt as to what to expect in this strange place. Indeed it turns out that there is a kind of relative morality between rival animals all the way from fish to primates based on comparative values of whatever is being defended, under which code not only does the older inhabitant morally dominate and usually defeat the newer inhabitant but the one with a nest or home has a moral advantage over the one without, the one with a better-kept home an edge over the more careless one, and the one who has eggs or young a definite prerogative over the one with none.

The territorial sense of fish in tanks has its own moral side and I was surprised to hear, from the late Christopher Coates, who ran the New York Aquarium, that, when you mix big fish with little fish, it makes a lot of difference whether you dump the little fish into the tank of the big fish or the other way around. Because, if you dump the little fish in with the big ones, the big fish assume they are being fed and just gobble them up. But if you pour water containing big fish into the tank of little ones, the big ones evidently feel they've entered a strange place and have no appetite for the little fish. Yet although any big fish dumped in with little fish may feel they're in the little fish's tank with no proprietary rights at the moment, time soon salves their unease (I wouldn't call it conscience), for within a few hours they are sufficiently "at home" to start feeding on the little fish. Mammals do not need any time at all, however, and probably due to superior intelligence, they resist this fishy code. Which explains why porpoises and whales, introduced into an unfamiliar tank, will devour fish without compunction.

The strongest animal property right of all, I think, must be the maternal instinct that arms a mother with a kind of divine fury when danger threatens her helpless young, even enabling her to successfully attack a marauder ten times bigger and stronger than herself. But many of nature's moral standards are very different from human ones, if no less rational, for she is broad-minded enough to try out an outrageous variety of plans, schemes and devices, including fraud, and with remarkable consistency she accepts the workable ones, no matter how disreputable, as virtuous. Lying is a common virtue among both animals and vegetables, where camouflage (page 192) is a highly developed art. The well-striped tiger is the good tiger in the same degree that his stripes deceive his prey. And the orchid that masquerades as a wasp (page 365) is the ambassador supreme between kingdoms.

There is no end to examples, like the ants who seem to feel a stigma associated with those who go to gather seeds but don't find any and return empty-handed. So such ants have been known to hide their shame by pretending they have a seed, often carrying a pebble or chip of wood instead. This substitute is almost always rejected by the inspector ants when they arrive back home but the unlucky gatherers are never admonished for their deceitfulness because, at this level, abstract honesty has not yet evolved a value.

I've heard it said that animals have no real moral choices in their behavior, but Konrad Lorenz has documented many cases of dogs telling lies. One involved his old dog Bully, partially blind and smuff, who one day failed to recognize him coming in the gate and rushed at him barking ferociously. But, on belatedly realizing who he was, Bully was so embarrassed he couldn't bring himself to admit his error or show he was sorry so, after a second of hesitation, he pushed past his master's legs, through the still-open gate and across the road to a neighbor's gate where he continued to bark furiously as if confronting a dangerous foe despite the absence of anything, man or beast, that could possibly justify it — except his obvious need to conceal his own inadequacy!

## Steps in Spiritual Evolution

How spirit begins and evolves is perhaps the Mystery of mysteries in this universe. I can only surmise that, as it works its way up from below the atom, diffusing imperceptibly into the molecule, gene, crystal, cell, organism, society, etc., it may arise in an animal or a human when he first begins to think beyond his immediate self, say in putting aside a present meal for the future, as when a dog stops gnawing a bone and decides to bury it for some hungrier day. Certainly the self of another time (or place) has less self in it than the self of here and now. To begin with, it is less real, because one cannot be sure it will actually come to pass. Death or disaster may intervene. Secondly, even if it is prompted more by shrewdness than anything consciously generous, it requires a least a slight degree of self-discipline, which is a vital antecedent of kindness.

In addition, it involves the insidious question of identity, which, I posit, is a key to comprehension of time. Have you ever considered the extent to which your "youth" exists when you are old? Materially of course you are now no longer composed of the same molecules and atoms that formed your body a decade ago, nor are your ideas and motivations quite the same as they were then. Shirley Temple, who was world-famous as a child movie star in the 1930s, said in the 1960s: "I've seen my old films on TV. I look on the 'little girl' which is what I call her, as somebody who isn't me." In a profound sense this is true, and the more removed one's past is in time and place, the more complete is the separation of identities. And what is true of the recordable past is at least as true of the unpredictable future, even though other parts of the future may be identified and forecast with some success and much of the past is already beyond reasonable hope of recall.

The next spiritual step, after the deferring of a present pleasure for a future one, must be one's first consideration of someone (or something) beyond oneself. If such a thought is for someone only slightly beyond oneself (say, one's Siamese twin or the baby one has just given birth to) it may not require much unselfishness to extend one's conscious care that far. Washing one's newborn son may be both easier and more personal than washing one's own foot. And saving the life of one's mate may be only half a step beyond preserving one's own life, depending on how closely entwined these lives may be. I heard an eyewitness account of a tiny female field mouse who was caught by a five-foot whip snake, which then glided up a tree holding her in its mouth, as she squeaked helplessly. A few seconds later the mouse's mate charged fearlessly to the rescue, going right up the tree

and sinking his little teeth in the snake's tail until it turned its head around to confront this menace. But, quickly realizing it could not defend itself with its mouth full, the snake dropped its prey, whereupon the gallant rescuer also let go and both mice scampered off to safety.

To what extent the whiskered hero could be said to be conscious of the risk he took is debatable, but he certainly cared as much for his mate as for himself and thereby displayed considerable spiritual maturity — indeed a good deal more of it, it would seem, than does the vilest of humans. Beyond the sensitive question of morality in a creature so different from ourselves, however, we may more confidently generalize that spirit progressively polarizes good and evil as the opposite sides of one potentiality. Which would explain why inorganic matter such as a rock practically has to be honest — having no real choice to be anything else — while organic life evolves more and more the freedom to cheat or be true. And that in turn says something about the nature of sin. It also explains why the cat is more honest than the dog. Or, in broader terms, why ignorance is more reliable than intelligence. For although the cat prowls and takes cover in order to ambush his prey (tactics that seem devious to a human) he plays a straightforward game according to well-known rules and always lets you know his mood. The dog, on the other hand, being more social, complex and intelligent, is capable of begging, apologizing, fawning, boasting, bluffing or otherwise concealing his true feelings, even from himself.

It is easy for relatively intelligent humans to misunderstand animals too, including their own pets, and the history of man-animal relations is full not only of sympathy and sentimentality but also of superstition, cruelty, strife, injustice and a surprising amount of bitter litigation in which offending animals have been hauled into court and solemnly charged with the crime of disobeying human laws. Have you heard of the famous trial of the caterpillars held in Valence, France, in A.D. 1585, involving a dozen lawyers and theologians? It lasted for months and ended only when the crawling defendants, who had been apprehended *in flagrante delicto*, were pronounced guilty and sentenced to banishment from the rich crop they had virtually eliminated to a piece of fallow land from which of course nature soon paroled them into butterflies.

Precedent for such judicial procedure is thought to date back at least to the Code of Hammurabi of Babylon in the eighteenth century B.C., the underlying concept being that all creatures were created by God for man's use and benefit, man being master of Earth, which was obviously the center of the universe. Most frequent challengers of this venerable code have been the pigs, presumably because they were customarily allowed to roam freely, which often led them into trouble. And when they killed anyone (most often a small child) they could be tried and executed, sometimes actually hanged on the gallows or burned at the stake. But the kinds of creatures prosecuted have ranged all the way from cattle and dolphins down to fish and ants — even occasionally vegetables, as when a priest in medieval Burgundy declared an anathema and, in the name of sanctity, excommunicated an orchard because its fruit had tempted children to stay away from mass. And as late as 1906 a "criminal" dog was sentenced to death in Switzerland. Yet none of all the bewildered defendants was ever granted the means to sue humanity for perpetrating on them such devilish cruelties as the wire noose and the steel trap, which seem to me morally worse than anything any animal could hope to devise.

Despite all man's offenses against other creatures and, it must be remembered, against his fellow man too, fairly continuous spiritual advances are being made around his lingering "blind spots" of pursuing and killing animals for sport, ritual slaughter, inhumane zoos, etc. The animals themselves also participate in this evolution as birdsong supplants pecking in territorial rivalry, victorious stags allow their defeated rivals to retire, chimpanzees are overheard lamenting the accidental death of one of their number and guide dogs daily risk their lives to lead blind humans through dangerous traffic for motives no one has yet satisfactorily explained.

Even the age-old barbarity of cannibalism includes some spiritual progress along with its religious formalities and taboos. For cannibals have been found to be relatively advanced, possibly eugenic-minded, people, some of whose tribes limit their population by eating babies while others control senility by consuming the elderly. And certain of their more restless societies specialize in eating enemies, while others, better settled, partake mainly of friends — the latter tradition explained by the fact that there is often a strong bond of love and faith between the cannibal and his designated victim. Indeed this relation is not nearly as analogous to the farmer's affection for the goose he is fattening for Christmas dinner as it is a mutual and profoundly religious acceptance of God's will that a sacrifice must be made for the sake of the world to come.

And still we are in the morning of the earth, the "Sun" of Spirit having just come up and our eyes barely begun to open to the growing light. Slowly a sense of creaturehood is leading to a feeling of brotherhood. It takes time more than anything, time which may be a greater barrier than space. For many mysteries are hidden in this world whose visible surface, after all, yields us only an elementary semblance of reality. Admittedly the planet, as a superorganism, is still nearly blind and numb, despite its having long since evolved eyes for itself and more recently begun to learn (through seismology) how to take its own pulse.

It also continues to harbor unknown numbers of stone-age tribes, like the gentle group called Tasaday, discovered only in 1971 in a Philippine jungle, whose people do not fight or hunt, but eat berries, fruit, worms, crabs and eggs, and possess neither a word for "anger" nor a desire for civilization. Which makes me suspect it was their kind Mark Twain was thinking of when he remarked that "soap and education are not as sudden as a massacre, but more deadly in the long run."

Even with soap, however, educated man on the average is learning to control his social aggression, at least as compared with hyenas, lions and langur monkeys, who, according to recent statistics from the field, engage in lethal fighting, infanticide and cannibalism at a much higher rate, leaving humans "well down the list of aggressive creatures." Man also can claim a conscious and rapidly evolving moral code which, in his peculiarly erratic way, he uses to preserve a semitolerable level of inconsistent harmony. I think it was Voltaire who cynically pointed out that "murderers are punished unless they kill in large numbers to the sound of trumpets," to which General Robert E. Lee countered a century later that "it is fortunate war is so terrible: otherwise men would love it too much." And, extending Lee's logic one step farther, we today might call ourselves more fortunate still to have invented nuclear war that is so much more terrible than ordinary war that no sane man could possibly love it. Indeed our present nuclear stand-off (which, I pray, will stay off) may be considered a mystic projection of the proverbial warning to tyrants that they are most liable to be assassinated by their own lieutenants. And that means that evil, in its very nature, is self-defeating.

## MYSTIC RELATIVITY

The signs of mystic potency in this universe appear not only in genes, seeds and eggs but in nearly everything, including such superorganisms as flocks of flying birds, gestating galaxies and even the unexplainable outpourings of the human mind. One sort of mind I am thinking of particularly here is that of the number prodigy, the child who can calculate at almost the speed of an electronic computer yet without knowing how or why. A six-year-old boy in Vermont named Zerah Colburn, for instance, was found able to factorize any number up to a million instantly. Given 171,395 he immediately knew it was the product of 5, 7, 59 and 83. For 247,483 he correctly blurted out 941 and 263. But he couldn't begin to explain his gift. He would only hold out his pudgy little hands, mumbling, "God does it." And the mysterious power soon started to fade and was completely gone by the time he finished school.

Even though modern computers are not yet comparable to human minds, except in regard to the speed and reliability of the particular calculations they're programmed for, they have begun to take on a certain mystique. For not only do they come down with undiagnosable ailments the engineers call glitches but they find themselves increasingly involved in weird warfare with a new type of human criminal specializing in getting around their protective codes and making secret unauthorized telephone calls to steal highly valuable information directly from their brains. The computers (conservative by nature) so far have nearly always managed to apprehend the crooks by signaling their owners when they sensed something irregular (therefore probably improper) going on, so investigators could check and see that matters were put right. But if ever in future a computer is found to be victimized not by a human but rather by another computer, that will call for new and serious rethinking in cybernetics and possibly some drastic evolutionary corrective measures. Surely any kind of automated competition involving morality would appear dangerous. For even the old traditional peer rivalry has long been a thorn in the side of spiritual progress, since, as I recall, it was not the peasants who refused to look through Galileo's telescope but the scholars with their learning from Aristotle. And it was not the laity who rejected Harvey's discovery of blood circulation so much as the almost unanimous consensus of doctors over forty.

Even in humble everyday matters, our spiritual struggle continues and evolves. I read a few years ago of a water pipe crossing a ravine in California that had been broken several times by the neighborhood children swinging on it. Because water is precious in that dry area, an indignation meeting was called in which angry men vowed to end the mischief by wrapping the pipe with barbed wire or, if need be, rigging it to give an electric shock. But in the midst of the wrangling, one man quietly arose with a suggestion. "Why don't we fix the pipe strong enough," he said, "so the kids *can* swing on it?"

Every now and then someone asks, "Who owns the tree that stands on the other side of your neighbor's fence?" Legally of course it is his. But it shades your lawn and beautifies your view — and you may well benefit from it more than he does. And what of the oriole who nests in it, the vine that clings to it and the wind that sways it? Truly it belongs in spirit to everyone and everything that touches or passes it. So really are all things owned by the world, or at least by any part of the world capable of owning them. And all things are yours as you are parcel of creation.

This realization, which may come late (if at all) to the ambitious property owner, is a broad equalizer of wealth as well as a great liberator from poverty. Indeed it makes everyone who sees a sunset or hears the song of a bird feel and be as rich as Croesus. It makes us all as free as thistledown or a puff of smoke whose freedom derives from having surrendered its will to a power greater than its own. And although King Croesus may have thought he inherited the earth, it must be a lot truer to say the earth inherited King Croesus among other things. For Croesus' little kingdom of Lydia was a drop in the bucket compared to the earth, then still largely unknown even if already rich enough in living forms to be considered lavishly extravagant from a human standpoint. I mean it was and is (as we have seen) a planet on which each individual of its trillions of creatures produces thousands of times more seeds and eggs than can possibly ever grow up — a world in which millions of people die of plague and war and starvation for no very obvious reason, yet which somehow grows and evolves and creates.

## THE LOGIC OF MYSTICISM

All in all we humans, who have made so much technological progress in recent centuries through our efforts to be practical, economical and efficient, find it very hard to understand such seeming wantonness. Perhaps that is because it is impertinent for the likes of us even to think of judging God's realm, for a part to doubt the whole. Besides, such a space-time universe as ours seems clearly finite. And finitude has a kind of built-in mystery to it, apparently due to the fundamental fact that light and radiation do not travel at infinite speed, while associated influences such as gravitation may be presumed just as slow. At best this apprises us finite mortals that we have no way of knowing what any distant place in the material universe was like when it influenced any other place a long distance away. If place A, for example, is a billion light-years from Earth in one direction and place B a billion light-years off in the opposite direction, then whatever influence A had upon B at the time (a billion years ago) that we now observe coming from B had to depend on conditions at A two billion years before that, assuming A and B stayed put all that time. However, not only do we have no way of knowing the influence of A upon B now (even if both are static), but we cannot possibly know much of conditions at A three billion years ago — not even if the universe had meantime been static, which it obviously hasn't been and presumably never will be as long as it is "alive," a state it may well be in forever.

When you add to this such ingredients as the question of whether

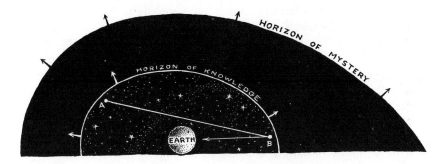

the universe is bounded or unbounded, whether anything exists beyond the horizon of knowledge (the range where galaxies recede faster than light), you might as well in the same wonder admit that the factor of mystery is enormous. In fact you could reasonably surmise a horizon of mystery bigger, farther and faster than the horizon of knowledge of which it is a function. Such a horizon of mystery also derives naturally from the fact that the most profound truths are the least definable and describable, largely because there is almost nothing to compare them to and, practically speaking, they are beyond the scope of language. What, after all, should be expected of a mind so limited it can visualize and accept neither an end to space, with absolutely nothing (not even emptiness) beyond it, nor space without end, with everything in all worlds inside it? Or even if somehow one could abandon both finitude and Infinitude, what (if anything) would remain?

Here we find ourselves wallowing in a field called mysticism, an elusive but wondrous subject that has had an immense if unfathomable influence on mankind for hundreds of millenniums. And it is easy to see why modern physicists, who have been pushing the frontier of knowledge into the unknown probably more profoundly than any other scientists in recent centuries, are ahead of most of their fellows in accepting that all-encompassing mystery of the universe commonly referred to as God.

Isaac Newton, for example, after formulating his famous laws of motion and gravitation, which stated precisely *how* the masses of the universe respond to each other, still wondered *why* they should respond to each other, particularly when separated by millions of miles of apparently empty space. Therefore he sought to trace cause and effect back to what he called the "first cause" behind all causes and effects, asking in his great book *The Opticks* (published in 1704), "Whence is it that the sun and planets gravitate towards one another without dense matter between them? Whence is it that nature doth nothing in vain; and whence arises all that order and beauty which we see in the world?"

In our own century Einstein declared that his most awe-inspiring experience was to see and contemplate the unknown, which taught him firsthand "that what is impenetrable to us really exists, manifesting itself as the highest wisdom and the most radiant beauty . . ." Although he never liked the traditional concept of a God in humanly recognizable form, he was profoundly impressed by what he described as "the harmony of natural law, which reveals an intelligence of such superiority that, compared with it, all the systematic thinking and acting of human beings is an utterly insignificant reflection."

Many others have voiced similar ideas, such as George Davis, the physicist, who suggested that God cannot be avoided by the common atheistic assumption that nature somehow sprang unassisted out of nothing and keeps on operating without conscious guidance because, "if a universe could create itself, it would embody the powers of a creator, and we should be forced to conclude that the universe itself is a God."

When a man does something at random like throwing a clock into the air while blindfolded, it is called entropy and is not likely to do the clock any good. Yet there are people, including plenty of scientists and perhaps the majority of atheists, who think nature's laws run the world this way: by chance. Presumably they regard Earth in her essence as a very probable world, random action being so casual and easy, and they can hardly be expected to accept the evidence I presented near the beginning of this chapter that no matter how many chimpanzees have been typing away at random since the universe began, they could not possibly, by chance alone, have made a single vitally significant contribution to this mysterious and extraordinary world.

The issue of course is fundamental to religion and philosophy. And it makes me think of a story about Charles Boyle, the fourth Earl of Orrery, who flourished in southern Ireland early in the eighteenth century — and of the theorem that bears his name. Having heard of Kepler's famous discovery of the laws of planetary motion and of Newton's recent work on gravitation, Lord Orrery had a working model of the solar system built inside his castle. It was an extraordinary, dynamic and up-to-date piece of clockwork with orbital hoops and a brass sun in the center plus smaller globes representing Mercury, Venus, Earth, Mars, Jupiter and Saturn slowly revolving around it, even a moon circling the Earth and four little ones going around Jupiter.

But it seems that Lord Orrery had an atheist friend who had an utterly materialistic outlook and thought of the universe as just an immense moving system of natural machinery that somehow coasts along, blindly but automatically maintaining itself without benefit of consciousness, mind or intelligence of any kind. So when the friend heard tell of Orrery's new and wonderful machine, he lost no time in going to the castle to see it. Entering the great hall where the model was in operation, the atheist's eyes widened with awe and the first question he asked Lord Orrery was: "Where did you get this magnificent thing? Who made it?"

But Orrery, remembering previous arguments with the atheist about creation, surprised him by replying, "Nobody made it. It just happened."

"How could that be?" retorted the atheist. "Surely these intricate gears and wheels couldn't create themselves. Who made them?"

Lord Orrery stood his ground, insisting that his model of the solar system had just happened by itself. Meantime the atheist worked himself into a state of hysterical frustration. Then at last, judging the time was ripe, Orrery let him have it. "Up to now," he declared, "I was testing you. Now I am going to offer you a bargain. I will promise to tell you truly who made my little sun and planets down here as soon as you tell me truly Who made the infinitely bigger, more wonderful and more beautiful real sun and planets up there in the heavens."

The atheist turned a little pale and, for the first time, began to wonder whether the Universe could really have made itself, or possibly be running all this time automatically and unguided by the slightest twinge of intelligence. And this was the origin of the Orrery Theorem which says: "If the model of any natural system requires intelligence for its creation and its working, the real natural system requires at least as much intelligence for its own creation and working."

## The Divine Hypothesis

There we have the hypothesis of God. It takes many forms. The ancient Chinese expressed it mildly in their saying that "If you keep a green bough in your heart, the singing bird will come." And I know a research biologist who feels he has a kind of divine right to experiment with large numbers of white mice but admits to wondering at times whether some sort of a superintelligence unknown to us isn't using us for a similar purpose.

Virtually every race of primitive people, so far as I can find out, recognizes some sort of mystic force, be it sun, moon, stars, lightning, volcanoes or a living creature. With the pygmies of central Africa, it is the forest they live in and commune with in their nightly chants of devotion for, as one old pygmy explained it, "We cannot see our god while we are alive, so we don't know exactly what he is like. We only know he is wise and good because he gives us everything we need, and we feel certain he is of the forest. So we sing to the forest and listen reverently to whatever he whispers in reply."

With the bushmen of the Kalahari Desert farther south, they say, "There is a great dream across the world that we are part of. It is not like any ordinary dream in sleep. For we do not dream this dream. Instead this dream dreams us. It dreams us all the time, even while we are awake, and we know that it must be lived out on this earth through everything we do."

## And Progressive Revelation

Such concepts may seem arbitrary in the apparently disconnected way they arise here and there among mankind, interspersed in recent millenniums with the widespread teachings of major prophets like Krishna, Buddha, Zoroaster, Moses, Christ, Muhammad and Baha'u'llah, each unique unto himself in his own place and time and all too often setting in motion differences of opinion that have led their believers into bitter "religious wars" from the Crusades to the recent fighting between Hindus and Muslims in India or between Arabs and Jews in the Near East. I submit, however, that in a larger perspective such differences are not fundamental. In fact I would say they are less fundamental than comparable differences in the revelations of science during the same time span. I expect some readers will deem it inappropriate thus to compare such disparate subjects as religion and science, but my own studies have persuaded me of the remarkable analogy between them. For science too grows by sequential revelation under the guidance of such great teachers as Aristotle, Ptolemy, Copernicus, Kepler, Galileo, Newton, Einstein — who, more obviously than the prophets, create a natural progression of evolving thought, each one figuratively standing on the shoulders of the one who preceded him and seeing farther than he saw.

In similar fashion then, let us look upon the great religious teachers not as opposing shafts in a confrontation but as connected cogs in a wheel, as branches of one tree, waves of one sea or notes of one melody. After all, they must commune with the same God, whether they call Him Jehovah, Allah or the Great Spirit. They are less enemies of each other than was Copernicus the enemy of Ptolemy or Einstein resentful toward Galileo or Newton. Did you ever hear of Buddha denying Krishna? Did Christ oppose Moses or revoke the Ten Commandments? On the contrary, he declared, "I came not to destroy the law and the prophets but to fulfill them."

The only reason progressive revelation is harder for us to accept in religion than in science, I suppose, is that religion is a more emotional, uncompromising and less rational realm than science, and its ardent adherents therefore more inclined to splinter into separate sects that pride themselves on their differences rather than their similarities. And that seems to be why western Protestantism, for example, has fragmented itself into more than a hundred sects and why the new ecumenical movement, in its serious current effort to reconcile such differences, is working only inside Christianity and, to my knowledge, has not even attempted to reconcile the seeming divergences between the major religions.

Yet a close study of history shows that the great religious teachers, as distinct from their apostles, are spiritually related in ways closely analogous to the relations between the different teachers in a school. The first-grade teacher of six-year-old children, for instance, can appear just as antediluvian to teen-age high school students as the high school teacher is incomprehensible to the six-year-olds. But the teachers of all grades are members of the same school system, hired by the same superintendent, who was appointed by the same school board. They are not opposed to each other and the main difference between them is that they are teaching children of different ages and different degrees of understanding. In other words, they are adapting their language, teaching techniques and discipline to the dissimilar needs of the pupils before them.

So has it ever been with the Prophets of God who teach mankind. Moses gave his crude followers the law now known as the Mosaic

613

Code, instructing them: "thou shalt give life for life, eye for eye, tooth for tooth," etc. (Exodus 21:23–24). It was of course a brutal and seemingly vengeful principle of justice in the light of present times. Yet, as the first moral code recorded by history and consigned to that primitive tribe of Jews more than three thousand years ago, it had the virtue of meting out swift, direct, clean-cut justice in an age before jails had been invented. And it gave the tribe a strict discipline for enforcing the highest standard of morality then known. It was both progressive and good in the time and place of Moses.

When Jesus appeared 1300 years later, however, western civilization was out of the cradle, having lived through the Golden Age of Greece, seen the rise of Rome and perfected writing and the pooling of knowledge on a large scale. Clearly man was ready for a more advanced spiritual code. So Christ taught him, instead of returning evil for evil, to turn the other cheek and love his enemy. And religion thereby took a giant step forward.

Then came Muhammad to the very backward land of Arabia 600 years later, advancing spiritual law another step by establishing reasonable property rights, limiting polygamy to four wives, prohibiting intoxicating drinks, etc.

And, in the nineteenth century, Baha'u'llah arose out of medieval Persia, offering the modern world the most enlightened teaching it has yet seen, with prime emphasis on unity, brotherhood, universal education, reasonableness, a global language and a world government — all urgent needs of the present day.

Each of these mystic figures, it is now becoming evident, serves as a vital link in the endless chain of progressive divine revelation, the sublimely integrated succession that goes back into the eons of long-forgotten prophets of prehistory and will continue into the unimaginable future ages of prophets still undreamed. It is a sequence remarkably parallel to the afore-mentioned chain of progressive revelation in science, although the major teachers in religion tend to be farther apart in time than the great scientists and, to a greater degree, they feel and express the mystic relationship between themselves, each not only honoring his predecessors but intuitively prophesying the advent of his still more exalted successors in millenniums to come.

Thus, for example, is Jesus reported to have told his disciples at the Last Supper (John 16:12–13): "I have yet many things to say unto you, but ye cannot bear them now. Howbeit when he, the Spirit of Truth, is come, he will guide you into all truth . . ." This prophecy, though variously interpreted, was fulfilled eighteen centuries later by Baha'u'llah, who specifically declared himself none other than the very Spirit of Truth referred to at the Last Supper, his life becoming his great testament to the fact. And the circumstance that he is the most recent of the great prophets, whose message was directed and timed to reach the majority of mankind at the height of Earth's germination period, seems to me reason enough to heed the promise of his still new, but steadily growing, Baha'i Faith.

## SCIENCE AND RELIGION

Not least of its teachings, I must not fail to mention, and one not only unique among world revelations but germane to our times, is the principle that everyone is entitled to his or her own independent investigation of truth, a human right that leads naturally into the singularly satisfying concept of the essential harmony of science and religion as integral parts of one whole. "Science and religion are as the two wings of a bird," wrote Baha'u'llah in one of his nearly two hundred books and tablets. "And the bird is mankind, which cannot fly on one wing alone. For the wing of science, if it lacks the insight of religion to balance it, leads to materialism. And the wing of religion, unguided by the reasonableness of science, leads to superstition."

I am reassured to find Einstein in agreement when he said that "science can only ascertain what *is*, but not what *should be*." And "though religion may determine the goal, it nevertheless learns from science, in the broadest sense, what means will contribute to the goal's attainment." In sum, "science without religion is lame, religion without science is blind." And I have no reason to doubt that Baha'u'llah was presciently honoring Einstein among others when he emphatically described the pioneers of science and philosophy as the "eyes" of humankind.

The idea that the invisible universe is more real than the visible one indeed has never been so widely accepted by practical scientists as now in this climactic century. But it is far from a new notion to seers and philosophers for, don't forget, Aristotle called life "spirit pervading matter," a concept all the great religions would heartily endorse. Some adherents of this doctrine may have been distracted, I surmise, by the perennial philosophy of solipsism which considers life nothing but a dream — not a world dream but a private dream dreamed all by oneself. If true, this would make both the visible, material universe and the invisible, nonmaterial one unreal and imaginary, something existing only in your mind or mine, yet as a concept, to my knowledge, it has never been successfully disproved. However it seems to be but a quibblous, semantic argument resolvable only arbitrarily according to whether you define this world as something subjective or objective. Whichever you call it, the world of course remains whatever it is, so whether you are a solipsist or not hardly affects the Universe and has no conclusive bearing on how you should deal with it.

The real Earth meanwhile spins on unheeding, inexorably tooling up for a maturity no man can fathom. And because its horizon of mystery continues to expand at a velocity much greater than that of knowledge (one proposed equation being $vm = vk^2$ in which the velocity of mystery equals that of knowledge squared) there is no more chance of eliminating it than of escaping one's own shadow under the sun. So the philosophy of mysticism emerges as eminently reasonable, even without considering its accordance with the beliefs of the greatest modern scientists and the profoundest prophets of God. And mysticism therefore would seem the proper approach for a humble and contemplative passenger on this solar vessel — pragmatic enough for daily use yet fully embracing the newly realized reality of the nonmaterial world, of fields that influence, of waves that convey, of minds that pervade.

## Spiral Grain of the Universe

If impalpable space thus turns out to be not nothingness but a mystic presence, we may ask of what it is made beyond the already mentioned sparsely strewn molecules of hydrogen, oxygen, carbon and dozens of life-harboring compounds. What is its structure? Does it have any definable grain or measurable texture that could be thought of as an integral part of it — as put there, say, when "God created the heaven" in the first verse of Genesis, while the earth was still "without form, and void; and darkness was upon the face of the deep" — a "deep," by the way, that must have been incomparably vaster than the unconceived, unborn oceans of the then nebulous, cometlike earth?

Yes, the firmament as we now know it in our finitude of space-time has a grain. Its grain is spiral with a propensity toward abstract nodes or waists as in an hourglass. A geometer might say it is singularly moniliform and plurally spiroid. The waistlike nodes are abstractions of relationship involving space and time and centering on all entities of life, on molecules, spores, seeds, eggs, germ cells and organisms, every one of which is descended from an ancestry spreading out behind it in a funnel shape toward the past of all life as well as potentially forward in a similar funnel into the future (page 358).

Spirality pervades the Universe concretely and abstractly, visibly and invisibly. Not only do the galaxies of stars tend to this rotary form, and possibly the supergalaxies or even the Universe itself, but the spiroid theme dominates the courses and shapes of bodies and substances all the way down into the atom. The elliptical orbit of any moving body or particle inevitably spirals in time through space, uttering the world lines of the sun, moon, stars and planets. And fluids both liquid and gaseous, as we have seen, corkscrew their way ahead — the great currents in the ocean, rivers, whirlpools, hurricanes and jet streams in the sky and all the smaller storms from thunder cells and tornadoes down to dust devils in the desert, snowdrifts around a post and smoke plumes out of chimneys — even the celestial streams of plasma with their dark vortex spots on the sun and stars. The shapes of trees (including family trees), vines and most plants and their roots also tend to the helix, as do the forms of shells, certain crystals such as stalagmites, bones, pine cones, flowers, seeds, muscle fibers, worms, fungus mycelium and the molecules within them, particularly protein, long nerve cells and the twisted double-helix DNA in genes — and so on and on to the abstract lines of all genetic connections in evolution, gyrating magnetic lines, invisibly curved space and, more than likely, the still undefined cores of atomic nuclei — all eventually transcending into *Urgrund*, the space-timeless abstraction of Infinitude.

## The Relativity of God

Just where or how any guiding mystic force or God fits into all this cosmic geometry is hard to say. But if "the world is a becoming," as Martin Buber said, and "we continually create it," all material structure may be considered a kind of script whose interrelations, like all geometric relations, depend on which of various possible viewpoints is used and are therefore, in essence, relative. So let us consider the relativity of God, relativity being a general principle that, in my view, almost surely pervades all things and non-things. When you were a baby, for instance, your mother presumably served as God in relation to you — in an inescapable if impersonal way before birth, and through the expression of an autocratic but very personal love after birth. Virtually everything good that came to you then, such as milk and love and coziness, seemed to be bestowed by her. And no doubt a few things "not so good," including an occasional scolding or a restraining slap, for of course she was sovereign and almighty — at least relatively speaking.

Comparable to this, more or less, may be the relation between a man and his dog. Or between a dog and his fleas, to whom the dog also represents God to the extent of being an incomprehensible source of sustenance, comfort and occasional discipline. To a flea, getting scratched or bitten by one's native habitat is a discipline akin, I suppose, to that experienced by a stone-age man getting battered by a thunderstorm, an attack he is well known to have attributed, when he first evolved the power to think, to the anger of a god. Moreover, if the dog symbolizes divinity to the flea, and the flea in turn to the microbes dwelling upon it, there may be no realistic limit to the relativity of God in the still unknown hierarchies of parasitism, neither deeper down in the microcosm, higher up in the macrocosm, nor anywhere in the mystic, intangible echelons of mind and spirit.

You may never have thought of it this way but, if you are accustomed to swatting mosquitoes, flies or other "bugs" that invade your home, and if you help birds, pets, plants or needy creatures of any kind, you are influencing evolution directly, even playing the part of a god in relation to the lives and deaths of some of these beings, to whom your motives may be just as inscrutable as your own deity's motives are inscrutable to you. At the very least this analogy is germane to the relativity of godliness and I think the Apostle Paul must have been acutely aware of such a principle when he wrote to the Corinthians that "the foolishness of God is wiser than men, and the weakness of God is stronger than men." If he was right to suggest God might appear foolish from any viewpoint, even from a divine one, who can say what profundities of wisdom and divinity may soar beyond Elysium? And who can be sure there is any finite number of the "many mansions" of the Lord?

## WHY THE WORLD?

It is obvious that many if not the majority of people regard the earth as basically a material world. And their number, I suppose, includes most astronauts, despite their rare advantage of an outside view. But at least one of them, Edgar D. Mitchell, saw more — for he wrote: "The first thing that came to mind as I looked at . . . planet Earth floating in the vastness of space . . . was its incredible beauty . . . a blue and white jewel suspended against a velvet black sky . . . The presence of divinity became almost palpable and I *knew* that life in the universe was not just an accident based on random processes. This knowledge came to me directly — noetically . . . an experiential cognition."

618    It made me think of the time many years earlier when philosopher

William Ernest Hocking asked his students at Harvard whether they would like to go to a remote but hospitable planet where they would be free to take advantage of any of innumerable opportunities, including living out their lives in any form of society they preferred with absolutely no interference from outside. When a few allowed it would be exciting, if true, he casually remarked, "Well, it *is* true. Right now you are on a remote planet, floating free and independent in the universe, and you actually have a chance to make any dream you dream come true — if you really want it badly enough!"

That indeed is the exciting circumstance in which man finds himself on Earth. Could anyone dream up a truer challenge? A more vital and precarious situation? It's almost as if the angels above were ripping out their feathers in frustration, imploring God, "Where can we find a place dangerous enough for those humans to live on? What do You say, Sir, to a tiny drop of splash balancing between a star too hot to touch and the cold surrounding ink of nothingness?"

That, for better or worse, is your Earth, dear reader. She is not exactly a picnic grove. Materially she is not even your home — but rather, in the long run, your tomb. Yet, if she ever proves unsuitable, might you not go elsewhere? In which case, I wonder if you would still be you. Or, regarding your body, is it you? Or yours? Or is it the world's?

## Evolution of the Impossible

Speaking of the world, now it comes to me that a real and critical portion of it is factually impossible. I mean its history literally is centered on an endless sequence of miracles that one by one come true. In 1,000,000 B.C. it was a miracle to kindle a fire. In 10,000 B.C. it was a miracle to grow one's own food. In 5000 B.C. a wheel was a miracle. So was metal. And riding a horse. And sailing the ocean. As recently as A.D. 1400 printing a book (on movable type) was a miracle. In 1700 an engine was a miracle. In 1780 human flight was a miracle. In 1835 instantaneous long-distance communication was a

miracle. In 1950 space travel was a miracle. And these miracles go on — and on — without end.

The biggest miracle of all, excitingly enough, may be that the world can exist under such an unbelievable system, that it actually operates without central senses amidst so much destruction, waste and woe — riding its inexplicable paradoxes.

"What have we done," cries mankind, "to deserve such treatment? Why does God try us so?"

The key philosophical question then boils down to: why the world? What are we here for? Specifically, why were you and I conceived such sorry worms upon a troublous mote named Earth?

It is a tough one. Scientists and philosophers have wrestled with it for millenniums to meager avail, while the world waxes not only much more complex, drastic and mysterious than we imagine but (as biologist J. B. S. Haldane once allowed) even much more so than we *can* imagine. Pondering why the Almighty — assuming there is One — would permit such imperfection in His realm, they have not come up with any completely convincing theories. But one of the better among recent ones is "process theology," which postulates that God, along with His universe, is in a perpetual process of development. This is based in large part on the natural science philosophy of Alfred North Whitehead, a mathematician who added the time dimension to static medieval concepts of absolutism so they would be more compatible with the findings of modern physics, biology, evolution and even psychology. Although it limits God, from a human viewpoint, by associating Him with a finite time field, process theology compensates, as I understand it, by permitting His all-knowingness to be explained by making all life an actual part of His experience. Thus what you and I do and think, God feels and knows eternally through our senses, our lives, our aspirations, our sacrifices, our creations, along with all such everywhere. In fact, if being the Creator means He has to create creatures, we creatures inevitably must be a vital part of Him, so that even our own small creations in turn become subreflections of His creatorship. And, to the extent that God's experience (essentially complete and therefore past in respect to Him) is also our experience, it is just as surely incomplete and therefore present in respect to us. In this way, even without overstepping the explored confines of our mortal space-time continuum, God's ancient suffering can be our present living, and our temporary pains and doubts and struggles and slow evolutions an indispensable jot of His eternal almightiness.

## THE SOUL SCHOOL

Although the promotion of earthly woe to an aspect of divine experience noticeably helps us in facing life's adversities, we need even more the principles of polarity and transcendence if we are really to explain why adversity is important and (as I believe) actually vital to our progress as spiritual beings. Thus at last we arrive at the only hypothesis for the nature of this troubled world that fits all the known facts — the hypothesis that planet Earth is, in essence, a Soul School.

Such is not what I'd call a solemn conclusion, even though fraught with meaning. For the soul is the part of us that is incorporeal, immortal and made of spirit, and it should be joyous news to mankind that a Soul School has been found, in which we are duly enrolled, in which raising souls is the prime goal. Indeed I feel like shouting: "Praise God, O humans, for your problems — the worse the better and look for more — because problems are what you are made of." For they are woven into the very texture of the world and every pain and trouble is food for the spirit without which it cannot grow.

If you doubt that Earth is in essence a Soul School, I suggest you just test the idea as a hypothesis. I mean: try to imagine that you are God. This may not come naturally to you. To be God of course you have to be a creator. And a creator, by definition, must create. So you, the creator, now find yourself creating creatures (a word meaning created beings) who have to have a world to live in. But what kind of a world should they live in? Or, specifically, what kind of a world will you decide to create for them?

This is a deeper question than at first you may have thought. For a mortal human it is certainly full of pitfalls and complexities. And you hardly can have ever devoted much time to pondering it. But what about checking into some known and already existing worlds around here? The moon, for a starter, is interesting, but airless, waterless, sterile and clearly unsuited at present for much life. Venus, on the other hand, is choked with clouds and hot as a stove at her solid surface. Mars is more promising, but seems mostly an endless, dusty, arctic desert with an oxygen-deficient atmosphere as vacuous as is air twenty miles above the earth. Jupiter and the giant planets are deathly cold on their visible surfaces of ammonic supertyphoons two thousand miles deep descending to probable oceans of methane. If there are any watery layers below the methane, it is hard to imagine them being an ideal abode for life, particularly the life of the mind and the spirit.

But now take a good look at Earth. Compared to those other worlds, she looks wonderful. Of course I'll admit I can't help having some bias in favor of this little third planet, seeing as I am human and not God, though I try mightily to overcome it. But all the same, even allowing for my humanness, I cannot help thinking that the virtues of Earth represent more than local vanity. In the first place, Earth's surface is extraordinarily varied and educational, not all ocean, not all desert, all lava, all cloudy nor all any one thing, like most other planets. Instead it is part sea, part land and part ice. Not only that, but the more habitable of the three, the land, is full of contrast, being part forest, part open plain, part desert, part mountains and part swamp or tundra with lots of lakes and rivers thrown in. And the sky, instead of being totally empty as on the moon and Mercury or completely cloud-blanketed like Venus, is almost exactly half clear and half cloud, so you can see the sun, moon and stars, yet also in your turn partake of rain, snow, hail, thunderstorms, hurricanes, tornadoes, even rainbows.

As for life and adventure, Earth is literally teeming with it. Creatures are not only walking, creeping and slithering all over the land, but burrowing under it, climbing the trees above it, swimming in the seas around it, flying through the air, even dwelling invisibly within each other, trying out, as we have seen, every imaginable mode of locomotion, of communication, of preying, eating, sheltering and propagating their kind. Earth is a place full of conflicts, surprises and surmises, multiple and bewildering revelations, evolving morals and heartrending struggles with adversity, of growing complexity, social uncertainty, political compromise, economic feedback and philosophical paradox. Earth provides the optimum, if not the maximum, in prolonged stimulation of body and mind and, most particularly, she excels in educating the spirit. In short, in the tiny portion of the universe thus far revealed to man, she is far and away the top-ranking Soul School available.

Honestly now, if you were God, could you possibly dream up any more educational, contrasty, thrilling, beautiful, tantalizing world than Earth to develop spirit in? If you think you could, do you imagine you would be outdoing Earth if you designed a world free of germs, diseases, poisons, pain, malice, explosives and conflicts so its people could relax and enjoy it? Would you, in other words, try to make the world nice and safe — or would you let it be provocative, dangerous and exciting? In actual fact, if it ever came to that, I'm sure you would find it impossible to make a better world than God has already created. Or even to free it from any basic limitation it has without doing more harm than good.

I know it seems almost blasphemous to associate pollution with spiritual beauty, yet overcoming such a prejudice is one of the first lessons of the Soul School, where decay is as much a part of life as growth and to be found in many of its loveliest features, from the smell of the rose to the flame of the maple. A rotting log, for example, may exhibit the graceful arboreal designs of bark beetles who tunneled blindly before the bark was shed. Even discarded tin cans on an unswept city street may be pressed into the abstract pose of blight. And, by mystic law, no depth of earthly imperfection but harbors as great a potential height of perfection. Even into the disciplined laboratories of science. In fact, I'm told, crystallographers see proof of this under the microscope, where a molecular imperfection must be present to initiate what they term a self-perpetuating, spiral dislocation (page 450), which alone permits the accretion of ions, layer upon layer, so they can grow into a single, whole gem-perfect crystal. Similar imperfection steers crystal growth in bones, aligns fibrils in muscles and seems to be a mysterious factor in the construction of everything from microdust to giant stars.

Mental and spiritual aspects of all protoplasm and organisms, including the Soul School's own special product, the soul, are even more mysterious. Essentially independent of space and time, the soul of course exists and reaches far beyond matter. Which explains Baha'u'llah's comment that "it is still, and yet it soareth; it moveth, and yet is still." Also his further description of the soul as "a sign of God, a heavenly gem . . . whose mystery no mind, however acute, can ever hope to unravel — even while, among all created things, it is the very first to realize something of the excellence of its Creator."

There are many people on Earth still of a mind to follow blindly the ancient superstition that all misfortune in life is meted out by God in His anger over the sins of man. But that is an error, says Baha'u'llah, for "tests in life are not punishment but rather serve to reveal the soul to itself. . . Neither need we dread the disasters that come to each individual life . . . according to station. For the earth in essence is a workshop, a crucible for the molding and refining of character." It is definitely not a global art gallery, nor a playground nor a torture chamber, though it may show temporary elements of all of these. Instead it is a Soul School, the perfection of which paradoxically is hidden within its imperfection.

I've heard it said that man's body needs the pig, as does his soul the eagle. If so, the Soul School is where he will find out how to reconcile the two. For this is a serious establishment in a venerable cosmos where we learn by trying and doing. Despite local appearances, ours is not a world composed entirely of neat three-acre lots, each sheltering a contented, well-fed, well-adjusted family that has never experienced mud, cancer, bugs, accidents, poverty, wars or rumors of wars. No, this is the place where a step is taken every day from thinking, "Someone ought to do it but why should I?" to "Someone ought to do it so why not I?" This is the planet where the bowel that issues entropy shares blood and nourishment with its neighbor, the womb, that issues negentropy. It is Saint Augustine's epic meeting ground between "Brother ass, the body, and his rider, the soul." It is where many a good man persists in denying his soul by telling himself it would be inhuman to deny his body — all because he has not yet discovered it is actually only his outdated animal body that is holding back the vast potential of his evolving human soul.

As spirit thus distributes itself through the world, obviously it will not treat all souls alike. For, in the service of justice, the Soul School must deal with us as individuals, making full allowance for the fact that the trials and lessons of one soul are rarely exactly appropriate for another. Thus arise the familiar and often puzzling disparities in

624

life's fortunes, like the exploding bombshell in a battlefield that inflicts cruel suffering upon one soldier, bestows heavenly relief in a hospital on another and grants a third his mystic release from life altogether. In a similar way Earth's approaching catastrophe of adjustment to germination may, for some souls, turn out instead to be a metastrophe of hope, a sort of musical beyond-beat or spiritual purge that will clear the way for general and joyous recognition of spiritual values, an aspect of the Soul School that I cannot hope to explain in any reasonable way because, quite simply, it is a matter of faith.

Faith of course is mystical and often a key in the struggles of mind and spirit — as when Jesus said to the father of the epileptic, "If thou canst believe: all things are possible to him that believeth." To which, paradoxically, the tearful man replied, "Lord, I believe; help thou mine unbelief" (Mark 9: 23-24).

For faith means more than holding something to be true. It requires action. It says: "I decide to do it. I stake my existence on it." Columbus did not just think he was right. He laid his life on the line. So did Lindbergh and Neil Armstrong.

Faith is likewise a spiritual form of vision. The Arabs said as much in their ancient proverb: "The eye is blind to what the mind does not see." Which really means: "Believing is seeing."

## THE AXIOMATIC PROOF

We hear a lot from scientists and others who maintain that the criterion of experience (say, a physical experiment) is superior to what they call the dogma of religion, but I wonder whether they remember or were ever made aware that science can experience only what it measures while religion may tune in on the even wider experience of things beyond measure.

"Is there any real evidence to support this suggestion?" you may ask. "No," I must reply. "I know of nothing likely to convince a skeptical modern scientist. Certainly there is no proof through logic." For religious experiences are spiritual truths, if they are anything, and beyond the reach of the thinking mind. You cannot get to them with common sense or by any deductive reasoning. You must go deeper, using what I would call the axiomatic proof.

Do you remember how you learned Euclidean geometry in school? You were presented first with an axiom such as: "A straight line is the shortest distance between two points." This was self-evident. It needed no logic. You didn't have to prove it but would simply accept it as true because it felt right in your bones. Matter of fact: being unprovable is the first criterion for an axiom. If it can be proved, it isn't an axiom but a theorem. If it cannot be proved, it may be an axiom and significantly may convey a feeling of absolute conviction through a knowledge deeper than thought. At the very least, an axiom is mysterious because it is made essentially of heart, not mind.

So hold fast to your axioms, brother man. Pay no money, pay only heed. For the axiomatic proof I am offering you knows no price. If it be overlooked by thinkers, it will be understood by seers. It is the realization that a profound, irrational quality in a concept, such as the timbre of its beauty, should be enough, unsupported and unpondered, to establish its truth. And who would doubt that John Keats fathomed as much when he wrote of his Grecian Urn, "Beauty is truth, truth beauty. That is all ye know on earth, and all ye need to know"?

For my part, I just cannot imagine anything being false if only it be beautiful enough. For beauty is its own proof, and it pertains very much to the world. I mean I don't think my dreams capable of portraying anything more beautiful than this world and universe — certainly not while I am in my current state of soul. Reason tells me that higher spiritual developments almost surely exist elsewhere in the firmament, but that does not necessarily place them in more beautiful worlds. Perhaps too my rhapsody with Earth is merely my personal limitation, though I suspect that those who think they can imagine greater beauty must be missing much of what is here. Of course I concede there is ugliness on Earth and misery and ignorance and hate, while peace often seems far away. But to me, beauty does not require tranquillity, nor even happiness, though it helps create both.

Contrast and struggle, as I've suggested, far from diluting beauty, only etch it deeper. For spiritual beauty of the highest order seems dependent on contrast most of all. By nature, it needs something to rise above, some darkness to illume, a negation to turn positive, a void to punctuate with islands, with lakes, with trees, with stars . . . Certainly supreme beauty needs more than a province, a sea or a species to dwell in. It needs a world!

And by such means I see the abstraction we have called Polarity transcending toward ultimate Divinity, which is thus revealed as the Mystery of mysteries, the great Mystery that embraces all our other mysteries of life, the unknowable Essence that many call by the name of God. But what matters it what we call It? It is abstruse, even bewilderingly abstruse, and remains so whether or no we accept that somehow by Its agency, out of utter nothingness is arisen everything in the Universe. Its station plainly implies intelligence, indeed Intelligence so far beyond the human as to justify the adjective "divine."

This, however, need not leave you or anyone behind — for, as Ali told a bewildered humanity in the seventh century, "Within thee the universe is folded." So be assured, dear earthling, that you are parcel of all mankind, of all life, of all matter, of all mind, of all spirit in the Universe. Even though the Mystery includes a veil to hide its awesome Glory from our feeble understanding, console yourself that your skin and senses are really less the boundary they always seemed than a bridge joining you to the world. And, as truly, the Universe is more than the pattern of matter we sense, for it is literally the greater aspect of one's own self. With profound confidence then, and lovingly, may we pray: "Thy will be done, O Universe!"

And as we drift wonderingly into transcendence, discarding at last our mundane selves to the tomb of Earth, gratefully receptive to the beauty of mysteries still hidden beyond the horizon of mystery, let us try not to forget that only God's Eye has the capacity to see God — yet, by His grace, that the stars can really smile, if we only knew. And that the veil of Glory, for all its blindingness, is woven of light.

# THE MEANING
# AND THE MELODY

# POSTLUDE

# The Meaning and the Melody

A S WE HAVE JUST CONCLUDED our main theme of the Seven Mysteries of Life, I shall briefly summarize them here with their meanings, so you may synthesize or digest the full philosophical message before tuning in to our final essay on the music of nature. The first mystery, Abstraction, shows our world to be basically intangible behind the seeming solidity of matter and life. Second is the Interrelatedness of all things, including animals, vegetables, minerals and stars. Third, the Omnipresence of life everywhere in every kingdom, even in the supersuperorganism of the Universe itself. Fourth, Polarity, the principle of symmetry expressed in polar opposites that interact and counterbalance throughout the worlds. Fifth, Transcendence through the natural law of progression from finitude toward Infinitude, using the tools of time, space and self. Sixth, the Germination of worlds, which happens once to every celestial organism and significantly is occurring now on Earth. And seventh, the greatest and ultimate mystery of Divinity, which includes all mysteries and remains the unknowable Essence behind Creation everywhere and forever.

Having now been in mental orbit through these many pages, scrutinizing Earth with one eye and Universe with the other, pondering their life and lives inside and out and between, there comes the time to call it a book. So let us hearken to what we have learned by this venture and see what melody it plays.

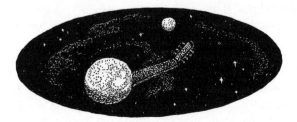

## MUSIC OF THE SPHERES

First we'll attend the seldom-heard lullaby of mellowing Earth, whom I behold hanging gourdlike upon the sparkling vine of space. On second glance though, the gourd looks more like a lute, that lilting contrivance first fashioned in Persia by stretching a sheepskin across a cut gourd and stringing it with silk. But what is our symbolic superlute singing as she spins? Such a question was only fit for poets or seers before this century, but now any competent geophysicist can answer it. The earth, as a musical instrument, knows two fundamental notes to which her body naturally oscillates: one with a vibration period of 53.1 minutes, the other of 54.7 minutes. Thus she sings a rather dissonant chord, like striking two adjacent keys on the piano. To be explicit, of course I should say Earth rather rings than sings, for solid planets are really more like gongs than lutes, and her ringing is sundered by her spinning, which perforce concentrates its energy around her swift equator while leaving her poles idling. She even dings forth a third and torsional note of higher pitch (42.3 minutes), not to mention a flight of girlish overtones all the way to a giddy 3.7 minutes.

Of course none of these planetary chimes can be heard by naked ears in their original slow frequency, since they are about twenty octaves lower in pitch than the tones we know and therefore, unless transposed, resemble silent periodic earthquakes. Yet they actually play the music of the spheres as divined by Pythagoras, and every living world in the heavens pours forth a comparable profundo harmony — even some pretty dead worlds like our barren moon. Do you remember the Apollo 12 astronauts who startled academe in 1970 by literally ringing the moon like a silver bell, hitting it with their discarded module so hard it reverberated for many minutes and probably (though unheard) for days?

The electronic ears "listening" to this celestial music are very sensitive seismographs that record both horizontal and vertical waves traveling outward as expanding spheres of compression through rock and magma. One of them was subtle enough, I'm told, to detect a squeeze of one sixteenth of an inch in the earth's crust between New York and California. But it was the violent Chilean earthquake of 1960 that provided the first clearly "heard" ringing of Earth, when scientists excitedly observed that the volume of the oscillations from each blow of the Andean "clapper" upon the Pacific "bell" diminished by almost half every two days, while continuing to fade more and more slowly for weeks, particularly in the very dense planetary core.

Out beyond the earth and moon a different and even more majestic kind of music is played by gravity and known to astronomers as interplanetary harmonics, also smaller resonances such as perturbations between the moons and rings of Saturn and newer-discovered relations involving scores of asteroids, like Alinda, Amor and Toro. Toro is an interesting example because it makes five orbits around the sun in the same time the earth makes eight and, being hardly a cubic mile in bulk, one would have thought it hopelessly dominated by our planet, yet its changing, lissajous, resonant orbit is so complex (and just dissonant enough) that every few millenniums it slips away from Earth and drifts into the gravitational clutches of Venus for a millennium or two before being tossed back to Earth in the first well-documented celestial game of catch.

THE FLUCTUATING AND TEMPORARY ORBITS OF THREE ASTEROIDS

Alinda 1600-1958          Amor 1800-1950          Toro 1800-2000

Now if the song of a planet is pitched twenty octaves below man's hearing, the song of the atom is sung a complementary twenty octaves above it, leaving us musically midway between the trebles and basses of the micro- and macrocosmic worlds. This to me is one of the most significant symmetries in nature, and we should not forget that atomic music is every bit as real as the planetary kind, in fact probably more accessible, as was charmingly demonstrated in April 1939 when my late friend Donald Hatch Andrews, professor of chemistry at Johns Hopkins University, put on a ballet for the national meeting of the American Chemical Society in Baltimore, transposing the vibration frequencies of many of the common atoms down to audible pitches, when, after trimming the overtones, ballerinas costumed to represent carbon (in black), hydrogen (in red), oxygen (blue), etc., pirouetted around on the stage. There was a water dance in which several blue girls, each flanked by two red ones, did an elaborate routine to barcarolle rhythm, wheeling around each other exactly like real $H_2O$    633 molecules so the audience could visualize the true rolling motion of water. Others later portrayed methane ($CH_4$), which, if the dissonant music was less than savory, at least let no one forget that it represented marsh gas. And the finale offered a series of alcohols ($CH_3OH$, etc.), some of which sounded to Dr. Andrews like Debussy and were, he hoped, plausibly intoxicating.

## MELODY OF LIFE

When we perceive the world thus as melody, I feel, we mortals are about as close as we can get to understanding its abstract essence, harking what Thoreau called that "most glorious musical instrument," whose generally unheard notes, grandnotes, great-grandnotes and succeeding harmonic posterity continuously join the chorus of evolution that may never end. I try not to assume the music of my native sphere is better than the music of other worlds unknown (even though it seems far too complex to have much likelihood of being identical to theirs just by chance) — so I listen to the saw-bows of insects and crustaceans as if they were unique to the universe. I marvel at the horns of earth fashioned by the mole cricket who stridulates underground (as loud as 90 decibels at one meter's range) to serenade his mate from the lonely sky. I hear the weedy warbling of the toad and the wistful whistling of the lovestruck turtle — even the almost inaudible whisper of the courting fruit fly, whose thousands of species are sorted almost entirely by the nuance of song. Diving off a reef, I discover the teeth-clicking conversations of fish, their oboeing of blown air and their drumming with special muscles upon their tuned air bladders (as were once the ancestors of our lungs).

Drumming is widely broadcast on land too, I note, by thumping rabbits and mice, by the wings of grouse and prairie chickens, by the fists of gorillas on their chests . . . Even leeches tap on leaves, as do termites on tunnel floors, while amorous earthworms beep faintly in their coded cadences. The ultrasonic twitter and aural "vision" of bats are well known (page 206), though not the haunting bell-like tones bats ring out while hanging at rest upside down in dark glades in the deep woods. And an old-time ship's doctor describes the shrill skirling of the humpback whale as "filled with tensions and resolutions" — all these but a meager sampling of the endless sequence of living, breathing parts of the song of Earth.

If such voices do not individually attain the full character of music, in concert they belong to something vastly greater than themselves.

And could we tune in on all of them at once, fully orchestrated as integral parts of the whole planet's symphony, we would surely hear sophisticated counterpoint, blended dissonant harmonies and all sorts of subtle sonorities just above the background pastiche of insignificant gabbing. A lot of small talk naturally has to be included in Earth's total chorus, because it comprises everything voiced by everybody from bacteria to whales.

Birds probably produce the most sophisticated music of any class of animals and put both meaning and emotion into many of their songs, which are much more individual than we usually think, a fact suggested by one ornithologist's count of 884 different songs sung by the American song sparrow and another's recording of chord variations including a four-note chord intoned repeatedly by a woodthrush. An Englishwoman named Len Howard, the professional musician and bird lover who wrote *Birds as Individuals*, says she can easily recognize dozens of individual birds by their voices, even though a song may change from day to day, partly through imitation. And she can tell whether the singer is happy, dejected or perhaps struggling with something unusual on his mind. At migration time the birds in her native Sussex, in effect, tell her not only when they are ready to take off for the winter but where they are going. They can't help it, for Spain, Italy, Morocco, East Africa or whatever place else is woven right into their music.

The English blackbird, called *Amsel* in German-speaking countries, sings songs that more closely resemble human compositions than does any other bird with the possible exception of the wattle-eyed flycatcher of East Africa, credited with a theme of *Salome* that presumably antedated Strauss. Miss Howard heard an *Amsel* sing a phrase from Bach one day that he may or may not have heard her playing a few weeks earlier on her violin. In any case, he had to work at it, making many attempts before he got a certain trill right. Then he began experimenting with doubling the length of the trill and adding new ones with a result Miss Howard called "flutelike" and "very lovely." Another young cock *Amsel* "actually composed the opening phrase of the rondo in Beethoven's violin concerto," something she could categorically "vouch for his not having heard." On the other hand, who knows what Beethoven may have picked up from *Amsels* in his day? Music is in the air and most certainly in the world's genes, and it was curiously reminiscent of Beethoven the way this bird composed the rondo by trial and error, pertinaciously trying numerous variations before reaching final satisfaction.

Many birds copy others or try to, even others of very different        635

families, while it has been estimated that a catchy tune may spread across the landscape in the mating season at a fairly constant mile and a half per day. Group singing is a factor in this process of course, being a common practice among certain species. Two dozen linnets, for example, will fly purposefully into a tree, all keyed up with anticipation. When one bursts into song, the others quickly join in, some twittering, some trilling, many slurring their notes up and down until all the voices unite in a great crescendo that can be heard across the fields, attracting other flocks of feeding birds, field by field, the effect sometimes traveling miles in a few minutes.

Other birds do antiphonal singing to maintain close communication in dense foiliage, different birds singing different notes of the same tune in responsive sequence. This has been reported of pairs of tawny-breasted wrens high in the jungled Andes of Colombia who sang their parts back and forth "like the twanging of liquid wires," diminuendoing then to gurgling babbles, which blended into the murmurs of a nearby brook. And there is the well-documented case of an "elegant trio" sung repeatedly by three boubou shrikes in overlapping territories on the shore of Lake Bunyoni in Uganda. The third bird was a male perched on a different branch from the first two, who may have been mates, but he inserted his note every time with professional precision.

1 sec. oscilloscope recording of birdsong

TRIO SUNG AT THE INTERSECTION OF THREE OVERLAPPING SHRIKE TERRITORIES IN UGANDA

A pair of these small shrikes has been heard by Dr. W. H. Thorpe of Cambridge University to sing as many as seventeen different duets on the same day. They use the orthodox western diatonic scale in at least five keys and know each other's parts. When the male shrike is away, the female will sing the complete duet by herself. But hearing her sing "his" part, even from a distance, often is enough to bring him flying back to reclaim his place in what seems to be a musical as well as a sexual marriage and which, should an outsider participate, can arouse musical jealousy, although trusted friends usually are more than musically welcome.

## HUMAN HARMONICS

I shall not invade the vast subject of human musical expression except to remind you that it has spread itself virtually everywhere on Earth from a sheep camp in lonely Australia to the streets of roaring New York. Indeed there is an unbelievable range of instruments and voices on some city streets and I vividly remember the haunting modulations of Arab vendors in the markets of Fez and Aleppo a few years ago, as well as the orchestrated medley of old Canton, where each tradesman would almost continuously broadcast his own trade's traditional signature in sound: the barber twanging a spiral wire, the bookseller ringing a bell, the butcher blowing a horn, and so on — so shoppers needed only to listen for what they wished and their ears would lead the way.

But while this sort of specialized polyphony may be expected in the exotic bazaars of the Orient, who would think to listen for anything like it among the grimy, can-whomping garbage men of New York? Yet the other day a musicologist of the Manhattoes for some reason did stop and listen and, to his surprise and delight, discovered that these burly characters not only bang their cans into the truck's grinder and do an incredible amount of clanking and thudding on the sidewalk but they also *call* — a call that reminded him of the whooping crane, though he later decided it is lower pitched and somehow different. What the garbage men actually call to of course is their truck, to let the driver know when to move to the next bunch of cans, but the call is very traditional and always the same: a sort of tenor "Yoh-hooo!" And the musicologist instantly recognized it to be the sol-mi interval — technically a minor third or, more descriptively, a descending minor third.

Why, you may be wondering, should garbage men always call in a descending minor third? Why not a descending *major* third? Or an

637

ascending fourth? Or a diminished fifth? The secret seems to be simply that those other intervals don't sound right. Surely Pythagoras would know what I mean. So would Beethoven, whose Fourth and Eighth Symphonies both begin on descending minor thirds. As does Mozart's Sonata in G Major for Violin and Piano, the march in Wagner's *Tannhäuser,* the first theme in Tchaikovsky's Piano Concerto No. 1 and a host of other major classical works. These include of course Rubenstein's *Kammenoi Ostrow* which, as it begins with fifty-four consecutive descending minor thirds, stands as the garbage man's rhapsody supreme. Garbage men, you see, low as they may be on the musical totem pole, and little though they may realize it themselves, are actually, like everyone else, in tune with every atom they heave into the grinder's maw and with every star that is churned by the Milky Way.

Among people who live out of town or at least in relatively quiet neighborhoods, many seem deceived by the apparent calmness of ordinary living. And they probably forget how drastic the world is, this world in which predation, parasitism and "wanton" extravagance are well established — for too often they attribute sudden illnesses and deaths to "infections" or "organic disorders," as if such things were rare, unaccountable disturbances on the otherwise serene sea of existence. Yet when I sit still and listen intently for a while, I can always hear a humming inside myself, often with overtones of singing that grow louder with awareness. This I take to be a natural manifestation of my body energy, perhaps residual molecular motion, the motor of organic matter engaged in the vital business of living. And to me there is no more reason to accept it as unmitigated calmness than I would assume absolute tranquillity in a beautiful, drifting thundercloud or a gently twinkling star.

## MUSIC OF THE ATOM

To come down to earth, a flame too has something of this deceptive calmness when in still air. Yet its volatile nature is immediately manifest when the right music gets to it, a phenomenon that, I'm told, was first noticed by physicist John Le Conte in the middle of the nineteenth century when at a concert he saw a gas light flare up every time a certain note was played on a cello. Since then researchers have gone on to discover that flames will actually accept and reproduce almost any kind of sound and, with the help of electricity, can be induced either to amplify or silence it. This occurs because the flames' heat

638

splits most of the gas molecules into negative electrons and positive ions, forming plasma (the stuff of stars) in the flame, and when any plasma is put into an electric field, its electrons naturally rush toward the field's positive pole and its ions toward the negative pole. But, as the ions are heavier than the electrons, there is a preponderant force toward the negative pole which bends the flame in that direction and, when this is controlled by an electric field shaped in a musical pattern, the fluctuating flame generates sound waves that exactly reproduce the music. It is a recent and quite extraordinary development and enough to make one wonder what kind of music the thousand-mile-high flares on the sun and other stars are dancing to, and what it will take to let us hear and understand them.

Now shifting to the microcosm, I hear that music, particularly ultrasonic music in the presence of heat, has been found capable of changing the chemical structure and strength of crystals by spreading imperfections at grain boundaries in the lattice. But more fundamental and presumably long ago surmised by Pythagoras is music's penetration to the very heart of the atom in the resonance principle, the revelation of which seems more and more to be establishing the concept that the smallest and most indivisible "particles" of matter may now realistically be considered nodes of resonance, which, in a sense, are poetically, if not scientifically, interpretable as living notes. Some physicists, I understand, are even hopeful that the dynamic school of physical research (which mostly studies what happens when such subatomic "particles" collide) and the theoretical school (which mostly tries to categorize the same "particles" relative to their presumed least common denominator, the quark) will get together in a harmony they have never known, through acceptance of something called "exotic resonance" (because it transcends quark harmonics), which now seems the most promising clue to the meaning of the complex, and seemingly ever more complex, symmetry of matter.

## The Vegetable Serenade

The vegetable kingdom has its musical side too, though not much is known about it among western scientists, the most successful researcher being Dr. T. C. N. Singh, head of the Department of Botany at Annamalai University in southern India. Singh evidently got his inspiration from none other than the Hindu Prophet Krishna, who, according to legend, used to play his flute in the Brindaban gardens near Mysore to make the flowers bloom. Singh realized that sound waves vibrate the molecules they encounter and that, since plants are made of molecules which must be in continuous motion for metabolism to take place, the right kind of music might just somehow stimulate them to accelerate their metabolism. So he spent years experimenting with young mimosa and pepper plants, marigolds, petunias, tobacco, sugar cane, sweet potatoes, onions, garlic, rice, tapioca and other vegetables in the university gardens, first subjecting certain ones to an electric tuning fork's hum for half an hour at dawn, then playing to some on a flute, to more on a violin and singing to still others to see how their growth would be affected compared to similar plants left in silence.

To his delight, in less than a week many of the plants that had "heard" the music, perhaps with the help of their earlike petals and leaves, began to outstrip their untreated companions and soon were growing at about twice the normal rate, particularly those that had been serenaded with high-pitched violin music and soprano voices. A few even outdid themselves, growing so fast they showed symptoms of a kind of hypersthenia and soon withered and died. But on the whole, as Singh matter-of-factly reported in 1959, "The treated seedlings were darker green, healthier, sturdier, with a more profuse root system than those derived from nurseries not excited by sound waves."

While the implications of this discovery do not seem to have been accepted without deep skepticism, research on similar lines has spread to Europe and America, where at least a few scientists have dared the heresy long enough to ask whether corn has ears — for music? And governments are getting interested, like Canada's, whose National Research Council recently financed research into music's influence on wheat seed and, in one experiment (repeated ten times so there could be no doubt about the results), found that seedlings of Rideau wheat (a winter variety) continuously exposed to 5000-cycle sound invariably exceeded the weight of control specimens by more than 250 percent and sprouted almost four times as many grain-bearing shoots. The researcher tentatively attributed this to the resonance produced by the high-pitched sound when it penetrated the wheat cells, which might thus conceivably store up enough energy to double their metabolism. And there is mounting evidence of genetic mutation in the fact that later generations of many of these plants repeated their musically inspired growth rates, even though only the first generation "heard" the music. Besides, due to the well-established fact that light of two different wavelengths or pitches is required in photosynthesis, which works through resonance, we can state categorically that energy in plants comes from the radiational equivalent of a musical chord. All of which fairly completes the botanical side of our thesis that life is physiologically, as well as abstractly, a melody.

## Musical Essence of the Universe

Winding up this final chapter on meanings to be gleaned from life's music, I would like to sneak in one last point about the musical character of the world, including the common dust we're made of. I have noticed that music, like solid matter, is essentially crystalline in structure. This is not exactly a recent discovery, being one of the latter-day surmises of the ancient art-science of harmonics founded by Pythagoras, but it is something everyone needs to understand if he would make himself aware of the seams where matter and spirit meet. Even though the music of matter is far too high-pitched for the human ear to hear either the individual notes sung by each element, the chords by each chemical compound or the intricate fugues reverberating from each type of crystal, matter's latticed waves are nevertheless spaced at intervals corresponding to the frets on a guitar or the holes in a flute, with analogous sequences of overtones arising from each fundamental tone. And, as Professor G. C. Amstutz, director of the Mineralogical and Petrographic Institute of the University of Heidelberg, said recently, "The science of harmony in music is, in these terms, practically identical with the science of symmetry in crystals. Indeed the crystals can now literally be seen to be the philosopher's stone, frozen music which presents to the eye . . . the dynamism of the molecules, atoms, particles and standing waves of which they are composed . . ."

Dante is remembered for having used harmonic haloes and symbolic crosses of light in arranging his blessed beings in a kind of rhythmic tick-tack-toe skeleton that formed what may be termed an ecclesiastical crystal. But a more perfect and more musical example is Bach's last work, *The Art of the Fugue*, which consists of a compound giant fugue elegantly fitted together from nineteen single ones, the core theme of which, after trimming all variations, bones down to a melodic, hexagonal skeleton of six symmetrical notes — the classic musical crystal.

DANTE'S BLESSED BEINGS          THE ART OF THE FUGUE

To Henry David Thoreau, living in his woodland cabin, a similar realization came in different guise with the erection of the first telegraph line past Walden Pond in 1851. For that August he heard the humming of the wire strung between the poles and wondered how much of it came from the wind and how much from the mysterious power of electricity to bind the world's thoughts together into a new and greater whole. "It was as the sound of a far-off glorious life," he wrote in his notebook, "a supernal life, which came down to us, and vibrated the latticework of this life of ours . . . every pore of the wood was filled with music . . . How this wild tree from the forest, stripped of its bark and set up here, rejoices to transmit this melody!" And on an occasion he "heard" the humming wire whisper, "Bear in mind, Child, and never for an instant forget, that there are higher planes of life than this thou are travelling on. Know that the goal is distant, and is upward, and is worthy all your life's efforts to attain to."

My thoughts return now to music, which seems to voice so perfectly the essence of the world I see below me. For your blue-swirled Earth down there is not merely a thing. It also lives through time. And in living, it is not so much an object, in three dimensions, as an event, in four. Its nature in fact resembles that of a melody which takes time to play and therefore exists not whole at any moment but rather strings itself out into a patterned sequence over a mortal span.

Any such span naturally has a beginning and an end, for that is the way with mortality, and mortality obviously is as much an attribute of music as of worlds in this pretranscendent phase of life. Indeed were any earthly melody to play unceasingly, whatever beauty it possessed would inexorably degrade into monotony and its once graceful form would bloat like a body with cancer. This rule is broad enough to include at least the vegetables and animals of Earth, who also need their gracious finales — as don't we all?

And in the larger view, beyond this nether finitude, we must ever remember that, as Buddha so succinctly put it, "It is not in the body of the lute that one finds the true abode of music." Almost any intuitive person who has had much experience riding a motorcycle, flying an airplane or handling some other sensitive vehicle knows that there comes a point where the driver begins to forget the mechanism and play it direct, as if the body of the machine were part of his own body and its limbs connected directly to his will. In music the great Arturo Toscanini was a supreme example of this transcendence of instrumentation when he bypassed all technique in exhorting his musicians to "Play not with your instruments but with your hearts!"

A seer like Thoreau would have known exactly what was meant, for he had more than ears to listen with and was always tuning in on, or wondering about, something new and beautiful. "As I climbed the hill again toward my old beanfield," he wrote, "I listened to the ancient, familiar, immortal, dear cricket sound under all others, hearing at first some distinct chirps; but when these ceased I was aware of the general earth-song, which my hearing had not heard . . . and I wondered if behind or beneath this there was not some other chant yet more universal . . ."

What chant, I wonder? Could it have been the forest air, whispering, "Breathe me and live." Or some gentle raindrops surrendering themselves with a sigh to the waters of the pond? Might it have been the Pythagorean octave, a withy consonance of small branches tossing among great trees to symbolize the man-woman, bass-soprano interval of life and harmony? Was it the moan of a blind planet groping for purpose in a boundless universe? Or a divine thought surging through the thresh of time?

All of us beings here are cells of the unknown essence of our world, nodes of flesh that could as well be notes of melody. We are part of something infinite and eternal. There is no boundary between us and the world. In a profoundly relative sense, each of us, as Alan Watts has suggested, may simultaneously occupy "that particular focal point through which the entire universe is singing at this moment."

Are we then God's dream set to music in the place where the sea and the wind have begun to awake and think? Grateful for our blessings, even when they hurt, we trust the world is not paining needlessly for our sweet incertitudes 'twixt desire and reason. We would wish to be wiser and more loving but, for good or ill, our memories are

643

young in this ancient oasis. And we comprehend little. How indeed could a part hear the Whole, or a note the Melody?

Yet the silence of space that enwombs the earth is not totally void. Indeed it is now revealed to be latent, pregnant, mystic — even as it was in the beginning that had no beginning — even as it will be in the end that can have no end. For this is the secret of the spirit that is the life of the form that is the language of the spirit — the eternal spirit that somewhere, somehow, found its voice, took wing and came alive.

# SUMMARY
# INDEX

# SUMMARY

# The Seven Mysteries of Life

### ABSTRACTION

*What's in an egg?*
  *A song is there, in chemical notation,*
    *Invisibly packed into the genes;*
      *Also detailed instructions for nest building,*
        *A menu or two, and a map of stars —*
*All in the one cell that multiplies into many,*
  *All put at the disposal*
    *Of the little feathered passenger*
      *So, once hatched and fledged,*
        *He will have more than a wishbone*
          *To launch his life.*

*What's an ocean wave made of?*
  *At first glance nothing but saltwater;*
    *But keep your eyes on it ten seconds . . . twenty seconds . . .*
      *You'll notice the water is roused*
        *Only momentarily by the wave*
          *Which passes it by,*
*That the wave leaves the molecules and bubbles behind,*
  *That the wave in essence is a kind of ghost*
    *Freed from materiality by the dimension of time.*
      *Made not of substance*
        *But energy.*

647

*And likewise with living bodies*
*And rocks, and all metabolizing matter*
*From atoms to stars,*
*Which all flow through space-time*
*Uttering the abstract nature*
*Of the Universe.*

# INTERRELATION

What relation is a white man
  To a black man?
    A yellow man to a red or brown?
      Closer maybe than you'd think,
        For all family trees meet and merge
          Within fifty generations, more or less —
            In round numbers a thousand years —
Which makes all men cousins,
  Brothers in spirit, if you will,
    Or, to be genetically precise,
      Within the range of fiftieth cousin.

But relations don't stop here:
  Man also has ancestors in common
    With the chimpanzee and other apes,
      Back twenty million years or so,
        Plus all the mammals further back —
          His ten millionth cousins
            If you'll abide my candor.
Still farther, the billionth cousin span
  Takes in the whole animal kingdom,
    And many vegetables, and trees;
      The trillionth must include rocks and worlds.
There is no line, you see, between these cousin kingdoms,
  No real boundary between you and the universe —
    For all things are related,
      Through identical elements in world and world,
        Even out to the farthest reaches
          Of space.

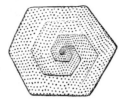

## OMNIPRESENCE

*Where did life begin?*
   *In the festering ooze of a primeval swamp?*
   *In a submicroscopic virus?*
   *In a stone? A star?*
*Strictly speaking, in none of these.*
   *For, truthfully, the question is wrong.*
   *Life did not literally begin. Life is.*
     *Life is everywhere everywhen,*
      *At least in essence,*
       *And of course*
        *It depends on your definition.*

*Did you ever meet a living stone,*
   *A stone that stirs, that travels,*
     *That eats, grows, heals its wounds,*
      *A stone that breeds its kind?*
*Yes, all stones are alive*
   *Essentially, potentially;*
     *At least they move around*
      *When weather and circumstances permit,*
      *Going mostly downhill,*
       *Sometimes waiting centuries*
        *In a deep pool in some stream*
         *For a torrent wild enough to drive them on.*
*And stones are crystals,*
   *Rock crystals that grow, molecule by molecule,*
     *Filling their own cracks or wounds,*
      *Reproducing themselves slowly*
       *But perfectly.*

*One kind is even magnetic and attracts iron.*
*The ancient Chinese called it*
*"The stone that loves."*

*Larger mineral-like organisms also live*
*In their patient, plodding way:*
*Dunes drift and glaciers creep,*
*As do mountains, islands, volcanoes and rivers —*
*That are born in the clouds and die in the sea —*
*And lakes and storms,*
*All moving as is their wont,*
*Even fires on Earth*
*And whirling spots on the sun.*
*In fact there is compelling evidence*
*That the earth lives as a superorganism,*
*Along with moons, planets, comets, stars, galaxies*
*And other celestial bodies,*
*And that, most of all,*
*The Universe itself*
*Is a growing, metabolizing supersuperBeing,*
*In very truth alive.*

## POLARITY

*Do you think matter is made of particles?*
  *Waves? Or what?*
    *Where is the line between body and mind?*
*How could God,*
  *Presumably the epitome of goodness,*
    *If He exists,*
      *Create a world harboring as much evil, pain, ugliness,*
        *Disease and war as we find in this world?*
          *How could He?*

*These are enigmas, paradoxes,*
  *Seemingly unsolvable;*
    *Yet somehow, if one relaxes one's heart*
      *And opens one's mind,*
        *And wonders the right wonders,*
          *They become resolvable.*
*Take Saint George and his dragon.*
  *If Earth is a good world, one asks oneself,*
    *Why the dragon?*
      *Obviously because he was needed.*
        *Can you, in fact, imagine*
          *Any way George could have made it to sainthood*
          *Without him?*

*There is a polarity about good and evil, you see.*
  *To a baby, getting spanked for trying to climb out of his cradle*
    *Is a dreadful experience: an "evil."*
      *But to his anxious mother, trying to tell him NO*
        *In sign language, it is a constructive deed and "good."*
          *The same act thus has two poles*

          *Expressing opposite aspects of good and ill.*

*Similarly to mankind as a whole, war is evil,*
  *A spanking of civilization,*
    *Something to be outlawed at all costs.*
*Yet, for all we know, in the perspective*
  *Of spiritual or cosmic forces far beyond man's understanding,*
    *War could possibly serve some useful*
      *Maternal purpose as a sign language,*
        *A challenge to try our souls —*
          *Even perhaps, relatively speaking,*
            *A constructive, spiritual purpose*
              *That is good.*

*For polarity is part of the symmetry of nature*
  *That brings a relativity, a complementarity*
    *To many qualities in this life,*
*To cause and effect, to predator and prey,*
  *Male and female, Creator and creature,*
    *Concrete and abstract, science and religion,*
      *Mortality and immortality, yin and yang,*
        *And other seeming opposites.*
*Free will, one of the most puzzling of these,*
  *Has for its counterpart predestination,*
    *Which turns out to be really its expanded aspect,*
      *A sort of bird's-eye view of the familiar scene*
        *Beheld from one dimension more.*
*And so it goes*
  *With body and mind,*
    *The first enmeshed in space, in time,*
      *The second free of both,*
        *Like poles of Earth and other paradoxes*
          *Which are, in a sense,*
            *Really just different sides*
              *Of the same thing.*

## TRANSCENDENCE

*Have you ever wondered*
  *Why each year you live*
    *Seems to pass faster than the year before?*
*There's a law at work here*
  *Called Transcendence,*
    *Influencing time and space and consciousness of self,*
      *For each year lived has to be a smaller portion*
        *Of one's experience to date.*
*To the year-old baby a year is a lifetime,*
  *To the ten-year-old a tenth as much,*
    *To the centenarian but one percent of his experience*
      *While people he knows appear, bloom and die*
        *Like flowers in a garden.*

*The same is as true of space as time.*
  *The baby learns the inch and foot*
    *Before he knows the yard,*
      *Then, as his horizon expands,*
        *The mile, the acre . . . the light-year . . .*
          *Progression from the finite*
            *Toward the Infinite, you see.*
*Yet, as you gain the mile, you do not lose the inch,*
  *Nor, as you gain the year, do you lose the minute or the hour,*
    *For finitude is a tool of learning,*
      *Learning the little before the big,*
        *The simple before the complex.*

*Transcendence affects the self too,*
  *For one begins as a fertile egg,*
    *The seed soul, stirring, seeking,*
      *Becoming a pupil in the Soul School of Earth,*

*Growing in consciousness,*
  *In awareness of other beings,*
    *Using the tools of finitude,*
      *The self in space and time,*
*The while developing spiritually*
  *Through life, through death —*
      *Death, which evolved only later in evolution because it had*
        *Survival value for the multicelled organisms —*
          *Death that we cannot live without.*

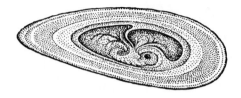

## GERMINATION

*A nova is an exploding star,*
  *Climax in the life of a blazing world,*
    *An example of the cyclic vitality of all worlds*
      *That grow and mature, ferment, germinate.*

*Germination happens only once per world,*
  *A crucial event amid the unfolding phases of life*
    *That develop mind, speech and spirituality*
      *In ways still scarcely known to history,*
        *To science, to philosophy.*

*Earth, for example, third planet*
  *Of a modest star called Sun,*
    *Is germinating right now.*
*After five billion years of slow, quiet evolvement*
  *Plus a few quick centuries of writing, printing,*
    *Industrial revolution, technological bloom*
      *And improved communication,*
        *Enabling her emerging mind for the first time*
          *To pool its knowledge,*
*Suddenly in the twentieth century Earth, with her human population,*
  *Is practically exploding!*
    *Man has won the planetary tournament of evolution*
      *By dominating all competing forms of life,*
        *Speed of travel has climbed a thousandfold*
          *From the gallop of the horse*
            *To the whoosh of the space rocket,*
*Revolutions have occurred in nearly every branch of learning,*
  *Man has explored not only his planet's entire surface*
    *But penetrated from the atom to the sky*
      *And into outer space.*

*He is now seriously trying to unite his home world*
  *Politically and culturally, through standardization,*
    *Liberalization, free compulsory education for all.*
*Even spiritual unity must soon loom as an attainable goal,*
  *An aspect of Earth's flowering into a mature superorganism —*
    *All this in fulfillment of the natural,*
      *The inevitable, evolutionary process*
        *Of planetary germination.*

## DIVINITY

Who or What runs the Universe?
  Is there a plan behind the daisy, the hummingbird,
    The whale, the world?
Who conceived the eye back in the primeval darkness
  Of early evolution?
    Who designed the fish's air bladder in the ancient deep
      As if foreseeing its future as a breathing lung
        Upon the dry land?
And out of what beginning evolved the mind?
  By any stretch could mind have been mindlessly created?
    Does science have an answer
      To the Voice out of the Whirlwind Which asked Job
        "Who hath put wisdom in the inward parts?"
Is the world really drifting along without pilot,
  Steering itself automatically,
    Running its own affairs at random?
      Could the Universe, just conceivably,
        Have created Itself?


Surely there is Mystery in this Universe,
  Not only somewhere and somewhen but everywhere everywhen
    And far, far beyond the scope of man's feeble
      Capacity to comprehend.
For man, puny, mortal and finite,
  As he is in this nether phase,
    Is permitted to visualize neither an end to space
      Nor space without end;
        Nor can he even grasp a start or a finish of time,
          Nor any sort of beginning that has no beginning
            Nor any end that has no end,
Hence the Mystery,
  The abiding, pervasive, universal Unknowability

*That many call by the name of God.*
*But what matters it what you call It?*
  *It is abstruse, bewilderingly abstruse, and remains so*
    *Whether or no we accept that somehow by Its agency*
      *Out of utter nothingness has arisen*
      *Everything in the Universe.*

*Its station plainly implies intelligence,*
  *Indeed Intelligence so far beyond the human*
    *As to justify the adjective "Divine,"*
*And this seems to be relative.*
  *If a human adult represents divinity to a baby or an animal,*
    *So must the animal be divine to a vegetable,*
      *The vegetable to a mineral . . .*
*Likewise, as wrote Paul to the Corinthians,*
  *"The foolishness of God is wiser than men,"*
    *And there is presumably a hierarchy in Divinity above*
      *As well as below us —*
        *Even as the doings and thoughts of humanity and of Earth*
        *Are but a negligible jot*
        *In the eternal consciousness of God,*
*Even as the horizon of knowledge expands outward from our planet*
  *Accompanied by the inexorable horizon of Mystery*
    *Which expands even faster and farther than knowledge,*
      *Leading man's consciousness*
      *To new dimensions.*

*Thus doth Divinity*
  *Embrace all the other six mysteries of life*
    *Even though callow man comprehendeth it not,*
      *Even though the Mystery remaineth*
        *So far beyond earthly finitude*
*That no eye but God's own Eye*
  *Hath the capacity to see*
    *GOD.*

# Index

This full index, made by the author, is especially suited for browsing. It lists abstract concepts and illustrations as well as proper names, and is liberally interlaced with suggestive references pertaining to the Seven Mysteries of Life.

Augustine (contd.)
383; on evolution, 541; on body and soul, 624
Auroville, future city, illus., 579
Aztecs of Mexico, 587

Babelnoah, future city, illus., 579
Bach, Johann Sebastian, on color, 235; melody of, 469; shared music with birds, 635
Backster, Cleve, lie detector, 307.
Baer, Karl Ernst von, embryologist, 405
Bagnold, Ralph A., on dunes, 418
Baha'i Faith, founded by Baha'u'llah, 372, 459, 492-93, 495, part of progressive revelation, 612-15. See also Baha'u'llah
Baha'u'llah, Prophet of God, ix, on eternity, 459; on polarity, 492-93; on God and transcendence, 495; part of progressive revelation, 612-15; his message, 614; his fulfillment of ancient prophecy, 614-15; on harmony between science and religion, 615-16; on the soul, 624; on Earth as Soul School, 624
Baldwin effect, in evolution, 552
Bancroft, Wilder D., biologist, 369
Barère, Simon, fast pianist, 230
Barnard, Dr. Christiaan, 351
Bascom, Willard, ix; on dynamics of the beach, 418-20, illus., 417
Bats, 39, mating of, 136; sonar of, 206-08, illus., 207; predation of, 206-08; sleep of, 295
Beach, dynamics of, 418-20, illus., 419. See also Dune, Analogies
Becker, Dr. Robert O., 202
Bee(s), pollination, 137-38; mind of, 247-49, illus., 247; compared to brain cells, 262; cross-fertilization rate of, 350; relationship with flowers, 363-66; with birds and other animals, 253; domestication of, 374; compared to mailman, 389; transcendence in, 501, 504-05, 591. See also Ant, Insect, Mind
Beethoven, Ludwig van, on deafness, 238-39; color of music, 235; hearing

through eyes, 265; works of, 540; shared composition with birds, 635-36, illus., 636; on harmony, 638. See also Music, Sense, Polarity, Abstraction
Beijerinck, Martinus Willem, on viruses, 102
Being, Earth as, 1-6; unit of, 327; forest as a, 382; abstract human being, 318-30; summary of abstraction, 340-43; Earth as a, 381-82, 388-95; the being of rocks, 383-88; universal being, 396-402; omnipresence of, 402-12; transcendence of consciousness, 515; of being, 517-19; afterlife, 534; supreme Being, chapter on, 596-627. See also Consciousness, Omnipresence, Life, God
Bergson, Henri, on growth-behavior polarity, 271
Bernoulli, James, on the spiral, 551
Besso, Michelangelo, heard Einstein, 517
Bhagavad-Gita, on death as dissolution, 521
Bible, Hebrew Tree of Knowledge, 43; Genesis of marriage "of one flesh," 127; John, 172, 359, and "Spirit of Truth," 614; Word made flesh, 533; Genesis, 541, 616; Job on interrelations, 542, 597; "time of the end," "the latter days," 563; Paul on relativity of God, 618; "many mansions," 618
Bird(s), 21-22; flight of, 23-24; at altitude, 28; navigation sense of, 40, 179-80, 316-17, illus., 316-17; vision of, 184-86, illus., 185-86; mating of, 132; eggs of, 141-43, illus., 142; language of, 252-55, 635-37, illus., 636; hearing and feeling his own song, 283; using tools, 285; communication with mammal, 253-54; pollination by, 363; domestication of, 373, 375; ecology of ducks, 376; parasitism in, 378, 513; helped by fire, 435, illus., 435; territory of, 477, illus., 477; transcendence in, 505-06, 508, 606, illus., 316, 505; singing bird divine,

677

Music (contd.)
633, illus., 633; of atom, 633-34, 638-39; of life, 634; of birds, 635-37, illus., 636; of man, 637-38; of vegetable, 639-41, illus., 640; of universe, 641-44; crystalline structure of, illus., 642. *See also* Song, Sound, Abstraction

*Music of the Spheres,* book, mentioned, 1, 272, 342

Myrberg, Arthur, zoologist, on fish language, 251

Mystery, mysteries, of life, 1, 6-7, 309; in photosynthesis, 50-53; in seeds, 74-78; in eggs, 141-42, 315-17, illus., 316; in the gene, 166-73; psychic senses, 235, 306-09; spiritual awareness, 240; of the atom, 273; of body and mind, 270-74; of Abstraction, 313-43; of identity, 329; of improbability, 329-30; of dimensions, 331-33, illus., 332; of mathematics, 334-39, illus., 335, 338; of universe, 343; of Earth's inner relatedness, 344; of parasitism, 378-80; of omnipresence of life, chapter on, 381-412; mechanism or vitalism concept of life?, 408; life in electron, 408; of electron as gene of matter, 412; mystery in life's analogies, 413; in hydraulics, 427; in life of crystals, 448-52; in life of bubbles, 461-62; dynamics of cells, 462-64; music of waves, 467-70; of location of beings, 466; of polarity principle, chapter on, 471-93; why evil?, 483-84; of the meaning of adversity, 484-90; of creative illnesses, 490; of transcendence, chapter on, 494-519; of creativity, 516; Infinitude of in Universe, 517; of death, chapter on, 520-38; of evolution, chapter on, 539-61; of the spiral, 551, illus., 551; of evolving language, 559; of sudden advent of world mind, 561; of world germination, chapter on, 562-95; of transcendence to superorganism mankind, 590-95; of abolition of slavery, 581; of divinity, chapter on, 596-627, illus., 596, 603, 609,

619, 623, 626; of spirit, 602; relativity, section on, 606-08; of number prodigy, 606-07; mysticism, section on, 608-11, illus., 609; horizon of, 609, 616, illus., 609; of God, 610-12; progressive revelation, 612-15; and science, 615-16; relativity in divinity, 617-18; why the world?, 618-19; evolution of the impossible, 619-20; process theology, 620; of soul school, 621-25; axiomatic proof, 625-27; Mystery of mysteries, 627; veil of, 627; mystics with visions, 236, 591-92; summary of The Seven Mysteries of Life, 6-7, 631, 645-54. *See also* Divinity, Spirit, God, Abstraction

Napoleon III, 289

Navigation sense, through light, 40; 179; of ants, 244; of bees, 247-50, illus., 247; birds, 316-17, illus., 316-17, illus., 316-17; of fish, 200-02, illus., 201; celestial, 360; of moth, illus., 551. *See also* Instinct

Nearchus, admiral, 64

Nebuchadnezzar, 289

Negation, ungrowth, 155; immunity, 158; eye of Nautilus, 188; invisibility sense, 192-94; absence of sense, 235-36, 238-39; smuffness, 212; in mathematics, 334; anti-matter in anti-universe, 401; as vital to life, 406; of memory, 406; half of quick clay, 429; negentrophy, 443-45; negative comparison, 489; uncertainty principle, 495; how positive is created by negative, 527; as trough of wave, 533; devolution, 489. *See also* Abstraction, Evolution, Mathematics, Polarity, Sense

Negentropy, law of order and life, 443-45, illus., 444; in evolution, 553. *See also* Order, Life, Symmetry, Mind, Spirit

Negroes, cousins of mankind, 351, ancestors of mankind, 355; slaves, 581

Nerve(s), section on, 116-17, mes-

Sound (contd.)
251; of mammals, 252; of birds, 252-55; of wolves, dogs, whales, 255; of human language, 287-91, 558-59; of sand dune, 417-18; acoustics, 460; of machine in distress, 225, 441-42; of bee hive, 505; evolution of voice, 558; of language, 558-59; cymatics, 468-70; of worlds, 632-33; of atom, 633-34, 638-39; of man, 637-38; of universe, 641-44. *See also* Music, Song, Hearing

Space, creature territories, 5; sense of, 179-80, 203, 208, 210; space-time concept, 291; in dreams, 301, 304; definition of, 331; field of force beyond, 412; infinite, 412; symmetry of, 412; cells that fill, 462-64, illus., 463; for wisdom to grow in, 496; territorial system, 476-79; transcendence of, 499-500; super-space, 511-12; spacelessness, 531; as mind, 533; explosion into, 568; travel in as a miracle, 619. *See also* Time, Speed, Abstraction, Sense, Universe

Spartacus, slave general, 580

Spassky, Boris, chess player, 270

Speed, law of, 20; of animals, 24-26; of vegetables, 55; of response in plants, 368-71, illus., 370; of minerals, 385-86; of supernova material, 397; of receding quasars, 399; rotation rate of electrons, 409-10; of moving sand, 384-86, 416, 419-20; of islands, 420-21; of glaciers, 422-23; of rivers, 426-27; of quick clay, 429-30; of storms, 431-32; of fire, 432-36; of lightning, 436-37; of atomic explosion, 437-38; of jet streams, 431; of magnetic plasma circulation on the sun, 431-32; of crystal growth, 447, illus., 448; of spin in liquid crystal, 454; of splash, 459; of bursting bubble, 460; of bubble adjustment, 461; of molecules in ice, 469; factor in happiness, 491; in transcendence, 498-99; of rotating ovum, 501; in school fish, 506; in fish response, 507; in

minerals, 384; of world germination, 562-95; of population explosion, 564; travel acceleration, 569; communication explosion, 570; knowledge explosion, 571-77. *See also* Interrelation, Dimension, Time, Space, Abstraction

Spemann, Hans, biologist, 153-56, 324

Spirality, in rattlesnake motion, illus., 28; in plants, 57-59, illus., 58; in bird sperm, illus., 136; in fruit fly eggs, illus., 142; in DNA, 166-70, illus., 169; in bioclock, illus., 232; in spider webs, 250; in shells, 318, illus., 596; of high dimensions, 332; of Golden Section, illus., 338; of cyclic interrelations, illus., 358; of deadly dodder, illus., 370; of galaxies, 398; of black holes, illus., 400; of paths of sand grains in dunes, 416; of water in rivers, 427; of crystals, illus., 439, 450, 454; of common animals and vegetables, illus., 551; grain of the Universe, 547, 616-17. *See also* Geometry, Abstraction, Mystery

Spirit, potential of Earth, 2; of the universe, 7; of the tree, 43; of the gene, 172-73; sense of, 180, 233, 239-40, 591; chapter on, 596-627, illus., 596, 603, 609, 619, 623, 626; in music, 641-44, illus., 642; of vegetable, 308-09, afterlife, 325-26, 532-38, illus., 326, 536; worlds beyond, 401; of universe, 412; immaturity of man's, 414; spectrum of, 483; polarity essential to its development, 484; meaning of thorn, tooth and claw, 488; boon of adversity, 489-91; transcendence, chapter on, 494-519; evolution of Golden Rule, 514, 555, evolution of language, 558-59; advent of human dominance and world mind, 560-61; germination of world, 562-95; transcendence to superorganism mankind, 590-95; spiritual evolution, 586-90, 602-09; history of vision, 573-74, illus., 574; source of abolition of slavery, 581; rise of hu-